Principles of Computational Biology and Genome Analysis

Principles of Computational Biology and Genome Analysis

Edited by Daniel McGuire

SYRAWOOD
PUBLISHING HOUSE

New York

Published by Syrawood Publishing House,
750 Third Avenue, 9th Floor,
New York, NY 10017, USA
www.syrawoodpublishinghouse.com

Principles of Computational Biology and Genome Analysis
Edited by Daniel McGuire

© 2017 Syrawood Publishing House

International Standard Book Number: 978-1-68286-407-4 (Hardback)

Cataloging-in-publication Data

Principles of computational biology and genome analysis / edited by Daniel McGuire.
    p. cm.
Includes bibliographical references and index.
ISBN 978-1-68286-407-4
1. Computational biology. 2. Genomes. 3. Bioinformatics. I. McGuire, Daniel.
QH324.2. P74 2017
570.28--dc23

Printed in the United States of America.

TABLE OF CONTENTS

PREFACE

Genome analysis has changed the way biological and anthropological evolution has been perceived. Computational analysis of genetic data has made it possible for the creation of speculative models that can predict possible evolutionary patterns while taking into account natural biological phenomena such as aging, disease and degeneration of the body. This book on computational biology and genome analysis contributes to the fields of computational neuroscience and computational evolutionary biology. The various studies that are constantly contributing towards advancing technologies and evolution of this field are examined in detail in this text. It elucidates new techniques and their applications in a multidisciplinary approach. This book is a vital tool for all researching or studying computational biology and genome analysis as it gives incredible insights into emerging trends and concepts.

The purpose of the book is to provide a glimpse into the dynamics and to present opinions and studies of some of the scientists engaged in the development of new ideas in the field from very different standpoints. This book will prove useful to students and researchers owing to its high content quality.

At the end, I would like to appreciate all the efforts made by the authors in completing their chapters professionally. I express my deepest gratitude to all of them for contributing to this book by sharing their valuable works. A special thanks to my family and friends for their constant support in this journey.

Editor

Somatic Mutations Favorable to Patient Survival Are Predominant in Ovarian Carcinomas

Wensheng Zhang[1], Andrea Edwards[1], Erik Flemington[2]*, Kun Zhang[1]*

1 Department of Computer Science, Xavier University of Louisiana, New Orleans, Louisiana, United States of America, **2** Tulane Cancer Center, Tulane School of Medicine, New Orleans, Louisiana, United States of America

Abstract

Somatic mutation accumulation is a major cause of abnormal cell growth. However, some mutations in cancer cells may be deleterious to the survival and proliferation of the cancer cells, thus offering a protective effect to the patients. We investigated this hypothesis via a unique analysis of the clinical and somatic mutation datasets of ovarian carcinomas published by the Cancer Genome Atlas. We defined and screened 562 macro mutation signatures (MMSs) for their associations with the overall survival of 320 ovarian cancer patients. Each MMS measures the number of mutations present on the member genes (except for TP53) covered by a specific Gene Ontology (GO) term in each tumor. We found that somatic mutations favorable to the patient survival are predominant in ovarian carcinomas compared to those indicating poor clinical outcomes. Specially, we identified 19 (3) predictive MMSs that are, usually by a nonlinear dose-dependent effect, associated with good (poor) patient survival. The false discovery rate for the 19 "positive" predictors is at the level of 0.15. The GO terms corresponding to these MMSs include "lysosomal membrane" and "response to hypoxia", each of which is relevant to the progression and therapy of cancer. Using these MMSs as features, we established a classification tree model which can effectively partition the training samples into three prognosis groups regarding the survival time. We validated this model on an independent dataset of the same disease (Log-rank p-value $<2.3\times10^{-4}$) and a dataset of breast cancer (Log-rank p-value $<9.3\times10^{-3}$). We compared the GO terms corresponding to these MMSs and those enriched with expression-based predictive genes. The analysis showed that the GO term pairs with large similarity are mainly pertinent to the proteins located on the cell organelles responsible for material transport and waste disposal, suggesting the crucial role of these proteins in cancer mortality.

Editor: Shannon M. Hawkins, Baylor College of Medicine, United States of America

Funding: Research reported in this publication was supported by a National Institutes of Health grant (NIGMS-2G12MD007595), an US Department of Army grant (W911NF-12-1-0066) and a seed grant from the Louisiana Cancer Research Consortium. The funders had no role in study design, data collection and analysis, decision to publish, or preparation of the manuscript.

Competing Interests: The authors have declared that no competing interests exist.

* Email: eflemin@tulane.edu (EF); kzhang@xula.edu (KZ)

Introduction

Ovarian cancer is the fifth-leading cause of cancer death among women in the United States [1]. The disease is often called a "silent killer" since its occurrence is usually not detected until an advanced stage. About 70% of the deaths occur in patients with advanced-stage, high-grade serous ovarian carcinomas [2]. The mortality has not been significantly improved in the past three decades [3]. Except for the detection delay and inaccessible location of the ovaries, other factors accounting for the persistent mortality include the poor understanding of the underlying biology and a lack of reliable biomarkers [4].

The formation of tumors largely results from cell growth that gets out of control [5]. In the human genome, there are many different types of genes that control cell growth in a very systematic, precise way. When these genes have an error in their DNA codes, the RNA or proteins that they encode may not function properly. Typically, a series of several mutations to certain classes of genes is usually required before a normal cell will transform into a cancer cell [6]. Nevertheless, some observed mutations may be neutral or even beneficial to patient survival. This perception can be considered from at least two perspectives. First, some mutations may be deleterious to the growth and proliferation of cancer cells, thus offering a protective mechanism to the patients. Second, some mutations may include the actual causal factors for relatively less-malignant subtypes of the same disease. For example, previous studies showed that cases with BRCA1/2 mutations have better overall survival than those with wild type BRCA1/2 in patients with ovarian carcinoma [7,8].

To date, the Cancer Genome Atlas (TCGA) [9] has generated and released comprehensive genomic, epigenomic and proteomic data of clinically annotated high-grade serous ovarian carcinomas (Ov-HGSCs). These rich data provide an unprecedented opportunity to investigate the genetic mechanisms underlying the variance in the survival of cancer patients and to advance the clinical prognosis and therapy of the disease. Besides the BRCA1/2 genotypes, the TCGA ovarian cancer paper [7] showed that gene expression-based sample clusters are also associated with the survival outcomes. Moreover, recent years have witnessed numerous studies that focus on the re-analysis of the TCGA data.

In these works, miscellaneous predictive signatures for survival outcomes have been identified. These signatures include the expression measures of coding and miRNA genes [10], genotypes of germline single-nucleotide polymorphisms (SNPs) [11], methylation patterns of genes in key cancer pathways [12], DNA copy number variations (CNV) [13] and the occurrences of chromosome aberrations [14].

As shown in [7], most of the Ov-HGSCs had 8 to 209 somatic mutations. These mutations, detected by exome sequencing, were present in 8945 genes, and 92% of them were validated by experiments using alternative technologies. However, most of the observed variants may be passenger mutations not involved in the formation and progression of ovarian cancer. Hidden among observed mutations are the individual-specific tumor drivers and the genetic alterations positively or adversely impacting the growth and survival of cancer cells. The identification of the clinically important mutations (genes) is far from completed. A major challenge impeding the effective statistical analysis of the somatic mutation spectrum (SMS) is the data sparseness issue. This is particularly implied by the fact that, among the 510 consensus cancer genes collected in the Catalogue Of Somatic Mutations In Cancer database [15], only six are significant in terms of their mutation frequencies over the 326 tumors. Nevertheless, two recent studies have demonstrated the potential to train a predictive model for survival outcomes of ovarian cancer patients using SMS [16,17]. In this study, we conducted a unique analysis of the recently updated TCGA's clinical and SMS datasets of ovarian cancer. Our study provides significant insights into the treatment of ovarian cancer and may open novel avenues for molecular prognosis and prediction.

Results

Predictive macro mutation signatures for patient survival

We developed a novel method to unravel the relationships between the somatic mutations and the survival time of cancer patients. First, by assuming that the DNA alterations on the genes of a similar function may have equivalent or complementary impacts on the growth and proliferation of cancer cells, we defined 562 macro mutation signatures (MMS), each of which corresponds to a highly-specific Gene Ontology (GO) term with 50 to 500 member genes. For each patient (i.e. a carcinoma sample), the MMS quantities were calculated as the number of the mutations on the genes (except for TP53) covered by the cognate GO term. When a gene involves in multiple GO terms, the mutation(s) present on each gene were counted with respect to each cognate MMS. In this way, we circumvented the sparsity issue inherent to the raw somatic mutation data (see Introduction section). After that, the MMSs were screened for their associations with the overall survival (OS) months of the cancer patients. More specifically, the associations were evaluated by performing the Log-rank test and Cox Proportional Hazards (Cox-PH) regression analysis on the mutation and clinical datasets of 320 training samples. In the implementation, quantities of the MMSs were capped by a ceiling value of 2, which represented that a tumor had at least two mutations present on the member genes covered by the corresponding GO term. Capping the MMS values was performed to alleviate the influence of leverage data points, which were related to un-ordinarily high MMS values and usually occurred in highly-specific GO terms. In the Cox-PH analysis, along with a focused MMS, the ages of the patients at the initial diagnosis and a binary measurement variable indicating the presence of somatic mutation on TP53 gene, which had a modestly significant (p<0.05) effect on the patient survival as

shown in our preliminary analysis of the same data, were included as covariates. In the Log-rank test, the three possible values (0, 1, 2) of a specific MMS were factorized as the indicators of three groups.

The analysis of the training set (N = 320) demonstrated strong evidence for the existence of an association between the MMSs and survival outcomes. As shown in Figure 1-A and 1-B, the distributional profiles of the p-values (from both the Log-rank test and Cox-PH regression) for the MMSs are deviated from a uniform distribution U (0, 1). Interestingly, most of the regression coefficients (i.e., beta values), especially those corresponding to small p-values, are negative (Figure 1-C). In the Cox-PH model, a negative regression coefficient indicates that the hazard function decreases (or equivalently, survival time increases) as the quantity of the corresponding predictive variable increases [18]. In this regard, we concluded that somatic mutations favorable to the survival of cancer patients are predominant in ovarian carcinoma compared to those indicating poor clinical outcomes. As shown in Figure 1-D and 1-E, this statement is also valid in terms of the number of the involved GO terms and the sizes of the relevant gene sets (Table S1).

Neither Log-rank test nor Cox-PH regression analysis are perfect for evaluating the associations between a MMS and the clinical outcome. The former ignores the patients' ages at the initial diagnosis, which intuitively influence survival time. The latter assumes that the quantity of the hazard functions is linearly dependent on the preprocessed MMS values, which is not true in many cases. Therefore, we determined the top significant MMSs (GO terms) by an alternative method. That is, we selected 20 MMSs if (1) their p-values from both the Log-rank test and Cox-PH analysis are less than 0.05 and (2) the resulting composite p-value (see Method section) is less than 0.025. Among those MMSs, 19 are "positive" predictors for survival time. Considering that the selection criteria may be too stringent for the potential MMSs adversely affecting overall survival outcomes, we chose another two MMSs. These two "negative" predictors, with Log-rank p-values less than 0.01, are relevant to two small patient sets of poor survival and correspond to GO:0045666 and GO:0042393, respectively. In this way, we established a predictor set consisting of 22 MMSs (Table 1).

We addressed the multiple-testing problem in the identification of predictive MMSs for patient survival by calculating false discovery rate (FDR) with a permutation-based algorithm (see Material and Methods section). In the implementation, we considered not only the skewness of the distribution of the effect parameters estimated from the original datasets (Figure 1C), but also the asymmetry of their null distribution established from the randomly permutated datasets (Figure S1). Because only one negative predictor (MMS) was rigorously selected, the analysis was focused on the 19 MMSs associated with good clinical outcomes. The result showed that, when those MMSs are stated to be significant, the FDR could be controlled at the level of 0.15.

Based on the definitions, we partition the cognate GO terms of the 19 positive predictors (MMSs) into six groups: (1) the gene products (proteins) locate in the cell organelles (membrane) responsible for waste disposal (lysosome) and material transport (recycling endosome); (2) the gene products locate in the sub-cellular structures playing roles in mitosis (nucleosome, spindle pole, centrosome); (3) the gene products perform function in cell division (mitosis and cytokinesis); (4) the gene products are involved in cellular responses to nutrient and hypoxia; (5) the gene products play roles in cancer pathways (integrin-mediated signaling pathway and positive regulation of ERK1 and ERK2 cascade); and (6) others. Numerous records regarding those GO

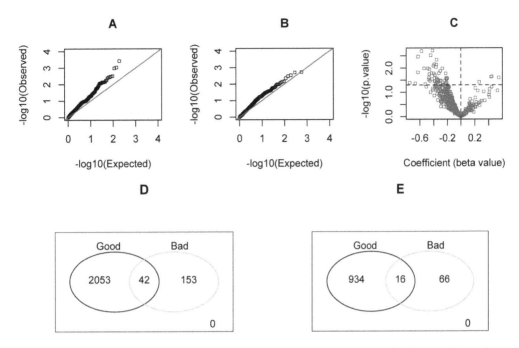

Figure 1. The profile for the associations between the somatic mutations and survival time of patients with ovarian cancer. A (B): The Q-Q plot of the p-values from Log-rank test (Cox-PH regression) for the 562 considered MMSs. C: The volcano plot of the Cox-PH p-values and regression coefficients for the 562 considered MMSs. The horizontal dot line marks p = 0.05. D: The Venn diagram for the entire set of genes covered by the 22 selected MMSs. Specifically, the good (bad) genes are the genes involved in the GO terms corresponding to the 19 (3) positive (negative) MMSs which predict good (poor) clinical outcomes. A gene can belong to both the positive and negative MMSs, therefore may be double counted. E: The Venn diagram for the subset of the genes which are covered by the 22 selected MMSs. Each of the genes has the mutation burden in at least one training sample.

terms' relevance to the formation, progression and therapy of tumors can be located in the literature (see Discussion section).

By looking into the Kaplan-Meier survival curves, we found a "dose-effect" relationship between the somatic mutations and survival outcomes. That is, for a specific GO term, a single mutation on the member genes usually does not make much difference to the patient survival time but two or multiple mutations do (Figure 2).

Robustness analysis of the selected predictors

In order to test the robustness of our main result, we randomly split the 320 training samples into two equal-size subsets, and estimated the effects of the 22 predictive signatures on each subset separately. The result showed that the sign (positive or negative) of the estimated regression coefficients of hazard functions on the MMSs were consistent with those estimated using the entire training set (upper left plot of Figure S2).

We further tested if each of the 22 selected MMSs can individually predict the survival of cancer patients in the validation set. The results showed that, just the predictors ranked at the second, third and fifth places had a marginally significant p-value (upper right and bottom plots of Figure S2). While this analysis only provided a minor support to our findings in the last subsection, the result is aligned to our expectation. This is because the insufficiencies of the training set, i.e. the small sample size (N = 140) and un-validated mutations, could lead to a lower statistical power.

A classification tree model for patient survival prediction

The findings presented above inspired us to build a classification tree to predict the patient survival using the 22 identified MMSs.

More specifically, based on the measures of all three negative predictors, we can separate a poor-prognosis group from the entire set of training samples whose members meet $\max_j (N_j^i) \geq 2$, where N_j^i indicates the value of the jth negative MMS on the ith sample. Then, based on the values of the top k positive predictors and the same criterion, i.e. $\max_k (P_k^i) \geq 2$ with P_k^i indicating the value of the jth positive MMS on the ith sample, a good-prognosis group can be split from the remaining samples that constitute an intermediate-prognosis group. See Figure S3 for an illustration. The threshold for the combined MMS values in the partition was heuristically chosen according to the pattern of dose-dependent effect of several MMSs of high interest to patient survival, as showed in Figure 2 and described in the ending paragraph of the first result subsection.

As shown in Figure 3, the patient groups generated by the tree model are significantly differentiated with respect to the times of overall survival (OS) and progression-free survival (PFS). Regardless of the k value (5 or 10), the Log-rank test p-value is less than 1.2×10^{-10} for OS and is less than 1.6×10^{-7} for PFS. From the Kaplan-Meier survival curves, we found that, for the poor-prognosis group, the upper limits of OS and PFS are 50 and 20 months, respectively. They are also the time points when the differences in the survival probabilities between the good-prognosis group and intermediate prognosis group become sharper. It is worth noting that, the choice of k value is somewhat arbitrary. The value determines the size of the predicted good-prognosis group that has a better survival curve compared to the intermediate-prognosis group. Therefore, a prior knowledge about the proportion of good-prognosis samples would help with the specification of k value.

Table 1. The summary of significant MMSs for the overall survival of patients with Ov-HGSCs.

GO (MMS) ID	β	Cox-PH p-value	Log-rank p-value	CP	N1	N2	GO Name
Positive predictors							
GO:0000786	−0.67	4.4E-02	4.3E-05	1.4E-03	65	25	nucleosome
GO:0005765	−0.28	1.1E-02	3.7E-04	2.0E-03	232	135	lysosomal membrane
GO:0050900	−0.42	1.9E-03	3.8E-03	2.7E-03	110	92	leukocyte migration
GO:0007229	−0.44	3.0E-03	8.0E-03	4.9E-03	77	80	integrin-mediated signaling pathway
GO:0010923	−0.54	1.1E-02	3.1E-03	5.7E-03	51	49	negative regulation of phosphatase activity
GO:0007584	−0.63	3.4E-03	1.1E-02	6.1E-03	67	54	response to nutrient
GO:0007067	−0.36	4.5E-03	8.5E-03	6.2E-03	245	126	mitosis
GO:0007160	−0.41	8.2E-03	7.0E-03	7.6E-03	79	75	cell-matrix adhesion
GO:0000910	−0.58	2.0E-03	3.0E-02	7.8E-03	64	54	cytokinesis
GO:0006928	−0.49	1.0E-02	5.9E-03	7.8E-03	97	60	cellular component movement
GO:0001666	−0.31	1.4E-02	7.4E-03	1.0E-02	153	115	response to hypoxia
GO:0000922	−0.39	7.0E-03	1.6E-02	1.1E-02	86	77	spindle pole
GO:0031965	−0.33	1.7E-02	8.6E-03	1.2E-02	162	108	nuclear membrane
GO:0005813	−0.21	2.6E-02	7.4E-03	1.4E-02	353	165	centrosome
GO:0004674	−0.23	1.7E-02	1.3E-02	1.5E-02	374	217	protein serine/threonine kinase activity
GO:0070374	−0.36	3.1E-02	9.1E-03	1.7E-02	97	57	positive regulation of ERK1 and ERK2 cascade
GO:0044325	−0.37	9.1E-03	3.5E-02	1.8E-02	71	87	ion channel binding
GO:0055037	−0.63	1.5E-02	2.7E-02	2.0E-02	60	37	recycling endosome
GO:0004843	−0.46	1.8E-02	2.9E-02	2.3E-02	55	39	ubiquitin-specific protease activity
Negative predictors							
GO:0051436	0.55	2.5E-02	1.0E-03	5.1E-03	65	27	negative regulation of ubiquitin-protein ligase activity involved in mitotic cell cycle
GO:0045666	0.15	4.0E-01	3.6E-03	3.8E-02	68	50	positive regulation of neuron differentiation
GO:0042393	0.17	3.6E-01	7.1E-03	5.0E-02	62	47	histone binding

β: the regression coefficients estimated by the Cox-PH model. **CP**: the composite p-value, which is the square root of the product of the Log-rank test p-value and the corresponding Cox-PH p-value. **N1**: the number of member genes covered by the corresponding MMS or GO term. **N2**: the number of mutated member genes present in at least one training sample. Note that a single gene can be covered by more than one GO term.

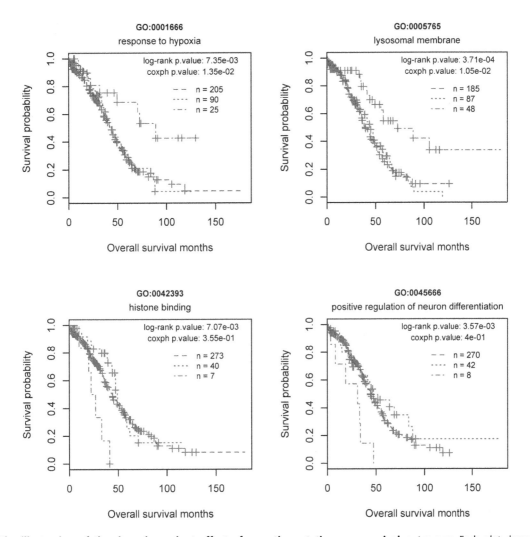

Figure 2. The illustration of the dose-dependent effect of somatic mutations on survival outcomes. Each plot demonstrates the relationship between the overall survival months and a specific macro mutation signature (MMS) that corresponds to a GO term. The purple curve represents the patients each of whom has at least two somatic mutations on the member genes of the indicated MMS (i.e., GO term). The red curve represents the patients each of whom has one somatic mutation on the member genes of the indicated MMS. The blue curve represents the patients without any somatic mutation on the member genes of the indicated MMS.

Model validation using independent datasets

We validated the tree model by applying it to an independent dataset of Ov-HGSCs. As shown in Figure 4, the survival curves of the patients in the three (good-, poor- and intermediate-prognosis) groups resemble those observed for the training set. The group effect on overall survival time is significant (Log-rank test p-value <0.001). When k is 5, the good prognosis group in this validation set has the same OS survival probability (~30%) as that in the training samples. Moreover, interestingly, although the underlying negative predictors are not defined on a stringent statistical criterion, both the survival profile of the poor-prognosis group and the patient percentage (5/140 = 3.8%) of in this group are similar to those (14/320 = 4.5%) of the training set. It is worthy of note that, in the TCGA database, the observed somatic mutations of the samples in the validation set have not been confirmed by other methods yet. The average number of mutations in this set is approximately 80, much higher than those (~50 observed and ~46 validated) of the training set. Hereby, the classification results are more sensitive to the number of used predictors.

Recent studies showed that the formation of ovarian tumors shares common cancer drivers with breast tumors. We assume that these two diseases may be similar regarding the biological mechanisms underlying the variance in the patient survival time. We look into this issue by applying the identified predictors for Ov-HGSCs to the TCGA data of invasive breast carcinomas. As shown in Figure 5, we can identify a good-prognosis group using the top positive predictors but cannot separate a poor-prognosis group via the negative predictors. The difference in the survival probability between the good-prognosis patents and other patients becomes evident at the point of 75 months, 25 months more than the time for ovarian carcinomas.

Comparison between mutation signatures and expression signatures

By analyzing the TCGA clinical and mRNA expression data of Ov-HGSCs, we identified 333 expression predictors (genes) for the overall survival time of patients with the p-values less than 0.01. 28 functionally specific non-redundant GO terms, either at level-4 or level-5 as categorized by DAVID [19], were over-represented

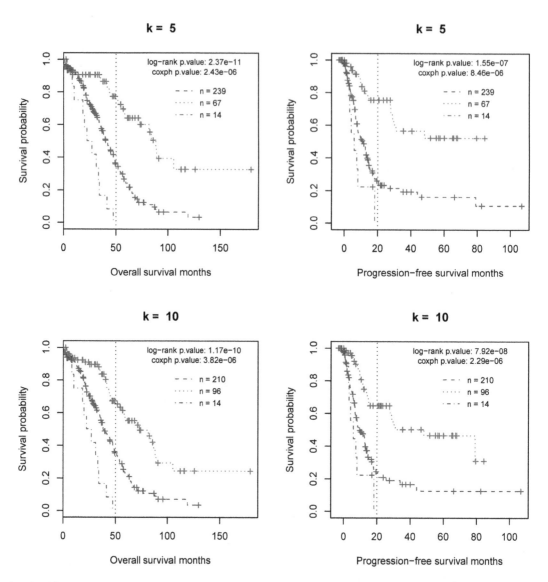

Figure 3. The classification of the training set of ovarian cancer patients by the proposed tree model. In each plot, the considered predictors include all three negative MMSs and the most significant (or top) k (5 or 10) positive MMSs as summarized in Table 1. The purple, red and blue curves represent the predicted poor, good, and intermediate-prognosis groups, respectively.

(FDR <0.1) by these genes. Hereafter, we named those 28 GO terms macro expression signatures (MESs). The matrix of the semantic similarity between the MESs and macro mutation signatures (MMSs), i.e. the GO terms corresponding to the 22 significant MMSs, was evaluated using the algorithm documented in [20]. As shown in Figure 6, the similarity coefficients are low in general. Four MES::MMS pairs have the coefficients over 0.5. They are: GO:0005788 (endoplasmic reticulum lumen) versus GO:0055037 (recycling endosome); GO:0005788 versus GO:0005665 (lysosomal membrane); GO:0051427 (human receptor binding) versus GO:00044325 (ion channel binding), GO:0051427 versus GO:0042393 (histone binding). Moreover, five MESs, relevant to the regulation of cellular process and cell death, show modest levels of similarity to seven MMSs, which correspond to some specific molecular functions and biological processes including integrin-mediated signaling pathway (GO:0007229) and positive regulation of ERK1 and ERK2 cascade (GO:0070374). These results suggest that: (1) only several survival-relevant somatic mutations impact the clinical outcomes

via the modification of the expression level of the host genes; and (2) the proteins located on the cell organelles responsible for material transport and waste disposal may be crucial for the survival of cancer patients in that both the modification of properties (due to a non-synonymous mutation) and the change of expression level in cancer cells can significantly influence the clinical outcomes.

Discussion

Over the last few decades, cancer researchers have pinpointed hundreds of cancer genes [21,22], including oncogenes and cancer suppressor genes, and established a number of DNA-alteration based theories for carcinogenesis [5,23]. Nevertheless, the genetic determination of survival outcomes for patients with malignant tumors has been less investigated yet. By analyzing 320 ovarian tumor samples, we found that somatic mutations favorable to the patient survival are predominant in ovarian carcinoma compared to those indicating poor clinical outcomes. This observation

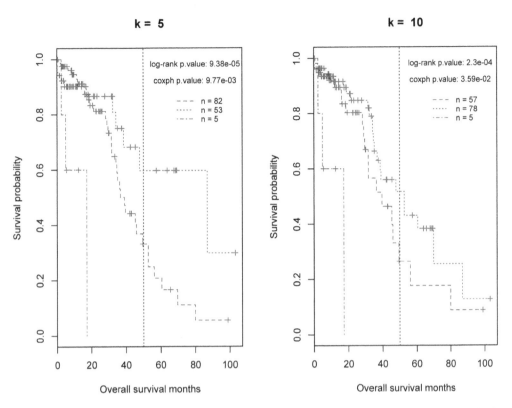

Figure 4. The classification of the validation set of ovarian cancer patients by the proposed tree model. In each plot, the considered predictors include all three negative MMSs and the most significant (or top) k (5 or 10) positive MMSs as summarized in Table 1. The purple, red and blue curves represent the predicted poor, good, and intermediate-prognosis groups, respectively.

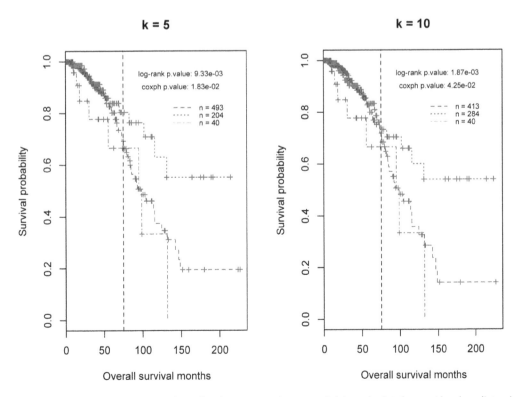

Figure 5. The classification of breast cancer patients by the proposed tree model. In each plot, the considered predictors include all three negative MMSs and the most significant (or top) k (5 or 10) positive MMSs as summarized in Table 1. The purple, red and blue curves represent the predicted poor, good, and intermediate-prognosis groups, respectively.

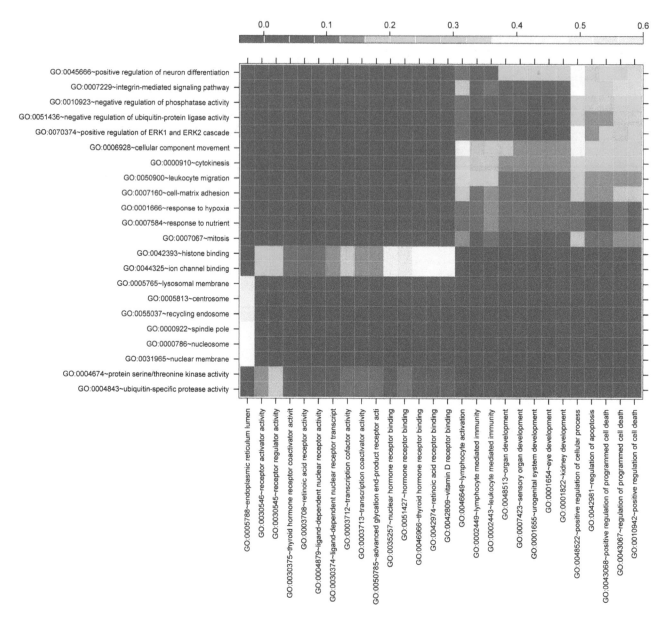

Figure 6. The visualization of the semantic similarity between the MESs and MMSs. The similarity is measured by a coefficient in the range of 0 to 1. 1 is the theoretical maximum of the similarity coefficient. For the GO term pairs considered here, the values are consistently less than 0.6.

highlights the vulnerability of cancer cells to "extra" mutations. That is, while the cancer-driver mutations prompt cancer cells to divide in an uncontrolled way or offer them selection advantage over the adjacent normal cells, the extra mutations may restrict the continuous proliferation in certain microenvironments. When the restriction occurs in some important organs such as liver and spleen, where ovarian metastases usually lead to mortality, the extra mutations may benefit the patient survival. Such a mechanism can be elucidated by a further scrutiny of our results. For example, among the predictive MMSs identified for good prognosis, there is one that measures the mutation events occurring on the genes involved in the biological process of "response to hypoxia" [24]. It is well known that the activation of anaerobic glycolysis (the Warburg effect) provides most of the building blocks required to duplicate the cellular components of a dividing cell; therefore, it is also essential for carcinogenesis [25,26]. If the properties of one or multiple protein(s) involved in

anaerobic glycolysis are altered, the tumors may lose the ability to produce enough energy for maintaining their growth. As a result, the carcinogenesis can be retarded.

On the other hand, in many cases, cancer cells acquire mutations to constitutively activate their survival pathway and to develop chemo-resistance. This mechanism seems to cast a shadow on our explanation to the main conclusion of this study. However, the dilemma could be resolved to some extent if we assume that only a few new driver mutations occur as the responses to the treatments. This assumption is supported by our preliminary analysis which showed that the average numbers of somatic mutations in Ov-HGSCs don't substantially increase across the development stages (from II to IV).

Our analysis suggests that the proteins located on the cell organelles responsible for material transport and waste disposal bear a special importance for cancer mortality since both the modification of properties (due to a non-synonymous mutation) and the change of expression level in cancer cells can significantly impact the clinical outcomes. In particular, the identified predictors for good clinical outcomes include the MMS corresponding to the cell component GO term "lysosomal membrane". This result provides the genetic insight into and clinical support for a promising cancer therapy strategy, in which the lysomoses of cancer cells can be treated as the drug targets. The strategy arose from the perception that the altered lysosomal trafficking and increased expression of the lysosomal proteases termed cathepsins may form an "Achilles heel" for cancer cells by sensitizing them to death pathways involving lysosomal membrane permeabilization and the release of cathepsins into the cytosol [27,28,29,30]. A recent study on the screening of a small molecule drug library provided strong evidence for this mechanism. The authors found that over half of the 11 compounds that induced significant cell death in p53-null colon cancer cells triggered lysosomal membrane permeabilization and cathepsin-mediated killing of tumor cells [31]. We speculate that these compounds may functionally resemble the mutations present on the genes related to lysosomal membrane. We further surmise that, for an ovarian cancer patient with a single mutation on the lysosomal membrane related genes, an additional functional disruption of these genes caused by the anti-cancer compounds (or by other treatments) may offer the patient a better chance for survival, which is similar to those patients denoted by the purple curve in the upper right plot of Figure 2.

Resistance to apoptosis and chemotherapy is a critical factor in cancer recurrence and patient relapse. Several studies over the last decade have demonstrated that ECM/integrin signaling provides a survival advantage to various cancer cell types against numerous chemotherapeutic drugs and against antibody/radiotherapy therapy [32,33,34]. Our result implies that such an advantage for cancer cells can be interrupted by the mutations occurred on the cognate genes. As shown in Table 1, the MMS corresponding to the biological process of "integrin-mediated signaling pathway" is a positive predictor for the survival time of Ov-HGSC patients. Furthermore, we find that the mutations on the genes that positively regulate ERK1/2 cascade [35] can be deleterious to the continuous proliferation of cancer cells in the sense that the patients with such mutations had a longer survival time. These observations suggest that even the mutations whose host genes play a role in a cancer pathway may benefit the survival of cancer patients.

Another novel finding from this study is the dose-dependent effect of somatic mutations on survival outcomes. In light of this observation, we established a classification tree model to predict the survival profiles of the Ov-HGSC patients. The model is robust and performs comparably to the classifiers created using gene expression and other –omic data [10,11,12,13,36]. The phenomenon that a single mutation does not make much difference to a biological process but two (or multiple) mutations do may be explained by genome evolution. That is, evolution often created "backup" genes (or gene fragments) that perform the normal functions of a specific gene (or gene fragment) and a biological aberration occurs only when both the gene (or fragment) and its backups are altered [37,38]. In fact, this mechanism may explain why the formation and malignancy of cancer require multiple mutations. To clarify this point, it is worth noting that a lethal biological aberration for tumor cells can imply a favorable change for the cancer patients and vice versa.

The proposed classification method can be implemented in a flexible way. For example, using the MMS corresponding to the GO term "histone binding" as the only predictor, a group of seven patients whose overall survival time is consistently less than 50 months can be identified from the Ov-HGSC training samples (Figure 2: bottom left plot). Each of these poor-survival patients has a short list of "lethal" mutations. Specifically, for the first sample of this group, there are three lethal mutations present on the genes NOC2L, CHD8 and CHAF1B. For the other six samples, the host genes of the lethal mutations are (L3MBTL2, L3MBTL2, L3MBTL2), (UIMC1, RNF20), (UIMC1, RNF20), (HJURP, NCAPD2), (NASP, PKN1), (MCM2, NCAPD2), respectively. Among the eleven member genes, three have been identified as prognostic indicators of breast or gastric cancers in previous studies. The evidences include: overexpression of MCM2 in gastric tumors predicted poor prognosis in the patients [39]; knock down of HJURP reduced the sensitivity of breast cancer patients to radiation treatment [40]; the loss of CHD8 may be an indicator for biological aggressiveness in gastric cancer [41]. Another two, i.e. UIMC1and CHAF1B, are cancer-relevant. The former codes BRCA1-A subunit RAP80 [42], a protein important for genomic stability [43]. The latter codes the chromatin assembly factor-1/p60, a proliferation marker in various malignant tumors with prognostic value in renal, endometrial and cervical carcinomas [44]. Therefore, further investigation on the functions and interaction of the proteins coded by these genes may facilitate the inference of the personalized mechanisms for the mortality of ovarian carcinomas.

Recent studies found cancer-driving changes shared across tumor types [45]. A well-known hallmark is the genetic similarity between breast cancer and ovarian cancer. For example, the major driver genes BRCA1/2 for breast cancer are frequently (10~20%) mutated in the cancer cells of the patients with ovarian tumors [7]. Moreover, somatic mutations on TP53 (a major cancer driver gene in Ov-HGSCs) have been observed in the breast cancer samples of all subtypes, including luminal A, B, basal-like, and Her2-enriched [46]. Interestingly, we found that the top predictive MMSs identified using the clinical data and SMS of Ov-HGSC can predict the survival time of breast cancer patients. However, the three predictors for poor-prognostic outcomes of ovarian cancer are invalid when applied to breast cancer. Intuitively, more significant predictive macro signatures for breast cancer could be identified using the information of the patients of the same disease but this work is out of our scope.

To date, survival prediction using the gene expression signatures for breast or ovarian cancer patients has been the subject of much research [47,48,49,50,51,52,53]. However, most of the reported predictive expression signatures cannot be consistently validated by the analysis on the independent datasets (cohorts) [54]. Our comparative analysis suggests that only a few survival-related somatic mutations impact the clinical outcomes by modifying the expression level of the host genes. A potential reason for the robustness deficiency in the expression-based prognostic signatures is the temporal and/or spatial gap between the sampling of the disease tissue and the occurrence of the lethal metastasis of cancer cells. We speculate that mutation prognostic signatures, such as those we identified, have an advantage over an expression-based signature in the sense that they are less likely subject to progression history and location transition of cancer cells.

At last, we note that there are some uncertainties in our results. First, a few genes (N = 16), such as FN1, are involved in both positive and negative predictors for patient survival. Those genes account for 1.6% (16/1016) of all the genes which have at least one mutation in the training set and are covered by the 22 significant MMSs. Second, the false discovery rate of the predictive MMSs is slightly high (at the level of 0.15 for the 19 MMSs associated with good clinical incomes). In other words, a small portion of those MMSs might be falsely identified.

Nevertheless, these issues are relatively minor to affect our conclusions regarding the predominance of somatic mutations favorable to patient survival and the prognostic usefulness of the identified predictive MMSs as a whole.

Material and Methods

Somatic mutation dataset for Ov-HGSC training samples (Data-1)

The dataset of 321 tumor samples was generated from three *mat*-format files (version 2.4)in the TCGA database [9]. The archives containing these files are "*broad.mit.edu_OV.Illumina-GA_DNASeq.Level_2.100.1.0*", "*hgsc.bcm.edu_OV.SOLiD_D-NASeq.Level_2.1.6.0*", and "*genome.wustl.edu_OV.Illumina-GA_DNASeq.Level_2.1.3.0*", respectively. Among the total 16306 mutations identified by exome-sequencing, 14960 have been validated using other methods and were used in our study. Most validated mutations belong to four single nucleotide mutation categories, namely missense_mutation (68.09%), silence (21.39%), nonsense_mutation (4.26%) and splice_site (2.20%). Among them, 257 validated mutations occurred on the gene TP53 of 225 samples. The cancer samples contained in this dataset were also used in [7]. There is a trivial difference between the SMS analyzed in our study and that used in [8].

Somatic mutation dataset for Ov-HGSC validation samples (Data-2)

This mat-format dataset (version 2.4) was obtained from the archive "*genome.wustl.edu_OV.IlluminaGA_DNASeq.Level_2.2.1.0*" at TCGA [9]. In total, there are 142 tumor samples and 11342 mutations, of which 111 are present on the gene TP53. None of these mutations has been validated yet. The mutation distribution over variant types is similar to that of the training set (Data -1). The entire mutation profiling was used in the study.

Somatic mutation dataset for breast invasive carcinoma samples (Data-3)

The dataset containing 776 tumor samples [55] was download-ed from TCGA [9]. The corresponding *mat*-format file is located in the archive "*genome.wustl.edu_BRCA.IlluminaGA_DNASeq.Level_2.5.3.0*". In total, there are 47243 mutations. The mutation distribution over variant types is similar to that of *Data-1*. Among these somatic mutations identified by exome-sequencing, only 6397 have been validated using other methods. The entire mutation profiling was used in the study.

Clinical dataset for Ov-HGSC training samples (Data-4)

This dataset is contained in the supplement, "*Copy of TCGA-OV-Clinical-Table_S1.2.xlsx*", of the TCGA paper [7]. We downloaded it from the Nature website. The dataset consists of the clinical information of 488 Ov-HGSC patients (samples), of which 320 had the somatic mutations collected in *Data-1*. This dataset was used because it contains the progression-free survival time (PFS) which are not present in the matrix data archive of [9]. While the tumor-stage and tumor-grade attributes are also available in the dataset, neither [7] nor our preliminary analysis showed that their effects on the survival time were statistically significant. Hereby, these two attributes were not considered as predictive variables in the study.

Clinical dataset for Ov-HGSC validation samples

This dataset was downloaded from [9]. Out of 573 patients in this set, 140 had the somatic mutations collected in *Data-2*.

Clinical dataset for breast invasive carcinoma samples

The dataset was downloaded from TCGA database. Out of 971 patients in this set, 737 had the somatic mutations collected in *Data-3*.

GO dataset

The gene function annotation data for human was downloaded (on Oct 8, 2013) from The Gene Ontology (GO) website [56]. In the dataset, 18920 genes (symbols) were annotated to 13863 GO terms. We used a heuristic method to select the GO terms considered in this study. That is, a GO term was selected if the number of genes annotated to this term was between 50 and 500. The reason for doing so is twofold. First, if a GO term has only a few genes, the values of its corresponding MMS may be too sparse to perform an efficient statistical inference. Second, if there are too many genes annotated to a GO term, the functional category can be rather broad to infer meaningful biological insight from the results. While this setting was somewhat arbitrary, it won't introduce the selection bias that might substantially impact the conclusion.

Gene expression dataset for Ov-HGSC training samples

The mRNA expression levels of the tumor sample contained in *Data-1* were measured on three different platforms, i.e. Affymetrix Human Exon1.0 ST Array, Agilent 244K Whole Genome Expression Array and Affymetrix HT-HG-U133A Array. In the study, the combined gene expression dataset of 11684 genes present on all three platforms was used. The dataset is a supplement of [7] and was downloaded from the Nature website.

Methods for survival analysis

Survival analysis was performed using the statistical functions included in R package "survival" [57,58]. For univariate survival analysis with a factorized MMS as the predictor, the function "*survdiff*" was implemented to generate the Log-rank test p-value. It worth noting that, when "*survdiff*" was applied to the breast cancer dataset in which the cases of death at an early stage are rare due to right censoring, we let the *rho* parameter equal to negative 2, *i.e.* assigned each death a weight of $S(t)^{-2}$, where $S(t)$ is the Kaplan-Meier estimate of survival. The Kaplan-Meier survival curves (in Figures 2, 3, 4, 5 and S1), with the censored observations being marked by a vertical tick, were obtained via the function "*survfit*". Multivariate survival analysis was conduct-ed using the function "*coxph*" which implements Cox PH regression.

Identification of MMSs for survival prediction

We identified the predictive MMSs for overall survival time using the procedure presented in the Result section, and ranked them according to the composite p-value *CP*. The *CP* value for a MMS was calculated as the square root of the product of the p-values obtained from the Log-rank test and the corresponding Cox-PH analysis.

Identification of expression predictors for survival time

The association between the patient survival time and the gene expression levels was evaluated by the Cox PH regression. Similar to the analysis for the association between a MMS and the survival time, the patient age at the initial diagnosis was included in the model as a covariate.

Comparison between macro mutation signatures and expression signatures

The similarity matrix for the macro mutation signatures (MMSs) and macro expression signatures (MESs) was calculated by the function "*goSim*" in the R package "GOSemSim" [59]. In the employed method [20], the semantics of GO terms are encoded into a numeric format and the different semantic contributions of the distinct relations are considered.

Estimation of FDR

By adapting the methods used in [60,61], we developed a permutation-based algorithm to estimate the false discovery rate (FDR) for the 19 predictive MMSs associated with good clinical outcomes. First, we generated 500 shuffled datasets via randomly permutating the clinical records of the 320 training samples while keeping their mutation profile untouched. Then, we repeated the survival analysis by the same method used in the identification of predictive MMSs, and recorded the Log-rank p-values (p_{rank}), Cox-PH p-values (p_{cox}), the complex p-values (p_{cp}) as well as the regression coefficients (i.e. the beta values c) for all the 562 addressed MMSs. By doing so, we established the null distributions for p_{rank}, p_{cox}, p_{cp} and c, respectively. Finally, we compared the true distributions of p-values and regression coefficients to the corresponding null distributions to estimate false discovery rate by the following equation.

$$FDR = \frac{P(p_{rank}^{(0)} < z_{rank}, p_{cox}^{(0)} < z_{cox}, p_{cp}^{(0)} < z_{cp}, c < 0)}{P(p_{rank}^{(1)} < z_{rank}, p_{cox}^{(1)} < z_{cox}, p_{cp}^{(1)} < z_{cp}, c < 0)} \quad (1)$$

In (1), $p_*^{(0)}$ is a p-value from the null distribution and the subscript index * represents "rank", "cox" or "cp"; $p_*^{(1)}$ is a p-value from the true distribution; z_* is the threshold specified for the identification of predictive MMSs, and it is set to be 0.05, 0.05 or 0.025 for p_{rank}, p_{cox} or p_{cp}, respectively. The numerator is the fraction of p-values from the null distributions that fall below the thresholds (z_*) with the cognate regression coefficients less than 0. The denominator is the corresponding fraction for the estimates of p-values and regression coefficients based on the original dataset.

Availability

R codes for the statistical analysis are available upon request.

Supporting Information

Figure S1 The asymmetry of the null distributions of the effect parameters. The volcano plot of the Cox-PH p-values and regression coefficients for the 562 considered MMSs is based on the results of five randomly shuffled datasets.

Figure S2 Robustness analysis of the predictive MMSs. Top-left: The scatter plot shows the regression coefficients estimated from the two equal-size subsets of 320 training samples using the same Cox-PH model in the identification of the predictors. The solid squares (triangle) represent the 19 (1) MMSs which were rigorously selected and associated with good (poor) clinical outcomes. The solid circles represent the two MMSs which were selected in a less-rigorous way and were associated with poor clinical outcomes. The MMSs focused in the top right and bottom plots of this figure are marked with red. **Top-right (bottom-left, bottom-right)**: The results were obtained by analyzing 140 training samples. Each plot demonstrates the relationship between overall survival months and a specific macro mutation signature (MMS) that corresponds to a GO term. The purple curve represents the patients each of whom has at least two somatic mutations on the member genes of the indicated MMS (i.e., GO term). The red curve represents the patients each of whom has one somatic mutation on the member genes of the indicated MMS. The blue curve represents the patients without any somatic mutation on the member genes of the indicated MMS.

Figure S3 An illustration of the proposed classification tree model for patient survival prediction. This sample tree is generated using the three negative predictors (N_j, $1 \leq j \leq 3$) and five positive predictors (P_k, $1 \leq k \leq 5$) as the features. S represents the entire sample (or patient) set. B represents the predicted poor-prognosis set of patients. $\bar{B}$ represents the remaining patient set after B is excluded. G represents the predicted patient set with good-prognosis. M represents the intermediate-prognosis set of patients, which is the remaining section of S after B and G are excluded. Note that in this sample tree, the feature tested at each internal node is a feature set instead of a single feature, which is different from the traditional classification/decision tree model.

Acknowledgments

The results presented here are based upon data published by The Cancer Genome Atlas managed by the NCI and NHGRI. Information regarding TCGA can be found at http://cancergenome.nih.gov. The authors are grateful to the four reviewers for their insightful comments which greatly improved this paper.

Author Contributions

Conceived and designed the experiments: WZ KZ. Performed the experiments: WZ. Analyzed the data: WZ KZ. Wrote the paper: WZ AE EF KZ. Helped with experiment design: EF AE.

References

1. Siegel R, Ma J, Zou Z, Jemal A (2014) Cancer statistics, 2014. CA Cancer J Clin 64: 9–29.
2. Goff BA, Mandel L, Muntz HG, Melancon CH (2000) Ovarian carcinoma diagnosis. Cancer 89: 2068–2075.
3. Bast RC Jr, Hennessy B, Mills GB (2009) The biology of ovarian cancer: new opportunities for translation. Nat Rev Cancer 9: 415–428.
4. Li J, Fadare O, Xiang L, Kong B, Zheng W (2012) Ovarian serous carcinoma: recent concepts on its origin and carcinogenesis. J Hematol Oncol 5: 8.
5. Bunz F (2008) Principles of cancer genetics. Dordrecht?: Springer. xi, 325 p. p.
6. Fearon ER, Vogelstein B (1990) A genetic model for colorectal tumorigenesis. Cell 61: 759–767.
7. TCGA (2011) Integrated genomic analyses of ovarian carcinoma. Nature 474: 609–615.

8. Birkbak NJ, Kochupurakkal B, Izarzugaza JM, Eklund AC, Li Y, et al. (2013) Tumor mutation burden forecasts outcome in ovarian cancer with BRCA1 or BRCA2 mutations. PLoS One 8: e80023.
9. TCGA website. Available: http://cancergenome.nih.gov/. Accessed 2013 Oct 22.
10. Delfino KR, Rodriguez-Zas SL (2013) Transcription factor-microRNA-target gene networks associated with ovarian cancer survival and recurrence. PLoS One 8: e58608.
11. Braun R, Finney R, Yan C, Chen QR, Hu Y, et al. (2013) Discovery analysis of TCGA data reveals association between germline genotype and survival in ovarian cancer patients. PLoS One 8: e55037.

12. Dai W, Zeller C, Masrour N, Siddiqui N, Paul J, et al. (2013) Promoter CpG island methylation of genes in key cancer pathways associates with clinical outcome in high-grade serous ovarian cancer. Clin Cancer Res 19: 5788–5797.

13. Engler DA, Gupta S, Growdon WB, Drapkin RI, Nitta M, et al. (2012) Genome wide DNA copy number analysis of serous type ovarian carcinomas identifies genetic markers predictive of clinical outcome. PLoS One 7: e30996.

14. Cope L, Wu RC, Shih Ie M, Wang TL (2013) High level of chromosomal aberration in ovarian cancer genome correlates with poor clinical outcome. Gynecol Oncol 128: 500–505.

15. COSMIC website. Available: http://cancer.sanger.ac.uk/cancergenome/projects/cosmic/. Accessed 2013 Oct 20.

16. Sohn I, Jung WY, Sung CO (2012) Somatic hypermutation and outcomes of platinum based chemotherapy in patients with high grade serous ovarian cancer. Gynecol Oncol 126: 103–108.

17. Sohn I, Sung CO (2013) Predictive modeling using a somatic mutational profile in ovarian high grade serous carcinoma. PLoS One 8: e54089.

18. Korosteleva O (2009) Clinical statistics: introducing clinical trials, survival analysis, and longitudinal data analysis. Sudbury, Mass.: Jones and Bartlett Publishers. vii, 120 p. p.

19. Huang da W, Sherman BT, Tan Q, Collins JR, Alvord WG, et al. (2007) The DAVID Gene Functional Classification Tool: a novel biological module-centric algorithm to functionally analyze large gene lists. Genome Biol 8: R183.

20. Wang JZ, Du Z, Payattakool R, Yu PS, Chen CF (2007) A new method to measure the semantic similarity of GO terms. Bioinformatics 23: 1274–1281.

21. Santarius T, Shipley J, Brewer D, Stratton MR, Cooper CS (2010) A census of amplified and overexpressed human cancer genes. Nat Rev Cancer 10: 59–64.

22. Futreal PA, Coin L, Marshall M, Down T, Hubbard T, et al. (2004) A census of human cancer genes. Nat Rev Cancer 4: 177–183.

23. Weinberg RA (2007) The biology of cancer. New York, NY: Garland Science.

24. Kunz M, Ibrahim SM (2003) Molecular responses to hypoxia in tumor cells. Mol Cancer 2: 23.

25. Lopez-Lazaro M (2010) A new view of carcinogenesis and an alternative approach to cancer therapy. Mol Med 16: 144–153.

26. Vander Heiden MG, Cantley LC, Thompson CB (2009) Understanding the Warburg effect: the metabolic requirements of cell proliferation. Science 324: 1029–1033.

27. Repnik U, Stoka V, Turk V, Turk B (2012) Lysosomes and lysosomal cathepsins in cell death. Biochim Biophys Acta 1824: 22–33.

28. Tardy C, Codogno P, Autefage H, Levade T, Andrieu-Abadie N (2006) Lysosomes and lysosomal proteins in cancer cell death (new players of an old struggle). Biochim Biophys Acta 1765: 101–125.

29. Guicciardi ME, Leist M, Gores GJ (2004) Lysosomes in cell death. Oncogene 23: 2881–2890.

30. Fehrenbacher N, Jaattela M (2005) Lysosomes as targets for cancer therapy. Cancer Res 65: 2993–2995.

31. Erdal H, Berndtsson M, Castro J, Brunk U, Shoshan MC, et al. (2005) Induction of lysosomal membrane permeabilization by compounds that activate p53-independent apoptosis. Proc Natl Acad Sci U S A 102: 192–197.

32. Aoudjit F, Vuori K (2001) Integrin signaling inhibits paclitaxel-induced apoptosis in breast cancer cells. Oncogene 20: 4995–5004.

33. Park CC, Zhang HJ, Yao ES, Park CJ, Bissell MJ (2008) Beta1 integrin inhibition dramatically enhances radiotherapy efficacy in human breast cancer xenografts. Cancer Res 68: 4398–4405.

34. Aoudjit F, Vuori K (2012) Integrin signaling in cancer cell survival and chemoresistance. Chemother Res Pract 2012: 283181.

35. Wortzel I, Seger R (2011) The ERK Cascade: Distinct Functions within Various Subcellular Organelles. Genes Cancer 2: 195–209.

36. Liu Y, Sun Y, Broaddus R, Liu J, Sood AK, et al. (2012) Integrated analysis of gene expression and tumor nuclear image profiles associated with chemotherapy response in serous ovarian carcinoma. PLoS One 7: e36383.

37. Kafri R, Bar-Even A, Pilpel Y (2005) Transcription control reprogramming in genetic backup circuits. Nat Genet 37: 295–299.

38. Ihmels J, Collins SR, Schuldiner M, Krogan NJ, Weissman JS (2007) Backup without redundancy: genetic interactions reveal the cost of duplicate gene loss. Mol Syst Biol 3: 86.

39. Liu M, Li JS, Tian DP, Huang B, Rosqvist S, et al. (2013) MCM2 expression levels predict diagnosis and prognosis in gastric cardiac cancer. Histol Histopathol 28: 481–492.

40. Hu Z, Huang G, Sadanandam A, Gu S, Lenburg ME, et al. (2010) The expression level of HJURP has an independent prognostic impact and predicts the sensitivity to radiotherapy in breast cancer. Breast Cancer Res 12: R18.

41. Sawada G, Ueo H, Matsumura T, Uchi R, Ishibashi M, et al. (2013) CHD8 is an independent prognostic indicator that regulates Wnt/beta-catenin signaling and the cell cycle in gastric cancer. Oncol Rep 30: 1137–1142.

42. Yan Z, Kim YS, Jetten AM (2002) RAP80, a novel nuclear protein that interacts with the retinoid-related testis-associated receptor. J Biol Chem 277: 32379–32388.

43. Wu J, Liu C, Chen J, Yu X (2012) RAP80 protein is important for genomic stability and is required for stabilizing BRCA1-A complex at DNA damage sites in vivo. J Biol Chem 287: 22919–22926.

44. Mascolo M, Vecchione ML, Ilardi G, Scalvenzi M, Molea G, et al. (2010) Overexpression of Chromatin Assembly Factor-1/p60 helps to predict the prognosis of melanoma patients. BMC Cancer 10: 63.

45. Weinstein JN, Collisson EA, Mills GB, Shaw KR, Ozenberger BA, et al. (2013) The Cancer Genome Atlas Pan-Cancer analysis project. Nat Genet 45: 1113–1120.

46. Silwal-Pandit L, Vollan HK, Chin SF, Rueda OM, McKinney S, et al. (2014) TP53 Mutation Spectrum in Breast Cancer Is Subtype Specific and Has Distinct Prognostic Relevance. Clin Cancer Res 20: 3569–3580.

47. van de Vijver MJ, He YD, van't Veer LJ, Dai H, Hart AA, et al. (2002) A gene-expression signature as a predictor of survival in breast cancer. N Engl J Med 347: 1999–2009.

48. Finak G, Bertos N, Pepin F, Sadekova S, Souleimanova M, et al. (2008) Stromal gene expression predicts clinical outcome in breast cancer. Nat Med 14: 518–527.

49. Li J, Lenferink AE, Deng Y, Collins C, Cui Q, et al. (2010) Identification of high-quality cancer prognostic markers and metastasis network modules. Nat Commun 1: 34.

50. Berchuck A, Iversen ES, Lancaster JM, Pittman J, Luo J, et al. (2005) Patterns of gene expression that characterize long-term survival in advanced stage serous ovarian cancers. Clin Cancer Res 11: 3686–3696.

51. Partheen K, Levan K, Osterberg L, Horvath G (2006) Expression analysis of stage III serous ovarian adenocarcinoma distinguishes a sub-group of survivors. Eur J Cancer 42: 2846–2854.

52. Bonome T, Levine DA, Shih J, Randonovich M, Pise-Masison CA, et al. (2008) A gene signature predicting for survival in suboptimally debulked patients with ovarian cancer. Cancer Res 68: 5478–5486.

53. Tothill RW, Tinker AV, George J, Brown R, Fox SB, et al. (2008) Novel molecular subtypes of serous and endometrioid ovarian cancer linked to clinical outcome. Clin Cancer Res 14: 5198–5208.

54. Venet D, Dumont JE, Detours V (2011) Most random gene expression signatures are significantly associated with breast cancer outcome. PLoS Comput Biol 7: e1002240.

55. TCGA (2012) Comprehensive molecular portraits of human breast tumours. Nature 490: 61–70.

56. GO website. Available: http://www.geneontology.org/. Accessed 2013 Oct 8.

57. Therneau T (2013) A Package for Survival Analysis in S. R package version 2.37–4. http://CRANR-projectorg/package=survival

58. Therneau TM, Grambsch PM (2000) Modeling survival data: extending the Cox model. New York: Springer. xiii, 350 p. p.

59. Yu G, Li F, Qin Y, Bo X, Wu Y, et al. (2010) GOSemSim: an R package for measuring semantic similarity among GO terms and gene products. Bioinformatics 26: 976–978.

60. Zhang W, Edwards A, Flemington EK, Zhang K (2013) Inferring polymorphism-induced regulatory gene networks active in human lymphocyte cell lines by weighted linear mixed model analysis of multiple RNA-Seq datasets. PLoS One 8: e78868.

61. Pickrell JK, Marioni JC, Pai AA, Degner JF, Engelhardt BE, et al. (2010) Understanding mechanisms underlying human gene expression variation with RNA sequencing. Nature 464: 768–772.

2

Development of a Gene-Centered SSR Atlas as a Resource for Papaya (*Carica papaya*) Marker-Assisted Selection and Population Genetic Studies

Newton Medeiros Vidal[1][*][¤], **Ana Laura Grazziotin**[1], **Helaine Christine Cancela Ramos**[2], **Messias Gonzaga Pereira**[2], **Thiago Motta Venancio**[1][*]

1 Laboratório de Química e Função de Proteínas e Peptídeos, Centro de Biociências e Biotecnologia, Universidade Estadual do Norte Fluminense Darcy Ribeiro, Campos dos Goytacazes, Rio de Janeiro, Brazil, **2** Laboratório de Melhoramento Genético Vegetal, Centro de Ciências e Tecnologias Agropecuárias, Universidade Estadual do Norte Fluminense Darcy Ribeiro, Campos dos Goytacazes, Rio de Janeiro, Brazil

Abstract

Carica papaya (papaya) is an economically important tropical fruit. Molecular marker-assisted selection is an inexpensive and reliable tool that has been widely used to improve fruit quality traits and resistance against diseases. In the present study we report the development and validation of an atlas of papaya simple sequence repeat (SSR) markers. We integrated gene predictions and functional annotations to provide a gene-centered perspective for marker-assisted selection studies. Our atlas comprises 160,318 SSRs, from which 21,231 were located in genic regions (i.e. inside exons, exon-intron junctions or introns). A total of 116,453 (72.6%) of all identified repeats were successfully mapped to one of the nine papaya linkage groups. Primer pairs were designed for markers from 9,594 genes (34.5% of the papaya gene complement). Using papaya-tomato orthology assessments, we assembled a list of 300 genes (comprising 785 SSRs) potentially involved in fruit ripening. We validated our atlas by screening 73 SSR markers (including 25 fruit ripening genes), achieving 100% amplification rate and uncovering 26% polymorphism rate between the parental genotypes (Sekati and JS12). The SSR atlas presented here is the first comprehensive gene-centered collection of annotated and genome positioned papaya SSRs. These features combined with thousands of high-quality primer pairs make the atlas an important resource for the papaya research community.

Editor: Chunxian Chen, USDA/ARS, United States of America

Funding: This work was funded by Coordenação de Aperfeiçoamento de Pessoal de Nível Superior (CAPES), Instituto Nacional de Ciência e Tecnologia - Entomologia Molecular (INCT-EM), Universidade Estadual do Norte Fluminense Darcy Ribeiro (UENF), and Fundação de Amparo à Pesquisa do Estado do Rio de Janeiro Carlos Chagas Filho (FAPERJ) (PENSARIO E-26/110.720/2012). The funders had no role in study design, data collection and analysis, decision to publish, or preparation of the manuscript.

Competing Interests: The authors have declared that no competing interests exist.

* Email: thiago.venancio@gmail.com (TMV); nwvidal@gmail.com (NMV)

¤ Current address: National Center for Biotechnology Information, National Library of Medicine, National Institutes of Health, Bethesda, Maryland, 20894, United States of America

Introduction

Papaya (*Carica papaya* Linneaus) is an economically and nutritionally important fruit tree of tropical and subtropical regions. Papaya is well known for its nutritional benefits [1], medical [2] and industrial [3] applications. Due to its commercial importance, papaya production is currently ranked as the third major global production among tropical fruits [4]. Notwithstanding the increased papaya trade, a limited number of cultivars are commercially available, hampering papaya production worldwide. Further, the low genetic diversity of selected *C. papaya* cultivars [5–7] makes the species susceptible to bacterial and viral infections [8,9]. To improve disease resistance, genetic diversity and productiveness, researchers have been using molecular marker-assisted selection (MAS), a well-established procedure employed in commercial breeding programs to enhance the gain from artificial selection.

Microsatellites, also known as simple sequence repeats (SSRs), are simple tandemly repeated DNA sequences, ranging from 2–6 base pairs per repeat unit [10]. These repeated sequences are highly variable in length, mainly due to unequal recombination events or DNA polymerase slippage. Microsatellite PCR amplification has been described as a reliable, rapid, and inexpensive technique. Combined with the highly polymorphic nature and co-dominant segregation of SSRs, PCR amplification of microsatellites is a powerful technique for plant breeding and genetic studies, such as MAS, population genetic analysis, quantitative trait locus (QTLs) mapping, DNA fingerprinting, and genome mapping [11–16].

Once a labor-intensive and time-consuming process, identification of new microsatellite markers became increasingly feasible with the improvement of molecular biology techniques and availability of genomic information for several plant species. Over the past decade, SSRs derived from expressed sequence tags (EST-

SSRs) markers emerged as a feasible alternative in marker development for several crop species [17]. EST-SSRs are transcribed from coding sequences (CDS), which tend to be conserved between species, high interspecies transferability rates can be achieved [18,19]. Moreover, CDS markers are more informative than intergenic 'anonymous' markers because they are more likely to be functional [20,21]. However, there are also a few disadvantages of using CDS markers. High conservation may result in low polymorphism rates, which are of limited use in MAS studies. In addition, primers designed for exon-intron junctions may result in PCR amplification failures.

With the increasing number of sequenced genomes, SSRs can be computationally detected and classified according to their genomic locations in 5′ untranslated regions (UTRs), exons, introns and 3′ UTRs. By using this strategy, exon-intron junctions can be avoided during primer design and fully exonic markers can be selected. Completely sequenced genomes also allow the selection of intronic markers which are more polymorphic than exonic markers and segregate with a particular gene that may be associated with a biochemical function or phenotype of interest [22,23].

In the present work we describe the analysis of papaya SSR markers in a genome-wide scale, integrating SSR positioning and functional annotation data. Stringent primer design criteria were used to allow better results in genetic studies. This complete catalog will be of great value for the papaya research community, especially for groups conducting MAS projects. Using this map, researchers will be able to filter and choose interesting markers according to SSR type, length, sequence, region location (exon, intron or intergenic), linkage group and gene annotation.

Materials and Methods

Genomic data and SSR annotation

Papaya genome assembly and annotation files were downloaded from Phytozome v7.0 FTP (http://www.phytozome.net/) [24]. The genome assembly 113 consists of 244.5 Mb of gapless sequences, distributed in 3,207 scaffolds and 2,693 contigs. Genomic locations of detected SSRs were integrated with gene and exon coordinates from the reference GFF file. Based on their genomic mapping, SSRs were categorized as exonic (entirely within the CDS), exon-intron (in exon-intron boundaries), intronic (within introns) and intergenic (outside of genic regions). Identifiers and annotations based on papaya-*Arabidopsis thaliana* homology were obtained for genic SSRs. Papaya genes were also annotated using Gene Ontology (GO) terms from the Plant Ontology Tool (http://www.arabidopsis.org/tools/bulk/po/index.jsp). In order to map scaffolds and contigs to the major nine linkage groups, all 47,483 papaya contigs [GenBank: ABIM01000001-

ABIM01047483] were downloaded from the National Center for Biotechnology Information (http://www.ncbi.nlm.nih.gov). Linkage group information was retrieved from GenBank files for each contig and scaffold [GenBank: DS981520-DS984726].

Simple Sequence Repeats Identification

Exact maximal repeats were detected in the papaya genome using *mreps* [25]. Perfect repeats with more than 12 nucleotides, motif lengths of 2–6 bp and at least 2 units of repetition were analyzed. Parameters were set as follows: *-r 0 -minsize 12 -minperiod 2 -maxperiod 6 -exp 2*. The *mreps* algorithm finds exact maximal repeats, removes redundancy by selecting the best period for each repeat, merges repeats with same period and eliminates statistically insignificant expected repeats [25].

For description of di- and trinucleotide motifs, circular permutations and complementary strand nucleotides were considered as equivalents and grouped in one class after determining the individual repeat frequencies. Thus, there are four possible di-nucleotide motifs and ten possible tri-nucleotide motifs. For example, motifs AG, GA, CT and TC are equivalent and grouped as AG/GA/CT/TC. Likewise, motifs ACG, CGA, GAC, CGT, TCG and GTC are also equivalents and represented as ACG/CGA/GAC/CGT/TCG/GTC.

Primer design

SSRs were retrieved from the genome with 250 bp upstream/downstream flanking regions and had their low complexity regions masked with DustMasker [26]. Primers were designed with the standalone version of Primer3 [27] using the following parameters: primer length between 18 and 25 nucleotides (optimal length = 20 nt), melting temperature between 57 and 63°C (optimal Tm = 60°C), PCR product size between 250 and 350 bp (optimal 300 bp), GC content of 20–60%, and PRIMER_MAX_HAIRPIN_TH = 24. All SSR sequences were also used as a repeat library (option PRIMER_MISPRIMING_LIBRARY) to avoid primer design within the SSRs. Primers with low complexity regions were discarded.

Analysis of genes involved in fruit ripening

As a proof of concept, our gene-centered SSR map were used to identify SSR markers for genes potentially involved in cell wall remodeling, transcriptional regulation and hormone signaling. Genes with differential expression in tomato ripening fruits (determined by RNA-Seq) [28] were used to identify homologous genes in papaya. Fifty-three cell wall and 222 transcription/ethylene proteins were used as BLASTP queries to search the papaya predicted proteins with the following criteria: E-value ≤ 1e–30, similarity of at least 50%, query and hit coverage of at least 75%. Tomato protein sequences and annotations release ITAG2.3

Table 1. Number of simple sequence repeats by genomic location.

Genomic location	SSR type					
	Di	Tri	Tetra	Penta	Hexa	Total
Exon	130	3588	117	206	1303	5344
Exon-intron	7	27	4	4	31	73
Intron	7158	1959	2343	1878	2476	15814
Intergenic	54853	24841	19584	19956	19853	139087
Total	**62148**	**30415**	**22048**	**22044**	**23663**	**160318**

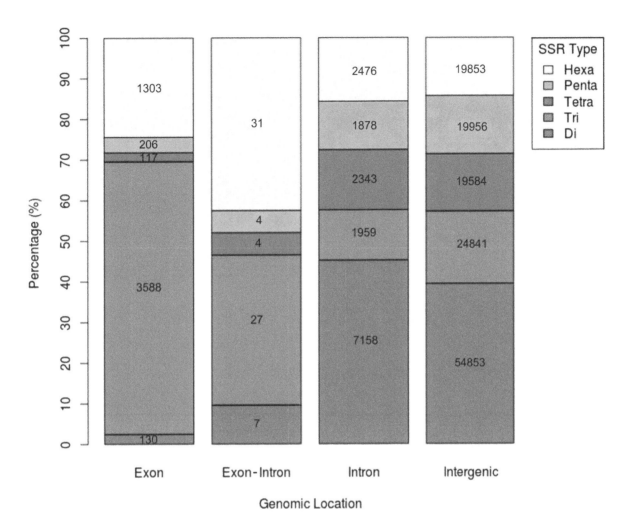

Figure 1. Distribution of SSR type according to genomic location.

were downloaded from the Sol Genomics Network FTP [29] (ftp://ftp.solgenomics.net/).

SSR screening and polymorphism survey

The papaya genotypes Sekati and JS12 were used for screening polymorphic genic SSRs. Total genomic DNA was extracted from young leaves according to the CBAB method [30]. A total of 73 primer pairs comprising ~8 genic SSR regions per chromosome were selected. PCR amplifications were performed in 15 μL reaction, containing 10 ng DNA, 10 mM Tris-HCl, pH 8.3, 50 mM KCl, 2 mM $MgCl_2$, 100 μM dNTPs, 0.2 μM of each primer, and 1 U Taq DNA polymerase. PCR cycling was performed in an Eppendorf thermal cycler, according to the following profile: 4 min of denaturation at 94°C, 35 amplification cycles (94°C at 30 s, 58°C at 1 min, 72°C at 3 min), followed by a final extension of 7 min at 72°C. In 4 of these cases, the primer annealing temperature was set to 65°C. Amplified products were separated in a 4% agarose gel Metaphor, stained by GelRedTM/ Blue Juice mixture (1:1) and visualized through the MiniBis Pro photodocumentation system (DNR Bio-Imaging Systems Ltd., Jerusalem, Israel).

Data Access and Retrieval

Information regarding SSR identifiers, genomic coordinates, motif sequence, period, size and exponent, genomic location (i.e. exon, intron, exon-intron, intergenic), linkage group, and gene annotations are fully available in two user-friendly spreadsheets (Table S1, Table S2).

Results and Discussion

SSR classification and genomic positioning

We identified 160,318 SSRs with a density of 656 SSR/Mb in the most recent version of the papaya genome (see methods for details). Previous studies of papaya SSRs reported densities of 1,340 SSR/Mb [31] and 746 SSR/Mb [32]. Such disparities in the number of identified microsatellites are usual among different reports, mainly due to differences in the algorithms, parameter settings, minimal repeat length and redundancy filtering [33–35]. Differently from other two previous studies [31,32], we used the *mreps* algorithm, which was ranked as the best algorithm for repeat detection in a systematic study [35]. Specifically, *mreps* does not report all the overlapping repeats, but efficiently retains only the most credible overlapping ones, giving more reproducible and reliable results.

We detected 160,318 perfectly matching, non-redundant SSRs. After integrating SSRs and gene coordinates, we found that 36%

Table 2. Distribution of SSRs by genomic location.

Motif	Exon		Exon-intron		Intron		Intergenic		Total	
	# SSRs	Percent (%)	# SSRs	Percent (%)	# SSRs	Percent (%)	# SSRs	Percent (%)	# SSRs	Percent (%)
Di-	130	2.4	7	9.6	7158	45.3	54853	39.4	62148	38.8
AT/TA	6	4.6	0	0.0	4554	63.6	38906	70.9	43466	69.9
AC/CA/GT/TG	8	6.2	1	14.3	1049	14.7	7307	13.3	8365	13.5
AG/GA/CT/TC	115	88.5	6	85.7	1546	21.6	8584	15.6	10251	16.5
CG/GC	1	0.8	0	0.0	9	0.1	56	0.1	66	0.1
Tri-	3588	67.1	27	37.0	1959	12.4	24841	17.9	30415	19.0
AAC/ACA/CAA/GTT/TGT/TTG	153	4.3	0	0.0	111	5.7	728	2.9	992	3.3
AAG/AGA/GAA/CTT/TCT/TTC	1236	34.4	7	25.9	546	27.9	5387	21.7	7176	23.6
AAT/ATA/TAA/ATT/TAT/TTA	78	2.2	0	0.0	913	46.6	13250	53.3	14241	46.8
GGA/GAG/AGG/TCC/CTC/CCT	406	11.3	3	11.1	44	2.2	537	2.2	990	3.3
GGC/GCG/CGG/GCC/CGC/CCG	102	2.8	1	3.7	10	0.5	260	1.0	373	1.2
GGT/GTG/TGG/ACC/CAC/CCA	316	8.8	7	25.9	49	2.5	350	1.4	722	2.4
ACG/CGA/GAC/CGT/TCG/GTC	89	2.5	2	7.4	7	0.4	107	0.4	205	0.7
ACT/CTA/TAC/AGT/TAG/GTA	225	6.3	1	3.7	88	4.5	2326	9.4	2640	8.7
AGC/GCA/CAG/GCT/TGC/CTG	372	10.4	4	14.8	26	1.3	278	1.1	680	2.2
ATC/TCA/CAT/GAT/TGA/ATG	611	17.0	2	7.4	165	8.4	1618	6.5	2396	7.9
Tetra-	117	2.2	4	5.5	2343	14.8	19584	14.1	22048	13.8
Penta-	206	3.9	4	5.5	1878	11.9	19956	14.3	22044	13.8
Hexa-	1303	24.4	31	42.5	2476	15.7	19853	14.3	23663	14.8
Total	5344	100.0	73	100.0	15814	100.0	139087	100.0	160318	100.0

Table 3. Distribution of Class I and Class II SSRs in different genomic regions.

	Exon		Exon-Intron		Intron		Intergenic	
SSR size (bp)	# SSRs	Percent (%)	# SSRs	Percent (%)	# SSRs	Percent (%)	# SSRs	Percent (%)
Class II								
12	1100	20.6	13	17.8	1450	9.2	13749	9.9
13	683	12.8	9	12.3	1883	11.9	16031	11.5
14	665	12.4	10	13.7	2237	14.1	19308	13.9
15	796	14.9	9	12.3	2474	15.6	20879	15.0
16	433	8.1	8	11.0	1542	9.8	13481	9.7
17	456	8.5	6	8.2	1280	8.1	11760	8.5
18	203	3.8	3	4.1	861	5.4	7970	5.7
19	158	3.0	2	2.7	763	4.8	6957	5.0
Subtotal	**4494**	**84.1**	**60**	**82.2**	**12490**	**79.0**	**110135**	**79.2**
Class I								
20	178	3.3	4	5.5	522	3.3	5275	3.8
21	104	1.9	2	2.7	449	2.8	4145	3.0
22	81	1.5	2	2.7	382	2.4	3223	2.3
23	91	1.7	1	1.4	340	2.1	3174	2.3
24	52	1.0	1	1.4	236	1.5	2095	1.5
25	54	1.0	1	1.4	213	1.3	1741	1.3
26	36	0.7	0	0.0	176	1.1	1725	1.2
27	30	0.6	0	0.0	169	1.1	1340	1.0
28	28	0.5	1	1.4	121	0.8	953	0.7
29	47	0.9	0	0.0	104	0.7	953	0.7
30	24	0.4	1	1.4	80	0.5	686	0.5
>30	125	2.3	0	0.0	532	3.4	3642	2.6
Subtotal	**850**	**15.9**	**13**	**17.8**	**3324**	**21.0**	**28952**	**20.8**
Total	**5344**	**100.0**	**73**	**100.0**	**15814**	**100.0**	**139087**	**100.0**

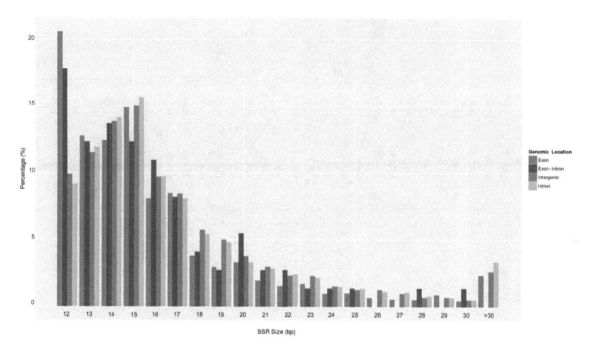

Figure 2. Percentage frequency of SSR according to SSR size.

of the papaya genes (9,992/27,769) have at least one SSR. A total of 21,231 SSRs were identified in genic regions, while 139,087 were intergenic (Figure 1). Because UTR annotations are not available for the papaya genome, SSRs located on these regions were not classified as such. As expected, most SSRs are intergenic (86.8%), followed by intronic (9.9%), exonic (3.3%), and only 73 (0.04%) SSRs in exon-intron boundaries (Table 1). Dinucleotide motifs were abundant in intergenic (39.4%) and intronic regions (45.3%), while tri- to hexanucleotides were uniformly distributed in such regions. On the other hand, exons and exon-intron boundaries were enriched in tri- (67.1% and 37%) and hexanucleotides (24.4% and 42.5%), which is expected due to the selective pressure against frameshift mutations in coding regions.

Sequence AT/TA was the most common dinucleotide motif (69.9%) in the papaya genome, corroborating previous results obtained from whole-genome shotgun sequences (WGS) and BAC End Sequences (BES) [7,31]. Nevertheless, when considering genomic locations, this motif was enriched only in intronic (63.6%) and intergenic (70.9%) regions, whereas AG/GA/CT/TC motifs were predominant in exons (88.5%) and exon-intron boundaries (85.7%) (Table 2). Among trinucleotide motifs, AAT/ATA/TAA/ATT/TAT/TTA has been described as most prevalent in papaya genome [7,31]. Although AAT/TTA sequence was also frequent (47%) in our study, it is mainly located in introns (46.6%) and intergenic (53.3%) regions. Conversely, the second most predominant trinucleotide motif, AAG/AGA/GAA/CTT/TCT/TTC, was more evenly distributed in exons (34.4%), exon-intron boundaries (25.9%), introns (27.9%), and intergenic (21.7%) locations in genomic context. Coherently to a previous study [31], dinucleotides CG/GC and trinucleotides CCG/GGC were rarely found (0.1% and 1.2% respectively) here.

Based on repeat length, papaya SSRs were defined as class I (≥ 20 nucleotides) and class II (between 12–19 nucleotides) (Figure 2). SSR lengths ranged from: 12–82 bases in exons; 12–30 bases in exon-intron boundaries; 12–201 bases in introns and 12–155 bases in intergenic regions. Most SSRs were classified as class II (79.0%–

84.1%), regardless of their genomic location (Table 3). However, 24,234 primer pairs were designed for class I SSRs (Table S3). Since these longer sequences are typically hypervariable and more likely to be polymorphic, they are the preferable choice as molecular markers for diversity studies.

All SSRs were assigned to chromosomes according to their scaffold or contig localization. A total of 116,453 SSRs (72.6%) could be mapped to one of the nine papaya linkage groups. The number of SSRs in each chromosome ranged from 10,566 (6.6%) in LG7 to 14,773 (9.2%) in LG9. The proportion of SSR types and motifs among different chromosomes was similar to the overall genomic distribution (Table S4). The proportion of SSR genomic locations in each chromosome was higher in intergenic regions, followed by introns, exons and exon-intron junctions. There was no bias for SSR types and SSR genomic locations among the chromosomes (Table S5), which is highly desirable for researchers aiming to develop a collection of polymorphic markers for genetic studies.

Design of high-quality primer pairs for papaya SSRs

Aiming to provide a comprehensive source of SSRs to be used in marker-assisted selection and population genetics studies, all 21,231 genic and 139,087 intergenic SSRs were submitted to primer design. All primer pairs were optimized for the same PCR conditions (see methods for details). In a preliminary analysis, 20,659 (97.3%) and 118,831 (85.4%) primer pairs were respectively designed for genic and intergenic SSRs using Primer3 with default parameters. Manual inspection of results revealed a significant number of primers containing repeated or low complexity sequences. Therefore, we decided to adopt more stringent criteria for primer design: 1) Primers within repetitive sequences were removed; 2) Primers in soft-masked low complexity 3′ end regions were not allowed (PRIMER_LOWERCASE_-MASKING = 1); and 3) A parameter to minimize hairpin formation (PRIMER_MAX_HAIRPIN_TH = 24) was employed. By using this stringent parameterization, we obtained a much more reliable primer set, although the overall success rate of

Table 4. Number of successfully designed primer pairs for SSR type and linkage group.

SSR Type	LG1 #	LG1 %	LG2 #	LG2 %	LG3 #	LG3 %	LG4 #	LG4 %	LG5 #	LG5 %	LG6 #	LG6 %	LG7 #	LG7 %	LG8+LG10[a] #	LG8+LG10[a] %	LG9 #	LG9 %	Un[b] #	Un[b] %	Total #	Total %
2	3669	41.3	4254	41.3	4212	40.4	3433	40.9	3154	39.5	4429	41.2	3221	41.7	3988	41.7	4446	41.1	10200	35.6	45008	39.7
3	1645	18.5	1922	18.7	1892	18.2	1565	18.6	1414	17.7	1871	17.4	1307	16.9	1639	17.1	1937	17.9	5483	19.2	20677	18.2
4	1233	13.9	1357	13.2	1351	13.0	1163	13.8	1068	13.4	1436	13.4	1003	13.0	1242	13.0	1458	13.5	3780	13.2	15093	13.3
5	1066	12.0	1271	12.4	1430	13.7	1048	12.5	1114	14.0	1342	12.5	1029	13.3	1252	13.1	1347	12.4	4539	15.9	15441	13.6
6	1264	14.2	1486	14.4	1531	14.7	1193	14.2	1225	15.4	1659	15.5	1158	15.0	1449	15.1	1639	15.1	4621	16.1	17227	15.2
Total	8877	100.0	10290	100.0	10416	100.0	8402	100.0	7975	100.0	10737	100.0	7718	100.0	9570	100.0	10827	100.0	28623	100.0	113446	100.0

LG represents the Linkage Groups associated to the nine papaya chromosomes.
[a]LG8 and LG10 are associated to the same chromosome.
[b]SSRs located in scaffolds/contigs were classified as unplaced.

Table 5. Number of successfully designed primer pairs for each genomic context and linkage group.

Genomic location	LG1 #	LG1 %	LG2 #	LG2 %	LG3 #	LG3 %	LG4 #	LG4 %	LG5 #	LG5 %	LG6 #	LG6 %	LG7 #	LG7 %	LG8+LG10[a] #	LG8+LG10[a] %	LG9 #	LG9 %	Un[b] #	Un[b] %	Total #	Total %
Exon	443	5.0	576	5.6	466	4.5	385	4.6	349	4.4	540	5.0	316	4.1	414	4.3	520	4.8	693	2.4	4702	4.1
Exon-intron	10	0.1	7	0.1	9	0.1	5	0.1	8	0.1	8	0.1	5	0.1	4	0.0	2	0.0	5	0.0	63	0.1
Intron	1457	16.4	1468	14.3	1442	13.8	1231	14.7	1140	14.3	1589	14.8	1029	13.3	1346	14.1	1440	13.3	2018	7.1	14160	12.5
Intergenic	6967	78.5	8239	80.1	8499	81.6	6781	80.7	6478	81.2	8600	80.1	6368	82.5	7806	81.6	8865	81.9	25907	90.5	94521	83.3
Total	8877	100.0	10290	100.0	10416	100.0	8402	100.0	7975	100.0	10737	100.0	7718	100.0	9570	100.0	10827	100.0	28623	100.0	113446	100.0

LG represents the Linkage Groups associated to the nine papaya chromosomes.
[a]LG8 and LG10 are associated to the same chromosome.
[b]SSRs located in scaffolds/contigs were classified as unplaced.

primer design dropped to 71%. A total of 18,925 primer pairs were successfully designed for 89.1% and 67.9% of all genic and intergenic SSR sequences, respectively (Table 4). Although such stringent settings prevented primer design for 8% genic and 17% intergenic SSRs, this methodology resulted in an extensive and more reliable list of primer pairs designed for distinct SSR types, genomic locations and chromosome linkage groups (Table S1, Table S2).

Designed primers were uniformly distributed among SSR types (Table 4) and chromosomes (Table 5). Regarding SSR type, most of the 113,446 primer pairs were designed for dinucleotide repeats (39.7%), followed by tri- (18.2%), hexa- (15.2%), penta- (13.6%) and tetranucleotide repeats (13.3%) (Table 4). As expected, the majority of exonic SSRs with designed primers are composed of trinucleotide repeats (Table S6).

Analysis of SSRs located in genes related to fruit ripening

Next, we aimed to use our SSR atlas to find genes related to fruit ripening, a trait of high agronomical interest in papaya. To achieve this goal, tomato gene expression and tomato/papaya orthology data were integrated. Tomato is the papaya's closest climacteric fleshy fruit with available genome-wide gene expression data [28]. Sato et al. reported significant differential expression of 53 cell wall and 222 transcription factors (TFs)/ethylene-related genes during fruit ripening [28]. Based on BLASTP searches (see methods for details), 175 cell wall and 319 transcription factor orthologous genes were found in papaya (Table S7, Table S8). Our atlas include SSRs with primer pairs for 113 cell wall-related (257 SSRs, 40 exonic and 217 intronic; 2.3 SSRs/gene) and 187 TF/ethylene-related genes (528 SSRs, 127 exonic and 400 intronic; 2.8 SSRs/gene). These two groups comprise very good candidate pulp softening and pigmentation control genes [36]. By integrating this information in our gene-centered SSR map, we provide an unprecedented list of markers for studying the genetic and functional variability of fruit ripening processes.

Fruit ripening is a developmental process characterized by remarkable changes related to flavor, sugar metabolism, color, aroma, texture, softening and nutritional content [37]. These metabolic and physical alterations are driven by genetically coordinated expression profiles in several metabolic pathways, such as cell wall disassembly, sugar hydrolysis, ethylene biosynthesis and pigmentation. Using KEGG Orthology (KO) we identified ripening-related pathways for cell wall genes and TFs harboring SSRs with primer pairs. Nine entries for cell wall genes were found in KO00050 - Starch and sucrose metabolism (5) and KO00040– Pentose and glucuronate interconversions (4). During climacteric fruit ripening, respiration rate increases and a series of enzymes degrade starch and synthesize sucrose (e.g. starch phosphorylase and sucrose synthase) [38]. The carboxylic acid glucuronate is the precursor of pectin, one of the main components of plant cell wall [39]. Pectin levels decrease during fruit maturation [39] typically due to the increased expression of pectin lyases [28]. However, since pectin lyases and methyltransferases are repressed during papaya ripening, pectin solubilization is probably catalyzed by polygalacturonases [40].

Among TFs/ethylene-related genes, 24 entries were found, mostly from the category KO04075 (Plant hormone signal transduction; 8 genes), such as auxin-responsive protein, ethylene receptor and responsive transcription factor 1 and ABA-responsive element binding factor. The phytohormone auxin plays important roles in fruit growth, regulating cell division, differentiation, lateral root formation and embryogenesis [41]. Particularly, genes involved in signaling and auxin response factors were up-regulated

during papaya [40] and peach [42] ripening. In addition, auxin can stimulate ethylene biosynthesis via transcription of acetyl-coenzyme A synthetase genes [42]. In turn, the involvement of ethylene response factor in climacteric fruit ripening is well-established and this regulator determines, for example, fruit firmness reduction and defense response to pathogens after ripe [36]. By aggregating these annotations, our atlas provides a rich resource from which the scientific community can rapidly draw genes and SSRs with designed primer pairs for genetic studies.

SSR screening and polymorphism survey

A 100% PCR amplification rate was achieved for the 73 genic SSRs (16 exonic and 57 intronic), allowing the identification of 19 polymorphic alleles (26%) (Table S9). Twenty five of such genes are orthologs of tomato genes with differential expression during fruit ripening (Table S7, Table S8). Among the polymorphic alleles we found 5 cell wall and 2 transcription factors/ethylene-related genes. Polymorphisms were also detected in genes from the cellulose synthase, pectin lyase-like and ethylene response factor families/superfamilies. Taken together, these results not only validate the use of our atlas as an efficient tool in papaya breeding projects, but also stimulate additional genetic and biochemical studies to detail the functions these polymorphic genes in papaya, as they may be useful in the production of fruits with increased shelf life.

Conclusion

Non-coding SSRs near genes or inside introns can directly affect gene expression [21,43–45]. On the other hand, SSRs within exons often result in amino acid changes that may affect protein function. For example, a study using genic SSRs derived from candidate genes involved in wood formation identified two SSR markers (one in coding and the other in non-coding region) explaining 13.5% of the lignin content variation in Chinese white poplar [46]. These results demonstrate the power of genic markers to identify genotype-to-phenotype associations and make these markers very useful in genetic improvement of desired characteristics.

In the present work we surveyed the papaya genome for the presence of perfect, non-redundant SSRs. We analyzed the distribution of SSR locations (exon, exon-intron, intron, intergenic) and established a comprehensive atlas of SSR markers with SSR type, motif sequence, SSR size, genomic location (exon, intron or intergenic), linkage group location and gene-centered information (gene annotation and GO assignments). The resource reported here is fully accessible through our supplementary material (Table S1, Table S2), allowing plant breeders and researchers to easily choose gene-centered markers to test their association with biological processes or phenotypes of agronomic interest. Moreover, we achieved a 100% PCR amplification rate during a genetic survey of 73 SSR markers, supporting the high quality of the predicted SSRs and designed primers. The atlas developed in this study will certainly serve as a toolbox to assist and improve the efficiency of marker-assisted selection in papaya breeding and population genetic studies.

Supporting Information

Table S1   Catalog of gene-centered SSR markers within genic regions (exon, exon-intron, intron).

Table S2   Catalog of SSR markers within intergenic regions.

Table S3   Class I SSR markers.

Table S4   Distribution of SSR motifs by linkage group.

Table S5   Distribution of SSR types by linkage group.

Table S6   Number of successfully designed primer pairs for each SSR type, genomic location and linkage group.

Table S7   SSR markers for cell wall genes.

Table S8   SSR markers for transcriptional/ethylene genes.

Table S9   Primer pairs used for polymorphism analysis. Genes related with cell wall metabolism and transcriptional regulation/ethylene signaling genes are highlighted in green and yellow, respectively.

Author Contributions

Conceived and designed the experiments: NMV ALG HCCR MGP TMV. Performed the experiments: NMV HCCR TMV. Analyzed the data: NMV ALG TMV. Contributed reagents/materials/analysis tools: NMV ALG HCCR MGP TMV. Contributed to the writing of the manuscript: NMV ALG TMV.

References

1. Marotta F, Catanzaro R, Yadav H, Jain S, Tomella C, et al. (2012) Functional foods in genomic medicine: a review of fermented papaya preparation research progress. Acta Biomed 83: 21–29.
2. Jimenez-Coello M, Guzman-Marin E, Ortega-Pacheco A, Perez-Gutierrez S, Acosta-Viana KY (2013) Assessment of the anti-protozoal activity of crude Carica papaya seed extract against Trypanosoma cruzi. Molecules 18: 12621–12632.
3. Tu T, Meng K, Bai Y, Shi P, Luo H, et al. (2013) High-yield production of a low-temperature-active polygalacturonase for papaya juice clarification. Food Chemistry 141: 2974–2981.
4. Evans EA, Ballen FH (2012) An overview of global papaya production, trade, and consumption. Gainesville: University of Florida. 7 p.
5. Kim MS, Moore PH, Zee F, Fitch MM, Steiger DL, et al. (2002) Genetic diversity of Carica papaya as revealed by AFLP markers. Genome 45: 503–512.
6. Ma H, Moore PH, Liu Z, Kim MS, Yu Q, et al. (2004) High-density linkage mapping revealed suppression of recombination at the sex determination locus in papaya. Genetics 166: 419–436.
7. Eustice M, Yu Q, Lai CW, Hou S, Thimmapuram J, et al. (2008) Development and application of microsatellite markers for genomic analysis of papaya. Tree Genetics and Genomes 4: 333–341.
8. Davis MJ, Ying Z, Brunner BR, Pantoja A, Ferwerda FH (1998) Rickettsial relative associated with papaya bunchy top disease. Current Microbiology 36: 80–84.
9. Gonsalves D (1998) Control of papaya ringspot virus in papaya: a case study. Annual Review of Phytopathology 36: 415–437.
10. Tautz D, Trick M, Dover GA (1986) Cryptic simplicity in DNA is a major source of genetic variation. Nature 322: 652–656.
11. Zhou H, Steffenson BJ, Muehlbauer G, Wanyera R, Njau P, et al. (2014) Association mapping of stem rust race TTKSK resistance in US barley breeding germplasm. Theoretical and Applied Genetics.
12. Carrillo E, Satovic Z, Aubert G, Boucherot K, Rubiales D, et al. (2014) Identification of quantitative trait loci and candidate genes for specific cellular resistance responses against Didymella pinodes in pea. Plant Cell Reports.
13. Wang Z, Kang M, Liu H, Gao J, Zhang Z, et al. (2014) High-level genetic diversity and complex population structure of Siberian apricot (Prunus sibirica L.) in China as revealed by nuclear SSR markers. PLoS One 9: e87381.
14. Yilancioglu K, Cetiner S (2013) Rediscovery of historical Vitis vinifera varieties from the South Anatolia region by using amplified fragment length polymorphism and simple sequence repeat DNA fingerprinting methods. Genome 56: 295–302.
15. Liu SR, Li WY, Long D, Hu CG, Zhang JZ (2013) Development and characterization of genomic and expressed SSRs in citrus by genome-wide analysis. PLoS One 8: e75149.
16. Xiao L, Hu Y, Wang B, Wu T (2013) Genetic mapping of a novel gene for soybean aphid resistance in soybean (Glycine max [L.] Merr.) line P203 from China. Theoretical and Applied Genetics 126: 2279–2287.
17. Kantety RV, La Rota M, Matthews DE, Sorrells ME (2002) Data mining for simple sequence repeats in expressed sequence tags from barley, maize, rice, sorghum and wheat. Plant Molecular Biology 48: 501–510.
18. Pashley CH, Ellis JR, McCauley DE, Burke JM (2006) EST databases as a source for molecular markers: lessons from Helianthus. Journal of Heredity 97: 381–388.
19. Victoria FC, da Maia LC, de Oliveira AC (2011) In silico comparative analysis of SSR markers in plants. BMC Plant Biology 11: 15.
20. Coulibaly I, Gharbi K, Danzmann RG, Yao J, Rexroad CE (2005) Characterization and comparison of microsatellites derived from repeat-enriched libraries and expressed sequence tags. Animal Genetics 36: 309–315.
21. Varshney RK, Graner A, Sorrells ME (2005) Genic microsatellite markers in plants: features and applications. Trends in Biotechnology 23: 48–55.
22. Zhang L, Zuo K, Zhang F, Cao Y, Wang J, et al. (2006) Conservation of noncoding microsatellites in plants: implication for gene regulation. BMC Genomics 7: 323.
23. Parida SK, Dalal V, Singh AK, Singh NK, Mohapatra T (2009) Genic non-coding microsatellites in the rice genome: characterization, marker design and use in assessing genetic and evolutionary relationships among domesticated groups. BMC Genomics 10: 140.
24. Goodstein DM, Shu S, Howson R, Neupane R, Hayes RD, et al. (2012) Phytozome: a comparative platform for green plant genomics. Nucleic Acids Research 40: D1178–1186.
25. Kolpakov R, Bana G, Kucherov G (2003) mreps: Efficient and flexible detection of tandem repeats in DNA. Nucleic Acids Research 31: 3672–3678.
26. Morgulis A, Gertz EM, Schaffer AA, Agarwala R (2006) A fast and symmetric DUST implementation to mask low-complexity DNA sequences. Journal of Computational Biology 13: 1028–1040.
27. Rozen S, Skaletsky H (2000) Primer3 on the WWW for general users and for biologist programmers. Methods in Molecular Biology 132: 365–386.
28. Sato S, Tabata S, Hirakawa H, Asamizu E, Shirasawa K, et al. (2012) The tomato genome sequence provides insights into fleshy fruit evolution. Nature 485: 635–641.
29. Bombarely A, Menda N, Tecle IY, Buels RM, Strickler S, et al. (2011) The Sol Genomics Network (solgenomics.net): growing tomatoes using Perl. Nucleic Acids Research 39: D1149–1155.
30. Doyle J, Doyle J (1990) Isolation of plant DNA from fresh tissue. Focus 12: 3.
31. Wang J, Chen C, Na J-K, Qingyi Y, Hou S, et al. (2008) Genome-wide comparative analyses of microsatellites in papaya. Tropical Plant Biology: 14.
32. Shi J, Huang S, Fu D, Yu J, Wang X, et al. (2013) Evolutionary dynamics of microsatellite distribution in plants: insight from the comparison of sequenced brassica, Arabidopsis and other angiosperm species. PLoS One 8: e59988.
33. Leclercq S, Rivals E, Jarne P (2007) Detecting microsatellites within genomes: significant variation among algorithms. BMC Bioinformatics 8: 125.
34. Merkel A, Gemmell N (2008) Detecting short tandem repeats from genome data: opening the software black box. Briefings in Bioinformatics 9: 355–366.
35. Lim KG, Kwoh CK, Hsu LY, Wirawan A (2013) Review of tandem repeat search tools: a systematic approach to evaluating algorithmic performance. Briefings in Bioinformatics 14: 67–81.
36. Li X, Zhu X, Mao J, Zou Y, Fu D, et al. (2013) Isolation and characterization of ethylene response factor family genes during development, ethylene regulation and stress treatments in papaya fruit. Plant Physiology and Biochemistry 70: 81–92.
37. Giovannoni JJ (2004) Genetic regulation of fruit development and ripening. Plant Cell 16: S170–S180.
38. Hubbard NL, Pharr DM, Huber SC (1990) Role of sucrose phosphate synthase in sucrose biosynthesis in ripening bananas and its relationship to the respiratory climacteric. Plant Physiology 94: 201–208.
39. Saito K, Kasai Z (1978) Conversion of labeled substrates to sugars, cell-wall polysaccharides, and tartaric acid in grape berries. Plant Physiology 62: 215–219.
40. Fabi JP, Seymour GB, Graham NS, Broadley MR, May ST, et al. (2012) Analysis of ripening-related gene expression in papaya using an Arabidopsis-based microarray. BMC Plant Biology 12.
41. Quint M, Gray WM (2006) Auxin signaling. Current Opinion in Plant Biology 9: 448–453.
42. Trainotti L, Tadiello A, Casadoro G (2007) The involvement of auxin in the ripening of climacteric fruits comes of age: the hormone plays a role of its own and has an intense interplay with ethylene in ripening peaches. Journal of Experimental Botany 58: 3299–3308.
43. Young ET, Sloan JS, Van Riper K (2000) Trinucleotide repeats are clustered in regulatory genes in Saccharomyces cerevisiae. Genetics 154: 1053–1068.

44. Li YC, Korol AB, Fahima T, Beiles A, Nevo E (2002) Microsatellites: genomic distribution, putative functions and mutational mechanisms: a review. Molecular Ecology 11: 2453–2465.

45. Li YC, Korol AB, Fahima T, Nevo E (2004) Microsatellites within genes: structure, function, and evolution. Molecular Biology and Evolution 21: 991–1007.

46. Du Q, Gong C, Pan W, Zhang D (2013) Development and application of microsatellites in candidate genes related to wood properties in the Chinese white poplar (Populus tomentosa Carr.). DNA Research 20: 31–44.

3

The Genome of the Generalist Plant Pathogen *Fusarium avenaceum* Is Enriched with Genes Involved in Redox, Signaling and Secondary Metabolism

Erik Lysøe[1]*, **Linda J. Harris**[2], **Sean Walkowiak**[2,3], **Rajagopal Subramaniam**[2,3], **Hege H. Divon**[4], **Even S. Riiser**[1], **Carlos Llorens**[5], **Toni Gabaldón**[6,7,8], **H. Corby Kistler**[9], **Wilfried Jonkers**[9], **Anna-Karin Kolseth**[10], **Kristian F. Nielsen**[11], **Ulf Thrane**[11], **Rasmus J. N. Frandsen**[11]

1 Department of Plant Health and Plant Protection, Bioforsk - Norwegian Institute of Agricultural and Environmental Research, Ås, Norway, **2** Eastern Cereal and Oilseed Research Centre, Agriculture and Agri-Food Canada, Ottawa, Canada, **3** Department of Biology, Carleton University, Ottawa, Canada, **4** Section of Mycology, Norwegian Veterinary Institute, Oslo, Norway, **5** Biotechvana, València, Spain, **6** Bioinformatics and Genomics Programme, Centre for Genomic Regulation, Barcelona, Spain, **7** Universitat Pompeu Fabra, Barcelona, Spain, **8** Institució Catalana de Recerca i Estudis Avançats, Barcelona, Spain, **9** ARS-USDA, Cereal Disease Laboratory, St. Paul, Minnesota, United States of America, **10** Department of Crop Production Ecology, Swedish University of Agricultural Sciences, Uppsala, Sweden, **11** Department of Systems Biology, Technical University of Denmark, Lyngby, Denmark

Abstract

Fusarium avenaceum is a fungus commonly isolated from soil and associated with a wide range of host plants. We present here three genome sequences of *F. avenaceum*, one isolated from barley in Finland and two from spring and winter wheat in Canada. The sizes of the three genomes range from 41.6–43.1 MB, with 13217–13445 predicted protein-coding genes. Whole-genome analysis showed that the three genomes are highly syntenic, and share>95% gene orthologs. Comparative analysis to other sequenced Fusaria shows that *F. avenaceum* has a very large potential for producing secondary metabolites, with between 75 and 80 key enzymes belonging to the polyketide, non-ribosomal peptide, terpene, alkaloid and indole-diterpene synthase classes. In addition to known metabolites from *F. avenaceum*, fuscofusarin and JM-47 were detected for the first time in this species. Many protein families are expanded in *F. avenaceum*, such as transcription factors, and proteins involved in redox reactions and signal transduction, suggesting evolutionary adaptation to a diverse and cosmopolitan ecology. We found that 20% of all predicted proteins were considered to be secreted, supporting a life in the extracellular space during interaction with plant hosts.

Editor: Yin-Won Lee, Seoul National University, Republic Of Korea

Funding: Nordisk komité for jordbruks- og matforskning (NKJ), Project number: "NKJ 135 - Impact of climate change on the interaction of Fusarium species in oats and barley", and Agriculture and Agri-Food Canada's Genomics Research & Development Initiative funded this work. The funders had no role in study design, data collection and analysis, decision to publish, or preparation of the manuscript.

Competing Interests: One of the co-authors in this manuscript, Carlos Llorens, is employed by the company Biotechvana. The authors have purchased some bioinformatic analysis from Biotchvana, and Biotechvana has no ownership to any results or material.

* Email: erik.lysoe@bioforsk.no

Introduction

Fusarium is a large, ubiquitous genus of ascomycetous fungi that includes many important plant pathogens, as well as saprophytes and endophytes. The genomes of sixteen *Fusarium* spp. have been sequenced during the past decade with a focus on species that either display a narrow host plant range or which have a saprophytic life style. *Fusarium avenaceum* is a cosmopolitan plant pathogen with a wide and diverse host range and is reported to be responsible for disease on>80 genera of plants [1]. It is well-known for causing ear blight and root rot of cereals, blights of plant species within genera as diverse as *Pinus* and *Eustoma* [2], as well as post-harvest storage rot of numerous crops, including potato [3], broccoli [4], apple [5] and rutabaga [6]. *Fusarium*

avenaceum has also been described as an endophyte [7,8] and an opportunistic pathogen of animals [9,10]. The generalist pathogen nature of *F. avenaceum* is supported by several reports on isolates that lack host specificity. One example of this is the report of *F. avenaceum* isolates from *Eustroma* sp. (aka Lisianthus) being phylogenetically similar to isolates from diverse geographical localities or which have been isolated from other hosts [11].

Fusarium avenaceum is often isolated from diseased grains in temperate areas, but an increased prevalence has also been reported in warmer regions throughout the world [12,13]. The greatest economic impact of *F. avenaceum* is associated with crown rot and head blight of wheat and barley, and the contamination of grains with mycotoxins [12]. Co-occurrence of multiple *Fusarium* species in head blight infections is often observed, and several studies covering the boreal and hemiboreal climate zones in the northern hemisphere have revealed that *F. avenaceum* is often among the dominating species [14]. Previously, *F. avenaceum* has been shown to produce several secondary metabolites, including moniliformin, enniatins, fusarin C, antibiotic Y, 2-amino-14,16-dimethyloctadecan-3-ol (2-AOD-3-ol), chlamydosporol, aurofusarin [12,15] and recently also fusaristatin A [16].

The genus *Fusarium* includes both broad-host pathogenic species, utilizing a generalist strategy, and narrow-host pathogenic species, which are specialized to a limited number of plant species. The *F. oxysporum* complex is a well-documented example of the specialist strategy, as each *forma specialis* displays a narrow host range. The genetic basis for this host specialization is dictated by a limited number of transferable genes, encoded on dispensable chromosomes [17]. However, the genetic foundation that allows *F. avenaceum* to infect such a wide range of host plant species and cope with such a diverse set of environmental conditions is currently not well understood. In an effort to shed light on the genetic factors that separates generalists from specialists within *Fusarium*, we sequenced the genomes of three different *F. avenaceum* strains isolated from two geographical locations, Finland and Canada, and from three small grain host plants: barley, spring and winter wheat. Comparison with existing *Fusarium* genomes would further explore pathogenic strategies.

Results and Discussion

Fusarium avenaceum genome sequences

We have sequenced three *F. avenaceum* genomes, one Finnish isolate from barley (Fa05001) and two Canadian isolates from spring (FaLH03) and winter wheat (FaLH27). Assembly of the 454 pyrosequencing based genomic sequence data from Fa05001 resulted in a total genome size of 41.6 Mb, while assembly of the Illumina HiSeq data for FaLH03 and FaLH27 resulted in genome sizes of 42.7 Mb and 43.1 Mb, respectively. Additional details on the assemblies can be found in (Table 1). Gene calling of the three *F. avenaceum* strains resulted in 13217 (Fa05001, gene naming convention *FAVG1_XXXXX*), 13293 (FaLH03, genes named *FAVG2_XXXXX*) and 13445 (FaLH27, genes named *FAVG3_XXXXX*) unique protein coding gene models. Previous comparative genomics studies of filamentous fungi have identified 69 core genes that are found ubiquitously across all fungal clades [18]. All three gene sets included the 69 core genes, suggesting a good assembly and reliable protein-coding gene prediction. Genome sequence data has been deposited at NCBI GenBank in the Whole Genome Shotgun (WGS) database as accession no. JPYM00000000 (Fa05001), JQGD00000000 (FaLH03) and JQGE00000000 (FaLH27), within BioProject PRJNA253730. The versions described in this paper are JPYM01000000, JQGD01000000, and JQGE01000000.

The mitochondrial genome sequence was contained within a single assembled contig for each strain (Fa05001, 49075 bp; FaLH03, 49402 bp; FaLH27, 49396 bp), supporting sufficient coverage and a high quality assembly. Prior to trimming, the FaLH03 and FaLH27 mitochondrial contigs contained 39 and 53 bp, respectively, of sequence duplicated at each end, as expected with the acquisition of a circular sequence. As found in other *Fusarium* mitochondrial genomes [19,20], the *F. avenaceum* mitochondrial genome sequences contain a low G+C content (about 33%) and encode 26 tRNAs and the ribosomal rRNAs *rnl* and *rns*. In addition, the 14 expected core genes (*cob, cox1, cox2, cox3, nad1, nad2, nad3, nad4, nad4L, nad5, nad6, atp6, atp8, atp9*) involved in oxidative phosphorylation and ATP production

are present and in the same order as other *Fusarium* mitochondrial genomes.

Genome structure in *F. avenaceum*

Electrophoretic karyotyping was performed to resolve the number of chromosomes in Fa05001. Previous karyotyping via fluorescence *in situ* hybridization has suggested that *F. avenaceum* isolated from wheat had 8–10 chromosomes [21]. Our attempt to determine the chromosome number in *F. avenaceum* Fa05001 strain by electrophoretic karyotyping was hampered due to the large size of several of the chromosomes. Southern analysis using a telomeric probe did however result in the detection of four distinct bands ranking from 1 to 5 MB, and several diffuse bands above the detection limit of the method (~5 Mb) (Figure S1 in File S1). A high order reordering of the scaffolds from the three sequenced genomes resulted in 11 supercontigs ranging from 0.8 Mb to 6.5 Mb in size, likely corresponding to entire chromosomes or chromosome arms (Figure S2 in File S1). The three genomes display a high level of microcolinearity and only a single putative large genome rearrangement was observed in an internal region of Supercontig 1 between Fa5001 and Canadian isolates (Figure S3 in File S1).

Sequence comparisons between the three genomes revealed a 91–96% nucleotide alignment, with the two Canadian isolates having the fewest unaligned bases. In addition, approximately 1.4–3.2% of the aligned nucleotides exhibited single nucleotide polymorphisms (SNPs), insertions, or deletions between isolates; these were also fewer between the Canadian isolates (Figure 1). These genetic differences were unevenly distributed across the genomes and were largely concentrated at the ends of the supercontigs, while centrally located regions remained relatively conserved. This is similar to what has been previously observed between chromosomes of other *Fusarium* spp. [22]. BLASTn analysis indicated that more than 95% of predicted genes had a significant hit within the two other *F. avenaceum* genomes (Figure S4 in File S1). Together, the results suggest that, despite the large geographical distance between the collection sites, there is a high level of similarity between the three *F. avenaceum* genomes, both in genome structure and gene content. However, some instances of poorly conserved or missing genes were observed in either one or two isolates out of the three (Figure S4 in File S1, File S2, File S3). For example, the three isolates contained some unique polyketide synthases and non-ribosomal peptide synthases. This suggests that there may be some differences in secondary metabolism between the isolates.

Comparison of genome structure to other *Fusarium* species

Phylogenetic analysis of genome-sequenced Fusaria based on *RPB1, RPB2*, rDNA cluster (18S rDNA, ITS1, 5.8S rDNA and 28S rDNA), *EF-1a* and *Lys2* suggest that *F. avenaceum* is more closely related to *F. graminearum*, with greater phylogenetic distance to *F. verticillioides, F. oxysporum* and *F. solani* [23,24]. Phylogeny using *β-tub* alone [24] suggested that *F. avenaceum* is more closely related to *F. verticillioides* than the other three species. The genome data for *F. avenaceum* allowed us to reanalyse the evolutionary history within the *Fusarium* genus based on the 69 conserved proteins, initially identified by Marcet-Houben and Gabaldón [18]. The Maximum Likelihood analysis was based on 25,535 positions distributed on six super-proteins and showed that *F. graminearum* and *F. avenaceum* clustered together in 93% of 500 iterations, with *F. oxysporum* and *F. verticillioides* as sister taxa in 100% of the cases (Figure 2). *Fusarium avenaceum* scaffolds have good alignment with super-

Table 1. Main assembly summary and annotation features of the three *F. avenaceum* genomes.

Strain	Fa05001	FaLH03	FaLH27
Sequencing technology	454	Hiseq	Hiseq
Genome size (Mb)	41.6	42.7	43.2
Sequencing coverage*	21.6x	426.6x	986.2x
Number of contigs	110	180	169
Number of scaffolds	83	104	77
Number of Large Scaffolds (>100 Kb)	40	22	18
Number of Large Scaffolds (>1 Mb)	17	14	11
N50 scaffold length (Mb)	1.43	4.11	4.14
L50 scaffold count	10	5	5
GC content (%)	48%	48%	48%
Number of predicted genes	13217	13293	13445
Average no of genes per Mb	317.7	311.6	311.8
Mean gene length (base pairs)	1554	1557	1552

*Post- removal of mitochondrial genome.

contigs of both *F. graminearum* and *F. verticillioides* with long, similar stretches of syntenic regions (Figure S5 in File S1). The synteny of *F. avenaceum* with *F. graminearum* is visualized in Figure 3, in which long stretches of genes from the same *F. avenaceum* supercontig have orthologs to neighbouring genes on *F. graminearum* chromosomes, indicating a shared genomic architecture. *F. graminearum* genes lacking orthologs in *F. avenaceum* are not distributed uniformly across the supercontig, and are mostly confined to telomeric regions, except for some at interstitial chromosomal sites. Such chromosome regions in *F. graminearum* have been shown to have a higher SNP density [22] and are influencing host specific gene expression patterns [25].

Fusarium avenaceum possesses the genetic hallmarks of a heterothallic sexual life cycle

The observation of two mating-type idiomorphs was another dissimilarity between the *F. avenaceum* isolates [26,27]. The Finnish *F. avenaceum* isolate is of the mating-type *MAT1-1*, possessing the three genes *MAT1-1-1*, *MAT1-1-2*, and *MAT1-1-3* (*FAVG1_07020*, *FAVG1_07021*, and *FAVG1_07022*), while the two Canadian isolates are of mating-type *MAT1-2*, containing the genes *MAT1-2-1* and *MAT1-2-3* (*FAVG2_03853* and *FAVG2_03854* or *FAVG3_03869* and *FAVG3_03870*). Such idiomorphs with different sets of genes has been observed previously in *Fusarium* spp. [26,27], and surveys of *F. avenaceum* populations often find isolates evenly split between mating types [28]. The sexual stage of *F. avenaceum* has been observed [29,30], and both *MAT1-1* and *MAT1-2* transcripts have been detected in this species under conditions favorable for perithecial production in other Fusaria [31], suggesting that *F. avenaceum* is likely capable of heterothallism. This is further supported by our data, in which a single mating-type is present in a given *F. avenaceum* isolate. This is characteristic for heterothallic fungal species, differing from homothallic species such as *F. graminearum*, which contain both mating-types in a single nucleus [32].

Occurrence of few repetitive elements supports the hypothesis that F. avenaceum is sexually active

A search for repetitive elements in the Fa05001 genome (using RepeatMasker [33] with CrossMatch) identified 1.0% of the Fa05001 genome as being repetitive or corresponding to transposable elements (Table S1 in File S1). This value is comparable to the 1.12% found for *F. graminearum*, although there were differences in the distributions of the various types of genetic elements. Tad1 and MULE-MuDR transposons were more enriched in *F. avenaceum* while *F. graminearum* had higher proportions of the TcMar-Ant1 transposon and small RNA. The low level of repeats supports the hypothesis that *F. avenaceum* is sexually active in nature, as such low levels are typical for species with an active sexual cycle, such as *F. graminearum*, *F. verticillioides* [1.21% repeats] and *F. solani* [3.8% repeats], while species which rely on asexual reproduction, such as *F. oxysporum*, have higher levels [10.6% repeats]. These results could be somewhat influenced by the fact that the Fusaria genomes are generated with different technologies.

Gene families enriched in F. avenaceum

Analysis of the predicted function of the 13217 Fa05001 gene models showed that Fa05001 contains a greater diversity of InterPro families than the other four analyzed Fusaria (Table 2, Figure 4, and File S4). Two highly enriched InterPro categories stand out in Fa05001; "Polyketide synthase, enoylreductase" (IPR020843) and "Tyrosine-protein kinase catalytic domain" (IPR020635), involved in secondary metabolism and signal transduction, respectively. With the exception of *F. solani*, Fa05001 also has the highest number of predicted transcription factors (Table S2 in File S1). Sixty-eight InterPro domains were predicted to be unique to Fa05001, including four tryptophan dimethylallyltransferase (IPR012148) proteins, commonly found in alkaloid biosynthesis. A comparison of Gene Ontology (GO) terms [34] indicated additional differences between the analyzed *Fusarium* species. Functional categories in which Fa05001 had higher numbers of proteins than the other genome-sequenced Fusaria were: "Cellular response to oxidative stress", "Branched-chain amino acid metabolic process", "Toxin biosynthetic process", "Oxidoreductase activity", "rRNA binding" and "Glutathione transferase activity" (Table S3, S4, S5 in File S1).

Reciprocal BLAST revealed that about ¾ of the predicted proteins in Fa05001 have orthologs in *F. graminearum* (76.7%), *F. verticillioides* (76.9%), *F. oxysporum* (78.8%) and *F. solani* (74.1%)

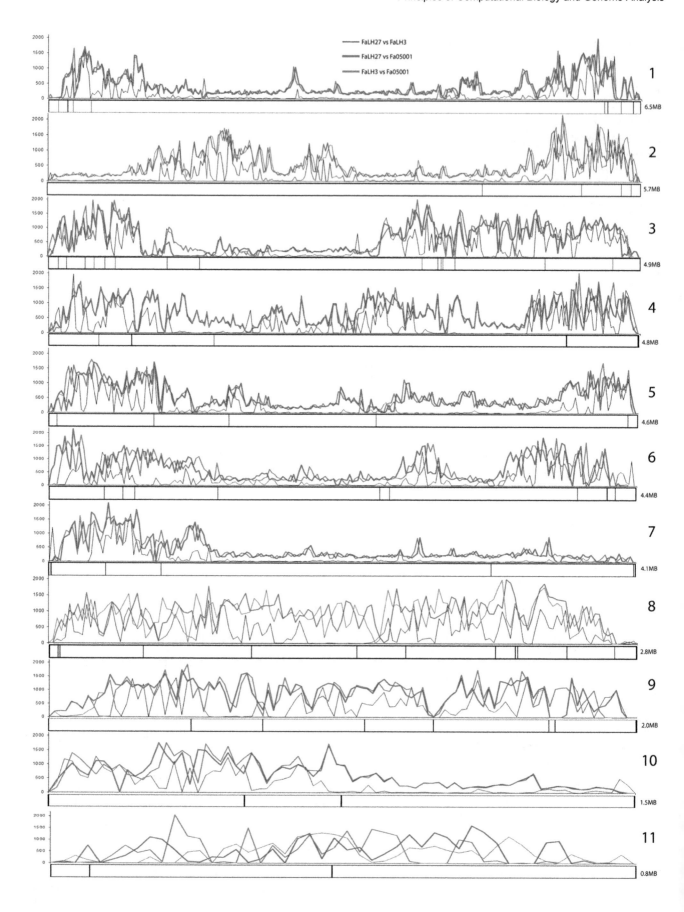

Figure 1. Sliding window map with numbers of SNP's and indels per 20 kb in the three *F. avenaceum* strains Fa05001, FaLH03 and FaLH27 on the 11 supercontigs. Locations of the polyketide synthase and non-ribosomal peptide synthetase genes in the strain FaLH27 are plotted on the supercontigs.

(File S5). The *F. avenaceum* proteins for which no ortholog (no hits or e-value>10^{-10}) was found in the other *Fusarium* genomes were especially enriched in GO biological categories "Oxidation-reduction process"; "Toxin biosynthetic process", "Alkaloid metabolic process"; "Cellular polysaccharide catabolic process" and "Transmembrane transport" (Table 3, Table 4, Figure S6 in File S1).

The secretome of *F. avenaceum*

The interplay between the invading fungus and the host plant occurs mainly in the extra-cellular space. The proportion of genes encoding predicted secreted proteins in the Fa05001 genome (File S6, Table S6 in File S1) is remarkably high (~20%; 2,580 proteins) as compared to plant pathogens such as *F. graminearum* (11%) and *Magnaporthe grisea* (13%), saprophytes such as *Neurospora crassa* (9%) and *Aspergillus nidulans* (9%) [22], and the insect pathogen *Cordyceps militaris* (16%) [35]. The secretome appears particularly enriched in proteins involved in redox reactions (Figure S7 in File S1). Inspecting the *F. avenaceum* proteins with no orthologs in other Fusaria (1223 proteins from Figure S6 in File S1), we found that 36% were predicted to be secreted.

Small cysteine-rich proteins (CRPs) can exhibit diverse biological functions, and some have been shown to play a role in virulence, including Avr2 and Avr4 in *Cladosporium fulvum* [36] and the Six effectors in *F. oxysporum* f. sp. *lycopersici* [37]. Other reported functions for CRPs have been adherence [38], anti-microbiosis [39] and carbohydrate binding activity that interferes with host recognition of the pathogen [40]. We found 19 candidates in Fa05001 containing more than four cysteine residues, and an additional 55 containing less than four cysteine residues, but with significant similarity to CRP HMM models (File S7, Table S7 in File S1). Of the predicted *F. avenaceum* CRPs, several are also found in other sequenced *Fusarium* species, but only two were noted with a putative function: CRP5760, with similarity to lectin-B, and CRP5810, a putative chitinase 3.

Metabolic profiling of *F. avenaceum*

A determination of secondary metabolites produced by these three *F. avenaceum* strains was performed on agar media, additionally Fa05001 was also grown *in planta* (barley and oat). Extraction of *F. avenaceum* cultures grown on PDA and YES solid

media revealed the presence of 2-amino-14,16-dimethyloctade-can-3-ol, acuminatopyrone, antibiotic Y, fusaristatin A, aurofu-sarin, butenolide, chlamydosporols, chrysogine, enniatin A, enniatin B, fusarin C, and moniliformin (Table 5). These metabolites have previously been reported from a broad selection of *F. avenaceum* strains [5,16,41,42] whereas the following metabolites previously reported from *F. avenaceum* were not detected in the present study: beauvericin [43], fosfonochlorin [44], diacetoxyscirpenol, T-2 toxin and zearalenone [45–50]. These reports are based on single observations using insufficient specific chemical methods that could yield false positive detection and/or poor fungal identification and no deposition of the strain in a collection for verification. The lack of zearalenone production agrees with the finding that none of the three *F. avenaceum* genomes described here contains the genes required for its production [51,52].

Preliminary genomic analysis had predicted the production of ferricrocin, malonichrome, culmorin, fusarinine and gibberellins [23], but chemical analysis only verified the production of the first two metabolites. The polyketide fuscofusarin (an aurofusarin analogue or intermediate in the biosynthetic pathway) was detected for the first time in *F. avenaceum*. Furthermore, fusaristatin A was found for the first time to be produced *in planta* during climate chamber experiments, and was generally found in *F. avenaceum*-infected barley, but only in trace amounts in oat. In addition, the tetrapeptide JM-47, an HC-toxin analogue reported from an unidentified *Fusarium* strain [53], was detected from all three strains in all cultures, including in barley and oats in climate chamber experiments. Apicidins have been detected in other strains of *F. avenaceum* cultured on YES agar (unpublished results), whereas these compounds were not detected in cultures of the three sequenced strains. Further characterization of the metabolomes on PDA and YES media revealed three major peaks in the ESI$^+$ chromatograms, which were also detected in the barley and oat extracts, that could not be matched to known compounds in Antibase [54]. Since these novel compounds are produced *in planta*, they are candidates for novel virulence factors.

Large potential for secondary metabolite production

The three *F. avenaceum* genomes encode 75 (Fa05001), 77 (FaLH03) and 80 (FaLH27) key enzymes for the production of

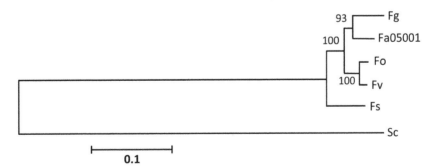

Figure 2. Molecular phylogenetic analysis of Fusarium species based on 69 orthologous proteins. The evolutionary history was inferred by using the Maximum Likelihood method and the tree with the highest log likelihood (−152577,9625) is shown. Bootstrap values, as percentages, are shown next to the individual branches. The tree is drawn to scale, with branch lengths measured in the number of substitutions per site. All positions containing gaps and missing data were eliminated prior to the ML analysis and the final data set contained 25535 positions.

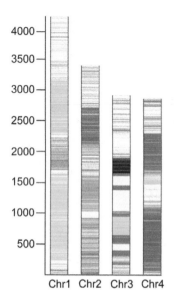

Figure 3. Shared gene homology map between Fa05001 and *F. graminearum* PH-1 has been created using the four defined *F. graminearum* chromosomes as templates. Genes in *F. graminearum* are coloured according to whether genes have corresponding orthologs in Fa05001, with one gene being one strip. Regions of the same color match to the same supercontigs (Figure S2 in File S1) in *F. avenaceum*. White regions represent lack of orthologs in Fa05001.

secondary metabolites, exhibiting a far greater biosynthetic potential than the known secondary metabolites produced by the species. These numbers include genes for 25–27 iterative type I polyketide synthases (PKSs), 2–3 type III PKSs, 25–28 nonribosomal peptide synthases (NRPSs), four aromatic prenyltransferases (DMATS), 12–13 class I terpene synthases (head-to-tail incl. cyclase activity), two class I terpene synthase (head-to-head) and four class II terpene synthases (cyclases for the class I terpene synthase head-to-head type) (Table 6).

The number of type I PKSs is surprisingly high, considering that the six other public *Fusarium* genomes (*F. graminearum*, *F. verticillioides*, *F. oxysporum*, *F. solani*, *F. pseudograminearum* and *F. fujikuroi*) only encode between 13 and 18 type I PKSs. The three *F. avenaceum* genomes share a core set of 24 type I PKS, and the individual isolates also encode unique type I PKS: oPKS47 (FAVG1_08496) is unique to Fa05001, while the two Canadian isolates share a single oPKS53 (FAVG2_01811, FAVG3_01846) and FaLH27 has two additional PKSs, oPKS54 (FAVG3_02030) and oPKS55 (FAVG3_06174). Of the 55 different type I PKSs described in the seven *Fusarium* genomes, only two (oPKS3 and oPKS7) are found in all species. Orthologs for 12 of the type I PKSs found in *F. avenaceum* could be identified in one or several of the publicly available *Fusarium* genomes, and hence 16 are new to the genus (Figure 5). None of these 16 have obvious characterized orthologs in other fungal genomes or in the GenBank database. The PKSs with characterized orthologs within the *Fusarium* genus includes the PKSs responsible for formation of fusarubin (oPKS3), fusaristatins (oPKS6), fusarins (oPKS10), aurofusarin (oPKS12) and fusaristatin A (oPKS6) [16,55–57]. Prediction of domain architecture of the 28 PKSs found in *F. avenaceum* showed that 17 belong to the reducing subclass, four to the non-reducing subclass, two are PKSs with a carboxyl terminal Choline/Carnitine O-acyltransferase domain, four are PKS-NRPS hybrids and one is a NRPS-PKS hybrid. The iterative nature of this enzyme class makes it

impossible to predict the products of these enzymes without further experimental data; this research has been initiated. The *F. avenaceum* genomes also encode type III PKSs, a class that among fungi was first described in *Aspergillus oryzae* [58], and recently in *F. fujikuroi* [59]. The FaLH03 isolate possesses three proteins (oPKSIII-1 to oPKSIII-3), while the two other isolates only have two (oPKSIII-1 and oPKSIII-3).

NRPSs provide an alternative to ribosomal-based polypeptide synthesis and in addition allow for the joining of proteinous amino acids, nonproteinous amino acids, α-hydroxy acids and fatty acids as well as cyclization of the resulting polypeptide [60]. The non-ribosomal peptide group of metabolites includes several well characterized bioactive compounds, such as HC-toxin (pathogenicity factor) and apicidin (histone deacetylase inhibitor) [61,62]. Of the 30 unique NRPSs encoded by the three *F. avenaceum* isolates, 16 are novel to the *Fusarium* genus, and include seven mono-modular and nine multi-modular NRPS, with between 2 and 11 modules. Of these, NRPS41 (FAVG1_08623, FAVG2_11354 and FAVG3_11434) is a likely ortholog to gliP2 (similar to gliotoxin synthetase) from *Neosartorya fischeri* (Genbank accession no. EAW21276), sharing 75% amino acid identity. The other 14 NRPSs are orthologs to previously reported proteins in other *Fusarium* species [63], and include the three NRPSs responsible for the formation of the siderophores malonichrome (oNRPS1), ferricrocin (sidC, oNRPS2) and fusarinine (sidA, oNRPS6).

All three strains encoded oNRPS31 which shows a significant level of identity to the apicidin NRPS (APS1) described in *F. incarnatum* and *F. fujikuroi* [59,64], and the HC-toxin NRPS (HTS1) from *Cochliobolus carbonum* SB111, *Pyrenophora tritici-repentis* and *Setosphaeria turcica* [65,66]. It has not yet been investigated whether NRPS31 is involved in the production of JM-47. Alignment of the genomic regions surrounding oNRPS31 with the APS1 and HTS1 clusters, showed that the three *F. avenaceum* strains encode nine of the twelve APS proteins found in the other two Fusaria, but lack clear orthologs encoding APS3 (pyrroline reductase), APS6 (O-methyltransferase) and APS12 (unknown function). The *APS*-like gene cluster in *F. avenaceum* has undergone extensive rearrangements, resulting in the loss of the three *APS* genes and gain of three new ones (*APS13-15*) (Figure 6 and Table S8 in File S1). Of these, *APS14* (*FAVG1_08581*, *FAVG2_02887* and *FAVG3_02926*) encodes a fatty acid synthase β subunit, which shares 60% identity with the FAS β subunit (encoded by *FAVG1_03575*, *FAVG2_06952* and *FAVG3_07032*) involved in primary metabolism. APS14 likely interacts with APS5 (fatty acid synthase α subunit) to form a functional fungal FAS (α6β6) responsible for the formation of the decanoic acid core of (S)-2-amino-8-oxodecanoic acid, proposed by Jin and co-workers [64] to be incorporated into apicidin by APS1. It has previously been hypothesized that APS5 (FAS α) interacts with the FAS α unit from primary metabolism to fulfill this role [64], however, the new model implies that the *F. incarnatum* genome encodes an unknown APS14 ortholog (no genome sequence available). The *F. fujikuroi* genome does not contain an APS14 ortholog (TBLASTn against the genome), but apicidin F has been detected in this species [59,67]. Alignment of the apicidin and HC-toxin gene clusters confirmed the observations made by Manning *et al.* [66] and Condon *et al.* [65], regarding the similarities of the two types of gene clusters, which yield very different products (Figure 6).

Dimethylallyltransferase and indole-diterpene biosynthesis proteins are common in the production of bioactive compounds in endophytes [68,69]. The three *F. avenaceum* genomes encoded four tryptophan dimethylallyltransferase (DMATS), aromatic

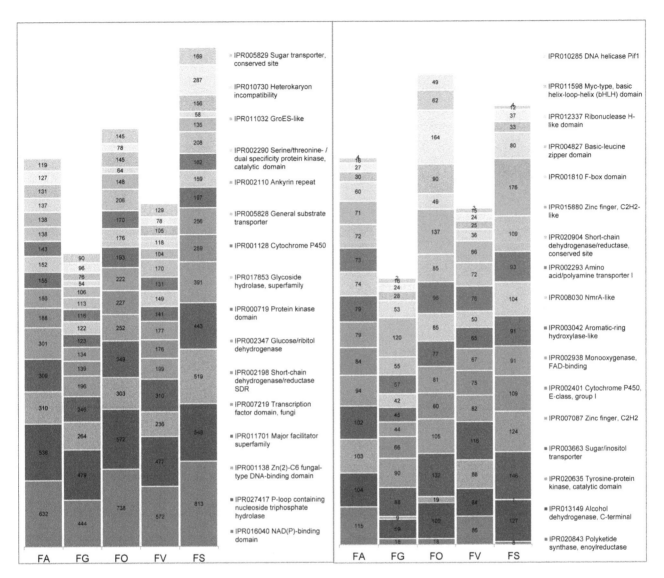

Figure 4. Functional analysis of Fa05001 and other sequenced Fusaria based on InterPro visualizing similarities and differences between the fungi. Categories with most differences between Fa05001 and others are presented with the number of proteins in each category. All details are listed in File S4. FA = *F. avenaceum* Fa05001, FG = *F. graminearum* PH-1, FO = *F. oxysporum* f. sp. *lycopersici* 4287, FV = *F. verticillioides*, FS = *F. solani*.

prenyltransferases, typically involved in alkaloid biosynthesis or modification of other types of secondary metabolites [68]. This is one more than is found in *F. verticillioides* and *F. oxysporum*, and two more than found in *F. fujikuroi*. One putative indole-

diterpene biosynthesis gene was found in all three *F. avenaceum* genomes (*FAVG1_08136*, *FAVG2_00906* and *FAVG3_00934*).

The three *F. avenaceum* genomes are also rich in genes involved in terpene biosynthesis. In the case of the class I terpene synthase,

Table 2. InterproScan analysis and comparison of Fa05001 with other Fusaria.

InterPro analysis	Families	Total domains
Fa05001	5,710	27,474
F. graminearum	5,602	23,560
F. oxysporum	5,609	29,873
F. verticillioides	5,545	25,393
F. solani	5,582	30,538

Table 3. Enriched biological processes in Fa05001 proteins without orthologs in other genome sequenced Fusaria (P<0.05).

F. avenaceum vs F. graminearum

GO-ID	Term	FDR	P-Value
GO: 0055114	Oxidation-reduction process	2.16E-06	9.72E-10
GO: 0009820	Alkaloid metabolic process	0.35	6.33E-04
GO: 0003333	Amino acid transmembrane transport	0.37	0.001
GO: 0009403	Toxin biosynthetic process	0.71	0.004
GO: 0016998	Cell wall macromolecule catabolic process	1	0.02
GO: 0006366	Transcription from RNA polymerase II promoter	1	0.03
GO: 0044247	Cellular polysaccharide catabolic process	1	0.03
GO: 0006573	Valine metabolic process	1	0.03

F. avenaceum vs F. verticillioides

GO-ID	Term		
GO: 0055114	Oxidation-reduction process	0.02	3.67E-05
GO: 0009403	Toxin biosynthetic process	0.98	0.003
GO: 0009820	Alkaloid metabolic process	1	0.01
GO: 0044247	Cellular polysaccharide catabolic process	0.02	1
GO: 0018890	Cyanamide metabolic process	1	0.02
GO: 0042218	1-aminocyclopropane-1-carboxylate Biosynthetic process	1	0.03
GO: 0006032	Chitin catabolic process	1	0.03
GO: 0006366	Transcription from RNA polymerase II promoter	1	0.04
GO: 0009636	Response to toxin	1	0.05

F. avenaceum vs F. oxysporum

GO-ID	Term		
GO: 0016106	Sesquiterpenoid biosynthetic process	0.44	0.002
GO: 0009403	Toxin biosynthetic process	0.44	0.002
GO: 0055114	Oxidation-reduction process	1	0.02
GO: 0046355	Mannan catabolic process	1	0.04
GO: 0006558	L-phenylalanine metabolic process	1	0.05

F. avenaceum vs F. solani

GO-ID			
GO: 0055114	Oxidation-reduction process	0.023	4.28E-05
GO: 0007155	Cell adhesion	0.13	5.08E-04
GO: 0045493	Xylan catabolic process	0.17	8.33E-04
GO: 0009403	Toxin biosynthetic process	0.8	0.005
GO: 0009820	Alkaloid metabolic process	1	0.017
GO: 0000162	Tryptophan biosynthetic process	1	0.033

See also Figure S6 in File S1.

responsible for the head-to-tail joining of isoprenoid and extended isoprenoid units and eventually cyclization, the three genomes encode 12–13 enzymes, of which three were putatively identified as being involved in primary metabolism (ERG20, COQ1, BTS1). In the case of the head-to-head class I systems, *F. avenaceum* encodes two enzymes, similar to the other fully genome-sequenced Fusaria, of which one gene encodes ERG9 and the other is involved in carotenoid biosynthesis. Cyclization of the formed head-to-head type product, if such a reaction occurs, is probably catalyzed by a class II terpene synthase/cyclase, of which *F. avenaceum* encodes four, including an ERG7 ortholog. This is more than the other Fusaria spp. genomes, as *F. graminearum, F.*

pseudograminearum and *F. verticillioides* only have ERG7, while *F. solani, F. oxysporum* and *F. fujikuroi* have two.

One of the four Type II terpene synthase encoding genes (TS_II_01: *FAVG1_10701, FAVG2_04190* and *FAVG3_04223*) shared by all three *F. avenaceum* strains was found to be orthologous to the gibberellic acid (GA) copalyldiphosphate/ent-kaurene synthase (cps/ks) from *F. fujikuroi* (Table S9 in File S1). Previously, the ability to synthesize the GA group of diterpenoid plant growth hormones in fungi has only been found in *F. fujikuroi* mating population A, *Fusarium proliferatum, Sphaceloma manihoticola* and *Phaeosphaeria sp.* strain L487 [70–74]. Biosynthesis of GA was first thoroughly characterized in *F.*

Table 4. Genes annotated as reduction-oxidation process in Fa05001 proteins without orthologs in other genome sequenced Fusaria (with expect>1e-10).

Reduction-oxidation function	Genes
4-carboxymuconolactone decarboxylase	FAVG1_07738
ABC multidrug	FAVG1_08208
Acyl dehydrogenase	FAVG1_08563, FAVG1_04763
Alcohol dehydrogenase	FAVG1_04699, FAVG1_12680
Aryl alcohol dehydrogenase	FAVG1_08747
Bifunctional p-450: nadph-p450 reductase	FAVG1_11632
Transcription factor	FAVG1_09453, FAVG1_07648
Choline dehydrogenase	FAVG1_12183
Cytochrome p450 monooxygenase	FAVG1_08576, FAVG1_13151, FAVG1_07923, FAVG1_10161, FAVG1_08627, FAVG1_02807, FAVG1_08721, FAVG1_10699
Delta-1-pyrroline-5-carboxylate dehydrogenase	FAVG1_04749
Dimethylaniline monooxygenase	FAVG1_11196
Ent-kaurene synthase	FAVG1_10701
Glutaryl- dehydrogenase	FAVG1_02842
Homoserine dehydrogenase	FAVG1_09776
l-lactate dehydrogenase a	FAVG1_08604
Mitochondrial 2-oxoglutarate malate carrier protein	FAVG1_08564
Monooxygenase fad-binding protein	FAVG1_10705
Nadp-dependent alcohol dehydrogenase	FAVG1_10174, FAVG1_12103, FAVG1_06950
Nitrilotriacetate monooxygenase component b	FAVG1_07690
Pyoverdine dityrosine biosynthesis	FAVG1_12703
Salicylate 1-monooxygenase	FAVG1_09825
Salicylaldehyde dehydrogenase	FAVG1_09676
Short-chain dehydrogenase	FAVG1_07710, FAVG1_12690, FAVG1_04239, FAVG1_10281, FAVG1_08519

See also Figure S6 in File S1.

fujikuroi and depends on the coordinated activity of seven enzymes, encoded by the GA gene cluster, that convert dimethylallyl diphosphate (DMAPP) to various types of gibberellic acids, with the main end-products being GA_1 and GA_3 [70]. *S. manihoticola*'s GA biosynthesis ends at the intermediate GA_4 due to the lack two of genes (DES and P450-3), compared to *F. fujikuroi*, responsible for converting GA_4 to GA_7, GA_3 and GA_1 [73]. Analysis of the genes surrounding the *F. avenaceum* CPS/KS encoding gene showed that six of the seven genes from the *F. fujikuroi* GA gene cluster are also found in all three *F. avenaceum* genomes, with only P450-3 (C13-oxidase) missing (Figure 7). The architecture of the GA cluster in *F. avenaceum* is highly similar to the *F. fujikuroi* cluster, and a single inversion in five of the six genes can explain the relocation of the desaturase (*des*) encoding gene. The inversion could potentially have involved the P450-3 gene and resulted in the disruption of its coding sequence, however the shuffle-LAGAN analysis (Figure 7) and dot plot showed that this has not been the case. The missing gene is not found elsewhere in the genome based on a tblastn search. The presence of six of the seven GA biosynthesis genes suggest that *F. avenaceum* has the potential to produce all the GA's up to G_4 and G_7, but lack the ability to convert these into the GA_1 and GA_3. None of the three *F. avenaceum* isolates have been reported to produce this plant growth hormone.

In summary, *F. avenaceum* has a very large potential for producing secondary metabolites belonging to the PKS, NRPS and terpene classes, with a total of 75–80 key enzymes, see File S8

that summarizes all orthology groups. However, it is expected that multiple enzymes will participate in a single biosynthetic pathway, thereby reducing the potential number of final metabolites. It is possible that some of the metabolites function as virulence factors during infection; however systematic deletion of all PKS encoding genes in *F. graminearum* showed that none of the 15 PKSs in this fungus had significant effect on virulence [75]. It is therefore more likely that at least some of the secondary metabolites function as antibiotics towards competing microorganisms in the diverse set of niches that the species inhabits. When plotting the PKSs and NRPSs on the 11 supercontigs, areas with higher numbers of SNPs, insertions, or deletions between the three strains, were also more populated with secondary metabolite genes (Figure 1), as seen in other Fusaria [22].

Transcriptomics of *F. avenaceum* in barley

To increase our understanding of *F. avenaceum* behaviour *in planta*, we performed RNA-seq on *F. avenaceum*-inoculated barley. Table S10 in File S1 shows a list of *F. avenaceum* genes with the most stable and significant expression (FDR<0.05). Due to putative false positive genes expressed in the host, we applied high stringency in the analysis, and only genes found expressed at 14 days post inoculation (dpi), and which were absent in control, were considered. Genes involved in stress related responses, especially oxidative stress (as defined in the fungal stress response database [76]) were highly represented. This strongly supports our hypothesis formulated from the comparative genomic analysis that

Table 5. Metabolic profiling of the three *F. avenaceum* strains.

	FaLH03		FaLH27		Fa05001				Other strains
	YES	PDA	YES	PDA	YES	PDA	Barley	OAT	*F. avenaceum*
PKS									
2-Amino-14,16-dimethyloctadecan-3-ol	+	+	+	+	+	+	ND	ND	+
Acuminatopyrone	+	+	ND	ND	+	+	ND	ND	+
Antibiotic Y	+	+	+	+	+	+	ND	ND	+
Aurofusarin	+	+	+	+	+	+	+	+	+
Fuscofusarin	+	+	+	+	+	+	ND	ND	NA
Moniliformin	+	+	+	+	+	+	NA	NA	+
Chlamydosporols	+	+	ND	ND	+	+	ND	ND	+
NRPS and mixed NRPS-PKS									
Butenolide	ND	ND	ND	ND	+	+	+	ND	+
Chrysogine	+	+	+	+	+	+	+	+ but 10 × lower than barley	+
Visoltricin	ND	ND	ND	ND	ND	ND	ND	ND	+ (by UV-Vis)
Fusarins C and A	+	ND	+	ND	+	+	ND	ND	+
Enniatins A's and B's	+	+	+	+	+	+	+	+ but 100 × lower than barley	+
Beauvericin	ND	ND	ND	ND	ND	ND	ND	ND	ND
Apicidin	ND	ND	ND	ND	ND	ND	ND	ND	+
Fusaristatins	+	+	+	+	+	+	+	Trace	+
Fusarielin A	ND	ND	ND	ND	ND	ND	ND	ND	ND
Ferricrocin	+	ND	+	ND	+	+	ND	ND	NA
Fusarinines	ND	ND	ND	ND	ND	ND	ND	ND	NA
JM-47	+	+	+	+	+	+	+	+	NA
Malonichrome	+	ND	+	ND	+	ND	ND	ND	NA
Other									
Fosfonochlorin	ND	ND	ND	ND	ND	ND	ND	ND	NA
Unknown 26 (NRPS)					+	+	+	+	
Fusarium unknown 31 (NRPS)					+	+	+	+	

ND not detected

NA not analyzed

Table 6. The number of identified signature genes for secondary metabolism in the three *F. avenaceum* strains compared to the other public *Fusarium* genome sequences.

	PKS I	PKS III	NRPS	TS I HT*	TS I HH**	TS II***	DMATS
Fa05001	25 (12)	2 (1)	25 (9)	13 (2)	2(0)	4(0)	4(2)
FaLH03	25 (12)	3 (2)	26 (10)	12 (1)	2(0)	4(0)	4(2)
FaLH27	27 (14)	2 (1)	28 (12)	13 (1)	2(0)	4(0)	4(2)

The number of genes that are unique for *F. avenaceum* is given in the parentheses. Gene classes: type I iterative PKS (PKS I), type III PKS (PKS III), non-ribosomal peptide synthetases (NRPS), terpene synthase class I head-to-tail type (TS I HT), terpene synthase class I head-to-head type (TS I HH), terpene synthases class II (TS II) and aromatic prenyltransferases (DMATS).
*incl. ERG20, COQ1, BTS1,
**incl. ERG9,
***incl. ERG7.

the broad host range of *F. avenaceum* is likely due to a generalized mechanism allowing it to cope with and overcome the innate immune response of plants, such as the generation of reactive oxygen species (ROS) [77].

Genes involved in signal transduction were also overrepresented in the transcriptome, including GO categories for GTP binding, ATP binding, calcium ion binding, and membrane activity (Figure S8 in File S1, File S8). Approximately 33% of all proteins predicted in the genome with Interpro "Tyrosine-protein kinase, catalytic domain" (IPR020635) were found in the transcriptome. This was one of the most highly enriched *F. avenaceum* categories when compared to other *Fusarium* genome sequences, and the *in planta* transcriptome hence supports the comparative results from the genome analysis. During plant infection, fungi need to monitor the nutrient status and presence of host defenses, and respond to or tolerate osmotic or oxidative stress, light and other environmental variables [78]. Stress-signaling/response genes of fungal pathogens are known to play important roles in virulence, pathogenesis and defense against oxidative burst (rapid production of ROS) from the host [79,80]. It is plausible to predict that tyrosine-protein kinases assist in the stress related response. There is a tendency that *F. avenaceum* isolates from one host can be pathogenic on other distantly related plants [11]. This is in contrast to, for example, *F. oxysporum*, a pathogen with a remarkably broad host range at the species level, but where individual isolates often cause disease only on one or a few plant species [81]. Our results support the chameleon nature of *F. avenaceum*, as it is capable of adapting to diverse hosts and environments. This lack of host specialization is likely to be a driving force in *F. avenaceum* evolution. Apart from the general functional categories, the *in planta* transcriptome of *F. avenaceum* also revealed many orthologs of *F. graminearum* pathogenicity and virulence factors, especially those involved in signal transduction and metabolism (Table S11 in File S1, [82]).

Conclusions

In summary, the comparative genomic analyses of *F. avenaceum* to other Fusaria point out several functional categories that are enriched in this fungal genome, and which indicate a great potential for *F. avenaceum* to sense and transduce signals from the surroundings, and to respond to the environment accordingly. *Fusarium avenaceum* has a large potential for redox, signaling and secondary metabolite production, and 20% of all predicted proteins were considered to be secreted. This could suggest that interaction with plant hosts is predominantly in the extracellular space. These genome sequences provide a valuable tool for the discovery of genes and mechanisms for bioactive compounds, and to increase our knowledge of the mechanisms contributing to a fungal lifestyle on diverse plant hosts and in different environments.

Materials and Methods

Sequencing, assembly, gene prediction and annotation

Fusarium avenaceum isolate Fa05001 (ARS culture collection: NRRL 54939, Bioforsk collection: 202103, DTU collection: IBT 41708) was isolated from barley in 2005 in Finland [83]. The strain was grown on complete medium [84] at room temperature for three days on a shaker, before mycelium was vacuum filtered and harvested for storage at −80°C. DNA was isolated using the Qiagen DNeasy Plant Maxi. The sequencing and assembly was performed by Eurofins MGW, using a combination of shotgun (1.5 plate) and paired-end (¼ plate of 3 kb, and ¼ plate of 20 kb) 454 pyrosequencing. Newbler 2.6 (www.roche.com) was used for

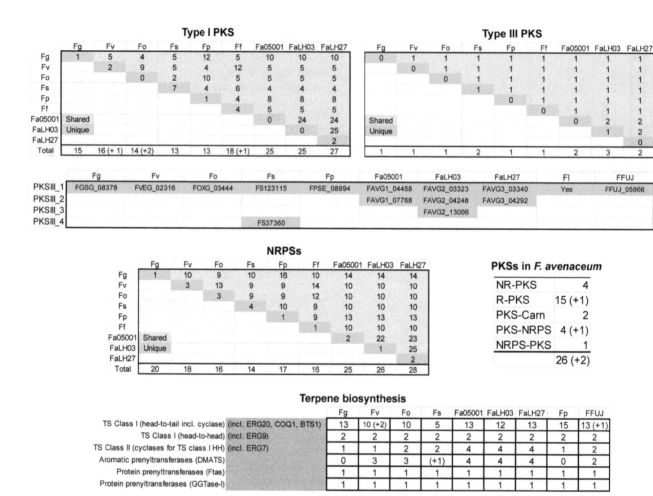

Figure 5. Shared and unique polyketide synthase (PKS), non-ribosomal peptide synthetases (NRPS) and terpene cyclase (TC) encoding genes in public available Fusaria genomes. Green and yellow boxes are the number of shared and unique genes, respectively. *Fg* = *F. graminearum*, *Fv* = *F. verticillioides*, *Fo* = *F. oxysporum*, *Fs* = *F. solani*, *Fp* = *F. pseudograminearum*, *Ff* = *F. fujikuroi* and *Fa05001, FaLH03* and *FaLH27* = *F. avenaceum*.

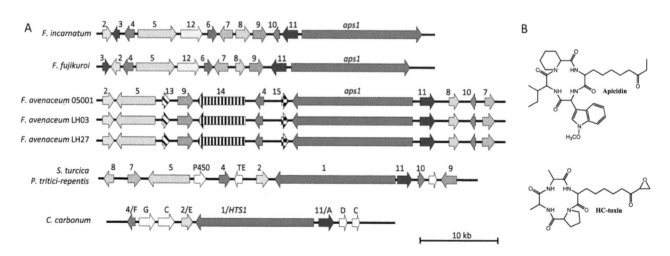

Figure 6. Apicidin-like gene cluster (oNRPS31) in the three *F. avenaceum* strains (Fa05001_Scaffold14, FaLH03_contig11, FaLH27_contig13) compared to the characterized apicidin gene cluster from *F. incarnatum* and the HC-toxin gene clusters from *Cochliobolus carbonum*, *Pyrenophora tritici-repentis* and *Setosphaeria turcica* **(A).** The genes are colored based on homology across the species. Chemical structure of apicidin and HC-toxin (B). See Table S8 in File S1 for further details.

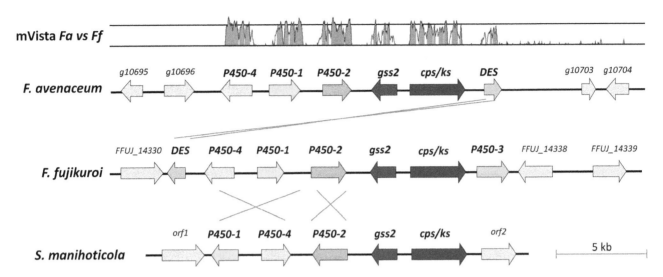

Figure 7. Architecture of the gibberellic acid (GA) gene clusters from *F. fujikuroi* **MP-A,** *F. avenaceum* **and** *Sphaceloma manihoticola.* The gene cluster and surrounding genes are identical in the three *F. avenaceum* strains and only Fa05001 is shown (FaLH03 cluster: *FAVG2_04186 - FAVG2_04192* and FaLH27 cluster: *FAVG3_04219 - FAVG3_04224*). The mVista trace shows the similarity over a 100 bp sliding windows (Shuffle-LAGAN plot) between the *F. avenaceum* and *F. fujikuroi* clusters, bottom line =50% and second line =75% identity. Genes: *gss2* = geranylgeranyldiphosphate synthase, *cps/ks* = copalyldiphosphate/ent-kaurene synthase, *P450-4* = ent-kaurene oxidase, *P450-1* = GA$_{14}$ synthase, *P450-2* = C20-oxidase, *P450-3* = 13-hydroxylase and *DES* = desaturase. Note that the intergenic regions are unknown for the *S. manihoticola* GA cluster, while the size of these regions is not to drawn to size.

automatic assembly, and gap closure and further assembly was performed using GAP4 (Version 4.4; 2011, http://www.gap-system.org), a total of 50 primer pairs, and manual editing. PCR products from gaps were Sanger sequenced in both directions, and 38 PCR products were successfully integrated into the assembly. These approaches significantly improved the results, starting from 502 contigs and 89 scaffolds after automatic annotation to 110 contigs and 83 scaffolds (Table 1).

Two Canadian *F. avenaceum* strains, FaLH03 (spring wheat, New Brunswick) and FaLH27 (winter wheat, Nova Scotia), were isolated from wheat samples harvested in 2011 (Canadian Grain Commission, Winnipeg, MB) and deposited in the Canadian Collection of Fungal Cultures (AAFC, Ottawa, ON) with the strain designations DAOM242076 and DAOM242378, respectively. Species identification was confirmed by sequencing the *tef1-α* gene [85]. Strains were single-spored prior to any analysis and confirmed to retain virulence towards durum wheat. After growth for 3 days in glucose-yeast extract-peptone liquid culture, mycelia was collected and freeze-dried. Genomic DNA was extracted using the Nucleon Phytopure genomic DNA extraction kit (GE Healthcare, Baie d'Urfe, Québec) and then used to prepare an Illumina TruSeq library. Each library was sequenced in a single lane on an Illumina HiSeq platform (100 bp paired-end) at the Génome Québec Innovation Centre (Montréal, Québec), yielding 99,386,445 and 233,211,138 reads for FaLH03 and FaLH27, respectively. Reads were assembled in CLC Genomics Workbench 6.0.1. The higher order of the obtained scaffolds in the three isolates was resolved through comparison between the strains. The contigs from the Canadian isolates were ordered to each other by ABACAS [86]. A reiterative reordering approach was used to generate a stable order of the contigs. The successful reordering is illustrated in the alignment of the ordered contigs by MUMMER [87] (Figure S2 in File S1). The scaffolds from the Fa05001 were then ordered according to the Canadian isolates by ABACAS, and aligned to the Canadian isolates by 'MUMMER' (Figure S2 in File

S1). Contig/scaffold overlaps and boundaries in the 'MUMMER' alignments were used to determine the higher order assembly.

Gene prediction was performed with Augustus v2.5.5 (Fa05001) and v2.6 (FaLH03, FaLH27) [88], using default settings and *F. graminearum* as a training set. Protein sequences were annotated and enrichment analysis of gene ontology categories were compared using Blast2GO [89]. Table 1 shows the general statistics of the three genome sequences.

The species phylogenies were constructed based on 69 orthologous proteins from the included species. The 69 protein sets were first aligned individually using MUSCLE with default parameters [18], then manually inspected and concatenated. These super-protein alignments were then analyzed using MEGA6.0 by first identifying the best substitution model and then inferring the evolutionary history of the species using the Maximum Likelihood method, Nearest-Neighbour-Interchange, [90] using the LG+(F) substitution model, uniform substitution rate, removal of all positions with gaps and missing data and bootstrapping with 500 iterations.

Functional analysis and orthology prediction

For comparative genomics analysis, the previously sequenced genomes of *F. graminearum*, *F. verticillioides*, *F. oxysporum* f. sp. *lycopersici* 4287 (*Fusarium* Comparative Sequencing Project, Broad Institute of Harvard and MIT, http://www. broadinstitute.org/) and *F. solani* [91] were re-annotated using Blast2GO [89] concurrently with Fa05001. A functional comparison was performed using the Gene Ontology (GO) categories Biological Process, Molecular Function, and Cellular Compartments at level six, and Interpro. To compare transcription factors, we used the Interpro list from the Fungal Transcription Factor Database [92]. To identify orthologs, protein sets were compared in both directions with BLASTp to identify the best reciprocal hit for each individual protein with expectation values less than 1e-10. RepeatMasker [11] was used on all species to find repetitive sequences in the *Fusarium* genomes, using CrossMatch (http://

www.phrap.org/phredphrap/general.html). BLASTn was used to compare gene sequences between the three *F. avenaceum* genomes.

Electrophoretic karyotyping

Plugs containing 4×10^8/ml protoplasts were loaded on a CHEF gel (1% FastLane agarose [FMC BioProducts, Rockland, Maine] in 0.5×TBE) and ran for 255 hours, using switch times between 1200–4800 s at 1.8 V/cm. Chromosomes of *Schizosaccharomyces pombe* and *Hansenula wingei* were used as a molecular size marker (BioRad). Chromosomes separated in the gel were blotted to Hybond-N+. Southern hybridization was done overnight at 65°C using the CDP star method (GE Healthcare). A 350 bp *Hind*III-*Eco*RI fragment from plasmid pNLA17 was used as a probe, containing the *F. oxysporum* telomeric repeat TTAGGG 18 times.

Prediction of putative secretome

To determine the putative secretome, we employed a pipeline consisting of the following: A combination of WolfPsort (http://wolfpsort.org/), IPsort (http://ipsort.hgc.jp/) and SignalP4.1 [93] to identify subcellular localization and/or signalP motifs of all *F. avenaceum* Fa05001 proteins. Then, we used TMHMM [94] to predict all transmembrane domains. The secretome fasta file with the sequences was created with GPRO [95], with proteins predicted by either the PSORT tools or SignalP that does not contain transmembrane domains. Of the whole set of 13217 predicted proteins in Fa05001, a subset of 2580 sequences has been determined as the putative secretome of *F. avenaceum* (File S6).

Prediction of cysteine-rich proteins

Prediction of putative cysteine-rich proteins CRPs was based on a previously described approach [96]. In brief, this method is based on their expected sequence characteristics, with predicted small open reading frames (ORFs) (20 to 150 amino acids), containing at least four cysteine residues and a predicted signal peptide from the secretory pathway. The 83 Fa05001 scaffolds were used as input against GETORF available in the EMBOSS package [97]. We obtained 565.652 ORFs, which were translated to proteins using the tool Transeq [97]. All proteins were separated into two files (more/less than four cysteine residues). We downloaded a collection of 513 HMM profiles based on CRP models [98], and then created a single HMM database file that was formatted with HMMER3 [99] and used as subject against a HMMER comparison with the two files obtained. Nineteen candidates with more than four cysteine residues were predicted as CRP, and 55 additional sequences with less than four cysteine residues but with significant similarity to particular CRP HMM models were also included (Table S7 in File S1, File S7). We compared these to the supercontigs and scaffolds of the genome sequenced *Fusarium* spp. using BLASTn and TBLASTx, bit-score>50, and BLASTp in NCBI.

Identification of secondary metabolites

The three *F. avenaceum* strains were grown at 25°C in darkness for 14 days as triple point inoculations on Potato Dextrose Agar (PDA, [100]) and Yeast Extract Sucrose agar (YES, [100]). The metabolites of the strains were extracted using a modified version of the micro-scale extraction procedure for fungal metabolites [101]. Six 5-mm plugs from each plate taken across the colonies were transferred to 2 mL HPLC vials and extracted with 1.2 mL methanol:dichloromethane:ethyl acetate (1:2:3 v/v/v) containing

1% (v/v) formic acid. After 1 hr in an ultrasonication bath, extracts were evaporated with nitrogen, the residue dissolved in 150 μL acetonitrile:water (3/2 v/v) and filtered through a standard 0.45-μm PTFE filter.

Barley and oat samples (all in biological triplicates), including none-inoculated samples from climate chamber experiments described below were ground in liquid nitrogen and 50 mg extracted with 1 mL of 50% (vol) acetonitrile in water in a 2 mL Eppendorf tube. Samples were placed in an ultrasonication bath for 30 min, centrifuged at 15,000 g, and the supernatant was transferred to a clean 2 mL vial that was loaded onto the auto sampler prior to analysis. UHPLC-TOFMS analysis of 0.3–2 μL extracts were conducted on an Agilent 1290 UHPLC equipped with a photo diode array detector scanning 200–640 nm, and coupled to an Agilent 6550 qTOF (Santa Clara, CA, USA) equipped with a dual electrospray (ESI) source [102]. Separation was performed at 60°C at a flow rate of 0.35 mL/min on a 2.1 mm ID, 250 mm, 2.7 μm Agilent Poroshell phenyl hexyl column using a water-acetonitrile gradient solvent system, with both water and acetonitrile containing 20 mM formic acid. The gradient started at 10% acetonitrile and was increased to 100% acetonitrile within 15 min, maintained for 4 min, returned to 10% acetonitrile in 1 min. Samples were analyzed in both ESI$^+$ and ESI$^-$ scanning m/z 50 to 1700, and for automated data-dependent MS/MS on all major peaks, collision energies of 10, 20 and 40 eV for each MS/MS experiment were used. An MS/MS exclusion time of 0.04 min was used to get MS/MS spectra of less abounded ions.

Data files were analyzed in Masshunter 6.0 (Agilent Technologies) in three different ways: i) *Aggressive dereplication* [103] using lists of elemental composition and the *Search by Formula* (10 ppm mass accuracy) of all described *Fusarium* metabolites as well as restricted lists of only *F. avenaceum* and closely related species; ii) Searching the acquired MS/MS spectra in an in-house database of approx. 1200 MS/MS spectra of fungal secondary metabolites acquired at 10, 20 and 40 eV [102]; iii) all major UV/Vis and peaks in the base peak ion chromatograms not assigned to compounds (and not present in the media blank samples) were also registered. For absolute verification, authentic reference standards were available from 130 *Fusarium* compounds and additional 100 compounds that have been tentatively identified based on original producing strains using UV/Vis, LogD and MS/HRMS [102–104].

Identification of secondary metabolite genes

Type I iterative polyketide synthase (PKS), type III PKS, non-ribosomal peptide synthase (NRPS), aromatic prenyltranferase (DMATS) and class I & II terpene synthase encoding genes were identified by BLASTp using archetype representatives for the six types of genes [105]. Identification of orthologous genes was further supported by comparison to the genomic DNA, using the shuffle LAGAN algorithm with default settings [106,107]. Functional protein domains were identified using the NCBI CDD and pfam databases [108]. Domains specific to non-reducing PKSs, e.g. 'Product template' (PT) and 'Starter Acyl-Transferase' (SAT), were inferred via multiple sequence alignment with the bikaverin PKS (PKS16), which was one of the founding members of the domain group [109]. The nomenclature for PKS and NPS follows that which was introduced by Hansen et al. [63], as indicated by the use of the oPKSx and oNRPSx name, where the prefix 'o' signals that it refers to orthology-groups rather than the original overlapping names schemes used previously in each species. Following the idea regarding transparency in the names, introduced by Hansen and co-works, we applied a similar nomenclature scheme to the type III PKSs (oPKSIII_x) and the various enzyme classes involved in terpene biosynthesis: Terpene Synthase class I head-to-tail (oTS-I-HT_x), Terpene Synthase class I head-to-head (oTS-II-HH_x) and Terpene Synthase class II (oTS-II-x).

Climate chamber infection experiment

Fa05001 was grown on mung bean agar [110] for three weeks at room temperature under a combination of white and black (UVA) light with a 12 h photoperiod. Macroconidia were collected by washing the agar plate with 5 mL sterile distilled water, and diluted with 1.5% carboxymethylcellulose solution to a concentration of 5×10^4 conidia/mL for inoculation. Conidial concentration was determined using a Bürker hemacytometer.

Barley (*Hordeum vulgare*), cultivar Iron, and oat (*Avena sativa*) cultivar Belinda were grown in a climate chamber under the following conditions: Two weeks at 10°C/8°C 17 h/7 h 70%RH/60%RH, two weeks 15°C/12°C 18 h/6 h 70%RH/60%RH, three weeks 18°C/15°C 18 h/6 h 70%RH/60%RH and three weeks 20°C/15°C 17 h/6 h 70%RH/60%RH. During anthesis, approximately 1 mL of conidial solution was sprayed on each panicle, a bag was placed over the panicle and removed after 4 days. We used 6 plants per pot, 2 panicles per plant and 3 replicate pots per treatment. At sampling, panicles were immediately stored at −80°C.

Transcriptomics of Fa05001 on barley heads

Panicles from one pot grown in climate chamber experiment were mixed and ground in liquid nitrogen. RNA was extracted from 50 mg subsample from three biological replicates (pots) of untreated control (0 dpi) and *F. avenaceum*-inoculated tissue (14 dpi), using Spectrum plant total RNA kit (Sigma-Aldrich, Steinheim, Germany), with slight modifications. Due to the high amount of starch in barley heads at 14 dpi, the volume of lysis buffer and binding solution were increased from 500 μL to 750 μL per sample, and samples were incubated for 5 min at room temperature and the lysates were filtered 2 times for 10 minutes. On-column DNase digestion (Sigma-Aldrich, Steinheim, Germany) was used.

PolyA purification and fragmentation, cDNA synthesis, library preparation and 1×100 bp single read module (half a lane Hi-seq 2500) sequencing were done by Eurofins MGW. The resulting fastaq files were trimmed (quality score limit: 0.05, maximum number of ambiguities: 2), and RNA-seq was performed with predicted *F. avenaceum* genes using CLC Genomics Workbench 6.05, with stringent settings (minimum similarity fraction: 0.95, minimum length fraction: 0.9, maximum number of hits for a read: 10) to subtract host-specific transcripts. Gene expression was calculated using reads per kilobase per million (RPKM) values. A T-test was used to determine significant expression levels in the biological replicates, comparing *F. avenaceum* inoculated samples against a control. Transcripts found solely in the *F. avenaceum*-inoculated plant were used to limit the amount of false positives coming from the host.

Supporting Information

File S1   Supplementary figures and tables.

File S2   BLASTn results of the three *F. avenaceum* isolates Fa05001, FaLH03 and FaLH27.

File S3   Venn diagram of BLASTn results corresponding to Figure S4.

File S4   Interpro results of *F. avenaceum* Fa5001, *F. graminearum*, *F. verticillioides*, *F. oxysporum* and *F. solani*.

File S5   Reciprocal blast of *F. avenaceum* Fa5001 vs *F. graminearum*, *F. verticillioides*, *F. oxysporum* and *F. solani*.

File S6   Secretome of *F. avenaceum* Fa5001.

File S7   Cysteine rich proteins in *F. avenaceum* Fa5001.

File S8   Summary of secondary metabolite genes.

File S9   Transcriptome of *F. avenaceum* Fa5001 on barley heads.

Acknowledgments

We thank Päivi Parikka, MTT, Finland for the *F. avenaceum* Fa05001 strain and Tom Gräfenhan, Canadian Grain Commission, for the two Canadian isolates FaLH03 and FaLH27 used for genome sequencing. We thank Danielle Schneiderman and Catherine Brown for technical support and Philippe Couroux for bioinformatics support.

Author Contributions

Conceived and designed the experiments: EL LH SW HD HCK WJ KN UT RF. Performed the experiments: EL LH SW RS HCK WJ AKK KN UT RF. Analyzed the data: EL LH SW RS CL TG KN UT RF. Contributed reagents/materials/analysis tools: EL LH SW RS CL TG AKK KN UT RF. Wrote the paper: EL LH SW HD ER TG HCK AKK KN UT RF.

References

1. Leach MC, Hobbs SLA (2013) Plantwise knowledge bank: delivering plant health information to developing country users. Learned Publ 26: 180–185.
2. Desjardins AE (2003) *Gibberella* from a (*venaceae*) to z (*eae*). Annu Rev Phytopathol 41: 177–198.
3. Satyaprasad K, Bateman GL, Read PJ (1997) Variation in pathogenicity on potato tubers and sensitivity to thiabendazole of the dry rot fungus *Fusarium avenaceum*. Potato Res 40: 357–365.
4. Mercier J, Makhlouf J, Martin RA (1991) *Fusarium avenaceum*, a pathogen of stored broccoli. Can Plant Dis Surv 71: 161–162.
5. Sørensen JL, Phipps RK, Nielsen KF, Schroers HJ, Frank J, et al (2009) Analysis of *Fusarium avenaceum* metabolites produced during wet apple core rot. J Agricult Food Chem 57: 1632–1639.
6. Peters RD, Barasubiye T, Driscoll J (2007) Dry rot of rutabaga caused by *Fusarium avenaceum*. Hortscience 42: 737–739.
7. Crous PW, Petrini O, Marais GF, Pretorius ZA, Rehder F (1995) Occurrence of fungal endophytes in cultivars of *Triticum aestivum* in South Africa. Mycoscience 36: 105–111.
8. Varvas T, Kasekamp K, Kullman B (2013) Preliminary study of endophytic fungi in timothy (*Phleum pratense*) in Estonia. Acta Myc 48: 41–49.
9. Yacoub A (2012) The first report on entomopathogenic effect of *Fusarium avenaceum* (Fries) Saccardo (Hypocreales, Ascomycota) against rice weevil (*Sitophilus oryzae* L.: Curculionidae, Coleoptera). J Entomol Acarol R 44: 51–55.
10. Makkonen J, Jussila J, Koistinen L, Paaver T, Hurt M, et al. (2013) *Fusarium avenaceum* causes burn spot disease syndrome in noble crayfish (*Astacus astacus*). J Invert Pat 113: 184–190.
11. Nalim FA, Elmer WH, McGovern RJ, Geiser DM (2009) Multilocus phylogenetic diversity of *Fusarium avenaceum* pathogenic on lisianthus. Phytopathology 99: 462–468.

12. Uhlig S, Jestoi M, Parikka P (2007) *Fusarium avenaceum* - The North European situation. Int J Food Microbiol 119: 17–24.

13. Kulik T, Pszczolkowska A, Lojko M (2011) Multilocus phylogenetics show high intraspecific variability within *Fusarium avenaceum*. Int J Mol Sci 12: 5626–5640.

14. Kohl J, de Haas BH, Kastelein P, Burgers SLGE, Waalwijk C (2007) Population dynamics of *Fusarium* spp. and *Microdochium nivale* in crops and crop residues of winter wheat. Phytopathology 97: 971–978.

15. Sørensen JL, Giese H (2013) Influence of carbohydrates on secondary metabolism in *Fusarium avenaceum*. Toxins 5: 1655–1663.

16. Sørensen L, Lysøe E, Larsen J, Khorsand-Jamal P, Nielsen K, et al. (2014) Genetic transformation of *Fusarium avenaceum* by *Agrobacterium tumefaciens* mediated transformation and the development of a USER-Brick vector construction system. BMC Mol Biol 15: 15. 10.1186/1471-2199-15-15.

17. Ma LJ, van der Does HC, Borkovich KA, Coleman JJ, Daboussi MJ, et al. (2010) Comparative genomics reveals mobile pathogenicity chromosomes in *Fusarium*. Nature 464: 367–373.

18. Marcet-Houben M, Gabaldón T (2009) The tree versus the forest: The fungal tree of life and the topological diversity within the yeast phylome. Plos One 4: e4357.

19. Al-Reedy RM, Malireddy R, Dillman CB, Kennell JC (2012) Comparative analysis of *Fusarium* mitochondrial genomes reveals a highly variable region that encodes an exceptionally large open reading frame. Fung Genet Biol 49: 2–14.

20. Fourie G, van der Merwe NA, Wingfield BD, Bogale M, Tudzynski B, et al. (2013) Evidence for inter-specific recombination among the mitochondrial genomes of *Fusarium* species in the *Gibberella fujikuroi* complex. BMC Genom 14: 1605.

21. Sato T, Taga M, Saitoh N, Nakayama T, Takehara T (1998) Karyotypic analysis of five *Fusarium* spp. causing wheat scab by fluorescence microscopy and fluorescence in situ hybridization. Int Con Plant Pat 2.2.48.

22. Cuomo CA, Guldener U, Xu JR, Trail F, Turgeon BG, et al. (2007) The *Fusarium graminearum* genome reveals a link between localized polymorphism and pathogen specialization. Science 317: 1400–1402.

23. O'Donnell K, Rooney AP, Proctor RH, Brown DW, McCormick SP, et al. (2013) Phylogenetic analyses of RPB1 and RPB2 support a middle Cretaceous origin for a clade comprising all agriculturally and medically important fusaria. Fung Genet Biol 52: 20–31.

24. Watanabe M, Yonezawa T, Lee K, Kumagai S, Sugita-Konishi Y, et al. (2011) Molecular phylogeny of the higher and lower taxonomy of the *Fusarium* genus and differences in the evolutionary histories of multiple genes. Bmc Evolutionary Biology 11: 332.

25. Lysøe E, Seong KY, Kistler HC (2011) The transcriptome of *Fusarium graminearum* during the infection of wheat. Mol Plant Microbe Interact 24: 995–1000.

26. Ma LJ, Geiser DM, Proctor RH, Rooney AP, O'Donnell K, et al. (2013) *Fusarium* pathogenomics. Annu Rev Microbiol 67: 399–416.

27. Martin SH, Wingfield BD, Wingfield MJ, Steenkamp ET (2011) Structure and evolution of the *Fusarium* mating type locus: New insights from the *Gibberella fujikuroi* complex. Fung Genet Biol 48: 731–740.

28. Holtz MD, Chang KF, Hwang SF, Gossen BD, Strelkov SE (2011) Characterization of *Fusarium avenaceum* from lupin in central Alberta: genetic diversity, mating type and aggressiveness. Can J Plant Pathol 33: 61–76.

29. Cook RJ (1967) *Gibberella avenacea* sp. n., perfect stage of *Fusarium roseum* f. sp. *cerealis* 'avenaceum'. Phytopathology 57: 732–736.

30. Booth C, Spooner BM (1984) *Gibberella avenacea*, teleomorph of *Fusarium avenaceum*, from stems of *Pteridium aquilinum*. T Brit Mycol Soc 82: 178–180.

31. Kerényi Z, Moretti A, Waalwijk C, Oláh B, Hornok L (2004) Mating type sequences in asexually reproducing *Fusarium* species. Appl Environ Microbiol 70: 4419–4423.

32. Lee J, Lee T, Lee YW, Yun SH, Turgeon BG (2003) Shifting fungal reproductive mode by manipulation of mating type genes: obligatory heterothallism of *Gibberella zeae*. Mol Microbiol 50: 145–152.

33. Tarailo-Graovac M, Chen N (2009) Using RepeatMasker to identify repetitive elements in genomic sequences. Curr Protoc Bioinformatics 4: 4–10.

34. Ashburner M, Ball CA, Blake JA, Botstein D, Butler H, et al. (2000) Gene Ontology: tool for the unification of biology. Nat Genet 25: 25–29.

35. Zheng P, Xia YL, Xiao GH, Xiong CH, Hu X, et al. (2011) Genome sequence of the insect pathogenic fungus *Cordyceps militaris*, a valued traditional chinese medicine. Genome Biol 12: R116.

36. Thomma BPHJ, Van Esse HP, Crous PW, De Wit PJGM (2005) *Cladosporium fulvum* (syn. *Passalora fulva*), a highly specialized plant pathogen as a model for functional studies on plant pathogenic *Mycosphaerellaceae*. Mol Plant Pathol 6: 379–393.

37. Rep M, van der Does HC, Meijer M, van Wijk R, Houterman PM, et al. (2004) A small, cysteine-rich protein secreted by *Fusarium oxysporum* during colonization of xylem vessels is required for I-3-mediated resistance in tomato. Mol Microbiol 53: 1373–1383.

38. Farman ML, Eto Y, Nakao T, Tosa Y, Nakayashiki H, et al. (2002) Analysis of the structure of the AVR1-CO39 avirulence locus in virulent rice-infecting isolates of *Magnaporthe grisea*. Mol Plant Microbe Interact 15: 6–16.

39. Marx F (2004) Small, basic antifungal proteins secreted from filamentous ascomycetes: a comparative study regarding expression, structure, function and potential application. Appl Microbiol Biotechnol 65: 133–142.

40. de Jonge R, Thomma BPHJ (2009) Fungal LysM effectors: extinguishers of host immunity? Trends Microbiol 17: 151–157.

41. Hershenhorn J, Park SH, Stierle A, Strobel GA (1992) *Fusarium avenaceum* as a novel pathogen of spotted knapweed and its phytotoxins, acetamido-butenolide and enniatin B. Plant Sci 86: 155–160.

42. Thrane U (1988) Screening for Fusarin C production by European isolates of *Fusarium* species. Mycotox Res 4: 2–10.

43. Morrison E, Kosiak B, Ritieni A, Aastveit AH, Uhlig S, et al. (2002) Mycotoxin production by *Fusarium avenaceum* strains isolated from Norwegian grain and the cytotoxicity of rice culture extracts to porcine kidney epithelial cells. J Agricult Food Chem 50: 3070–3075.

44. Takeuchi M, Nakajima M, Ogita T, Inukai M, Kodama K, et al. (1989) Fosfonochlorin, a new antibiotic with spheroplast forming activity. J Antibiot (Tokyo) 42: 198–205.

45. Hussein HM, Baxter M, Andrew IG, Franich RA (1991) Mycotoxin production by *Fusarium* species isolated from New Zealand maize fields. Mycopathologia 113: 35–40.

46. Chelkowski J, Manka M (1983) The ability of Fusaria pathogenic to wheat, barley and corn to produce zearalenone. Phytopathol Z 106: 354–359.

47. Chelkowski J, Visconti A, Manka M (1984) Production of trichothecenes and zearalenone by *Fusarium* species isolated from wheat. Nahrung 28: 493–496.

48. Chelkowski J, Golinski P, Manka M, Wiewiórowska M, Szebiotko K (1983) Mycotoxins in cereal grain. Part IX. Zearalenone and Fusaria in wheat, barley, rye and corn kernels. Die Nahrung 27: 525–531.

49. Ishii K, Sawano M, Ueno Y, Tsunoda H (1974) Distribution of zearalenone-producing *Fusarium* species in Japan. Appl Microbiol 27: 625–628.

50. Marasas W. F O., Nelson P E., Tousson T A. (1984) Toxigenic *Fusarium* species. Identity and mycotoxicology. University Park, Pennsylvania, U.S.A., 328p: Pennsylvania State University Press.

51. Lysøe E, Klemsdal SS, Bone KR, Frandsen RJN, Johansen T, et al. (2006) The *PKS4* gene of *Fusarium graminearum* is essential for zearalenone production. Appl Environ Microbiol 72: 3924–3932.

52. Kim YT, Lee YR, Jin JM, Han KH, Kim H, et al. (2005) Two different polyketide synthase genes are required for synthesis of zearalenone in *Gibberella zeae*. Mol Microbiol 58: 1102–1113.

53. Jiang Z, Barret MO, Boyd KG, Adams DR, Boyd ASF, et al. (2002) JM47, a cyclic tetrapeptide HC-toxin analogue from a marine *Fusarium* species. Phytochemistry 60: 33–38.

54. Laatch H. (2012) Antibase 2012: The natural compound identifier. Wiley-VCH Verlag GmbH.

55. Studt L, Wiemann P, Kleigrewe K, Humpf HU, Tudzynski B (2012) Biosynthesis of fusarubins accounts for pigmentation of *Fusarium fujikuroi* perithecia. Appl Environ Microbiol 78: 4468–4480.

56. Song ZS, Cox RJ, Lazarus CM, Simpson TJ (2004) Fusarin C biosynthesis in *Fusarium moniliforme* and *Fusarium venenatum*. Chembiochem 5: 1196–1203.

57. Malz S, Grell MN, Thrane C, Maier FJ, Rosager P, et al. (2005) Identification of a gene cluster responsible for the biosynthesis of aurofusarin in the *Fusarium graminearum* species complex. Fung Genet Biol 42: 420–433.

58. Seshime Y, Juvvadi PR, Fujii I, Kitamoto K (2005) Discovery of a novel superfamily of type III polyketide synthases in *Aspergillus oryzae*. Biochem Biophys Res Commun 331: 253–260.

59. Wiemann P, Sieber CMK, Von Bargen KW, Studt L, Niehaus EM, et al. (2013) Deciphering the cryptic genome: genome-wide analyses of the rice pathogen *Fusarium fujikuroi* reveal complex regulation of secondary metabolism and novel metabolites. Plos Pathog 9: e1003475.

60. von Döhren H (2004) Biochemistry and general genetics of nonribosomal peptide synthetases in fungi. Adv Biochem Engin/Biotechnol 88: 217–264.

61. Panaccione DG, Scottcraig JS, Pocard JA, Walton JD (1992) A cyclic peptide synthetase gene required for pathogenicity of the fungus *Cochliobolus carbonum* on maize. Proc Natl Acad Sci U S A 89: 6590–6594.

62. Jose B, Oniki Y, Kato T, Nishino N, Sumida Y, et al. (2004) Novel histone deacetylase inhibitors: cyclic tetrapeptide with trifluoromethyl and pentafluoroethyl ketones. Bioorg Med Chem Lett 14: 5343–5346.

63. Hansen FT, Sørensen JL, Giese H, Sondergaard TE, Frandsen RJN (2012) Quick guide to polyketide synthase and nonribosomal synthetase genes in *Fusarium*. Int J Food Microbiol 155: 128–136.

64. Jin JM, Lee S, Lee J, Baek SR, Kim JC, et al. (2010) Functional characterization and manipulation of the apicidin biosynthetic pathway in *Fusarium semitectum*. Mol Microbiol 76: 456–466.

65. Condon BJ, Leng YQ, Wu DL, Bushley KE, Ohm RA, et al. (2013) Comparative genome structure, secondary metabolite, and effector coding capacity across *Cochliobolus* pathogens. Plos Genet 9: e1003233.

66. Manning VA, Pandelova I, Dhillon B, Wilhelm LJ, Goodwin SB, et al. (2013) Comparative genomics of a plant-pathogenic fungus, *Pyrenophora tritici-repentis*, reveals transduplication and the impact of repeat elements on pathogenicity and population divergence. G3-Genes Genom Genet 3: 41–63.

67. Niehaus EM, Janevska S, von Bargen KW, Sieber CMK, Harrer H, et al. (2014) Apicidin F: Characterization and genetic manipulation of a new secondary metabolite gene cluster in the rice pathogen *Fusarium fujikuroi*. Plos One 9: e103336. doi: 10.1371/journal.pone.0103336.

68. Lee SL, Floss HG, Heinstein P (1976) Purification and properties of dimethylallylpyrophosphate - tryptophan dimethylallyl transferase, first enzyme of ergot alkaloid biosynthesis in *Claviceps*. sp. SD 58. Arch Biochem Biophys 177: 84–94.

69. Young CA, Tapper BA, May K, Moon CD, Schardl CL, et al. (2009) Indole-diterpene biosynthetic capability of *Epichloe* endophytes as predicted by *ltm* gene analysis. Appl Environ Microbiol 75: 2200–2211.

70. Bomke C, Tudzynski B (2009) Diversity, regulation, and evolution of the gibberellin biosynthetic pathway in fungi compared to plants and bacteria. Phytochemistry 70: 1876–1893.

71. Rim SO, You YH, Yoon H, Kim YE, Lee JH, et al. (2013) Characterization of gibberellin biosynthetic gene cluster from *Fusarium proliferatum*. J Microbiol Biot 23: 623–629.

72. Malonek S, Rojas MC, Hedden P, Gaskin P, Hopkins P, et al. (2005) Functional characterization of two cytochrome P450 monooxygenase genes, P450-1 and P450-4, of the gibberellic acid gene cluster in *Fusarium proliferatum* (*Gibberella fujikuroi* MP-D). Appl Environ Microbiol 71: 1462–1472.

73. Bomke C, Rojas MC, Gong F, Hedden P, Tudzynski B (2008) Isolation and characterization of the gibberellin biosynthetic gene cluster in *Sphaceloma manihoticola*. Appl Environ Microbiol 74: 5325–5339.

74. Kawaide H (2006) Biochemical and molecular analyses of gibberellin biosynthesis in fungi. Biosci Biotechnol Biochem 70: 583–590.

75. Gaffoor I, Brown DW, Plattner R, Proctor RH, Qi WH, et al. (2005) Functional analysis of the polyketide synthase genes in the filamentous fungus *Gibberella zeae* (anamorph *Fusarium graminearum*). Eukaryot Cell 4: 1926–1933.

76. Karányi Z, Holb I, Hornok L, Pócsi I, Miskei M (2013) FSRD: fungal stress response database. Database (Oxford) 2013: bat037.

77. Plancot B, Santaella C, Jaber R, Kiefer-Meyer MC, Follet-Gueye ML, et al. (2013) Deciphering the responses of root border-like cells of *Arabidopsis* and flax to pathogen-derived elicitors. Plant Physiol 163: 1584–1597.

78. Kosti I, Mandel-Gutfreund Y, Glaser F, Horwitz BA (2010) Comparative analysis of fungal protein kinases and associated domains. BMC Genom 11: 1133.

79. Hamilton AJ, Holdom MD (1999) Antioxidant systems in the pathogenic fungi of man and their role in virulence. Med Mycol 37: 375–389.

80. de Dios CH, Roman E, Monge RA, Pla J (2010) The role of MAPK signal transduction pathways in the response to oxidative stress in the fungal pathogen *Candida albicans*: Implications in virulence. Curr Protein Pept Sci 11: 693–703.

81. Dean R, van Kan JAL, Pretorius ZA, Hammond-Kosack KE, Di Pietro A, et al. (2012) The Top 10 fungal pathogens in molecular plant pathology. Mol Plant Pathol 13: 414–430.

82. Urban M, Hammond-Kosack KE (2013) Molecular genetics and genomic approaches to explore *Fusarium* infection of wheat floral tissue. In: Brown DW, Proctor RH, editors.*Fusarium*: Genomics, Molecular and Cellular Biology.- Norfolk, UK: Caister Academic Press. pp.43–79.

83. Kokkonen M, Ojala L, Parikka P, Jestoi M (2010) Mycotoxin production of selected *Fusarium* species at different culture conditions. Int J Food Microbiol 143: 17–25.

84. Harris SD, Morrell JL, Hamer JE (1994) Identification and characterization of *Aspergillus nidulans* mutants defective in cytokinesis. Genetics 136: 517–532.

85. Geiser DM, Jimenez-Gasco MD, Kang SC, Makalowska I, Veeraraghavan N, et al. (2004) FUSARIUM-ID v. 1.0: A DNA sequence database for identifying Fusarium. Eur J Plant Pathol 110: 473–479.

86. Assefa S, Keane TM, Otto TD, Newbold C, Berriman M (2009) ABACAS: algorithm-based automatic contiguation of assembled sequences. Bioinformatics 25: 1968–1969.

87. Kurtz S, Phillippy A, Delcher AL, Smoot M, Shumway M, et al. (2004) Versatile and open software for comparing large genomes. Genome Biol 5.

88. Stanke M, Diekhans M, Baertsch R, Haussler D (2008) Using native and syntenically mapped cDNA alignments to improve de novo gene finding. Bioinformatics 24: 637–644.

89. Conesa A, Gotz S, Garcia-Gomez JM, Terol J, Talon M, Robles M (2005) Blast2GO: a universal tool for annotation, visualization and analysis in functional genomics research. Bioinformatics 21: 3674–3676.

90. Le SQ, Gascuel O (2008) An improved general amino acid replacement matrix. Mol Biol Evol 25: 1307–1320.

91. Coleman JJ, Rounsley SD, Rodriguez-Carres M, Kuo A, Wasmann CC, et al. (2009) The genome of *Nectria haematococca*: contribution of supernumerary chromosomes to gene expansion. Plos Genet 5: e1000618.

92. Park J, Park J, Jang S, Kim S, Kong S, et al. (2008) FTFD: an informatics pipeline supporting phylogenomic analysis of fungal transcription factors. Bioinformatics 24: 1024–1025.

93. Petersen TN, Brunak S, von Heijne G, Nielsen H (2011) SignalP 4.0: discriminating signal peptides from transmembrane regions. Nature Methods 8: 785–786.

94. Krogh A, Larsson B, von Heijne G, Sonnhammer ELL (2001) Predicting transmembrane protein topology with a hidden Markov model: Application to complete genomes. J Mol Biol 305: 567–580.

95. Futami R, Muñoz-Pomer L, Viu JM, Dominguez-Escriba L, Covelli L, et al. (2011) GPRO: The professional tool for annotation, management and functional analysis of omics databases. Biotechvana Bioinformatics 2011-SOFT3 2011.

96. Marcet-Houben M, Ballester AR, de la Fuente B, Harries E, Marcos JF, et al. (2012) Genome sequence of the necrotrophic fungus *Penicillium digitatum*, the main postharvest pathogen of citrus. BMC Genom 13: 646.

97. Rice P, Longden I, Bleasby A (2000) EMBOSS: the European Molecular Biology Open Software Suite. Trends Genet 16: 276–277.

98. Silverstein KAT, Moskal WA, Wu HC, Underwood BA, Graham MA, et al. (2007) Small cysteine-rich peptides resembling antimicrobial peptides have been under-predicted in plants. Plant J 51: 262–280.

99. Finn RD, Clements J, Eddy SR (2011) HMMER web server: interactive sequence similarity searching. Nucleic Acids Res 39: W29–W37.

100. Samson RA, Houbraken J, Thrane U, Frisvad JC, Andersen B (2010) Food and indoor fungi. Utrecht: CBS-KNAW Fungal Biodiversity Centre.

101. Smedsgaard J (1997) Micro-scale extraction procedure for standardized screening of fungal metabolite production in cultures. J Chromatogr A 760: 264–270.

102. Kildgaard S, Månsson M, Dosen I, Klitgaard A, Frisvad JC, et al. (2014) Accurate dereplication of bioactive secondary metabolites from marine-derived fungi by UHPLC-DAD-QTOFMS and a MS/HRMS Library. Mar Drugs 12: 3681–3705.

103. Klitgaard A, Iversen A, Andersen MR, Larsen TO, Frisvad JC, et al. (2014) Aggressive dereplication using UHPLC-DAD-QTOF: screening extracts for up to 3000 fungal secondary metabolites. Anal Bioanal Chem 406: 1933–1943.

104. Nielsen KF, Månsson M, Rank C, Frisvad JC, Larsen TO (2011) Dereplication of microbial natural products by LC-DAD-TOFMS. J Nat Prod 74: 2338–2348.

105. Altschul SF, Madden TL, Schaffer AA, Zhang JH, Zhang Z, et al. (1997) Gapped BLAST and PSI-BLAST: a new generation of protein database search programs. Nucleic Acids Res 25: 3389–3402.

106. Frazer KA, Pachter L, Poliakov A, Rubin EM, Dubchak I (2004) VISTA: computational tools for comparative genomics. Nucleic Acids Res 32: W273–W279.

107. Brudno M, Malde S, Poliakov A, Do CB, Couronne O, et al. (2003) Global alignment: finding rearrangements during alignment. Bioinformatics 19: i54–i62.

108. Marchler-Bauer A, Lu SN, Anderson JB, Chitsaz F, Derbyshire MK, et al. (2011) CDD: a conserved domain database for the functional annotation of proteins. Nucleic Acids Res 39: D225–D229.

109. Crawford JM, Dancy BCR, Hill EA, Udwary DW, Townsend CA (2006) Identification of a starter unit acyl-carrier protein transacylase domain in an iterative type I polyketide synthase. Proc Natl Acad Sci U S A 103: 16728–16733.

110. Dill-Macky R (2003) Inoculation methods and evaluation of *Fusarium* head blight resistance in wheat. In: Leonard KJ, Bushnell WR, editors.Fusarium head blight of wheat and barley. pp.184–210.

4

Sequencing, Annotation and Analysis of the Syrian Hamster (*Mesocricetus auratus*) Transcriptome

Nicolas Tchitchek[1], David Safronetz[2], Angela L. Rasmussen[1], Craig Martens[3], Kimmo Virtaneva[3], Stephen F. Porcella[3], Heinz Feldmann[2], Hideki Ebihara[2*⌕], Michael G. Katze[1,4*⌕]

1 Department of Microbiology, University of Washington, Seattle, Washington, United States of America, 2 Laboratory of Virology, Division of Intramural Research, National Institute of Allergy and Infectious Diseases, National Institutes of Health, Rocky Mountain Laboratories, Hamilton, Montana, United States of America, 3 Genomics Unit, Research Technologies Section, National Institute of Allergy and Infectious Diseases, National Institutes of Health, Rocky Mountain Laboratories, Hamilton, Montana, United States of America, 4 Washington National Primate Research Center, University of Washington, Seattle, Washington, United States of America

Abstract

Background: The Syrian hamster (golden hamster, *Mesocricetus auratus*) is gaining importance as a new experimental animal model for multiple pathogens, including emerging zoonotic diseases such as Ebola. Nevertheless there are currently no publicly available transcriptome reference sequences or genome for this species.

Results: A cDNA library derived from mRNA and snRNA isolated and pooled from the brains, lungs, spleens, kidneys, livers, and hearts of three adult female Syrian hamsters was sequenced. Sequence reads were assembled into 62,482 contigs and 111,796 reads remained unassembled (singletons). This combined contig/singleton dataset, designated as the Syrian hamster transcriptome, represents a total of 60,117,204 nucleotides. Our *Mesocricetus auratus* Syrian hamster transcriptome mapped to 11,648 mouse transcripts representing 9,562 distinct genes, and mapped to a similar number of transcripts and genes in the rat. We identified 214 quasi-complete transcripts based on mouse annotations. Canonical pathways involved in a broad spectrum of fundamental biological processes were significantly represented in the library. The Syrian hamster transcriptome was aligned to the current release of the Chinese hamster ovary (CHO) cell transcriptome and genome to improve the genomic annotation of this species. Finally, our Syrian hamster transcriptome was aligned against 14 other rodents, primate and laurasiatheria species to gain insights about the genetic relatedness and placement of this species.

Conclusions: This Syrian hamster transcriptome dataset significantly improves our knowledge of the Syrian hamster's transcriptome, especially towards its future use in infectious disease research. Moreover, this library is an important resource for the wider scientific community to help improve genome annotation of the Syrian hamster and other closely related species. Furthermore, these data provide the basis for development of expression microarrays that can be used in functional genomics studies.

Editor: Vincent Laudet, Ecole Normale Supérieure de Lyon, France

Funding: This work was supported by the Intramural Research Program of the National Institute of Allergy and Infectious Diseases, National Institutes of Health and the NIAID Regional Centers of Excellence (U54 AI081680) to MGK. The funders were not involved in the study design, data collection and analysis, or preparation of the manuscript. NIAID was solely involved in the decision to publish.

Competing Interests: The authors have declared that no competing interests exist.

* Email: ebiharah@niaid.nih.gov (HE); honey@uw.edu (MK)

⌕ These authors contributed equally to this work.

Introduction

The Syrian hamster (golden hamster, *Mesocricetus auratus*) has recently been used as an experimental rodent model for important infectious diseases including Ebola and other viral hemorrhagic fevers [1–8]. For instance, Syrian hamsters infected with mouse-adapted Ebola virus (EBOV) manifest many of the clinical and pathological findings observed in EBOV-infected non-human primates (NHPs) and humans, including systemic viral replication, suppression of the innate immune response, an uncontrolled inflammatory response, and disseminated intravascular coagulation syndrome [9]. The Syrian hamster is emerging as a promising model for leishmaniasis [10] and dyslipidaemia research [11,12]. The Syrian hamster is also an important animal model in neurosciences research [13,14]. For instance, this species has been widely used in the studies of circadian rhythms [15], cardiomyopathy [16], aggression [17], reproduction [18], and sensory systems [19].

Genotyping of *Mesocricetus auratus* is currently under way at the Broad Institute (NCBI-BioProject accession: PRJNA77669) but not yet published. So far, only 860 cDNA sequences from the Syrian hamster are available in the NCBI-dbEST database [20], where 728 sequences have been collected in the context of testis organs [21] and 125 sequences have been collected in the context of embryonic cells [22]. More recently, while Schmucki et al. analyzed the liver transcriptome of the Syrian hamster with a focus on lipid metabolism [23] the data is not publicly available as of this writing.

Drafts of the genome and transcriptome of Chinese hamster ovary (CHO) cells have recently been published [24,25], although it should be noted that CHO cells represent cells in an immortalized condition and therefore will likely contain genetic mutations not present in natural conditions. The current release of the CHO cell draft genome is composed of 109,152 scaffolds and 265,786 contigs representing a total length of 2,318,115,958 nucleotides. Preliminary gene annotation of the CHO cell genome was performed using vertebrate experimental data and cross-species comparisons. The current release of the CHO cell transcriptome comprises 121,636 transcript fragments representing a total length of 179,731,611 nucleotides. More recently, Lewis et al. compared the genome of CHO cells and the genome of the Chinese hamster obtained from tissues, and they showed a significant proximity between these different conditions [26]. Further efforts will be continued regarding the update of the CHO and Chinese hamster genomes and transcriptomes.

The aims of our study were: (i) to provide to the scientific community a large panel of annotated mRNA sequences from the *Mesocricetus auratus* transcriptome; (ii) to provide new biological insights and knowledge about the *Mesocricetus auratus* species; and (iii) to use this data to allow the design of a future gene expression microarray. Here we sequenced a normalized 3′ mRNA fragment primed cDNA library produced from pooled RNA isolated from the major organs of adult female Syrian hamsters following strategies in common-use described elsewhere [27,28]. We reasoned that pooling a large variety of different organs of animals will provide a large pool of mRNA fragments to sequence and annotate. Sequencing reads were de novo assembled into contigs. The combined contig and unassembled read (singleton) dataset, designated as the Syrian hamster transcriptome, was annotated based on the mouse and rat transcriptomes. We identified the most highly covered and the most highly expressed transcripts in our Syrian hamster transcriptome and performed a functional enrichment analysis to identify which canonical pathways and biological functions were most significantly represented. In order to contribute to the annotation efforts of the Chinese hamster species, we aligned our Syrian hamster transcriptome to the current version of the CHO cell genome and transcriptome. Finally, we aligned our Syrian hamster transcriptome to 14 other primate species and analyzed the genomic divergence of our transcripts in order to gain insights into the genomic evolution of the Syrian hamster.

Results

Sample collection and sequencing of a cDNA library produced from female Syrian hamster organs

The brains, lungs, spleens, kidneys, livers, and hearts were collected from three adult female Syrian hamsters. Total RNAs were isolated, pooled, and contaminating genomic DNA removed. Following adaptor ligation, cDNAs were 3′ fragment-sequenced on a Roche 454 GS FLX Titanium instrument. The sequencing generated 1,283,840 reads with an average length of 344 bases. Reads were trimmed for quality and reads shorter than 40 bases were discarded, resulting in 1,212,395 sequence reads available for further assembly and analysis. **Figure 1A** shows the length distribution of reads before assembly. Consistent with most of the publicly available transcriptome libraries [29], we observed that our reads ranged between 200 and 600 nucleotides in length.

Library assembly

Quality-filtered reads were assembled into contigs. Resulting contigs and unassembled reads (singletons) were quality filtered and contigs or singletons shorter than 50 bases were discarded. Among the 1,212,395 reads, 62,482 contigs and 111,796 singletons were generated. **Figure 1B** shows the length distributions of the 174,278 combined contig/singleton dataset. The lengths of the singletons ranged from 50 to 614, with a median length of 187.50 bases. The lengths of the contigs ranged from 50 to 4,054, with a median length of 473.50 bases. We observed that most of the reads ranging between 75 and 400 nucleotides were assembled. Short reads are subject to noise and have low quality scores, making them more difficult to assemble. On the other hand, larger reads are difficult to assemble in this context because our library was targeted against 3′ mRNA priming. The final dataset (contigs plus singletons) represents a total of 60,117,204 nucleotides and is designated as the Syrian hamster transcriptome.

Library annotation

The Syrian hamster transcriptome was aligned to the mouse and rat transcriptome references (**Table 1**). Amongst the 174,278 contigs and singletons, 41,651 (23.90%) were significantly aligned (expected value cutoff of 10) to the mouse transcriptome and 26,258 (15.07%) were significantly aligned to the rat transcriptome. Of these, 11,648 transcripts (representing 9,562 genes) contained functional annotation in the mouse transcriptome, and 7,223 transcripts (representing 7,137 genes) were functionally annotated in the rat transcriptome (**Table 1**). Therefore, 11,648 Syrian sequence fragments or transcripts are now annotated by way of homology with the mouse genome.

We also investigated the positioning of the mRNA encoded contigs and singletons of our Syrian hamster transcriptome against other species' different transcript regions such as, 5′ untranslated regions (5′ UTR), coding regions, or 3′ untranslated regions (3′ UTR). With respect to the mouse transcriptome reference, 4,314 fragments of our Syrian hamster transcriptome (10.36%) aligned to 5′ UTRs, while 6,493 fragments of our dataset (15.59%) aligned to coding regions. In addition, 26,764 fragments of the Syrian hamster transcriptome (64.26%) aligned to 3′ UTRs (**Figure 2A**). A further 4,080 fragments of the Syrian hamster transcriptome (9.80%) aligned between 5′ UTRs, coding regions, and 3′ UTRs of the mouse transcriptome. Based on the rat transcriptome reference, 521 fragments of the Syrian hamster transcriptome (1.98%) aligned to 5′ UTRs while 5,568 fragments of the Syrian hamster transcriptome (21.20%) aligned to coding regions. In addition, 13,371 fragments of the Syrian hamster transcriptome (50.92%) aligned to 3′ UTRs (**Figure 2B**). Finally, a total of 6,798 fragments of the Syrian hamster transcriptome (25.89%) aligned between 5′ UTRs, coding regions, and 3′ UTRs of the rat transcriptome. As expected from the experimental design of our library, the majority of our Syrian hamster transcriptome sequences aligned to 3′ UTRs of mouse and rat annotated transcripts.

Regarding the publicly available mouse and rat genome datasets, 45,804 of our Syrian hamster transcriptome fragments aligned to either the mouse or rat transcriptomes, and 22,105 of these same sequences aligned to both transcriptomes simultaneously, suggesting commonly occurring transcripts. Our Syrian hamster transcriptome dataset was 65.10% similar to the mouse transcriptome and 64.46% similar to the rat transcriptome. These similarities increased to 74.48% and 74.26% for the mouse and rat respectively, when comparisons were restricted to coding regions-only within those two reference genomes.

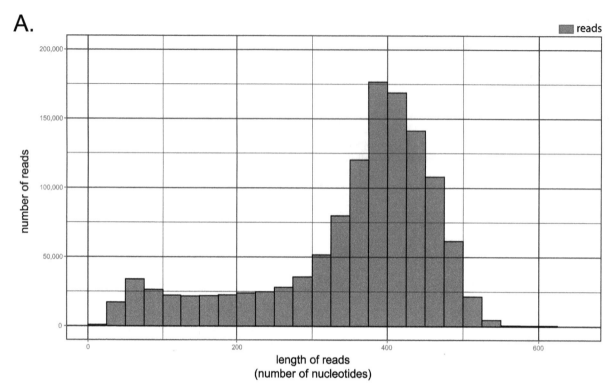

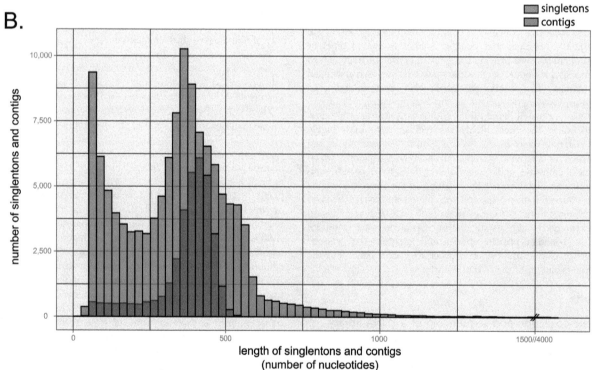

Figure 1. Histograms showing the length distribution of the reads and the length distribution of the singletons and contigs. (A) The length distribution of the reads is shown in a gray histogram. Bins of the histogram have been set to 50 nucleotides. The lengths of the reads range from 40 to 631, with a median length of 387 and a mean length of 352. The reads represents a total of 426,683,712 nucleotides bases. (B) The length distribution of the 111,796 singletons is shown in a red histogram while the length distribution of the 62,482 contigs is shown in a blue histogram. Bins of the histograms have been set to 25 nucleotides. The lengths of the singleton sequences range from 50 to 614, with a median length of 187 and a mean length of 265. The lengths of the contig sequences range from 50 to 4,054, with a median length of 473 and a mean length of 487. Our Syrian hamster transcriptome represents a total of 60,117,204 nucleotides bases.

Table 1. Transcriptome references and alignment statistics.

Species name	# of genes	# of transcripts	# and % of alignments	# and % of mapped genes	# and % of mapped transcripts
Mouse (Mus musculus)	38,293	92,484	41,651 (23.90%)	9,562 (22.96%)	11,648 (12.59%)
Rat (Rattus norvegicus)	26,405	29,189	26,258 (15.07%)	7,137 (27.18%)	7,223 (24.75%)
Chinese Hamster Ovary cells (Cricetulus griseus)	NA	121,636*	7,845 (4.50%)	NA	4,390 (3.61%)
Chimpanzee (Pan troglodytes)	28,012	29,160	2,884 (1.65%)	1,631 (56.55%)	1,643 (5.63%)
Ferret (Mustela putorius furo)	23,811	23,963	16,169 (9.28%)	4,169 (25.78%)	4,187 (17.47%)
Gorilla (Gorilla gorilla gorilla)	29,216	35,727	8,319 (4.77%)	2,733 (32.85%)	2,735 (7.66%)
Guinea pig (Cavia porcellus)	25,028	26,129	15,014 (8.61%)	4,050 (26.97%)	4,155 (15.90%)
Human (Homo sapiens)	62,316	213,551	23,020 (13.21%)	5,409 (23.50%)	7,254 (3.40%)
Kangaroo rat (Dipodomys ordii)	26,405	29,189	2,103 (1.21%)	1,252 (59.53%)	1,252 (4.29%)
Macaque (Macaca mulatta)	30,246	44,725	13,792 (7.91%)	3,804 (27.58%)	4,163 (9.31%)
Orangutan (Pongo abelii)	28,443	29,447	15,331 (8.80%)	3,929 (25.63%)	3,952 (13.42%)
Pig (Sus scrofa)	25,322	30,586	10,910 (6.26%)	2,978 (27.30%)	3,051 (9.98%)
Pika (Ochotona princeps)	23,028	23,028	1,575 (0.90%)	989 (62.79%)	989 (4.29%)
Rabbit (Oryctolagus cuniculus)	23,394	28,188	4,344 (2.49%)	1,946 (44.80%)	2,007 (7.12%)
Shrew (Sorex araneus)	19,134	19,139	1,330 (0.76%)	759 (57.07%)	759 (3.97%)
Squirrel (Ictidomys tridecemlineatus)	22,398	23,572	7,730 (4.44%)	2,723 (35.23%)	2,733 (11.59%)
Tree Shrew (Tupaia belangeri)	20,820	20,824	1,786 (1.02%)	1,091 (61.09%)	1,091 (5.24%)

For each transcriptome reference used in this study, the name of the species, the number of genes available, and the number of transcripts available are indicated. *The number of available transcripts indicated for the Chinese hamster ovary cells represents the number of available transcript fragments available and not the number of distinct transcripts. Moreover, for each transcriptome reference used in this study, the number of aligned contigs and singletons, the number of mapped transcripts and the number of mapped genes are indicated. The percentages of mapped transcripts and mapped genes relative to the total number of transcripts and genes available on the transcriptome references are provided. Moreover the percentage of alignments relative to the total number of contigs and singletons in our library (174,278) is also provided.

In the mouse genome, we found that 214 of those transcripts mapped at 90% of their lengths to either contigs or singletons in the Syrian hamster transcriptome (**Table S1**). Among these highly covered mouse transcripts were genes associated with a range of cellular activities involving, but not limited to inflammation, cell death, metabolism, and initiation of translation. These results

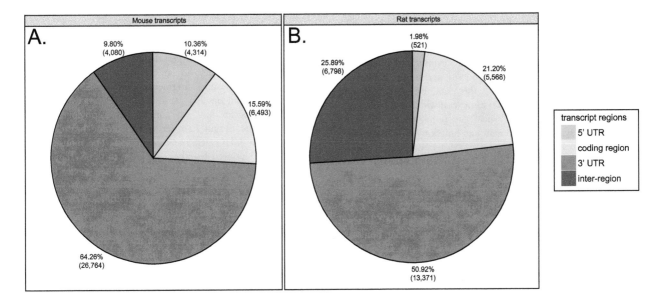

Figure 2. Pie diagrams showing the alignment positions of the contigs and singletons on the mouse and rat transcript regions. (A) Pie diagram showing the distribution of alignment positions of the 41,651 contigs and singletons on the mouse transcripts regions (5′ UTR, coding region, 3′ UTR, or inter-region). (B) Pie diagram showing the distribution of alignment positions of the 26,258 contigs and singletons on the rat transcripts regions. For each species and transcript region the number and percentage of aligned sequences are indicated.

suggest that highly covered transcripts, representative of a wide variety of cellular processes, were obtained through our methodology.

Over-expressed sequence reads and over-represented canonical pathways

In order to obtain further biological insight into our Syrian hamster transcriptome, we next identified over-expressed genes based on the number of individual reads that mapped to mouse-annotated genes (**Table 2**). We found that 20 mouse genes contained at least 600 x read depth, and 49 mouse genes contained at least 500 fold read depth.

Most of the mouse genes showing high read depth were annotated as being involved in fundamental cellular processes such as cell morphology and organization, cell cycle progression, cell function and maintenance, transcription, protein synthesis and turnover, cell death, and molecular transport. Genes associated with cell type or tissue-specific functions were not significantly over-represented, consistent with our method of generating cDNA reads from pooled, multiple organ tissues. Our aim in this study was to sequence and annotate a large number of hamster mRNA 3′ fragments as a preliminary effort towards generation of an expression array, our observation that the distribution of reads were across common cellular functions, suggests our assembly is not overly biased against a specific cell or tissue type.

We also performed a functional enrichment analysis of our Syrian hamster transcriptome. Based on the list of 9,562 mouse genes that were mapped to our contigs and singletons, we identified the over-represented canonical pathways in our library (**Table 3**). "Protein ubiquitination" (**Figure 3A**, p-value = 1.99E-18) and "molecular mechanisms of cancer" (**Figure 3B**, p-value = 5.01E-14) were the two most over-represented canonical pathways. However, there was also significant enrichment of many other canonical pathways related to biochemical, cellular, and disease-associated cellular processes. These included a multitude of signaling pathways, including RhoGTPase, protein kinase A, integrin, Rac, ERK/MAPK, mTOR, PI3K/Akt, PTEN, insulin, WNT/b-catenin, growth factor (VEGF, NGF, HGF, FGF, GM-CSF), and cellular junction signaling pathways. All of these pathways are biologically essential for intra- and intercellular communication and have known pleiotropic effects on transcription and translation, cellular proliferation, development, differentiation, cytoskeletal dynamics, cellular morphology, cell death, metabolism, and host responses to stress or infection. Consistent with this data, we also observed enrichment of functional categories associated with these biological activities (**Table 3**). The biological functions associated with "cardiovascular system development and function" (p-values range from 1.05E-03 to 4.15E-17) and "nervous system development and function" (p-values range from 1.29E-03 to 1.46E-19) were statistically over-represented.

Comparison with the Chinese hamster species

In order to contribute to the annotation efforts for the Chinese hamster (*Cricetulus griseus*) species, we aligned our Syrian hamster transcriptome to the current draft versions of the CHO cell genome and its transcriptome (**Table 1**).

We found that 7,845 fragments in our Syrian hamster transcriptome aligned to the CHO cell transcriptome (**Table 1**) and 85,652 aligned to the CHO cell genome (**Table S2**). On the other hand, 4,390 transcript fragments from the CHO cell dataset mapped to the Syrian hamster transcriptome (**Table 1**). Our aligned Syrian hamster transcriptome showed 85.14% similarity with the CHO cell transcriptome, an expectedly higher value than

what we saw for the same comparison with the mouse and rat transcriptomes.

Cross-species comparison

In order to obtain further insights about the genomic evolution of the Syrian hamster we aligned our Syrian hamster transcriptome to 14 other transcriptomes, all of which are publicly available on the Ensembl database [30] (**Table 1**). This compendium of transcriptome references included the human (*Homo sapiens*), chimpanzee (*Pan Troglodytes*), gorilla (*Gorilla gorilla gorilla*), macaque (*Macaca mulatta*), and orangutan (*Pongo abelii*) sequences, as well as the ferret (*Mustela putorius furo*), guinea pig sequences (*Cavia porcellus*), and pig (*Sus scrofa*). As expected, the greatest number of aligned sequences occurred with the mouse and rat species transcriptomes (**Table 1**). The human and the non-human primate species also showed high numbers of aligned sequences, possibly due to the current high quality assembly and annotation of those genomes. The CHO, ferret, pig, rabbit (*Oryctolagues cuniculus*), and squirrel (*Ictidomys tridecemlineatus*) species showed intermediary numbers of aligned sequences, while the guinea pig, kangaroo rat (*Dipodomys ordii*), pika (*Ochotona princeps*), shrew (*Sorex araneus*) and tree shrew (*Tupaia belangeri*) had the lowest numbers of aligned sequences. Of 174,278 Syrian hamster transcriptome fragments 50,433 aligned to at least one transcript reference while 61 fragments from our dataset aligned in common across all of these transcriptome references. Importantly, 76,175 of our Syrian hamster transcriptome fragments did not align to any of the 17 transcriptomes tested, nor to the CHO cell genome. It is important to note that some of the variability seen in our transcriptome comparisons may be due to differences in genome quality, assembly and annotation for the reference genomes tested.

Figure 4A is a distogram showing the results of our analysis of transcript sequences shared in common. The kangaroo rat, pika, shrew, and tree shrew had the lowest amount of commonly aligned sequences, amongst themselves and with the other species. The mouse and rat species showed the highest number of aligned sequences, presumably because of both their relatedness and genome quality/completeness.

We then investigated the evolutionary divergence between the Syrian hamster and the 13 species with the largest numbers of mapped sequences and the largest degrees of shared sequences (i.e. excluding the pika, kangaroo rat, shrew, tree shrew). We found that 611 transcriptomic fragments (**Table S3**) have been significantly aligned on the transcriptome references of these 13 most related species and we constructed a phylogenetic tree (**Figure 4B**). The Syrian hamster transcriptome branched most closely with the CHO genome as expected. The mouse and rat transcriptome clustered together and close to the Syrian hamster and CHO cluster, as expected. All the primate species formed a super group, while the ferret and pig transcriptomes clustered together as the rabbit and squirrel transcriptomes. Consistent with a recently published study [31], we observed that the genomic divergence between the Syrian and Chinese hamsters is comparable to the divergence seen between the rat and mouse. Also, as expected, we observed that the Guinea pig does not cluster with the rodent species [32,33].

Discussion

Here we present the assembly and analysis of a Syrian hamster transcriptome derived from the pooled RNAs from brains, lungs, spleens, kidneys, livers, and hearts of three adult females. The 3′ poly-T primed cDNAs that were sequenced on a long read-format

Table 2. List of the top 50 expressed genes in the library.

Ensembl Gene ID	Associated Gene Name	Description	Count
ENSMUSG00000028647	Mycbp	c-myc binding protein	1120
ENSMUSG00000020594	Pum2	pumilio 2 (Drosophila)	1017
ENSMUSG00000008575	Nfib	nuclear factor I/B	945
ENSMUSG00000022010	Tsc22d1	TSC22 domain family, member 1	895
ENSMUSG00000062078	Qk	quaking	861
ENSMUSG00000078578	Ube2d3	ubiquitin-conjugating enzyme E2D 3	795
ENSMUSG00000026621	Mosc1	MOCO sulphurase C-terminal domain containing 1	710
ENSMUSG00000028161	Ppp3ca	protein phosphatase 3, catalytic subunit, alpha isoform	707
ENSMUSG00000028790	Khdrbs1	KH domain containing, RNA binding, signal transduction associated 1	695
ENSMUSG00000006740	Kif5b	kinesin family member 5B	684
ENSMUSG00000031627	Irf2	interferon regulatory factor 2	682
ENSMUSG00000036781	Rps27l	ribosomal protein S27-like	660
ENSMUSG00000026655	Fam107b	family with sequence similarity 107, member B	658
ENSMUSG00000006373	Pgrmc1	progesterone receptor membrane component 1	652
ENSMUSG00000060961	Slc4a4	solute carrier family 4 (anion exchanger), member 4	641
ENSMUSG00000024750	Zfand5	zinc finger, AN1-type domain 5	639
ENSMUSG00000028788	Ptp4a2	protein tyrosine phosphatase 4a2	634
ENSMUSG00000019943	Atp2b1	ATPase, Ca++ transporting, plasma membrane 1	605
ENSMUSG00000097347	AC121292.1		603
ENSMUSG00000004980	Hnrnpa2b1	heterogeneous nuclear ribonucleoprotein A2/B1	600
ENSMUSG00000093904	Tomm20	translocase of outer mitochondrial membrane 20 homolog (yeast)	593
ENSMUSG00000068823	Csde1	cold shock domain containing E1, RNA binding	586
ENSMUSG00000020315	Spnb2	spectrin beta 2	579
ENSMUSG00000068798	Rap1a	RAS-related protein-1a	579
ENSMUSG00000020390	Ube2b	ubiquitin-conjugating enzyme E2B	570
ENSMUSG00000026064	Ptp4a1	protein tyrosine phosphatase 4a1	570
ENSMUSG00000020053	Igf1	insulin-like growth factor 1	569
ENSMUSG00000027706	Sec62	SEC62 homolog (S. cerevisiae)	553
ENSMUSG00000064373	Sepp1	selenoprotein P, plasma, 1	549
ENSMUSG00000014956	Ppp1cb	protein phosphatase 1, catalytic subunit, beta isoform	538
ENSMUSG00000007850	Hnrnph1	heterogeneous nuclear ribonucleoprotein H1	536
ENSMUSG00000031207	Msn	moesin	518
ENSMUSG00000020152	Actr2	ARP2 actin-related protein 2	515
ENSMUSG00000022261	Sdc2	syndecan 2	514
ENSMUSG00000047187	Rab2a	RAB2A, member RAS oncogene family	512
ENSMUSG00000004936	Map2k1	mitogen-activated protein kinase kinase 1	510
ENSMUSG00000026576	Atp1b1	ATPase, Na+/K+ transporting, beta 1 polypeptide	506
ENSMUSG00000022234	Cct5	chaperonin containing Tcp1, subunit 5 (epsilon)	504
ENSMUSG00000001175	Calm1	calmodulin 1	502
ENSMUSG00000069662	Marcks	myristoylated alanine rich protein kinase C substrate	490
ENSMUSG00000017776	Crk	v-crk sarcoma virus CT10 oncogene homolog (avian)	484
ENSMUSG00000038014	Fam120a	family with sequence similarity 120A	484
ENSMUSG00000036478	Btg1	B cell translocation gene 1, anti-proliferative	483
ENSMUSG00000027177	Hipk3	homeodomain interacting protein kinase 3	478
ENSMUSG00000043991	Pura	purine rich element binding protein A	474
ENSMUSG00000022283	Pabpc1	poly(A) binding protein, cytoplasmic 1	471
ENSMUSG00000031342	Gpm6b	glycoprotein m6b	471
ENSMUSG00000050608	Minos1	mitochondrial inner membrane organizing system 1	471
ENSMUSG00000018446	C1qbp	complement component 1, q subcomponent binding protein	469

Table 2. Cont.

Ensembl Gene ID	Associated Gene Name	Description	Count
ENSMUSG00000026568	Mpc2	mitochondrial pyruvate carrier 2	461

For each of the top 50 expressed genes in the library, based on the mouse annotations, the Ensembl mouse gene identified, the associated gene name, description, and the number of count (number of time that the genes have been mapped by the reads) are indicated.

Roche 454 were assembled into contigs, such that 22,105 of these contigs or singletons were annotated based on homology with both the mouse and rat transcriptomes, while 45,804 contigs or singletons were annotated based upon homology to one or the other mouse or rat transcriptomes. We identified 214 quasi-complete transcript sequences based on homology with mouse mRNAs and their annotations. In addition, we aligned our Syrian hamster transcriptome to the CHO cell transcriptome in order to further annotate our hamster species, and we observed a transcriptome similarity of 85.14% between the two.

When compared to a large compendium of transcriptome references, comprised of rodent, primate, and laurasiatheria species, using 661 Syrian hamster transcriptome fragments that aligned in common, the Syrian hamster transcriptome was found to be evolutionarily closest to the CHO genome and in close proximity to the mouse and rat species. The branch pattern and branch length between the Syrian and Chinese hamster transcriptomes was found to be similar to that observed between the mouse and the rat species. This observation was also described

by Ryu et al. [31], but those previous efforts focused on mitochondrial gene sequences for their phylogeny analysis.

In the Syrian hamster transcriptome, we were able to identify a number of genes involved in a broad spectrum of fundamental biological processes. In addition to the 214 quasi-complete transcripts, identified based on mouse annotations and the most highly expressed transcripts, functional analysis of the entire set of sequence fragments in the Syrian hamster transcriptome that mapped to mouse genes revealed that a number of critical biological pathways are well-represented, including many related to key processes that are potentially perturbed or induced during infection. Among the most significantly enriched canonical pathways were several involved with protein synthesis, turnover, and antigen processing (protein ubiquitination, EIF2 signaling), metabolism and stress responses (mitochondrial dysfunction, NRF2-mediated oxidative stress response, PI3K/Akt, and mTOR signaling), and inflammatory and immune responses (production of NO and reactive oxygen species by macrophages, CXCR4 signaling, IL-1 signaling, and IL-3 signaling). The aim of this study

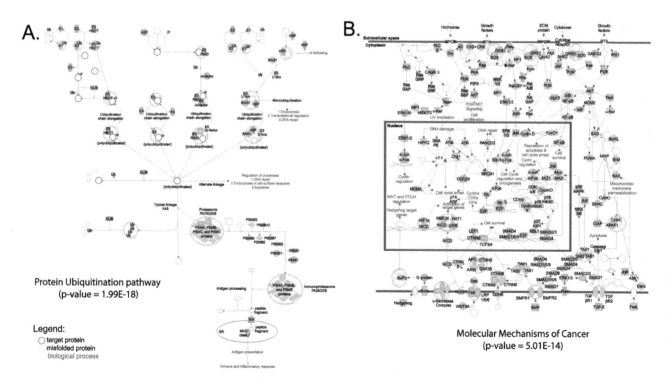

Figure 3. Schematic representation of the top two over-represented canonical pathways in our transcriptome assembly. (A) Representation of the "Protein Ubiquitination" canonical pathway. (B) Representation of the "Molecular Mechanisms of Cancer" canonical pathway. Both pathways have been generated based on mouse annotations. Transcripts involved in these pathways are indicated by different node shapes and associations are indicated by different edge shapes. Legends for the different nodes and edges are given in **Figure S1**. For both pathways, transcripts present in our library are indicated in gray. Associated p-values showing the statistical over-representation significance of the canonical pathways are also indicated.

Table 3. Functional enrichment of the mouse genes mapped by our transcriptome assembly.

Rank	Biological Function [p-value range]	Canonical pathway (p-value)
1	Organismal Surviva [1.11E-03 – 4.03E-26]	Protein Ubiquitination Pathway (1.99E-18)
2	Nervous System Development and Function [1.29E-03 – 1.46E-19]	Molecular Mechanisms of Cancer (5.01E-14)
3	Organ Morpholog [1.32E-03 – 4.20E-19]	Integrin Signaling (3.16E-13)
4	Tissue Morphology [1.08E-03 – 1.07E-18]	EIF2 Signaling (3.98E-12)
5	Cardiovascular System Development and Function [1.05E-03 – 4.15E-17]	Epithelial Adherens Junction Signaling (2.51E-11)

List of the top 5 biological functions and the top 5 canonical pathways found as statistically over-represented based on the list of 9,546 mouse genes mapped by our transcriptome assembly. The range of p-values is indicated for the biological functions and the p-value is indicated for each canonical pathways.

was to collect and annotate a large panel of transcripts regardless of tissue origin. These observations suggest that we have generated a representative transcriptome of the Syrian hamster. Therefore this transcriptome data could be used to generate a biologically meaningful first-generation expression DNA microarray for analysis of Syrian hamster response to disease, including those infectious agents known to alter immune and pro-inflammatory responses. Mechanisms of transcriptome regulation in the Syrian hamster, by way of these important pathways can now be monitored and analyzed further.

Only ~20% of the fragments in the Syrian hamster transcriptome aligned to the mouse and rat transcriptomes and even less aligned to the CHO cell transcriptome. This low percentage is due in part to species specificity, alignment stringency, but also to the fact that transcriptome references are far from being completely known and annotated. For instance, some classes of non-coding transcripts are now increasingly recognized as major components of regulation, and are widely expressed, but are poorly characterized and annotated. The transcriptome references that we used mainly contain known and annotated transcripts and our assembly may contain many expression contigs and singletons currently unknown and un-annotated in these other genomes.

The CHO cell genome is a useful tool for further improving the quality of our Syrian hamster transcriptome annotation for functional genomics work [24,25]. CHO cells have been used in a variety of genetic, cell biology, and pharmacology studies. They also are the mammalian cell line of choice for producing large quantities of recombinant proteins in large amounts or in or

industrial laboratory settings. Although Chinese and Syrian hamsters are phylogenetically distinct within the rodent subfamily *Cricetinae* [34,35], our data confirm that they are more closely related to one another as compared to other muroid rodents.

Through our work, we have increased the number of contig sequences available in the public domain for the Syrian hamster from 860 to 174,278, where 50,433 (28.93% of the Syrian hamster transcriptome) aligns to at least one transcriptome reference. Moreover, 85,652 (49.14% of the Syrian hamster transcriptome) fragments have aligned to the draft CHO genome, leading to an overall total of 98,103 (56.29%) annotated Syrian hamster transcripts. As a note, the work performed by Schmucki et al in [23] focused on transcriptome analysis of lipid metabolism in the golden hamster liver, and no contigs or other sequences have been released to the public domain to date.

With additional funding, future plans are in place for Illumina-based RNA sequencing using paired-end technology to add and improve on our current contig assembly. These efforts will improve our coverage of the Syrian hamster transcriptome, as well as permit more comprehensive and robust phylogenetic comparisons with other species. These combined efforts will lead to a better understand of the Syrian hamster transcriptome under a variety of infectious agent models related to human disease and pathogenesis.

Conclusions

The Syrian hamster is becoming an increasingly popular model for a variety of diseases, in particular, diseases known to infect non-

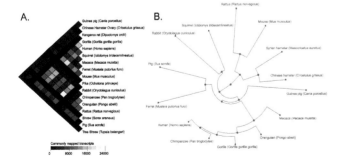

Figure 4. Distogram showing the commonly mapped transcripts and phylogenetic tree showing the divergences amongst the different species. (A) Distogram showing the number of transcripts commonly mapped by the Syrian hamster transcriptome between the different species used in this study. Each cell of the distogram represents the number of transcripts commonly mapped by two different species using a gradient color. (B) Phylogenetic tree showing the genomic divergence between a subset of the different species used in this study. Each leaf of the tree represents a different species and the distances of the edges are proportional to the genomic distances between the species. Genomic distances have been calculated based on the list of 611 Syrian hamster contigs and singletons that have been commonly aligned on the transcriptome references of the 13 species having the highest number of commonly aligned sequences.

human primates and humans. This Syrian hamster transcriptome discussed here represents a critical step forward in providing the tools necessary for advancing functional genomics in this important animal model.

Material and Methods

Animal housing

All hamsters were housed in individually ventilated cages (IVCs). All hamsters are co-housed, unless scientifically justified and approved by the Institutional Animal Care and Use Committee (IACUC) or deemed necessary for veterinary reasons. Housing density is determined by the guidelines outlined in the Guide for the Care and Use of Laboratory Animals and the Association for the Assessment and Accreditation of the Laboratory Animal Care, International (AAALAC). Food and sterile or acidified water were provided ad libitum. Hamster diets were consist of pellets containing a variety of foods such as grains and dried vegetables along with some seeds. Water was provided by either water bottles or water pouches. The light/dark cycle was 14 hours light, 10 hours dark.

RNA extraction

Three adult female Syrian hamsters were euthanized (exsanguinated while under isoflurane sedation) and six tissues – liver, lung, heart, brain, kidney, and spleen – were harvested from each hamster. All animal studies conformed to the guidelines set forth by the National Institutes of Health (NIH) and were reviewed and approved by the Institutional Animal Care and Use Committee (IACUC) at Rocky Mountain Laboratories, Division of Intramural Research, National Institute of Allergy and Infectious Diseases, NIH. One hundred mg of hamster tissue was homogenized with a Qiagen TissueLyzer II (Qiagen, Valencia, CA) in 1 mL Trizol (Invitrogen, Carlsbad, CA) following manufacturer's recommendations. To each aliquot 200 µL of 1-bromo-3-chloropropane (Sigma-Alrich) was added, the mixture was vortexed for 15 seconds and centrifuged at 4°C at 16,000x for 15 minutes. The aqueous phase was removed and passed through a Qiagen QiaShredder column to fragment remaining gDNA in the sample. The Qiagen AllPrep DNA/RNA 96 method was then performed including on-column Dnase 1 treatment to obtain high quality RNA with no genomic DNA contamination (Qiagen, Valencia, CA). RNA yield was determined by spectrophotometry (A260/A280) and RNA quality was determined using an Agilent 2100 Bioanalyzer (Agilent Technologies, Santa Clara, CA). The average RNA integrity number (RIN) for all 18 RNAs (3 animals times 6 tissues) was 6.4. An RNA aliquot from each organ of each animal was pooled and a total of 170 µg of RNA was prepared for sequencing.

Library construction and 454 sequencing.

The eighteen total RNA samples (6 tissues times 3 animals) were pooled equally into one pool. The total RNA pool underwent additional cleaning using the mirVana isolation kit following manufacturer's recommendations (Ambion). Poly A RNA cDNA was synthesized according to a standard protocol using an oligo(dT)-linker primer for first strand synthesis. The N0 cDNA was PCR amplified during 18 cycles using a high fidelity DNA polymerase. Normalization was carried out by one cycle of denaturation and renaturation of the cDNA, resulting in N1-cDNA. Reassociated ds-cDNA was separated from the remaining ss-can (normalized cDNA) by passing the mixture over a hydroxylapatite column. After hydroxylapatite chromatography, the ss-cDNA was amplified with 15 PCR cycles. For 454

sequencing, cDNA in the size range of 500–700 bps was eluted from a preparative agarose gel. An aliquot of the size fractionated cDNA was analyzed on a 1.5% agarose gel. 454 adaptors were ligated to the size fractionated N1 cDNA and 3′ fragment sequenced on a Roche 454 using GS FLX technology with Titanium series chemistry following manufacturer's recommendations.

GS FLX sequencing generated 1,283,840 reads with an average length of 344 bases. Raw reads were trimmed for quality and reads shorter than 40 bases were discarded. The sequencing resulted in 1,212,395 reads of a total length 426,683,712 bases.

Library assembly

The trimmed and filtered reads was assembled using MIRA [36] (version 3) with the following parameters: mira –job = denovo,est,accurate,454 454_SETTINGS -CL:qc = no:cpat = yes -AL:mo = 40:mrs = 90. MIRA assembly produced 62,567 contigs and 125,228 singletons. There were 85 contigs and 13,432 singletons discarded due to poor quality (repetitive or poly-T sequence) or short read length (<50 bases), resulting in 62,482 contigs and 111,796 singletons for a total of 174,278 Syrian hamster transcriptome sequences totaling 60,117,204 bases.

Transcriptome and genome references

The transcriptome references used in this study were retrieved from the Ensembl Database [30] via the Biomart interface. Transcriptome references used in this study were obtained from the release 71 of the Ensembl database. The draft version of the CHO genome and transcriptome were retrieved from the Pre Ensembl website.

Alignments

Syrian hamster transcriptome sequences were aligned to transcriptome references using BLAST [37]. An Expect value cutoff parameter of 10 was used and alignments results were filtered in order to only keep sequences aligned at least at 80%.

Similarities of the assembled library with the transcriptome references

The similarities to the mouse, rat and other transcriptome references were calculated based on BLAST results. For all Syrian hamster transcriptome sequences that aligned to the transcriptomes, we calculated the ratio between the total number of correct nucleotide matches and the total combined length of our Syrian hamster transcriptome, which is 60,117,204 bases.

Identification of over-represented canonical pathways and biological functions

Functional enrichment of canonical pathways and biological functions was performed using Ingenuity Pathways Analysis (Ingenuity Systems, Inc.). Canonical pathways refer to pathways curated by Ingenuity as part of its knowledgebase, based on extensive characterization in the peer-reviewed literature published using human, mouse, and rat experimental models. These typically represent common properties of a particular signaling module, mechanism, or pathway. IPA examines differentially expressed transcripts in the context of known biological functions, mapping each gene identifier to its corresponding molecule in the Ingenuity Pathways Knowledge Base (IPKB). For all analyses, the p-values – representing the statistical over-representation significance – were generated using the right-tailed Fisher's Exact Test [38] and were adjusted using the Benjamini-Hochberg Multiple Testing correction [39].

Distrogram construction

The distogram represented in figure 4A was constructed using the "squash" package [40] of the R suite [41]. This representation is a color-coded, rotated triangular matrix indicating the distance between every pair of species in term of number of aligned sequences shared.

Phylogenetic tree construction

To construct the phylogenetic tree, we used the 611 transcriptomic sequences that have been significantly aligned on the transcriptome reference of the 13 species having the largest numbers of mapped sequences and the largest degrees of shared sequences. Sequences and matched transcripts were aligned using the Needleman-Wunsch multiple alignment algorithm [42], using the multialign function in MATLAB (open and extend gap penalties have been taken into consideration). Genomic divergences between sequences were calculated using the Jukes-Cantor method [43], based on the 'NUC44' scoring matrix. Indel mismatches have not been taken into consideration for the computation of genomic divergences. The phylogenetic tree was constructed by using the neighbor-joining method (NJ) [44].

Supporting Information

Figure S1. Legend for the IPA canonical pathways representations. Figure showing the annotations of the different node and edge shapes in the representations of the canonical pathways obtained from Ingenuity Pathway Analysis (IPA).

Table S1 List of highly-covered transcripts by the Syrian hamster transcriptome. Table showing the 214 highly-covered transcripts (transcripts mapped at least at 90% by the contigs and singletons) based on the mouse annotations. For each highly-covered transcript, the Ensembl Mouse gene identified, the associated gene name as well as the gene description is indicated.

Table S2 Alignment positions of the Syrian hamster transcriptome over the Chinese Hamster Ovary cell genome. Table showing the alignment positions of the Syrian hamster transcriptome sequences to the draft of the Chinese hamster genome. We found that 85,652 contigs and singletons of our library have been aligned on the Chinese Hamster Ovary cell genome draft. For each aligned Syrian hamster transcriptome sequence, the CHO genome segment, the start alignment position, the end alignment position, and the strand are indicated.

Table S3 List of sequences used to infer the phylogenic tree. Table providing the list of contigs and singletons used to construct the phylogenic tree shown in figure 4. All the 611 contigs and singletons in this table have been significantly aligned on the transcriptomes of species having the highest number of commonly aligned sequences.

Acknowledgments

The authors thank Marcus J. Korth for valuable feedback on the manuscript.

Author Contributions

Conceived and designed the experiments: DS ALR HF HE MGK. Performed the experiments: KV DS. Analyzed the data: NT CM. Contributed reagents/materials/analysis tools: DS KV SFP. Wrote the paper: NT DS ALR CM SFP HE.

References

1. Wahl-Jensen V, Bollinger L, Safronetz D, de Kok-Mercado F, Scott DP, et al. (2012) Use of the Syrian hamster as a new model of ebola virus disease and other viral hemorrhagic fevers. Viruses 4: 3754–3784. doi:10.3390/v4123754.
2. Xiao SY, Guzman H, Zhang H, Travassos da Rosa AP, Tesh RB (n.d.) West Nile virus infection in the golden hamster (Mesocricetus auratus): a model for West Nile encephalitis. Emerg Infect Dis 7: 714–721. doi:10.3201/eid0704.010420.
3. Tesh RB, Guzman H, da Rosa AP, Vasconcelos PF, Dias LB, et al. (2001) Experimental yellow fever virus infection in the Golden Hamster (Mesocricetus auratus). I. Virologic, biochemical, and immunologic studies. J Infect Dis 183: 1431–1436. doi:10.1086/320199.
4. Aguilar P V, Barrett AD, Saeed MF, Watts DM, Russell K, et al. (2011) Iquitos virus: a novel reassortant Orthobunyavirus associated with human illness in Peru. PLoS Negl Trop Dis 5: e1315. doi:10.1371/journal.pntd.0001315.
5. Shinya K, Makino A, Tanaka H, Hatta M, Watanabe T, et al. (2011) Systemic dissemination of H5N1 influenza A viruses in ferrets and hamsters after direct intragastric inoculation. J Virol 85: 4673–4678. doi:10.1128/JVI.00148-11.
6. Roberts A, Lamirande EW, Vogel L, Jackson JP, Paddock CD, et al. (2008) Animal models and vaccines for SARS-CoV infection. Virus Res 133: 20–32. doi:10.1016/j.virusres.2007.03.025.
7. Safronetz D, Zivcec M, Lacasse R, Feldmann F, Rosenke R, et al. (2011) Pathogenesis and host response in Syrian hamsters following intranasal infection with Andes virus. PLoS Pathog 7: e1002426. doi:10.1371/journal.ppat.1002426.
8. Zivcec M, Safronetz D, Haddock E, Feldmann H, Ebihara H (2011) Validation of assays to monitor immune responses in the Syrian golden hamster (Mesocricetus auratus). J Immunol Methods 368: 24–35. doi:10.1016/j.jim.2011.02.004.
9. Ebihara H, Zivcec M, Gardner D, Falzarano D, LaCasse R, et al. (2013) A Syrian golden hamster model recapitulating ebola hemorrhagic fever. J Infect Dis 207: 306–318. doi:10.1093/infdis/jis626.
10. Gomes-Silva A, Valverde JG, Ribeiro-Romão RP, Plácido-Pereira RM, DA-Cruz AM (2013) Golden hamster (Mesocricetus auratus) as an experimental model for Leishmania (Viannia) braziliensis infection. Parasitology: 1–9. doi:10.1017/S0031182012002156.
11. Briand F (2010) The use of dyslipidemic hamsters to evaluate drug-induced alterations in reverse cholesterol transport. Curr Opin Investig Drugs 11: 289–297.
12. Castro-Perez J, Briand F, Gagen K, Wang S-P, Chen Y, et al. (2011) Anacetrapib promotes reverse cholesterol transport and bulk cholesterol excretion in Syrian golden hamsters. J Lipid Res 52: 1965–1973. doi:10.1194/jlr.M016410.
13. Morin LP, Wood RI (2001) A stereotaxic atlas of the golden hamster brain. Academic Press San Diego:
14. Van Hoosier GL, McPherson CW (1987) The Laboratory Hamsters. Access Online via Elsevier.
15. Monecke S, Brewer JM, Krug S, Bittman EL (2011) Duper: a mutation that shortens hamster circadian period. J Biol Rhythms 26: 283–292. doi:10.1177/0748730411411569.
16. Nigro V, Okazaki Y, Belsito A, Piluso G, Matsuda Y, et al. (1997) Identification of the Syrian hamster cardiomyopathy gene. Hum Mol Genet 6: 601–607.
17. Ricci LA, Schwartzer JJ, Melloni RH (2009) Alterations in the anterior hypothalamic dopamine system in aggressive adolescent AAS-treated hamsters. Horm Behav 55: 348–355. doi:10.1016/j.yhbeh.2008.10.011.
18. Chelini MOM, Palme R, Otta E (2011) Social stress and reproductive success in the female Syrian hamster: endocrine and behavioral correlates. Physiol Behav 104: 948–954. doi:10.1016/j.physbeh.2011.06.006.
19. Rawji KS, Zhang SX, Tsai Y-Y, Smithson IJ, Kawaja MD (2013) Olfactory ensheathing cells of hamsters, rabbits, monkeys, and mice express α-smooth muscle actin. Brain Res 1521: 31–50. doi:10.1016/j.brainres.2013.05.003.
20. Boguski MS, Lowe TM, Tolstoshev CM (1993) dbEST–database for "expressed sequence tags". Nat Genet 4: 332–333. doi:10.1038/ng0893-332.
21. Oduru S, Campbell JL, Karri S, Hendry WJ, Khan SA, et al. (2003) Gene discovery in the hamster: a comparative genomics approach for gene annotation by sequencing of hamster testis cDNAs. BMC Genomics 4: 22. doi:10.1186/1471-2164-4-22.
22. Landkocz Y, Poupin P, Atienzar F, Vasseur P (2011) Transcriptomic effects of di-(2-ethylhexyl)-phthalate in Syrian hamster embryo cells: an important role of early cytoskeleton disturbances in carcinogenesis? BMC Genomics 12: 524. doi:10.1186/1471-2164-12-524.

23. Schmucki R, Berrera M, Küng E, Lee S, Thasler WE, et al. (2013) High throughput transcriptome analysis of lipid metabolism in Syrian hamster liver in absence of an annotated genome. BMC Genomics 14: 237. doi:10.1186/1471-2164-14-237.

24. Xu X, Nagarajan H, Lewis NE, Pan S, Cai Z, et al. (2011) The genomic sequence of the Chinese hamster ovary (CHO)-K1 cell line. Nat Biotechnol 29: 735–741. doi:10.1038/nbt.1932.

25. Hammond S, Swanberg JC, Kaplarevic M, Lee KH (2011) Genomic sequencing and analysis of a Chinese hamster ovary cell line using Illumina sequencing technology. BMC Genomics 12: 67. doi:10.1186/1471-2164-12-67.

26. Lewis NE, Liu X, Li Y, Nagarajan H, Yerganian G, et al. (2013) Genomic landscapes of Chinese hamster ovary cell lines as revealed by the Cricetulus griseus draft genome. Nat Biotechnol 31: 759–765. doi:10.1038/nbt.2624.

27. Liu H, Wang T, Wang J, Quan F, Zhang Y (2013) Characterization of liaoning cashmere goat transcriptome: sequencing, de novo assembly, functional annotation and comparative analysis. PLoS One 8: e77062. doi:10.1371/journal.pone.0077062.

28. Ji P, Liu G, Xu J, Wang X, Li J, et al. (2012) Characterization of common carp transcriptome: sequencing, de novo assembly, annotation and comparative genomics. PLoS One 7: e35152. doi:10.1371/journal.pone.0035152.

29. Nagaraj SH, Gasser RB, Ranganathan S (2007) A hitchhiker's guide to expressed sequence tag (EST) analysis. Brief Bioinform 8: 6–21. doi:10.1093/bib/bbl015.

30. Flicek P, Ahmed I, Amode MR, Barrell D, Beal K, et al. (2013) Ensembl 2013. Nucleic Acids Res 41: D48–55. doi:10.1093/nar/gks1236.

31. Ryu SH, Kwak MJ, Hwang UW (2013) Complete mitochondrial genome of the Eurasian flying squirrel Pteromys volans (Sciuromorpha, Sciuridae) and revision of rodent phylogeny. Mol Biol Rep 40: 1917–1926. doi:10.1007/s11033-012-2248-x.

32. D'Erchia AM, Gissi C, Pesole G, Saccone C, Arnason U (1996) The guinea-pig is not a rodent. Nature 381: 597–600. doi:10.1038/381597a0.

33. Graur D, Hide WA, Li WH (1991) Is the guinea-pig a rodent? Nature 351: 649–652. doi:10.1038/351649a0.

34. Romanenko SA, Volobouev VT, Perelman PL, Lebedev VS, Serdukova NA, et al. (2007) Karyotype evolution and phylogenetic relationships of hamsters (Cricetidae, Muroidea, Rodentia) inferred from chromosomal painting and banding comparison. Chromosome Res 15: 283–297. doi:10.1007/s10577-007-1124-3.

35. Trifonov VA, Kosyakova N, Romanenko SA, Stanyon R, Graphodatsky AS, et al. (2010) New insights into the karyotypic evolution in muroid rodents revealed by multicolor banding applying murine probes. Chromosome Res 18: 265–275. doi:10.1007/s10577-010-9110-6.

36. Chevreux B, Pfisterer T, Drescher B, Driesel AJ, Müller WEG, et al. (2004) Using the miraEST assembler for reliable and automated mRNA transcript assembly and SNP detection in sequenced ESTs. Genome Res 14: 1147–1159. doi:10.1101/gr.1917404.

37. Altschul SF, Gish W, Miller W, Myers EW, Lipman DJ (1990) Basic local alignment search tool. J Mol Biol 215: 403–410. doi:10.1016/S0022-2836(05)80360-2.

38. Fisher RA (1922) On the Interpretation of χ2 from Contingency Tables, and the Calculation of P. J R Stat Soc 85: 87–94. doi:10.2307/2340521.

39. Benjamini Y, Hochberg Y (1995) Controlling the False Discovery Rate: A Practical and Powerful Approach to Multiple Testing. J R Stat Soc Ser B 57: 289–300. doi:10.2307/2346101.

40. Eklund A (2012) squash: Color-based plots for multivariate visualization. Available: http://cran.r-project.org/package=squash.

41. R Development Core Team R (2011) R: A Language and Environment for Statistical Computing. R Found Stat Comput 1: 409. doi:10.1007/978-3-540-74686-7.

42. Needleman SB, Wunsch CD (1970) A general method applicable to the search for similarities in the amino acid sequence of two proteins. J Mol Biol 48: 443–453.

43. Jukes TH, Cantor CR (1969) Evolution of Protein Molecules. Munro HN, editor Academy Press.

44. Saitou N, Nei M (1987) The neighbor-joining method: a new method for reconstructing phylogenetic trees. Mol Biol Evol 4: 406–425.

45. Kodama Y, Shumway M, Leinonen R (2012) The Sequence Read Archive: explosive growth of sequencing data. Nucleic Acids Res 40: D54–6. doi:10.1093/nar/gkr854.

5

Marked Microevolution of a Unique *Mycobacterium tuberculosis* Strain in 17 Years of Ongoing Transmission in a High Risk Population

Carolina Mehaffy[1,2*¤], Jennifer L. Guthrie[1], David C. Alexander[3], Rebecca Stuart[4], Elizabeth Rea[4], Frances B. Jamieson[1,2]

1 Public Health Ontario, Toronto, Canada, 2 University of Toronto, Toronto, Canada, 3 Saskatchewan Disease Control Laboratory, Regina, Canada, 4 Toronto Public Health, Toronto, Canada

Abstract

The transmission and persistence of *Mycobacterium tuberculosis* within high risk populations is a threat to tuberculosis (TB) control. In the current study, we used whole genome sequencing (WGS) to decipher the transmission dynamics and microevolution of *M. tuberculosis* ON-A, an endemic strain responsible for an ongoing outbreak of TB in an urban homeless/under-housed population. Sixty-one *M. tuberculosis* isolates representing 57 TB cases from 1997 to 2013 were subjected to WGS. Sequencing data was integrated with available epidemiological information and analyzed to determine how the *M. tuberculosis* ON-A strain has evolved during almost two decades of active transmission. WGS offers higher discriminatory power than traditional genotyping techniques, dividing the *M. tuberculosis* ON-A strain into 6 sub-clusters, each defined by unique single nucleotide polymorphism profiles. One sub-cluster, designated ON-A[NM] (Natural Mutant; 26 isolates from 24 cases) was also defined by a large, 15 kb genomic deletion. WGS analysis reveals the existence of multiple transmission chains within the same population/setting. Our results help validate the utility of WGS as a powerful tool for identifying genomic changes and adaptation of *M. tuberculosis*.

Editor: Igor Mokrousov, St. Petersburg Pasteur Institute, Russian Federation

Funding: This study was supported by a grant-in-aid from The Lung Association/Ontario Thoracic Society (http://www.on.lung.ca) (CM, FBJ). The funders had no role in study design, data collection and analysis, decision to publish, or preparation of the manuscript.

Competing Interests: The authors have declared that no competing interests exist.

* Email: Carolina.mehaffy@colostate.edu

¤ Current address: Colorado State University, Fort Collins, Colorado, United States of America

Introduction

Globally, tuberculosis (TB) is an important cause of morbidity and mortality. World-wide, *M. tuberculosis* is responsible for more than 1 million deaths per year. In low incidence countries, homeless/under-housed individuals represent one of the groups at greater risk for TB infection and disease [1–5]. TB outbreaks within homeless settings have been documented throughout North America [6–9].

One endemic strain, designated Ontario A (ON-A), has been circulating since at least 1997 in the urban homeless/under-housed population of Toronto, Canada and has been responsible for TB outbreaks in 2001 and 2004 with new cases identified every year [5,10]. Genotyping is an essential component of epidemiological investigations. However, the *M. tuberculosis* ON-A isolates are defined by a unique combined spoligotype and 24-locus MIRU-VNTR (24-MIRU) profile, while IS*6110* RFLP generates pseudo-clusters among these strains [10].

Although several reports have highlighted the epidemiological value of whole genome sequencing (WGS) over traditional genotyping techniques [11–17], presently only a few studies have evaluated the utility of WGS to study whole genome changes of *M. tuberculosis* during long-term continuous active transmission [15].

We evaluated the usefulness of WGS to retrospectively validate and identify transmission events associated with TB cases due to *M. tuberculosis* ON-A over the last 17 years. We used a phylogenetic analysis based on single nucleotide polymorphisms (SNP) to portray the microevolution of this *M. tuberculosis* strain during almost two decades of on-going transmission in a high risk population. Our analysis revealed the presence of six independent transmission chains and the presence of an ON-A natural mutant, defined by a large genomic deletion that most likely emerged during the first ON-A TB outbreak in 2001.

Methods

M. tuberculosis clinical isolates

M. tuberculosis isolates were obtained from clinical specimens routinely received at the Public Health Ontario Laboratories for TB diagnosis. All available isolates from 1997–2013 with genotypes consistent with the ON-A strain [6,18] were selected. All isolates were susceptible to all first line drugs.

The work described in this manuscript relates directly to improvement of routine TB surveillance and outbreak management, therefore research ethics board (REB) approval was not required.

DNA extraction

Genomic DNA (gDNA) was extracted as previously described [19] with minor modifications [20].

Genotyping

24-locus MIRU-VNTR [21], spoligotyping [22] and IS6110 RFLP [23] were performed using standard methods, and data were analyzed with BioNumerics v6.1 (Applied Maths, St-Martin Latem, Belgium). RFLP patterns were compared as previously described [10].

Whole genome sequencing

DNA was prepared for sequencing as described elsewhere [20]. Illumina paired-end reads were trimmed using quality scores and then aligned to *M. tuberculosis* H37Rv reference genome (NC_000962.2) using the CLC Genomics workbench (v.6.0.2) software. For 5 of the 61 isolates, quality and/or quantity of the DNA were not suitable for WGS and therefore these were not included in the analysis.

Accuracy of WGS assembly and analysis workflow was assessed by sequencing the H37Rv reference strain that is used in our clinical lab (Material S1).

Variant calling

Single nucleotide polymorphisms (SNPs) and small insertion-deletion (indel) events were identified using a probabilistic variant detection with cutoffs of a minimum read depth of 20X and a variant frequency of at least 75. Indels were not considered for any further analyses. SNPs were further filtered by removing positions associated with PE, PPE and PE_PGRS gene families which have been previously shown to represent false positives and due to their high variation are not suited for phylogenetic analysis [17]. SNPs unique to any of the fifty-six high quality whole genome sequences were manually inspected in each individual alignment for accuracy and all ambiguous results were discarded.

Phylogenetic analysis

A concatemer of the SNPs was generated and then used to reconstruct the phylogeny of ON-A using SplitsTree v.4 software [24] and the BioNJ algorithm [25]. Trees were then re-constructed using the Equal Angle algorithm [26] with equal-daylight and box opening optimization [27] available in SplitsTree v.4 (Figure 1).

Demographic and clinical data

Demographic characteristics and clinical information for each TB case was obtained from Ontario's integrated Public Health Information System (iPHIS) as well as responsible Public Health Units (Table 1). This information is routinely recorded by Public Health Units for all laboratory and clinically confirmed TB cases. Data was anonymized and all personal information removed from the final data set. Time lapse between onset of symptoms and diagnosis as well as treatment start date were not available for most cases and therefore were not included in this study.

Social network visualization

Known epidemiological/social connections identified during routine public health contact investigation (i.e roommates, close friends, used same drop-in, etc.) among forty-seven ON-A cases were available in iPHIS. The igraph [28] package of R (v3.0.2) was used to generate the analysis. Transmission events were defined based on the genomic and epidemiological information (i.e. SNP pattern, contact information, year of diagnosis and infectiousness based on smear and chest x-ray results).

Results

In this study we performed the whole genome sequencing of 61 *M. tuberculosis* isolates identified during routine genotyping as members of a large cluster denominated ON-A and spanning 17 years (1998–2013). The 61 isolates corresponded to 57 patients, most of whom were homeless/under-housed individuals (75.5%). The epidemiological and molecular typing characteristics of this cluster up to 2008 have been published elsewhere [5,6]. Table 1 summarizes the demographic characteristics of all 57 patients.

Isolates in this cluster are characterized by a highly similar 24MIRU and spoligotype patterns while RFLP analysis of the ON-A strain results into three pseudo-clusters defined by RFLP types A, B and C. The main 24MIRU pattern has not been reported in the international MIRU database (http://www.miru-vntrplus.org/MIRU/index.faces) and personal communication with the corresponding health authorities of the other Canadian provinces indicates that this genotype is unique to Toronto's inner city population.

Of the 61 *M. tuberculosis* isolates, WGS was successfully performed in 56 isolates, corresponding to 53 TB cases and further sub-divided the large ON-A group into 6 different sub-clusters (SC1-6) (Figure 1 and 2).

Figure 3 combines the RFLP and SNP-clustering results for all 53 cases with final WGS data. Most isolates belong to RFLP type A (n = 34) followed by RFLP B (n = 5) and RFLP C (n = 3). Nine isolates were not clustered by RFLP and 1 presented a mixed pattern corresponding to an infection with strain ON-A and strain ON-B, which is also commonly found in the homeless/under-housed population. Figure 3 illustrates the lack of correlation between RFLP and SNP-clustering with the exception of RFLP type B which corresponded to SC4.

Sub-cluster classification was based on SNP analysis using as reference, the genome of laboratory strain H37Rv. In summary, 722 SNPs were identified in ON-A isolates when compared with H37Rv. Of these, 641 SNPs, including 333 (52%) non-synonymous, 224 (35%) synonymous, and 84 (13%) non-coding SNPs, were conserved in all sequenced isolates and only served to differentiate H37Rv from the ON-A strain. The remaining 81 SNPs were variable among the sequenced isolates and were used to identify sub-cluster associations.

Most isolates belonged to the most recently emerged sub-cluster SC6 (n = 24), followed by SC4 (n = 10), SC1 (n = 8), SC5 (n = 5) and SC2 (n = 2). One isolate was not placed in any of the 6 sub-clusters.

Of the 81 SNPs, 2 were present in the majority of ON-A isolates, except for most isolates in SC1. One SNP was only present in SC-6 and one was present in both SC4 and SC5 isolates. The remaining 77 SNPs were either sub-cluster associated (i.e. present in two or more isolates) (17/77) or singletons (60/77).

Sub-clusters, SNPs and transmission events

Phylogenetic analysis identified 6 ON-A sub-clusters (Figure 1). One isolate did not group with any of the identified sub-clusters.

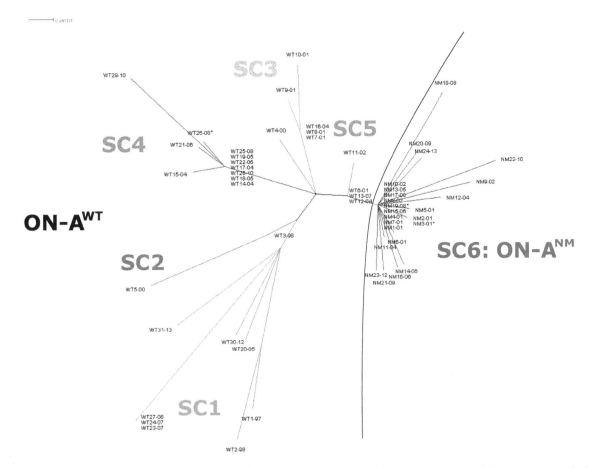

Figure 1. ON-A Phylogenetic tree constructed in Splitstree ver. 4. All 722 polymorphic sites were included. Phylogenetic tree was built using BioNJ algorithm and then filtered for parsimony-informative sites.

Phylogenetic analysis and temporal association of isolates from the ON-A strain suggests the independent emergence of SC1–SC5 prior to 1997 and subsequent clonal expansion of sub-clusters SC3, SC4 and SC5 (Figures 1 and 2). Contrary to this, SC6 appears to have emerged in 2001 during the first TB outbreak associated with ON-A.

SNP patterns of isolates belonging to SC1 group suggest they represent infections acquired from a common ancestor(s) prior to 1997 and most likely represent re-activation of latent TB infection rather than immediate secondary cases. The exceptions were two isolates, (WT24-07 and WT27-08) with SNP patterns identical to the source isolate WT23-07. The two individuals, from whom the isolates WT23-07 and WT24-07 were obtained, lived in the same rooming house, and both used the same drop-in location which was associated with the two ON-A TB outbreaks. Interestingly, the individual associated with isolate WT27-08 was diagnosed in a different jurisdiction north of Toronto, but is clearly related to the other two isolates and demonstrates the high mobility of this population.

Contrary to SC1, all other sub-clusters had a clear source case. In group SC2, isolate WT5-00 gained 4 SNPs in approximately 2 years after infection from WT3-98. In group SC3, the source case (WT07-01) resulted in two other primary TB cases in the same year and a later reactivation case in 2004.

In SC4, the source case (WT17-04) resulted in secondary cases in shelter workers (both in 2004). Five additional cases with an identical SNP pattern as the suspected source case were also identified in subsequent years (2005–2010). In addition, the source was also genetically related to a case in 2006 with 1 SNP gained and another case in 2010 with 4 additional SNPs. It is not possible to determine if the case in 2010 (WT29-10) originated from the source case or from one of the five cases with identical SNP pattern to the source. However, the large number of SNPs suggests that WT29-10 corresponds to a re-activation from an infection acquired in 2004 or 2005. This is also supported by epidemiological data that indicates WT29-10 may have shared a common workplace with WT17-04 at the time this source case was ill.

24-MIRU, spoligotype and RFLP indicated that A/B-01 represented a mixed infection with two strains commonly found in this population (ON-A and ON-B) [6] and therefore its WGS data was not included in the phylogenetic analysis. However, manual inspection of all 81 variable SNP positions in the genome alignment generated for this isolate demonstrated the presence of all SC5 SNPs, allowing us to include the A/B-01 case within the SC5 group.

The two suspected source cases in group SC5 (WT6-01 and A/B-01) resulted in two subsequent cases in 2004 and 2007, both of which have identical SNP patterns to the suspected source(s). In addition, these two cases were the most likely source for the emergence of SC6 in 2001.

It is possible that the high bacteria burden present in WT6_01 or the A/B-01 mixed case (smear 2+ and 3+ respectively) contributed to the emergence of SC6. Both patients also had abnormal chest x-rays and together with their smear results

Table 1. Demographic characteristics of ON-A TB cases.

Variable	All Individuals (n = 57)	ON-A^{WT} Individuals (n = 33)	ON-A^{NM} Individuals (n = 24)
Age (years)			
Mean age (± std dev)	50.0 (±12.9)	51.4 (±11.6)	48.2 (±14.4)
Range	20.3–74.3	23.8–74.3	20.3–72.5
Gender			
Female	5	3	2
Male	52	30	22
Birthplace			
Born in Canada	40	22	18
Born outside Canada	11	5	6
Unknown	6	6	0
Disease Classification			
Pulmonary	49	27	22
Extra-pulmonary	2	1	1
Pulmonary with extra-pulmonary involvement	6	5	1
Mortality (TB as cause of death$^{\psi}$)			
Yes	3	3	0
No	10	5	5
Unknown	6	5	1
Housing Status			
Under-housed	43	23	20
Housed	4	2	2
Unknown	10	8	2
Risk Factors			
HIV/AIDS	8	4	4
Injectable drug use	10	4	6
Alcohol abuse	31	17	14

$^{\psi}$ Number of deaths during or shortly after completing TB treatment.

suggest they were highly infectious. The chest x-ray of the mixed A/B case also showed cavitation. Social network analysis demonstrated the large number of contacts shared by these two patients, including individuals in both ON-A^{WT} and ON-A^{NM} groups (Figure S1).

The SC6 group was very homogeneous. SNPs within this group were mostly singletons and represented temporal acquisition of substitutions in single isolates. Nine SC6 isolates (37.5%) had no accumulation of SNPs in an 8 year time frame (Figure 2).

Because of the high number of isolates with identical SNP patterns in SC6, it was not possible to determine the exact transmission chain in this group. Multiple control measures to reduce spread of TB in Toronto's homeless shelters were implemented as a consequence of the outbreaks in 2001 and 2004. This provides support to the idea that most cases after 2005 correspond to reactivation of infections acquired during one of the two TB outbreaks. The only exception is case NM17-08 which is an individual outside of the high risk group for which contact investigations indicate the source case for that individual was NM15-06. For the remaining cases, the most likely index case was NM7-01 who was highly infectious, with smear +3 and abnormal chest x-ray.

WGS revealed a large genomic deletion in SC6

We discovered a large genomic deletion of more than 15 kb, comprising 12 genes in 26/61 (43.5%) ON-A isolates (Table 2), dividing the ON-A strain into two groups, ON-A^{WT} and ON-A^{NM} representing presence or absence of the 15 kb genomic region, respectively (Figure 1, Table 2). SC1–SC5 belong to the ON-A^{WT} group while isolates in SC6 are all ON-A^{NM}. In contrast to ON-A^{NM}, the ON-A^{WT} variant was very heterogeneous, characterized by the presence of multiple sub-clusters. In addition, thirty-two ON-A^{WT} SNPs were singletons while 16 were sub-cluster associated. ON-A^{WT} isolates also presented 3 additional SNPs within the deletion region. These SNPs were not included in the phylogenetic analysis. The genomic deletion was confirmed by a custom PCR in all isolates, including 5 for which WGS data was not available (Material S1). Sanger sequencing of the ON-A^{NM} PCR amplicons demonstrated that the insertion sequence IS*6110* is present and flanked by an incomplete Rv1358 at the 5'-end and by an incomplete Rv1371 at the 3'-end. Sanger sequencing of the flanking areas of the 15 kb deletion region in ON-A^{WT} also confirmed the presence of IS*6110* at the 3' end of the region (upstream of Rv1371).

Heterogeneity resulting from IS*6110*-mediated deletion events during active TB infection has previously been reported in an individual with a very high bacteria burden [29] and it is possible that the high bacteria burden present in the two possible source

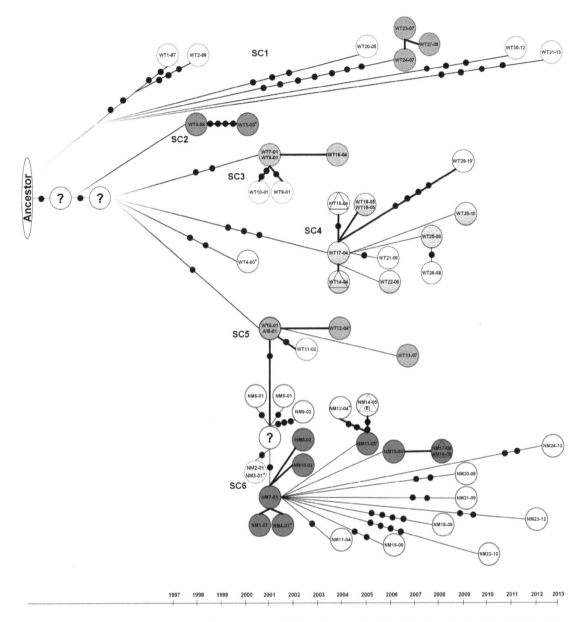

Figure 2. Transmission events of ON-A TB cases based on WGS information and epidemiologic analysis. Colored solid circles represent TB isolates that are identical to the suspected primary case (0 SNPs). Open circles represent related cases, separated by black solid dots representing SNPs acquired over time. Triangles represent TB cases outside of the homeless/under-housed group. Isolates from the same TB patient are represented by circles with dots in their background: WT25-08 and WT26-08; NM2-01 and NM3-01; and NM17-08 and NM19-08. *Isolates that had more than one passage prior to WGS.

Discussion

Microevolution and adaptation events of *M. tuberculosis* during active TB transmission

We analyzed a cluster of *M. tuberculosis* isolates (ON-A) associated with on-going TB transmission in a large urban setting and its inner-city homeless/under-housed population. In order to evaluate the evolution of this *M. tuberculosis* strain during a 17 year period, we performed WGS analysis for all available ON-A isolates. We discovered a large genomic deletion (>15 kb) in a

cases (WT6-01 and A/B-01) as described previously could have contributed to the emergence of the deletion and origin of the ON-A^{NM} variant.

subset of ON-A isolates that divided the ON-A strain into two strain variants (ON-A^{WT} and ON-A^{NM}). This large genomic deletion, which is present in 43% of the ON-A isolates demonstrates the presence of two distinct groups independent of the RFLP pattern.

Our phylogenetic and epidemiological analyses suggest this deletion emerged in 2000 during the first TB outbreak caused by ON-A.

The 15 kb deletion comprises a cluster of 12 genes, of which at least 5 code for regulatory proteins. Regulatory proteins govern the expression of clusters of genes involved in specific molecular pathways and therefore are ultimately responsible for the ability of the cell to adapt and survive in new environments. Two of the deleted genes identified in the ON-A^{NM} correspond to regulators

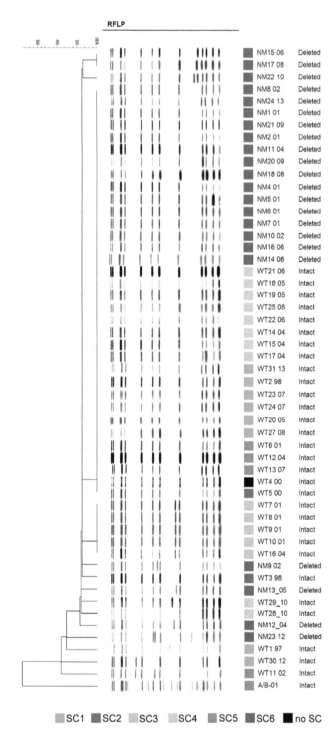

Figure 3. Dendrogram of IS6110 RFLP showing the 15 kb deletion distribution. Branches indicate the clusters with identical RFLP patterns. Colored squares represent each of the six sub-clusters identified by WGS. Information regarding the presence or absence of the 15 kb deletion for each isolate is shown as "deleted" or "intact".

deleted genes (Rv1364c) strongly interacts with sigF and also with the sigF antagonist RsbW [32]. Other regulatory genes identified in the 15 kb deletion are the LuxR family regulator (Rv1358) which is also regulated by SigF [33], a signal transduction regulatory gene (Rv1359) and a conserved gene (Rv1366) in which we identified a domain representative of RelA-SpoT superfamily, responsible for the regulation of ppGpp concentration. ppGpp is a modified nucleotide used by bacteria as intracellular messengers that respond to different environmental stresses [34,35]. In *M. tuberculosis*, ppGpp responses are associated with virulence and long-term survival [35,36].

The presence of the insertion sequence IS6110 at the site of the deletion suggests this event was IS6110 mediated. Reports elsewhere have shown that IS6110 mediated deletion events are common in *M. tuberculosis* [29,37,38] and even though the identification of a large genomic deletion in this highly related *M. tuberculosis* strain is very interesting, it should not be surprising given the high IS6110 mediated mutation rate (transposition/generation) [39]. It has been proposed that although deleterious IS6110 mediated mutations may occur frequently, these mutants are rapidly purged from the population by purifying selection [39]. Contrary to this, the 15 kb deletion seems to have emerged during the first TB outbreak in 2001; and since then the ON-A^NM has established itself among the homeless/under-housed population in Toronto.

Other examples of non-deleterious deletions exist. An *M. tuberculosis* isolate from the CAS family bearing a chromosomal deletion, was responsible for a large TB outbreak in Leicester, UK and was shown to have a lower inflammatory phenotype and higher intracellular growth similar to the hypervirulent strain HN878 [40]. An IS6110-mediated deletion was also reported elsewhere in an individual with disseminated TB and a high bacterial burden [29]. The deleted isogenic strain in this individual was only present in the lymph nodes and although transmission events were not identified, it is clear that the deletion did not impair the bacilli to replicate and cause disease.

Although the physiological effect of the ON-A^NM 15 kb deletion is presently unknown, based on the information available for the deleted genes, five of which have some role in gene regulation under environmental stresses; this deletion may have a direct impact on several molecular pathways and could enhance the capacity of this variant to spread and cause disease.

Conversely, the 15 kb deletion observed in ON-A^NM is in a hotspot position for IS6110 transposition events [41,42] and the genes located in this region may not be essential for the bacteria to transmit and cause disease. This would support the theory that the ON-A^NM was randomly fixed in the homeless/under-housed population, possibly supported by higher transmission rates in congregate setting such as shelters and a high prevalence of co-morbidities). It is possible that a weakened immune system allows for the appearance of potentially "unfit" mutations or large deletion events such as those reported here.

Studies focusing on fitness assessment and mining of the proteome and transcriptome of ON-A variants to determine the role of the deleted genes in the physiology and virulence of *M. tuberculosis* are currently underway.

WGS as tool to determine TB transmission dynamics

Although RFLP, MIRU-VNTR and spoligotype are all powerful genotyping techniques, they cannot establish chronology of infection and lack resolution in long-term transmission events [43–46]. In contrast WGS has the potential for resolving the transmission dynamics of TB infection and when complemented with epidemiological data, it is an excellent tool to identify

of the Sigma-factor F (SigF). Sigma factors bind to RNA polymerase and alter its promoter preference resulting in subsets of genes that are differentially expressed upon environmental stresses to which a particular sigma factor responds. *M. tuberculosis* SigF is induced by nutrient starvation [30] and it is involved in the late stages of the disease [31]. One of the anti-sigF

Table 2. Genes present in the 15 kb deleted region.

Deleted gene	Function	Annotations	Reference
Rv1358	Probable transcriptional regulatory protein	LuxR family signature, expression regulated by **SigF**; deleted in some clinical isolates	[33,49]
Rv1359	Probable signal transduction regulatory protein	Adenylate cyclase, family 3 (Sensory pathways)	
Rv1360	Probable oxidoreductase	Expressed during GP infections; mRNA down-regulated in starvation	[50,51]
Rv1361	PPE19 - Unknown	Expressed during GP infections; mRNA down-regulated in starvation; regulated by PhoP	[50–52]
Rv1362c	Unknown	Expressed during GP infections; membrane protein	[50]
Rv1363c	Unknown	None	
Rv1364c	Possible sigma factor F regulatory protein	May be a capreomycin target; responds to heat stress	[53]
Rv1365	Anti-anti sigma factor F	Regulated by Redox potential	
Rv1366	Unknown	Homology to RelA-SpoT superfamily (ppGpp synthesis)	
Rv1367c	Possibly involved in cell wall metabolism	Homology to PBPs, B-lactamases	
Rv1368	Lipoprotein	None	
Rv1371	Unknown	Probably conserved membrane protein	

individual transmission events, and its use is very valuable in highly mobile communities such as the homeless/under-housed.

In addition to providing relevant and detailed information that can be used to construct and/or validate the transmission dynamics of a given strain, WGS also provides important information regarding SNPs, indels and larger genomic structural variants that can provide insights into the physiological characteristics of *M. tuberculosis* as a whole as well as particular characteristics of the strain of interest.

Recent genomics studies have highlighted the benefits of whole genome sequencing over traditional genotyping to uncover the transmission dynamics and molecular-guided TB control and surveillance [11,12,14–17]. For instance, Gardy et al. [12] demonstrated the usefulness of overlapping WGS data and social network analyses during outbreak investigations. The authors were able to determine transmission dynamics of 32 *M. tuberculosis* outbreak isolates in a period of two and half years. Similar to our findings, their SNP analysis revealed two lineages suggesting not one, but two simultaneous chains of transmission. The SNP calling algorithm in Gardy's study did not include filtering out PE/PPE genes. This resulted in a higher genetic diversity than expected, but it is remarkable the similar findings with our study vis-à-vis the identification of multiple strain variants and clusters with otherwise identical RFLP and MIRU-VNTR patterns in a high risk population. Unlike the study by Gardy et al. [12] in which the two strain variants diverged and circulated in the community well before the outbreak, the ON-A^{WT} variant appears to have evolved during the 2001 TB outbreak and the emergent ON-A^{NM} was rapidly fixed in Toronto's homeless/under-housed population.

Most WGS studies related to TB transmission dynamics have been focused on TB outbreaks in short time frames (1–30 months) [11–14,16]. In all cases WGS represented an invaluable tool to either support or complement directionality of transmission based on epidemiological contact studies. A few long-term studies have also been conducted. Roetzer *et al.* [17] performed a prospective study to evaluate the correlation of WGS analyses with contact tracing data. Eighty-six *M. tuberculosis* isolates from a 13 year time frame were sequenced. WGS analysis confirmed the clonal expansion of an outbreak strain that was initially seen in a specific social setting in Hamburg, Germany, but slowly spreading outside of this particular setting [17]. In contrast to our findings, SNPs were the primary source of genome evolution during transmission and all small deletions and insertions detected were constant in all isolates.

Walker et al. [15] also confirmed the high power and resolution provided by WGS and determined the genomic diversity between patients in MIRU-VNTR community clusters in a period spanning 1–12 years.

Here, we studied the dynamics of TB disease over 17 years of on-going transmission in a homeless/under-housed population. Previous studies have demonstrated that large deletions and hundreds of polymorphic sites can exist in isolates with identical IS6110 and MIRU-VNTR patterns [15]. As expected, we confirmed that clusters obtained by IS6110 RFLP typing were not in agreement with the WGS-based division of ON-A isolates into two major groups based on the 15 kb large genomic deletion).

In addition, we identified 81 SNPs that are variable within the ON-A strain and further subdivide the strain into 6 distinct sub-clusters.

The large heterogeneity observed in our study group resulted in an average of 1.5 SNPs/case (81 SNPs/55 *M. tuberculosis* isolates) is slightly larger than those reported in the majority of *M. tuberculosis*-WGS studies which ranged from 0–1.3. However, in those studies, the larger variation rates were associated to community clusters (as opposed to household groups) spanning more than 10 yr of transmission [13,15,17]. Only two studies have shown larger variability. Cluster 9 in the study reported in Walker et al. (2013) [15] which was associated with substance abuse and presented an average of 2 SNPs/case in a 9 yr span; and the large cluster in Gardy et al. (2011) with a rate of more than 5 SNPs/case likely due to the inclusion of hypervariable PPE/PE genes [12].

Our rate of 1.5 SNPs/case may be a result from the long time span as well as the likely possibility of on-going transmission prior to 1997 and thus potential for missed links/cases. Although information regarding delay in diagnosis and adherence to treatment was not available, these are conditions likely to be encountered in a high risk setting such as the homeless/under-housed and could increase the possibility of bacilli evolution. For instance, SNP patterns of isolates belonging to SC1 group suggest they all represent infections acquired from a common ancestor prior to 1997 and most likely represent re-activation of latent TB infection with independently gained substitutions during latency ranging from 3–7 SNPs per event.

This hypothesis is supported by a study of TB latency in macaques that suggests M. tuberculosis mutation rate is fairly similar in active and latent TB, with the rate in latent TB being slightly higher [47]. In contrast, a recent study of latency in humans suggests a lower mutation rate during latency when compared to active TB [48]. However, this study only included two subjects with a long period (>10 yr) of latent infection, and as the authors suggested, it is possible that a broader spectrum of mutation rates exists during latent TB.

Certainly, our study supports this idea, as we were able to show both high heterogeneity as observed in SC1, but also several cases with a low SNP fixation (0–2 SNPs/yr) in the other ON-A groups particularly SC4–SC6 in which latency periods of more than 5 years and zero SNP changes were observed. For instance, in the ON-A^{NM} (SC6), some of the isolates obtained in 2008 did not present any additional SNPs when compared to isolates obtained in 2001. Similarly, group SC4 of ON-A^{WT} was represented by identical isolates from 2004 to 2010 and SC5 included isolates with identical SNP pattern from 2001 to 2007.

Although most SC1 cases are probably reactivations, we strongly believe that the high heterogeneity is not only due to latency but also to missing links within the group. This could be due to unrecognized/undiagnosed TB cases, TB cases ultimately diagnosed outside the province of Ontario, or to TB cases diagnosed prior to 1997 for which no isolates are available. Gaps in the PHOL genotyping database are another factor. Our laboratory implemented IS6110 RFLP in 1997 but, until the introduction of a semi-automated MIRU-VNTR and spoligotyping program in 2007, Ontario did not have universal genotyping and only a subset of strains were analyzed or archived. Currently, the PHOL database includes patterns for >4000 isolates, but data for hundreds of strains obtained between 1997 and 2007 were never obtained.

In summary, the large number of singletons observed in our study, and the absence of progenitors for some SCs, are most likely due to gaps in our strain collection and genotyping database. However, SNPs present in some isolates may represent substitutions that originated during latency. In group SC6, the large majority of cases resulted from the presence of a super-spreader (NM7-01), and this suggestion is supported by clinical findings (e.g. smear 3+ and abnormal chest x-rays) consistent with a highly infectious state. Although it is possible that the high bacterial burden in this case could have contributed to the emergence and spread of isolates with differences of 2–4 SNPs, the large heterogeneity observed in these isolates is also compatible with a more plausible explanation such as mutations originating after secondary infection and long latency periods.

Conclusion

We performed a large retrospective and longitudinal study which resulted in the characterization of the microevolution of a unique M. tuberculosis strain associated with a high-risk group population. Despite a clustered genotype pattern based on the combination of 24-MIRU and spoligotype, we identified a large genomic deletion in nearly half of the sequenced isolates and were able to identify the emergence of this deletion event to have occurred during the first TB outbreak caused by ON-A in 2001. Even though loss of large genomic regions is a major source of variation in M. tuberculosis [49], to the best of our knowledge, this is the first study in which identification of such a region was pinpointed during active TB transmission. Furthermore, we identified a larger than expected heterogeneity resulting from the microevolution of ON-A in 17 years of transmission and further delineated this strain into 6 distinct sub-clusters.

WGS has been proposed as a "gold standard" for strain typing in M. tuberculosis [12,13,15]. Our study confirms the value of WGS to determine transmission dynamics and isolate relatedness in a large cluster of on-going TB transmission extending over many years in a high risk population. Nonetheless, the large number of shared contacts between TB cases, the mobility of homeless/under-housed individuals, and TB latency contributed to the complexity of determining individual events of TB transmission in this population.

The peculiarities of our study cohort are noticeable: high risk cases, on-going transmission despite control measures, high mobility of cases and likely missing links. All of these factors could be associated with a higher than average microevolution dynamic resulting in high variability and multiple transmission chains. Our intention is not to make general conclusions adapted to other transmission environments but to decipher the high clonal complexity and microevolution rates that could be expected in a transmission event in a complex population.

Supporting Information

Figure S1 Social network analysis of ON-A subjects. Social network analysis was performed using R statistical software (v3.0.2) with the igraph package. Each large circle represents a single individual colored by their sub-cluster as determined by WGS SNP analysis. Grey lines represent common contacts between study individuals and thick black lines represent direct epidemiological/social-connections between study individuals.

Material S1 Supplementary methods and results section. The supplementary methods describe the PCR amplification of regions flanking the 15 kb deletion. The supplementary results describe the WGS results of our laboratory strain H37Rv and how these results were use to evaluate the accuracy of our SNP calling algorithm. Table S1 in Material S1 shows the primers used for PCR amplification of regions flanking the 15 kb deletion.

Acknowledgments

We would like to thank the Public Health Ontario TB and Mycobacteriology Laboratory and research staff, responsible for the initial isolation, cultivation, and susceptibility testing of the clinical isolates used in the study, deletion-PCR and genotyping.

Author Contributions

Conceived and designed the experiments: CM FBJ DCA. Performed the experiments: CM JLG. Analyzed the data: CM DCA JLG RS ER. Contributed reagents/materials/analysis tools: CM FBJ. Contributed to the writing of the manuscript: CM. Review and edit the manuscript: JLG DCA RS ER FBJ.

References

1. Feske ML, Teeter LD, Musser JM, Graviss EA (2013) Counting the homeless: a previously incalculable tuberculosis risk and its social determinants. Am J Public Health 103: 839–848. doi:10.2105/AJPH.2012.300973.

2. Haddad MB, Wilson TW, Ijaz K, Marks SM, Moore M (2005) Tuberculosis and homelessness in the United States, 1994–2003. JAMA J Am Med Assoc 293: 2762–2766. doi:10.1001/jama.293.22.2762.

3. Bamrah S, Yelk Woodruff RS, Powell K, Ghosh S, Kammerer JS, et al. (2013) Tuberculosis among the homeless, United States, 1994–2010. Int J Tuberc Lung Dis Off J Int Union Tuberc Lung Dis 17: 1414–1419. doi:10.5588/ijtld.13.0270.

4. McAdam JM, Bucher SJ, Brickner PW, Vincent RL, Lascher S (2009) Latent tuberculosis and active tuberculosis disease rates among the homeless, New York, New York, USA, 1992–2006. Emerg Infect Dis 15: 1109–1111. doi:10.3201/eid1507.080410.

5. Khan K, Rea E, McDermaid C, Stuart R, Chambers C, et al. (2011) Active tuberculosis among homeless persons, Toronto, Ontario, Canada, 1998–2007. Emerg Infect Dis 17: 357–365. doi:10.3201/eid1703.100833.

6. Adam HJ, Guthrie JL, Bolotin S, Alexander DC, Stuart R, et al. (2010) Genotypic characterization of tuberculosis transmission within Toronto's under-housed population, 1997–2008. Int J Tuberc Lung Dis Off J Int Union Tuberc Lung Dis 14: 1350–1353.

7. Lofy KH, McElroy PD, Lake L, Cowan LS, Diem LA, et al. (2006) Outbreak of tuberculosis in a homeless population involving multiple sites of transmission. Int J Tuberc Lung Dis Off J Int Union Tuberc Lung Dis 10: 683–689.

8. Centers for Disease Control and Prevention (CDC) (2012) Tuberculosis outbreak associated with a homeless shelter - Kane County, Illinois, 2007–2011. MMWR Morb Mortal Wkly Rep 61: 186–189.

9. Centers for Disease Control and Prevention (CDC) (2013) Notes from the Field: Outbreak of Tuberculosis Associated with a Newly Identified Mycobacterium tuberculosis Genotype - New York City, 2010–2013. MMWR Morb Mortal Wkly Rep 62: 904.

10. Alexander DC, Guthrie JL, Pyskir D, Maki A, Kurepina N, et al. (2009) Mycobacterium tuberculosis in Ontario, Canada: Insights from IS6110 restriction fragment length polymorphism and mycobacterial interspersed repetitive-unit-variable-number tandem-repeat genotyping. J Clin Microbiol 47: 2651–2654. doi:10.1128/JCM.01946-08.

11. Bryant JM, Schürch AC, van Deutekom H, Harris SR, de Beer JL, et al. (2013) Inferring patient to patient transmission of Mycobacterium tuberculosis from whole genome sequencing data. BMC Infect Dis 13: 110. doi:10.1186/1471-2334-13-110.

12. Gardy JL, Johnston JC, Sui SJH, Cook VJ, Shah L, et al. (2011) Whole-Genome Sequencing and Social-Network Analysis of a Tuberculosis Outbreak. N Engl J Med 364: 730–739. doi:10.1056/NEJMoa1003176.

13. Kato-Maeda M, Ho C, Passarelli B, Banaei N, Grinsdale J, et al. (2013) Use of whole genome sequencing to determine the microevolution of Mycobacterium tuberculosis during an outbreak. PloS One 8: e58235. doi:10.1371/journal.pone.0058235.

14. Schürch AC, Kremer K, Daviena O, Kiers A, Boeree MJ, et al. (2010) High-resolution typing by integration of genome sequencing data in a large tuberculosis cluster. J Clin Microbiol 48: 3403–3406. doi:10.1128/JCM.00370-10.

15. Walker TM, Ip CLC, Harrell RH, Evans JT, Kapatai G, et al. (2013) Whole-genome sequencing to delineate Mycobacterium tuberculosis outbreaks: a retrospective observational study. Lancet Infect Dis 13: 137–146. doi:10.1016/S1473-3099(12)70277-3.

16. Török ME, Reuter S, Bryant J, Köser CU, Stinchcombe SV, et al. (2013) Rapid whole-genome sequencing for investigation of a suspected tuberculosis outbreak. J Clin Microbiol 51: 611–614. doi:10.1128/JCM.02279-12.

17. Roetzer A, Diel R, Kohl TA, Rückert C, Nübel U, et al. (2013) Whole Genome Sequencing versus Traditional Genotyping for Investigation of a Mycobacterium tuberculosis Outbreak: A Longitudinal Molecular Epidemiological Study. PLoS Med 10: e1001387. doi:10.1371/journal.pmed.1001387.

18. Alexander DC, Guthrie JL, Pyskir D, Maki A, Kurepina N, et al. (2009) Mycobacterium tuberculosis in Ontario, Canada: Insights from IS6110 Restriction Fragment Length Polymorphism and Mycobacterial Interspersed Repetitive-Unit-Variable-Number Tandem-Repeat Genotyping. J Clin Microbiol 47: 2651–2654. doi:10.1128/JCM.01946-08.

19. Van Soolingen D, Hermans PW, de Haas PE, Soll DR, van Embden JD (1991) Occurrence and stability of insertion sequences in Mycobacterium tuberculosis complex strains: evaluation of an insertion sequence-dependent DNA polymorphism as a tool in the epidemiology of tuberculosis. J Clin Microbiol 29: 2578–2586.

20. Jamieson FB, Guthrie JL, Neemuchwala A, Lastovetska O, Melano RG, et al. (2014) Profiling of rpoB Mutations and MICs to Rifampicin and Rifabutin in Mycobacterium tuberculosis. J Clin Microbiol. doi:10.1128/JCM.00691-14.

21. Supply P, Allix C, Lesjean S, Cardoso-Oelemann M, Rüsch-Gerdes S, et al. (2006) Proposal for standardization of optimized mycobacterial interspersed repetitive unit-variable-number tandem repeat typing of Mycobacterium tuberculosis. J Clin Microbiol 44: 4498–4510. doi:10.1128/JCM.01392-06.

22. Cowan LS, Diem L, Brake MC, Crawford JT (2004) Transfer of a Mycobacterium tuberculosis genotyping method, Spoligotyping, from a reverse line-blot hybridization, membrane-based assay to the Luminex multianalyte profiling system. J Clin Microbiol 42: 474–477.

23. Van Embden JD, Cave MD, Crawford JT, Dale JW, Eisenach KD, et al. (1993) Strain identification of Mycobacterium tuberculosis by DNA fingerprinting: recommendations for a standardized methodology. J Clin Microbiol 31: 406–409.

24. Huson DH, Bryant D (2006) Application of phylogenetic networks in evolutionary studies. Mol Biol Evol 23: 254–267. doi:10.1093/molbev/msj030.

25. Gascuel O (1997) BIONJ: an improved version of the NJ algorithm based on a simple model of sequence data. Mol Biol Evol 14: 685–695.

26. Dress AWM, Huson DH (2004) Constructing splits graphs. IEEEACM Trans Comput Biol Bioinforma IEEE ACM 1: 109–115. doi:10.1109/TCBB.2004.27.

27. Gambette P, Huson DH (2008) Improved layout of phylogenetic networks. IEEEACM Trans Comput Biol Bioinforma IEEE ACM 5: 472–479. doi:10.1109/tcbb.2007.1046.

28. Csardi G, Nepusz T (2006) The igraph software package for complex network research. InterJournal Complex Systems: 1695.

29. Sampson SL, Richardson M, Van Helden PD, Warren RM (2004) IS6110-mediated deletion polymorphism in isogenic strains of Mycobacterium tuberculosis. J Clin Microbiol 42: 895–898.

30. Chen P, Ruiz RE, Li Q, Silver RF, Bishai WR (2000) Construction and characterization of a Mycobacterium tuberculosis mutant lacking the alternate sigma factor gene, sigF. Infect Immun 68: 5575–5580.

31. Geiman DE, Kaushal D, Ko C, Tyagi S, Manabe YC, et al. (2004) Attenuation of late-stage disease in mice infected by the Mycobacterium tuberculosis mutant lacking the SigF alternate sigma factor and identification of SigF-dependent genes by microarray analysis. Infect Immun 72: 1733–1745.

32. Parida BK, Douglas T, Nino C, Dhandayuthapani S (2005) Interactions of anti-sigma factor antagonists of Mycobacterium tuberculosis in the yeast two-hybrid system. Tuberculosis 85: 347–355. doi:10.1016/j.tube.2005.08.001.

33. Hartkoorn RC, Sala C, Uplekar S, Busso P, Rougemont J, et al. (2012) Genome-Wide Definition of the SigF Regulon in Mycobacterium tuberculosis. J Bacteriol 194: 2001–2009. doi:10.1128/JB.06692-11.

34. Pesavento C, Hengge R (2009) Bacterial nucleotide-based second messengers. Curr Opin Microbiol 12: 170–176. doi:10.1016/j.mib.2009.01.007.

35. Klinkenberg LG, Lee J, Bishai WR, Karakousis PC (2010) The Stringent Response Is Required for Full Virulence of Mycobacterium tuberculosis in Guinea Pigs. J Infect Dis 202: 1397–1404. doi:10.1086/656524.

36. Primm TP, Andersen SJ, Mizrahi V, Avarbock D, Rubin H, et al. (2000) The stringent response of Mycobacterium tuberculosis is required for long-term survival. J Bacteriol 182: 4889–4898.

37. Fang Z, Doig C, Kenna DT, Smittipat N, Palittapongarnpim P, et al. (1999) IS6110-mediated deletions of wild-type chromosomes of Mycobacterium tuberculosis. J Bacteriol 181: 1014–1020.

38. Ho TB, Robertson BD, Taylor GM, Shaw RJ, Young DB (2000) Comparison of Mycobacterium tuberculosis genomes reveals frequent deletions in a 20 kb variable region in clinical isolates. Yeast Chichester Engl 17: 272–282. doi:10.1002/1097-0061(200012)17:4<272::AID-YEA48>3.0.CO;2-2.

39. Pepperell CS, Casto AM, Kitchen A, Granka JM, Cornejo OE, et al. (2013) The Role of Selection in Shaping Diversity of Natural M. tuberculosis Populations. PLoS Pathog 9: e1003543. doi:10.1371/journal.ppat.1003543.

40. Newton SM, Smith RJ, Wilkinson KA, Nicol MP, Garton NJ, et al. (2006) A deletion defining a common Asian lineage of Mycobacterium tuberculosis associates with immune subversion. Proc Natl Acad Sci 103: 15594–15598. doi:10.1073/pnas.0604283103.

41. Yesilkaya H, Dale JW, Strachan NJC, Forbes KJ (2005) Natural transposon mutagenesis of clinical isolates of Mycobacterium tuberculosis: how many genes does a pathogen need? J Bacteriol 187: 6726–6732. doi:10.1128/JB.187.19.6726-6732.2005.

42. Reyes A, Sandoval A, Cubillos-Ruiz A, Varley KE, Hernández-Neuta I, et al. (2012) IS-seq: a novel high throughput survey of in vivo IS6110 transposition in multiple Mycobacterium tuberculosis genomes. BMC Genomics 13: 249. doi:10.1186/1471-2164-13-249.

43. Benedetti A, Menzies D, Behr MA, Schwartzman K, Jin Y (2010) How close is close enough? Exploring matching criteria in the estimation of recent transmission of tuberculosis. Am J Epidemiol 172: 318–326. doi:10.1093/aje/kwq124.

44. De Boer AS, Kremer K, Borgdorff MW, de Haas PE, Heersma HF, et al. (2000) Genetic heterogeneity in Mycobacterium tuberculosis isolates reflected in

IS6110 restriction fragment length polymorphism patterns as low-intensity bands. J Clin Microbiol 38: 4478–4484.

45. Glynn JR, Vynnycky E, Fine PE (1999) Influence of sampling on estimates of clustering and recent transmission of Mycobacterium tuberculosis derived from DNA fingerprinting techniques. Am J Epidemiol 149: 366–371.

46. Schürch AC, Kremer K, Hendriks ACA, Freyee B, McEvoy CRE, et al. (2011) SNP/RD typing of Mycobacterium tuberculosis Beijing strains reveals local and worldwide disseminated clonal complexes. PloS One 6: e28365. doi:10.1371/journal.pone.0028365.

47. Ford C, Yusim K, Ioerger T, Feng S, Chase M, et al. (2012) Mycobacterium tuberculosis – Heterogeneity revealed through whole genome sequencing. Tuberculosis 92: 194–201. doi:10.1016/j.tube.2011.11.003.

48. Colangeli R, Arcus VL, Cursons RT, Ruthe A, Karalus N, et al. (2014) Whole Genome Sequencing of Mycobacterium tuberculosis Reveals Slow Growth and Low Mutation Rates during Latent Infections in Humans. PLoS ONE 9: e91024. doi:10.1371/journal.pone.0091024.

49. Tsolaki AG, Hirsh AE, DeRiemer K, Enciso JA, Wong MZ, et al. (2004) Functional and evolutionary genomics of Mycobacterium tuberculosis: Insights from genomic deletions in 100 strains. Proc Natl Acad Sci 101: 4865–4870. doi:10.1073/pnas.0305634101.

50. Kruh NA, Troudt J, Izzo A, Prenni J, Dobos KM (2010) Portrait of a Pathogen: The Mycobacterium tuberculosis Proteome In Vivo. PLoS ONE 5: e13938. doi:10.1371/journal.pone.0013938.

51. Betts JC, Lukey PT, Robb LC, McAdam RA, Duncan K (2002) Evaluation of a nutrient starvation model of Mycobacterium tuberculosis persistence by gene and protein expression profiling. Mol Microbiol 43: 717–731.

52. Walters SB, Dubnau E, Kolesnikova I, Laval F, Daffe M, et al. (2006) The Mycobacterium tuberculosis PhoPR two-component system regulates genes essential for virulence and complex lipid biosynthesis. Mol Microbiol 60: 312–330. doi:10.1111/j.1365-2958.2006.05102.x.

53. Arnvig KB, Comas I, Thomson NR, Houghton J, Boshoff HI, et al. (2011) Sequence-Based Analysis Uncovers an Abundance of Non-Coding RNA in the Total Transcriptome of Mycobacterium tuberculosis. PLoS Pathog 7: e1002342. doi:10.1371/journal.ppat.1002342.

6

Comparative Analysis of Predicted Plastid-Targeted Proteomes of Sequenced Higher Plant Genomes

Scott Schaeffer[1,2], Artemus Harper[1], Rajani Raja[3], Pankaj Jaiswal[3], Amit Dhingra[1,2]*

1 Department of Horticulture, Washington State University, Pullman, WA, United States of America, **2** Molecular Plant Science Graduate Program, Washington State University, Pullman, WA, United States of America, **3** 2082 Cordley Hall, Department of Botany and Plant Pathology, Oregon State University, Corvallis, OR, United States of America

Abstract

Plastids are actively involved in numerous plant processes critical to growth, development and adaptation. They play a primary role in photosynthesis, pigment and monoterpene synthesis, gravity sensing, starch and fatty acid synthesis, as well as oil, and protein storage. We applied two complementary methods to analyze the recently published apple genome (Malus × domestica) to identify putative plastid-targeted proteins, the first using TargetP and the second using a custom workflow utilizing a set of predictive programs. Apple shares roughly 40% of its 10,492 putative plastid-targeted proteins with that of the Arabidopsis (Arabidopsis thaliana) plastid-targeted proteome as identified by the Chloroplast 2010 project and ~57% of its entire proteome with Arabidopsis. This suggests that the plastid-targeted proteomes between apple and Arabidopsis are different, and interestingly alludes to the presence of differential targeting of homologs between the two species. Co-expression analysis of 2,224 genes encoding putative plastid-targeted apple proteins suggests that they play a role in plant developmental and intermediary metabolism. Further, an inter-specific comparison of Arabidopsis, Prunus persica (Peach), Malus × domestica (Apple), Populus trichocarpa (Black cottonwood), Fragaria vesca (Woodland Strawberry), Solanum lycopersicum (Tomato) and Vitis vinifera (Grapevine) also identified a large number of novel species-specific plastid-targeted proteins. This analysis also revealed the presence of alternatively targeted homologs across species. Two separate analyses revealed that a small subset of proteins, one representing 289 protein clusters and the other 737 unique protein sequences, are conserved between seven plastid-targeted angiosperm proteomes. Majority of the novel proteins were annotated to play roles in stress response, transport, catabolic processes, and cellular component organization. Our results suggest that the current state of knowledge regarding plastid biology, preferentially based on model systems is deficient. New plant genomes are expected to enable the identification of potentially new plastid-targeted proteins that will aid in studying novel roles of plastids.

Editor: Steven M. Theg, University of California – Davis, United States of America

Funding: This research and support for AH was funded in part by United States Department of Agriculture National Institute of Food and Agriculture Grant # 2009-05031 to AD. Support for SS was provided by National Institutes of Health Training Grant number T32 GM008336 to AD as co-applicant. The funders had no role in study design, data collection and analysis, decision to publish, or preparation of the manuscript.

* Email: adhingra@wsu.edu

Introduction

The plastid is an intracellular organelle derived from an endosymbiotic event wherein a free-living autotrophic photosynthetic bacterium was phagocytized by a separate heterotrophic organism [1]. These organelles have since become essential to plant survival and have been documented to participate in numerous biological processes including photosynthesis, storage of oils, and proteins, pigment synthesis and storage, monoterpene synthesis [2], gravity sensing [3], and starch and fatty acid synthesis [4]. Over an extensive period of evolution, large parts of the plastid genome are hypothesized to have integrated into the nuclear genome [5]. In higher plants, the vast majority of proteins constituting the plastid proteome are encoded by genes physically resident in the nuclear genome, with about 120 genes retained in the plastid genome, a number which varies between species [6]. Comparative genomic analysis between Arabidopsis (*Arabidopsis thaliana*) and cyanobacteria indicates that 18% of the Arabidopsis protein–coding genes were derived from events involving transfer of genetic material from the plastid to the nucleus [7]. In part, exchange of genetic material and related biological functionality has necessitated an orchestration of processes between the plastid and nucleus where the nucleus actively exerts control on all aspects of plastid function.

Plant cells have developed intricate mechanisms to import nuclear-encoded proteins to or across the three plastid membranes (outer, inner plastid envelope, thylakoid). The presence of multiple protein transport pathways has been shown to play a role in aiding protein transport across the inner and outer plastid envelopes; however, the vast majority of the plastid proteome is transported via the tic/toc pathway [8]. In order to utilize this pathway, most proteins possess a signal peptide which interacts with chaperones and is later cleaved. Stromal-targeting peptide sequences, while not conserved, possess some similarities in amino acid composition. These targeting peptides are typically comprised of a relatively high abundance of serine and threonine residues [9,10] and are positively charged [11]. There are also some proteins that do not have any canonical signaling peptides and yet localize to plastids [12,13,14]. Therefore, the signaling prediction programs provide a good reference point to initiate an understanding of the plastid-targeted proteome for any new species, but as predictions, they do require experimental validation.

Due largely to the technical complexity with whole plastid proteome characterization, transcriptome or genome sequences have become a widely used dataset to predict plastid-targeted motifs. Such an approach also enables the identification of plastid-targeting proteins in a spatial and temporal context. Prediction of subcellular localization has been reported to be performed with software such as PCLR [15], iPSORT [16], TargetP [17,18], and PREDOTAR [19] amongst many other programs. Most prediction methods exploit the presence of an N-terminal signal sequence to predict cellular localization. Of these, TargetP was recommended to be most successful in prediction and was comparable to PCLR, with each having sensitivity values of 0.72 [20].

The Rosaceae family represents a unique diversity in fruit development, which is unrealized in the many model plants whose genomes have been sequenced [21,22,23,24,25,26]. Pomes (apples and pears), stone fruits (cherry and peach), and aggregate fruits (strawberry and raspberry) display a diversity that suggests the presence of novel metabolic processes, and is supported by a large number of genes which far exceeds the number of genes in Arabidopsis. While fruit development in these Rosaceae species differs vastly, the ubiquitous process of the plastidial transition from a chloroplast to chromoplast is often assumed to be conserved. Within Rosaceae fruit, plastids play extremely important roles in determining fruit quality and organoleptic appeal as they are the site for synthesis of carotenoids [27,28], monoterpenes [2], fatty acids [4] and aromatic amino acids. Many of these compounds have been linked to human health and nutrition [29,30]. Plastids are also important in converting starch into various types of carbohydrate and sugars in developing fruits [31].

The plastid structure has also been reported to differ between different tissues of the fruit. Phan [32] reported the presence of a large single granum comprised only of stacked thylakoid membranes in the plastids of the endocarp tissue of apple. In addition,

chloroplasts with leaf-like thylakoid and grana organization in the outer six cell layers of mature apple fruit and presence of globular chromoplasts in epidermal cells were described. It is expected that differences in structure, physiology and biochemistry in the Rosaceae fruit plastids as well as other non-model systems will assist in identifying novel processes associated with plastids in plants.

In this study we tested the three primary hypotheses using a bioinformatics approach, (1) The total number and composition of the plastid-targeted protein coding genes in apple, a model representative of Rosaceae, that is taxonomically different from Arabidopsis, (2) The plastid-targeted protein coding genes are under transcriptional control during apple fruit development and (3) There is a subset of unique plastid-targeted protein coding genes that are unique and novel to each plant species.

In order to test the first hypothesis, we performed an in-depth computational analysis predicting the plastid-targeted proteome of apple and compared it with Arabidopsis resulting in the identification of a much larger number of plastid-targeted genes with nearly 4000 plastid-targeted protein coding genes being unique to apple. The second hypothesis was tested by reanalyzing publically available apple transcriptome data which revealed the presence of co-expression profiles of plastid-targeted genes and their association to development and metabolism. Finally, the third hypothesis was tested by extending the custom analysis workflow to an inter-genera comparison between six published genomes: *Arabidopsis thaliana*, *Vitis vinifera*, *Prunus persica*, *Populus trichocarpa*, *Fragaria vesca*, and *Solanum lycopersicum* resulting in the identification of plastid-targeted proteins unique to each species. A core set of 737 *Arabidopsis thaliana* proteins, highly enriched in photosynthesis and primary metabolism gene ontology (GO) terms, were identified to have homologous plastid-targeted proteins in all investigated species.

Materials and Methods

TargetP-based prediction of *Malus × domestica* plastid proteome

The *Malus × domestica* predicted protein set was obtained from the apple genome sequencing project [26]. Protein sequences were analyzed using TargetP using plant networks with default parameters [17,18]. All sequences with predicted chloroplast transit peptides were compiled into a new dataset and were sorted based on length using USEARCH [33].

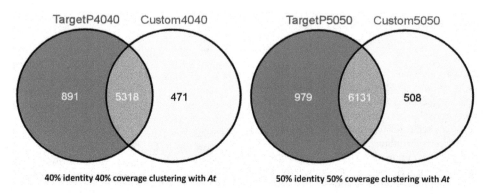

Figure 1. Venn diagrams displaying the predicted plastid-targeting proteins unique to apple compared to *Arabidopsis*. Two plastid-targeting methods, TargetP and a custom analysis method, were used to predict genes encoding plastid localized proteins. Sequences in these data sets were compared to *Arabidopsis* plastid-targeted proteins from the Chloroplast2010 project using USEARCH. Genes not clustered to *Arabidopsis* were compared between prediction methods displaying a high agreement between the methods. Venn diagrams were constructed using Venny (Oliveros, 2007).

Table 1. Enriched GO terms associated with co-expressed genes encoding plastid-targeted proteins in developing apple.

	GO term	Ontology	Description	p-value	FDR
Cluster 1	GO:0015979	P	photosynthesis	7.80E-08	2.60E-05
	GO:0009579	C	thylakoid	5.50E-16	9.10E-14
	GO:0009536	C	plastid	4.70E-06	3.90E-04
	GO:0016020	C	membrane	1.10E-05	6.20E-04
	GO:0043231	C	intracellular membrane-bounded organelle	3.20E-05	1.10E-03
	GO:0043227	C	membrane-bounded organelle	3.20E-05	1.10E-03
	GO:0043229	C	intracellular organelle	6.90E-05	1.60E-03
	GO:0043226	C	organelle	6.90E-05	1.60E-03
	GO:0005623	C	cell	8.30E-05	1.70E-03
	GO:0044464	C	cell part	1.30E-04	2.30E-03
	GO:0044424	C	intracellular part	6.50E-04	1.10E-02
	GO:0044444	C	cytoplasmic part	7.00E-04	1.10E-02
	GO:0005622	C	intracellular	1.60E-03	2.20E-02
Cluster 2	GO:0006629	P	lipid metabolic process	8.40E-10	2.60E-07
	GO:0019748	P	secondary metabolic process	8.70E-05	1.30E-02
	GO:0009058	P	biosynthetic process	1.60E-04	1.60E-02
	GO:0009056	P	catabolic process	4.40E-04	3.40E-02
	GO:0006810	P	transport	9.20E-04	4.00E-02
	GO:0051234	P	establishment of localization	9.20E-04	4.00E-02
	GO:0051179	P	localization	9.20E-04	4.00E-02
	GO:0003824	F	catalytic activity	5.60E-04	3.80E-02
	GO:0005737	C	cytoplasm	7.00E-12	1.40E-09
	GO:0044464	C	cell part	1.80E-11	1.80E-09
	GO:0005623	C	cell	4.90E-11	3.20E-09
	GO:0044444	C	cytoplasmic part	7.30E-11	3.50E-09
	GO:0044424	C	intracellular part	1.40E-10	4.50E-09
	GO:0005622	C	intracellular	1.30E-10	4.50E-09
	GO:0043229	C	intracellular organelle	1.40E-09	3.10E-08
	GO:0009536	C	plastid	1.30E-09	3.10E-08
	GO:0043226	C	organelle	1.40E-09	3.10E-08
	GO:0043231	C	intracellular membrane-bounded organelle	1.80E-09	3.20E-08
	GO:0043227	C	membrane-bounded organelle	1.80E-09	3.20E-08
Cluster 3	No significant term				
Cluster 4	No significant term				
Cluster 5	No significant term				
Cluster 6	No significant term				
Cluster 8	No significant term				
Cluster 9	No significant term				
Cluster 10	No significant term				
Cluster 11	No significant term				

Expression data from Janssen et al. [49] was mined to identify the expression profiles for any genes encoding plastid-targeted protein. Clustering of genes was performed based upon expression profile revealing a set of 11 significant clusters. GO term enrichment was performed to identify significantly enriched terms associated with each cluster. Gene Ontology terms are provided for biological process (P), molecular function (F), and cellular component (C).

Custom protein targeting analysis

A part of the functional annotation pipeline was applied to identify organelle 'plastid' targeted gene products encoded by the apple genome [26]. The peptide sequences were analyzed first through InterProScan [34] results provided by the genome consortium [26], followed by in-house analysis using the SignalP [17], Predotar [19] and TMHMM [35]. InterPro provided the domain annotations, and any genes/peptides with transposable element/domain annotations were filtered out for further analysis. The next steps of the pipeline employed: (1) SignalP to predict localization to the mitochondrial or plastid or secretion pathway, plus providing signal peptide cleavage sites, (2) Predotar to predict localization to either or both the mitochondrion or plastid, and (3)

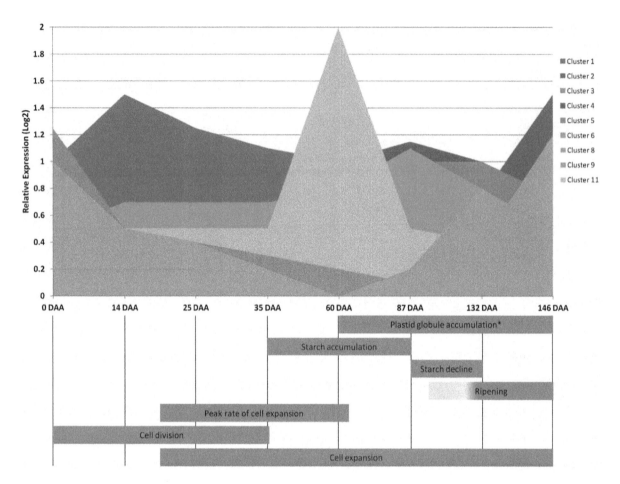

Figure 2. Overlay of apple gene expression clusters with distinct fruit development events. Relative expression of the co-expressed genes encoding plastid-targeted proteins were merged and displayed along significant events occurring within the developmental continuum of apple fruits, adapted from Janssen et al. 2008. An additional event unique to apple fruit plastids, globule accumulation was added as described by Clijsters in 1969.

TMHMM to identify predicted transmembrane domains in the protein sequences. After collecting these annotations, standardized protocols for assigning the annotations were adopted [24]. The higher quality scores with reviewed after computational analyses

(RCA) were selected if the scores of 0.75 and greater were predicted for TargetP and Predotar and two or more transmembrane annotations were predicted by the TMHMM. The parameters selected for inferred by electronic annotation (IEA)

Table 2. Results of putative plastid-targeted protein prediction.

Species	Reference	Unique predicted protein-coding genes	Predicted to be plastid-targeted
Arabidopsis thaliana	[40]	27,416	5,382* (19.6%)
Malus × *domestica* (Apple)	[26]	57,386	10,492 (18.3%)
Vitis vinifera (Grapevine)	[22]	26,346	3,015 (11.4%)
Prunus persica (Peach)	[21]	28,689	3,860 (13.5%)
Populus trichocarpa (Black cottonwood)	[25]	45,555	4,515 (9.9%)
Fragaria vesca (Woodland strawberry)	[24]	34,809	4,922 (14.1%)
Solanum lycopersicum (Tomato)	[23]	33,926	4,009 (11.8%)

Predicted transcripts were selected from 7 sequenced genomes representing both model organisms as well as agriculturally important fruit-producing crops. Translated sequences were analyzed using TargetP to predict cellular localization. Sequences were clustered with 40% coverage and 40% identity to predict proteins unique to the plastid proteome of each species. *Represents sequences from Chloroplast 2010 project and sequences associated with predicted plastid-targeted embryo lethal mutant.

include the scores of 0.5–0.749 for TargetP and Predotar and one/single transmembrane domain suggested by the TMHMM. The majority of these annotations were IEA evidence codes. If the annotations overlapped for gene products that had plastid-targeting predicted from TargetP and Predotar and membrane spanning domains identified by the TMHMM, then the suggested location of the targeted protein was 'plastid membrane'.

The Inparanoid algorithm [36] was used to find orthologous genes and paralogous genes that arise by duplication events. The pipeline was discussed in the *Fragaria vesca* genome paper [24]. For this study, the analysis included the peptide sequences from 22 species, including *Arabidopsis thaliana, Brachypodium distachyon, Caenorhabitis elegans, Chlamydomonas reinhardtii, Danio rerio, Eschericia coli, Fragaria vesca, Glycine max, Homo sapiens sapiens, Zea mays, Malus × domestica, Mus musculus, Neurospora crassa, Oryza sativa, Physcomitrella patens, Populus trichocarpa, Saccharomyces cerevisiae and pombe, Selaginella moellendorffii, Sorghum bicolor, Synechosystis, and Vitis vinifera* to cover the tree of life with emphasis on fully/nearly complete and published genomes. The peptide sequences were downloaded from Phytozome.net for grapevine, *Selaginella, Physcomitrella, Chlamydomonas, Glycine, Populus, and Malus* from the genome portal [26], and *Gramene* [37] for rice, sorghum, maize and Arabidopsis. The remaining sequences were downloaded from Ensembl [38,39].

Identification of sequences unique to apple datasets compared to Arabidopsis

The *Arabidopsis thaliana* plastid-targeted gene set was obtained from the Chloroplast 2010 project website (www.plastid.msu.edu) [40]. Arabidopsis embryo lethal mutants were analyzed using TargetP [17,18] and any chloroplast targeted proteins were added to the aforementioned dataset, as these were omitted from the Chloroplast 2010 database. Proteins predicted to target the apple plastid were then compared to plastid-targeted proteins from *Arabidopsis thaliana* using USEARCH [33]. Predicted plastid-targeted proteins were compared using two conditions: first, a global USEARCH was performed using 40% amino acid identity and 40% coverage (40/40), and a second global comparison was performed using 50% amino acid identity with 50% coverage (50/50). Header files of proteins unique to the *M. × domestica* dataset were compared between the TargetP-based method as well as the custom analysis to investigate any bias associated with either respective prediction technique.

USEARCH-based multispecies comparative analysis of predicted plastid-targeted proteomes

Predicted coding sequences were collected from the genomes of *Fragaria vesca* (Woodland Strawberry) [24], *Vitis vinifera* (Grapevine) [22], *Solanum lycopersicum* ITAG1 release (Tomato) [23], *Prunus persica* (Peach) [21] and *Populus trichocarpa* (Black Cottonwood) [25]. Protein sequences were analyzed with TargetP [17,18] using default parameters to predict localization. Sequences predicted to be plastid-targeted were organized into new files for each species. Comparisons were performed for each plastid-targeted dataset using USEARCH [33] 40/40 and 50/50 global parameters against the *Arabidopsis thaliana* plastid-targeted dataset, the entire Arabidopsis TAIR V10 protein set (Arabidopsis.org) [41], the predicted proteins from *Solanum lycopersicum*, as well as a file comprised of the sequences of the predicted plastid-targeted protein sequences of the other six species. All datasets were first sorted by length using USEARCH.

Further analysis was performed with USEARCH to identify those proteins present in the predicted plastid proteomes of all investigated species. To perform this analysis, the *Arabidopsis thaliana* putative plastid-targeted protein set was compared using USEARCH 40/40 global parameters separately against the plastid-targeted proteins from woodland strawberry, grapevine, tomato, peach, black cottonwood, and apple. Output files were then analyzed to identify those Arabidopsis sequences which had a match in the plastid-targeted proteomes of all species.

UCLUST-based multispecies comparative analysis of predicted plastid-targeted proteomes

A second comparative analysis was performed using the clustering feature of the USEARCH package, UCLUST. In this analysis, the plastid-targeted protein sequences from the seven examined species were compiled into a single file and sorted by length. UCLUST was performed at 50% identity. The output was parsed to identify protein clusters with members from all seven species, as well as those clusters containing sequences from only one species.

Determination of Jaccard's similarity coefficients

Two separate techniques were used to create the similarity matrices based upon Jaccard's coefficients. To calculate the value of an individual cell (the distance between species A and species B) we first determined if two genes were considered homologous. If a

Table 3. Clustering of predicted plastid proteomes with *Arabidopsis* datasets.

Species	Predicted plastid-targeted proteins	Percent clustered with *At* chloroplast (40/40)	Percent clustered with all *At* proteins (40/40)	Percent clustered with *At* chloroplast (50/50)	Percent clustered with all *At* proteins (50/50)
Arabidopsis thaliana	5,382*	100%	100%	100%	100%
Malus × domestica	10,492	40.7%	56.7%	32.2%	44.5%
Vitis vinifera	3,015	57.3%	77.0%	49.5%	65.8%
Prunus persica	3,860	59.9%	76.9%	52.0%	67.3%
Populus tichocarpa	4,515	51.8%	68.3%	44.5%	58.4%
Fragaria vesca	4,922	40.9%	56.1%	33.3%	45.0%
Solanum lycopersicum	4,009	55.2%	69.7%	45.7%	57.0%

Predicted plastid-targeted proteins unique to each species were clustered against the peptide sequences derived from the Chloroplast 2010 dataset with the addition of embryo lethal sequences (*At* chloroplast) as well as the entire *Arabidopsis* protein set from TAIR V10 (all *At* proteins) using USEARCH. Clustering was performed using the parameters of 40% coverage with 40% identity (40/40) as well as 50% coverage with 50% identity (50/50).

Table 4. USEARCH-based matching of unique plastid-targeted proteins.

Species	Predicted plastid-targeted Proteins	Number Clustered with *At* Chloroplast (40/40)	Number Clustered with *Sl* Chloroplast (40/40)	Proteins clustered with Cp Proteins from 6 Other Species	Unique Plastid-targeted Proteins
Arabidopsis thaliana	5,382*	5,382 (100%)	2,503 (46.5%)	4,099 (76.2%)	1,446 (26.9%)
Malus × domestica	10,492	4,265 (40.7%)	3,442 (32.8%)	6,235 (59.4%)	4,257 (40.6%)
Vitis vinifera	3,015	1,729 (57.3%)	1,396 (46.3%)	2,265 (75.1%)	750 (24.9%)
Prunus persica	3,860	2,312 (59.9%)	1,857 (48.1%)	3,232 (83.7%)	628 (16.3%)
Populus tichocarpa	4,515	2,340 (51.8%)	1,949 (43.2%)	3,207 (71.0%)	1,308 (29.0%)
Fragaria vesca	4,922	2,014 (40.9%)	1,584 (32.2%)	2,880 (58.5%)	2,042 (41.5%)
Solanum lycopersicum	4,009	2,213 (55.2%)	4,009 (100%)	2,799 (69.8%)	1,210 (30.2%)

Datasets comprised of putative plastid-targeted proteins unique to each species were clustered against the predicted plastid-targeted protein sequences of *Arabidopsis thaliana* (*At*) from Chloroplast2010 and TAIR V10, *Solanum lycopersicum* (*Sl*), and a database consisting of the plastid-targeted proteins from all 6 species using USEARCH. Sequences were clustered globally with 40% coverage with 40% identity (40/40). Results suggest that a large portion of the plastid-targeted proteins may be unique to each respective species.

gene from the species A matches with another gene in species B, then both A and B are included in the intersection set. The result in the cell is the count of the intersection set divided by the sum of all the genes in both species. For the USEARCH-based approach, two genes match if they align to each other using 40/40 parameters. Alternatively, in the UCLUST-based approach two genes match if they belong to the same cluster.

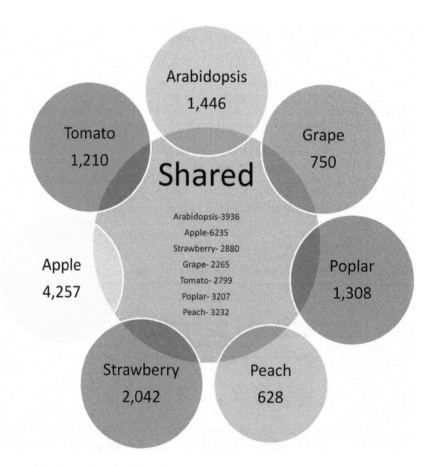

Figure 3. Comparison of predicted plastid-targeted proteomes across seven sequenced genomes. Protein sequences generated from the genomes of *Arabidopsis thaliana, Vitis vinifera, Populus trichocarpa, Prunus persica, Fragaria vesca, Malus × domestica,* and *Solanum lycopersicum* were clustered at 40 percent identity and 40 percent coverage using USEARCH. Comparison reveals large subsets for each species appearing to be unique to each species' respective plastids.

Table 5. Uniquely targeted plastid-targeted protein sequences.

Species	Predicted plastid-targeted Proteins	Unique to Species (USEARCH 40/40)	Unique to Species (UCLUST50)	Number Clustered with All Proteins from 6 Other Species (USEARCH 40/40)
Arabidopsis thaliana	5,382*	1,446 (26.9%)	2,154 (40.0%)	4,928 (91.6%)
Malus × domestica	10,492	4,257 (40.6%)	4,787 (45.6%)	7,911 (76.2%)
Vitis vinifera	3,015	750 (24.8%)	976 (32.3%)	2,561 (84.9%)
Prunus persica	3,860	628 (16.3%)	795 (20.6%)	3,506 (90.8%)
Populus tichocarpa	4,515	1,308 (29.0%)	1,732 (38.4%)	3,630 (80.4%)
Fragaria vesca	4,922	2,042 (41.5%)	2,305 (46.8%)	3,338 (67.8%)
Solanum lycopersicum	4,009	1,210 (30.2%)	1,654 (41.3%)	3,122 (77.9%)

A USEARCH comparison of plastid-targeted protein datasets was performed at 40% identity and 40% coverage against a database containing the chloroplast protein sequences from all six other species investigated in this study. A second comparison was performed against a database containing the entire protein set from the other species. An increase in matching suggests the presence of differentially localized homologues in other systems. Additionally, these results suggest a sizeable number of plastid-targeted proteins may be unique to each species.

Blast2GO Gene Ontology analysis and GO term enrichment analysis

Sequences for all genes encoding unique or shared plastid-targeted proteins in the investigated apple, Arabidopsis, grapevine, peach, strawberry, black cottonwood, and tomato datasets were analyzed via Blast2GO [42,43]. BLASTP was performed using the NCBI nr database with Blast2GO default parameters. Gene ontology mapping and annotation were also performed using default parameters with the August 2012 database. Following GO annotation, an Interpro scan [34,44] was performed and results were merged with the GO annotations. Annotation augmentation was performed using ANNEX [45], followed by GO-slim with the goslim_plant.obo database. Kyoto Encyclopedia of Genes and Genomes (KEGG) information was downloaded from the KEGG Pathway Database [46,47]. Datasets comprising those unique to each plastid-targeted proteome, as well as those shared between all

seven species were investigated using Single Enrichment Analysis with agriGO [48] to identify enriched GO terms. Analysis was performed using the Fisher test for significance and adjusted using the Yekutieli multi test adjustment with the minimum mapping entries set to three. A significance level was set at 0.01 and all terms GO-terms with p-values lower than this cutoff were reported as enriched.

Analysis of apple fruit gene expression

In order to ascertain if genes encoding plastid-targeted proteins in apple are expressed in fruits, as well as to identify co-expressed gene sets, microarray data from a previously published experiment were used [49]. The Janssen study measured the relative expression of about 13,000 features designed from apple fruit expressed sequence tags (ESTs) at 8 time points ranging from 0 days after anthesis (DAA) to 146

Table 6. GO terms enriched in uniquely targeted proteins as determined by USEARCH 40/40 method.

	GO term	Ontology	Description	p-value	FDR
Apple	GO:0006259	P	DNA metabolic process	8.30E-60	3.30E-57
	GO:0044260	P	cellular macromolecule metabolic process	4.20E-14	8.40E-12
	GO:0006807	P	nitrogen compound metabolic process	1.60E-12	1.60E-10
	GO:0006139	P	nucleobase, nucleoside, nucleotide and nucleic acid metabolic process	1.60E-12	1.60E-10
	GO:0043170	P	macromolecule metabolic process	2.90E-11	2.30E-09
	GO:0003676	F	nucleic acid binding	1.30E-53	1.60E-51
	GO:0005488	F	Binding	2.80E-24	1.70E-22
	GO:0003677	F	DNA binding	8.30E-05	3.30E-03
Arabidopsis	GO:0005634	C	Nucleus	8.30E-16	1.70E-13
Grapevine	No significant term				
Peach	GO:0016787	F	hydrolase activity	1.60E-06	1.10E-04
Black cottonwood	GO:0003676	F	nucleic acid binding	2.90E-06	3.40E-04
Strawberry	No significant term				
Tomato	No significant term				

GO terms were determined for all predicted plastid-targeted proteomes. Enrichment of GO terms was determined with agriGO for the subset of plastid-targeted proteins unique to each species compared to the entire plastid proteome. Gene Ontology terms are provided for biological process (P), molecular function (F), and cellular component (C).

DAA. All EST sequences utilized in the microarray experiment were retrieved from NCBI and a BLASTX was performed against the predicted apple protein set generated from the apple genome [26]. The EST expression data were then assigned to the top protein hit. Sequences which were previously found to be plastid-targeted were extracted and their respective expression data were analyzed by determining relative expression to the lowest measured mean expression value. The Log2 of relative expression data were imported into and analyzed with Multi-Experiment Viewer [50,51]. Sequences were clustered using Cluster Affinity Search Technique [52] using Pearson Correlation and a threshold of 0.8. Blast2GO [42] was used to assign annotation to those proteins with associated gene expression data. Single Enrichment Analysis was performed with AgriGO [48] as previously described, however a chi-square test was used instead to determine statistical significance.

Results

Predicted Plastid-targeted Proteomes of *Malus* × *domestica*

The apple genome has a total of 57,386 predicted genes [26] nearly 30,000 more genes than Arabidopsis [41,53]. We analyzed the complete apple gene set for cellular localization using two approaches, namely TargetP [17,18] and a custom prediction method (see materials and methods section for details). TargetP predicted the presence of 10,492 plastid-targeted proteins in the apple genome, while the custom gene ontology-based analysis predicted 9,882 genes, with an overlap of 9,256. Each data set was then clustered with the Arabidopsis plastid-targeted protein set using USEARCH [33] with 40% identity and 40% coverage (40/40 parameters) to identify homologous protein sequences. The TargetP method and custom analysis predicted 6,209 and 5,789 plastid-targeted proteins respectively to be unique to the apple dataset. The two methods agreed upon 5,318 proteins (86% and 92% respectively) uniquely targeted to apple chloroplasts and absent from those of *Arabidopsis thaliana* (Figure 1). Alternative clustering using 50% identity and 50% coverage (50/50 parameters) resulted in less clustering with *Arabidopsis* sequences and, consequently, increased the number of proteins predicted to be unique to apple. Using these parameters 7,110 sequences were predicted to be unique to the apple plastid proteome by TargetP, 6,639 with the custom analysis, and a set of 6,131 agreed upon by the two methods.

In order to identify prediction biases between the custom analysis and TargetP, an agriGO [48] GO term enrichment was performed on the proteins predicted to be differentially targeted. No significant GO terms were found to be enriched in the 1,236 proteins predicted to target the plastid with TargetP. However, agriGO identified the GO terms oxygen binding (GO:0019825, p-value 3.2e-05), hydrolase activity (GO:0016787 p-value 6e-05) and catalytic activity (GO:0003824 p-value 3.3e-04) are enriched in the plastid-targeted-proteins unique to the custom analysis.

Expression analysis of genes encoding plastid-targeted proteins in *Malus* × *domestica*

In order to test the hypothesis that plastid-targeted protein coding genes are under transcriptional control as the apple fruit develops, we reanalyzed data from a previously published microarray-based analysis of developing apple fruit [49]. Of the 13,000 unigene microarray probe sets studied, 2,698 were determined to map back to putative plastid-targeted proteins identified in this study, and represent a total of 2,224 unique sequences. Clustering of expression data using MultiExperiment

Viewer [50,51] identified 92 different expression clusters, however, only 64 of these had 5 or more members. Over 50% of the genes fit into 9 co-expression clusters. These co-expressed genes were annotated using Blast2GO to infer their functions. Expression data for each cluster are provided in File S1 as well as their associated GO term information (File S2).

Of the 11 main clusters investigated, only the two most populous clusters of co-expressed genes contained GO terms which were determined to be significantly enriched by agriGO analysis (Table 1). Cluster 1 had a single enriched biological process GO term of photosynthesis (GO:0015979) along with 12 enriched cellular component GO terms with thylakoid (GO:0009579) having the lowest p-value. Cluster 2 was enriched in the biological process GO terms lipid metabolic process (GO:0006629), secondary metabolic process (GO:0019748), biosynthetic process (GO:0009058), catabolic process (GO:0009056), transport (GO:006810), establishment of localization (GO:0051234), and localization (GO:0051179). Additionally, one molecular function GO term was enriched in cluster 2, catalytic activity (GO:0003824), along with 11 cellular component GO terms.

In order to determine if genes encoding plastid-targeted proteins were indeed expressed within the fruit of apple, data from a previous study were analyzed [49]. The initial microarray experiment was a large scale analysis representing 13,000 of the ~57,000 apple genes, and was designed around many significant physiological events occurring during apple fruit development. These 13,000 genes were compared to the genes encoding predicted plastid targeted proteins described earlier in this study. About 20% of the genes (2,224 genes) encoding predicted plastid-targeted proteins mapped back to genes represented in the Janssen study. Analysis with MultiExperiment Viewer revealed that the majority of these genes were co-expressed in 9 clusters. To show how these expression profiles may relate to important fruit developmental events, expression profiles for the co-expressed genes were overlaid with those events described in Janssen et al. (Figure 2). An additional event, plastid globule accumulation, was also added, as it was noted in developing apple fruits alongside the unstacking of photosynthetic membranes [54]. As seen in Figure 2, the gene expression of these clusters and their GO terms coincide to some extent with the processes occurring within the apple fruits. Many of the biological process GO terms and KEGG pathways associated with each gene expression cluster suggest that expression of genes encoding plastid-targeted proteins may coincide with these important events. The expression of Cluster 1 greatly mirrors the photosynthetic activity of apple fruit tissue, with highest expression occurring in young, photosynthetically-capable fruit, and expression lowering as the fruit matures and has a reduction in photosynthetic capabilities. Additionally, the expression of those genes in Cluster 2 appear to mirror the development of carotenoids, volatile compounds, and maturation of fruit, with expression lowest in young fruit and increasing as the fruit reaches maturity. In particular the expression of genes whose products are involved in lipid metabolic processes, secondary metabolic processes, biosynthetic processes, and catabolic processes, as determined via GO term enrichment would be great candidates for further study in their participation in apple fruit volatile production. Cluster 11 is particularly interesting as it is comprised of genes whose expression peaks at a single time point (60 DAA), however, the associated KEGG pathways and GO terms do not suggest a connection to the significant fruit processes of cell expansion and starch accumulation occurring at that time point. Blast2GO analysis revealed that 15.2% of the entire plastid-targeted proteome of apple lacked GO term information.

Table 7. GO terms enriched in uniquely plastid-targeted proteins identified with UCLUST 50% method.

	GO term	Ontology	Description	p-value	FDR
Apple	GO:0006259	P	DNA metabolic process	4.60E-08	1.80E-05
	GO:0016265	P	death	2.90E-04	3.80E-02
	GO:0008219	P	cell death	2.90E-04	3.80E-02
	GO:0003676	F	nucleic acid binding	3.90E-15	4.90E-13
	GO:0003677	F	DNA binding	2.30E-12	1.40E-10
	GO:0005488	F	binding	2.20E-10	9.30E-09
Arabidopsis	GO:0030528	F	transcription regulator activity	7.20E-06	3.60E-04
	GO:0003700	F	transcription factor activity	6.90E-06	3.60E-04
	GO:0003677	F	DNA binding	9.10E-05	3.00E-03
	GO:0005634	C	nucleus	1.30E-25	2.70E-23
	GO:0005654	C	nucleoplasm	6.60E-06	6.70E-04
	GO:0031981	C	nuclear lumen	3.90E-05	1.30E-03
	GO:0031974	C	membrane-enclosed lumen	3.90E-05	1.30E-03
	GO:0043233	C	organelle lumen	3.90E-05	1.30E-03
	GO:0070013	C	intracellular organelle lumen	3.90E-05	1.30E-03
	GO:0044428	C	nuclear part	1.10E-04	2.60E-03
	GO:0044446	C	intracellular organelle part	1.10E-04	2.60E-03
	GO:0044422	C	organelle part	1.10E-04	2.60E-03
Grapevine	No significant term				
Peach	GO:0016787	F	hydrolase activity	1.20E-04	8.90E-03
Black cottonwood	GO:0003676	F	nucleic acid binding	3.60E-04	4.40E-02
Strawberry	No significant term				
Tomato	GO:0003677	F	DNA binding	8.60E-05	6.60E-03
	GO:0030528	F	transcription regulator activity	2.60E-04	1.00E-02
	GO:0003700	F	transcription factor activity	4.70E-04	1.20E-02

Blast2GO was used to determine GO terms associated with all predicted plastid-targeted proteins. Enrichment analysis was performed with agriGO to identify significant enriched GO terms. Gene Ontology terms are provided for biological process (P), molecular function (F), and cellular component (C).

However, the set of 2,224 genes represented in this study reveals that this subset is better characterized as it contains only 78 (3.5%) sequences with no associated GO terms. Of course the mere expression of a gene does not indicate that a functional protein is present within the fruit plastids as this process could be affected or controlled at a number of levels including translation, interaction with chaperone proteins, redox state of the plastid, presence of appropriate translocation proteins, protein and mRNA stability and turnover, and likely many other factors. Regardless, the data presented in this study indicate that the expression of genes encoding plastid-targeted proteins is dynamic in the fruit of *Malus*

Table 8. Percentage of unique plastid proteome containing GO information.

Species	Plastid-targeted Proteins	Number w/o GO information	Unique to Species (USEARCH)	Number w/o GO information	Unique to Species (UCLUST)	Number w/o GO information
Arabidopsis thaliana	5,382	10 (0.2%)	1,446	6 (0.4%)	2,154	8 (0.4%)
Malus × domestica	10,492	1,592 (15.2%)	4,257	1,348 (31.7%)	4,817	1,386 (28.8%)
Vitis vinifera	3,015	542 (18.0%)	750	415 (55.3%)	976	455 (46.6%)
Prunus persica	3,860	719 (18.6%)	628	379 (60.4%)	795	450 (56.6%)
Populus tichocarpa	4,515	527 (11.7%)	1,308	425 (32.5%)	1,732	462 (26.6%)
Fragaria vesca	4,922	1,539 (31.3%)	2,042	1,350 (66.1%)	2,305	1,385 (60.1%)
Solanum lycopersicum	4,009	873 (21.8%)	1,210	693 (57.3%)	1,654	750 (45.3%)

Plastid-targeted proteins were analyzed using Blast2GO to identify GO terms associated with each protein sequence. With the exception of *Arabidopsis,* significant proportions of chloroplast-targeted proteins datasets lack GO term information. This further increases in the datasets comprised of chloroplast-targeted proteins unique to each investigated species.

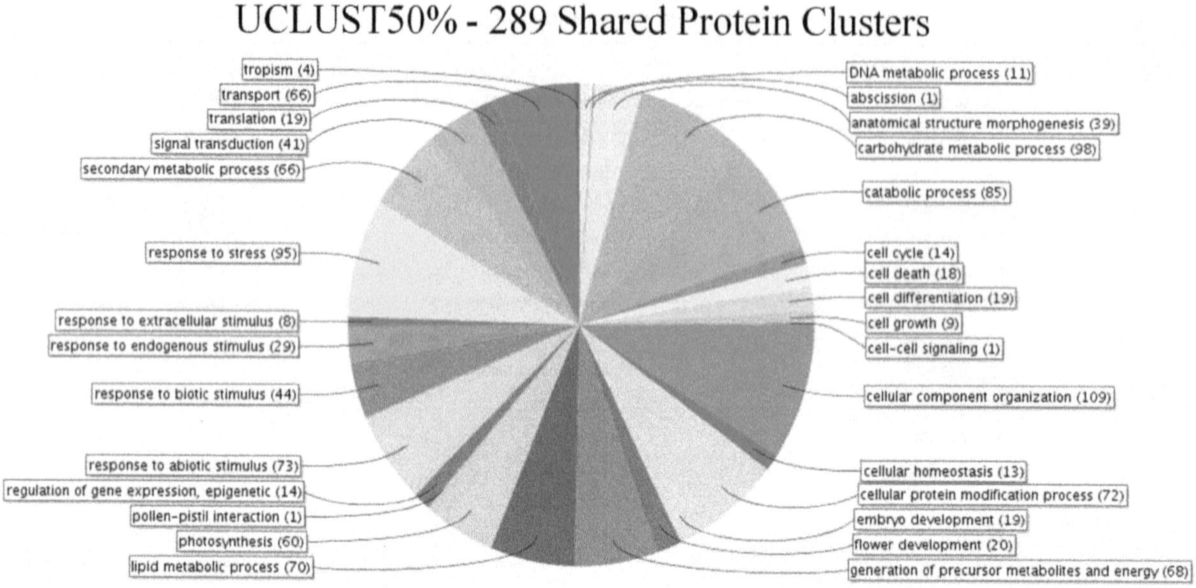

UCLUST50% - 289 Shared Protein Clusters

Figure 4. Biological process GO term composition of proteins shared between the 7 investigated plastid-targeted proteomes using two separate techniques. Two separate analyses were performed to identify proteins within the predicted plastid-targeted proteomes of seven species. The first, UCLUST with 50% identity, generated 15,750 clusters, 289 of which contained a member from all 7 species. USEARCH comparison performed at 40% identity and 40% coverage identified 737 sequences in the Arabidopsis thaliana putative plastid-targeted dataset which had a match in a protein sequence from the plastid-targeted sequences from all of the other 6 species. Sequences were analyzed via Blast2GO to determine the biological processes in which they partake.

× *domestica* and may play key roles in the development and quality of apple fruit.

Prediction and comparative analysis of plastid-targeted proteomes

In order to identify plastid targeted-proteins in seven species of interest (*Arabidopsis thaliana*, *Prunus persica*, *Malus × domestica*, *Populus trichocarpa*, *Fragaria vesca*, *Solanum lycopersicum*, and *Vitis vinifera*), plastid-targeting predictions were primarily performed using Target P. TargetP was selected in order to be consistent with previously published work in this area and because previous studies have found TargetP to be the most reliable single prediction program [17,20]. TargetP analysis revealed a large variance in the percentage of total transcripts encoding putative plastid-targeted proteins between the investigated species. The largest of these datasets belonged to *Malus × domestica* with 18.3% of its nuclear-encoded proteins predicted to be plastid-targeted, while the lowest was that of *Populus trichocarpa* with only 9.9% (Table 2). Header information for the predicted plastid-targeted datasets is provided in File S3.

Comparison of plastid-targeted proteomes with model systems

The predicted plastid proteomes for *Malus × domestica*, *Fragaria vesca*, *Populus trichocarpa*, *Prunus persica*, *Vitis vinifera*, and *Solanum lycopersicum* were independently compared with the *Arabidopsis* plastid proteome dataset as well as the entire *Arabidopsis* protein set using USEARCH. In *Arabidopsis*, the Chloroplast2010 project (www.plastid.msu.edu) [40] identified 5,181 unique genes encoding plastid-targeted proteins using software predictions and direct experimental evidence. Since this dataset did not represent embryo lethal mutants, an additional 201 genes predicted to encode embryo lethal plastid-targeted proteins from the SeedGenes database [55] were added to the dataset used

in this study for comparative analysis to bring it to a total of 5,382 sequences. Comparison at 40% identity and 40% coverage (40/40) reveals that about 50% of each predicted plastid proteome has a likely homolog in the *Arabidopsis* plastid-targeted proteome subset, while about 60–70% of the proteins have likely homologs in the entire *Arabidopsis* protein subset (Table 3). Further comparison with 50/50 clustering parameter lowers these estimates significantly.

Additional comparison of putative plastid-targeted protein sequences for all six species was performed against the plastid-targeted proteome of *Solanum lycopersicum*, another model system for plastid biology research. This analysis showed that smaller proportions of the plastid proteomes had homologs in the tomato plastid proteome (ranging from 32–48%) than they had in the Arabidopsis proteome (40–60%) (Table 4). Strawberry and apple had the lowest similarity with the predicted plastid-proteomes of both *Arabidopsis* and tomato, while peach had the highest.

Identification of unique plastid-targeted proteins

Two separate analyses were performed to identify the plastid-targeted proteins in each of the seven species examined. The first consisted of a USEARCH-based comparison of predicted plastid-targeted proteins against the plastid-targeted protein sequences from the other six species. A second comparison utilized a clustering technique with UCLUST [33]. In this analysis the plastid-targeted proteins from all species were clustered together and clusters containing singletons or sequences from a single species were identified and further analyzed. In both the USEARCH and UCLUST-based analyses, a significant proportion of each predicted plastid proteome was found to be unique to that species (Table 5). The proportion of uniquely targeted proteins ranges from 16.3% in *Prunus persica* to 41.5% in *Fragaria vesca* in the USEARCH method and 20.6% in *Prunus persica* to 46.8% in *Fragaria vesca* in the UCLUST method.

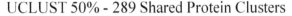

UCLUST 50% - 289 Shared Protein Clusters

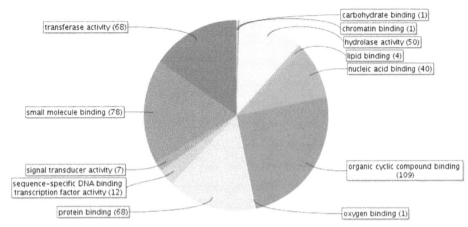

USEARCH 40/40 – 737 Shared Protein Sequences in *Arabidopsis thaliana*

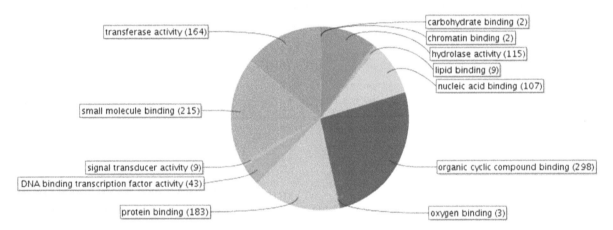

Figure 5. Molecular Function GO term composition of proteins shared between the 7 investigated plastid-targeted proteomes using two separate techniques. Two separate analyses were performed to identify proteins within the predicted plastid-targeted proteomes of seven species. The first, UCLUST with 50% identity, generated 15,750 clusters, 289 of which contained a member from all 7 species. USEARCH comparison performed at 40% identity and 40% coverage identified 737 sequences in the Arabidopsis thaliana putative plastid-targeted dataset which had a match in a protein sequence from the plastid-targeted sequences from all of the other 6 species. Sequences were analyzed via Blast2GO to determine the molecular functions of proteins at level 3.

However, the majority of the protein sequences have a homolog in at least one of the investigated species (Figure 3). In order to determine if uniquely-targeted proteins were in fact completely unique to each species, as opposed to alternatively targeted, a comparison with USEARCH 40/40 parameters was performed against datasets comprising the entire predicted proteomes of the other 6 species (Table 5). This investigation revealed thata significant number of proteins that lacked homology with interspecies plastid-targeted proteins, in fact have alternatively targeted homologs. This difference was most significant in *Arabidopsis* and *Malus × domestica* increasing the percentage of homology by 15.3% and 16.8%, respectively. Roughly 7–10% more proteins had homologs when using this matching scheme in the other investigated species. Information on these uniquely plastid-targeted proteins is provided in File S4.

GO term enrichment analysis was performed on the species-specific plastid-targeted protein sequences for both UCLUST and USEARCH-based analyses. Comparisons were performed using a Fisher's test with the entire predicted plastid proteome as a

reference. This analysis revealed significantly enriched GO terms for both comparative techniques. In the USEARCH based technique, the majority of enriched GO terms were present in the unique apple plastid-targeted sequences with the most significant GO terms were DNA metabolic process (GO:006259, p-value 8.30E-60) and cellular macromolecule metabolic process (GO:0044260) (Table 6). Additionally, the GO term nucleic acid binding (GO:0003676) was enriched in poplar as well as apple. No significant GO terms were found for grape, strawberry or tomato. More GO terms were found to be enriched in the UCLUST-based comparison (Table 7). In this analysis transcription regulator activity (GO:0030528) and transcription factor activity (GO:0003700) were enriched in the *Arabidopsis* and tomato datasets. DNA binding (GO:0003677) was enriched in apple, *Arabidopsis* and tomato. Cell death (GO:0008219) and death (GO:0016265) were also enriched in the apple UCLUST50 dataset.

While a significant amount of functional annotation was performed with Blast2GO, a large proportion of the proteins

Table 9. GO terms enriched in *Arabidopsis thaliana* members of the 289 plastid-targeted protein clusters shared between all species investigated.

GO term	Ontology	Description	Number in UCLUST50	Number in At Cp-targeted Proteome	p-value	FDR
GO:0015979	P	photosynthesis	96	362	1.30E-22	5.10E-20
GO:0005975	P	carbohydrate metabolic process	167	911	3.70E-19	7.40E-17
GO:0043412	P	macromolecule modification	138	718	2.10E-17	2.10E-15
GO:0006464	P	protein modification process	138	718	2.10E-17	2.10E-15
GO:0008152	P	metabolic process	419	3515	1.90E-16	1.50E-14
GO:0006091	P	generation of precursor metabolites and energy	108	515	2.40E-16	1.60E-14
GO:0044238	P	primary metabolic process	388	3152	4.60E-16	2.60E-14
GO:0009987	P	cellular process	436	3796	9.90E-15	4.90E-13
GO:0044267	P	cellular protein metabolic process	157	925	1.40E-14	6.70E-13
GO:0016265	P	death	54	203	7.10E-13	2.60E-11
GO:0008219	P	cell death	54	203	7.10E-13	2.60E-11
GO:0019748	P	secondary metabolic process	112	624	8.60E-12	2.80E-10
GO:0009628	P	response to abiotic stimulus	148	940	6.80E-11	2.10E-09
GO:0019538	P	protein metabolic process	178	1238	8.30E-10	2.40E-08
GO:0006629	P	lipid metabolic process	116	703	1.00E-09	2.70E-08
GO:0016043	P	cellular component organization	181	1274	1.50E-09	3.30E-08
GO:0006519	P	cellular amino acid and derivative metabolic process	130	824	1.50E-09	3.30E-08
GO:0044281	P	small molecule metabolic process	130	824	1.50E-09	3.30E-08
GO:0050896	P	response to stimulus	220	1657	3.90E-09	8.10E-08
GO:0009058	P	biosynthetic process	293	2394	4.40E-09	3.80E-08
GO:0044237	P	cellular metabolic process	316	2637	4.20E-09	3.80E-08
GO:0006950	P	response to stress	177	1274	1.60E-08	2.80E-07
GO:0023052	P	signaling	99	636	5.40E-07	9.40E-06
GO:0007165	P	signal transduction	86	534	8.80E-07	1.30E-05
GO:0050794	P	regulation of cellular process	86	534	8.80E-07	1.30E-05
GO:0009056	P	catabolic process	146	1056	8.80E-07	1.30E-05
GO:0065007	P	biological regulation	135	959	1.00E-06	1.50E-05
GO:0023046	P	signaling process	87	545	1.10E-06	1.50E-05
GO:0023060	P	signal transmission	87	545	1.10E-06	1.50E-05
GO:0050789	P	regulation of celbiological process	102	674	1.40E-06	1.80E-05
GO:0044260	P	cellular macromolecule metabolic process	194	1530	2.60E-06	3.30E-05
GO:0006810	P	transport	136	1010	1.10E-05	1.30E-04
GO:0051234	P	establishment of localization	136	1010	1.10E-05	1.30E-04
GO:0051179	P	localization	136	1010	1.10E-05	1.30E-04
GO:0009605	P	reponse to external stimulus	57	342	3.50E-05	4.00E-04
GO:0006807	P	nitrogen compound metabolic process	203	1686	4.70E-05	5.00E-04
GO:0006139	P	nucleobase, nucleoside, nucleotide and nucleic acid metabolic process	203	1686	4.70E-05	5.00E-04
GO:0009607	P	response to biotic stimulus	88	618	1.10E-04	1.10E-03
GO:0043170	P	macromolecule metabolic process	216	1842	1.20E-04	1.20E-03
GO:0009719	P	reponse to endogenous stimulus	69	484	7.40E-04	7.40E-03
GO:0003824	F	catalytic activity	291	2207	4.00E-13	4.40E-11
GO:0000166	F	nucleotide binding	139	934	1.80E-08	1.00E-06
GO:0016740	F	transferase activity	115	760	2.20E-07	4.80E-06
GO:0016301	F	kinase activity	56	286	1.80E-07	4.80E-06

Table 9. Cont.

GO term	Ontology	Description	Number in UCLUST50	Number in At Cp-targeted Proteome	p-value	FDR
GO:0016772	F	transferase activity, transferring phosphorus-containing groups	56	286	1.80E-07	4.80E-06
GO:0005488	F	binding	327	2878	9.30E-07	1.70E-05
GO:0016817	F	hydrolase activity, acting on acid anhydrides	11	31	1.30E-04	1.30E-03
GO:0016818	F	hydrolase activity, acting on acid anhydrides, in phosphorus-containing anhydrides	11	31	1.30E-04	1.30E-03
GO:0003774	F	motor activity	11	31	1.30E-04	1.30E-03
GO:0016462	F	pyrophosphatase activity	11	31	1.30E-04	1.30E-03
GO:0017111	F	nucleoside-triphosphatase activity	11	31	1.30E-04	1.30E-03
GO:0009579	C	thylakoid	111	524	3.50E-17	6.90E-15
GO:0009536	C	plastid	337	2749	1.10E-11	1.10E-09
GO:0016020	C	membrane	228	1712	1.10E-09	7.10E-08
GO:0044444	C	cytoplasmic part	394	3569	4.70E-08	2.30E-06
GO:0005737	C	cytoplasm	409	3823	1.10E-06	4.40E-05

GO terms from the 497 *Arabidopsis thaliana* proteins present within the 289 shared clusters were analyzed by agriGO to identify enriched GO terms. Chi-square test was performed with a p-value cutoff of 0.01.

predicted to be unique to each species lack any associated GO term (Table 8). Proteins predicted to be unique to the plastids of *Arabidopsis* appear to be the best characterized with 98% containing some form of GO information, followed by those of *Prunus persica* with 81.9%. *Solanum lycopersicum* displays the least amount of GO information with only 56.5% of these uniquely plastid-targeted proteins having associated GO terms.

Analysis of proteins conserved between all plastid-targeted proteomes

Two separate analyses were performed to identify proteins which were predicted to be targeted to the plastids of all seven angiosperms studied. First analysis workflow utilized a semi-global approach where UCLUST [33] was used to cluster the proteins of all predicted plastid proteomes at 50% identity. The second approach utilized a global approach where the plastid proteome of *Arabidopsis thaliana* was compared with every other species' predicted plastid proteome at 40% identity and 40% coverage. Those proteins which have a matched protein from all other six species were then determined to be conserved across the plastid proteomes.

The first analysis using UCLUST at 50% identity identified 289 clusters of proteins which had at least one member from all seven species. These 289 clusters contain 497 unique sequences from *Arabidopsis thaliana*, 773 from *Malus × domestica*, 384 from *Vitis vinifera*, 392 from *Fragaria vesca*, 545 from *Populus trichocarpa*, 439 from *Prunus persica*, and 478 from *Solanum lycopersicum*. Blast2GO analysis reveals that these proteins are involved in a large number of biological processes (Figure 4) with the most populous being cellular component organization (GO:0016043, 109 proteins), carbohydrate metabolic process (GO:0005975, 98 proteins), and response to stress (GO:0006950, 95 proteins). The electron transport chain-related molecular functions for this data set reveals that the majority of proteins fit into six main categories; organic cyclic compound binding (GO:0097159 109 proteins), small molecule binding (GO:0036094 78 proteins), transferase

activity (GO:0016740 68 proteins), protein binding (GO:0005515 68 proteins), hydrolase activity (GO:0016787 50 proteins), and nucleic acid binding (GO:0003676 40 proteins) (Figure 5). GO term enrichment was performed by selecting a single *Arabidopsis thaliana* protein sequence and utilizing agriGO to compare with those GO terms from the entire *Arabidopsis thaliana* predicted plastid-targeted proteome. This dataset contains 56 GO terms which were enriched with a p-value cut-off of 0.01 (Table 9). The lowest p-values are associated with the GO terms photosynthesis (GO:0015979), carbohydrate metabolic process (GO:0005975), macromolecule modification (GO:0043412), protein modification process (GO:0006464) and thylakoid (GO:0009579) respectively.

Analysis with USEARCH 40/40 identified the presence of 737 unique protein sequences from Arabidopsis thaliana with a matching protein in the predicted plastid-targeted proteomes of *Solanum lycopersicum, Prunus persica, Vitis vinifera, Malus × domestica, Fragaria vesca*, and *Populus trichocarpa*. As in the UCLUST50 analysis, the top three biological processes in this dataset as determined by Blast2GO are cellular component organization (GO:0016043, 254 proteins), response to stress (GO:0006950, 251 proteins), and carbohydrate metabolic process (GO:0005975, 210 proteins) (Figure 4). Again, the most populous molecular function GO terms mirror those in the UCLUST 50% with the majority falling into the six categories of: organic cyclic compound binding (GO:0097159 298 proteins), small molecule binding (GO:0036094 215 proteins), transferase activity (GO:0016740 164 proteins), protein binding (GO:0005515 183 proteins), hydrolase activity (GO:0016787 115 proteins), and nucleic acid binding (GO:0003676 107 proteins) (Figure 5). GO term enrichment using agriGO identified 59 GO terms to be enriched with a p-value cut-off of 0.01 (Table 10). The lowest p-values are associated with the GO terms photosynthesis (GO:0015979), thylakoid (GO:0009579), macromolecule modification (GO:0043412), protein modification process (GO:0006464), and generation of precursor metabolites and energy respectively (GO:0006091).

Table 10. GO terms enriched in *Arabidopsis thaliana* members of the 737 proteins with a match in all 6 species as determined by USEARCH4040.

GO Term	Ontology	Description	Number in USEARCH 4040	Number in At Cp-targeted Proteome	p-value	FDR
GO:0015979	P	photosynthesis	127	362	8.90E-22	3.50E-19
GO:0043412	P	macromolecule modification	190	718	7.70E-18	1.00E-15
GO:0006464	P	protein modification process	190	718	7.70E-18	1.00E-15
GO:0006091	P	generation of precursor metabolites and energy	146	515	4.70E-16	3.70E-14
GO:0009987	P	cellular process	634	3796	4.50E-16	3.70E-14
GO:0008152	P	metabolic process	598	3513	1.80E-15	1.20E-13
GO:0044238	P	primary metabolic process	551	3152	4.30E-15	2.40E-13
GO:0044267	P	cellular protein metabolic process	218	925	7.60E-15	3.80E-13
GO:0005975	P	carbohydrate metabolic process	210	911	3.00E-13	1.30E-11
GO:0044237	P	cellular metabolic process	471	2637	1.50E-12	5.90E-11
GO:0009628	P	response to abiotic stimulus	212	940	1.90E-12	6.90E-11
GO:0006519	P	cellular amino acid and derivative metabolic process	191	824	3.60E-12	1.10E-10
GO:0044281	P	small molecule metabolic process	191	824	3.60E-12	1.10E-10
GO:0050896	P	response to stimulus	316	1657	6.00E-10	1.70E-08
GO:0019538	P	protein metabolic process	249	1238	1.40E-09	3.70E-08
GO:0016043	P	cellular component organization	254	1274	2.00E-09	4.60E-08
GO:0065007	P	biological regulation	203	959	2.00E-09	4.60E-08
GO:0006950	P	response to stress	251	1274	8.20E-09	1.80E-07
GO:0044260	P	cellular macromolecule metabolic process	288	1530	2.70E-08	5.70E-07
GO:0016265	P	death	61	203	6.30E-08	1.20E-06
GO:0008219	P	cell death	61	203	6.30E-08	1.20E-06
GO:0019748	P	secondary metabolic process	137	624	3.00E-07	5.40E-06
GO:0009058	P	biosynthetic process	407	2394	5.50E-07	9.50E-06
GO:0048856	P	anatomical structure development	155	749	1.40E-06	2.40E-05
GO:0065008	P	regulation of biological quality	84	344	2.00E-06	3.10E-05
GO:0006629	P	lipid metabolic process	146	703	2.70E-06	4.10E-05
GO:0043170	P	macromolecule metabolic process	318	1842	1.40E-05	2.00E-04
GO:0050789	P	regulation of biological process	134	674	6.70E-05	9.50E-04
GO:0009607	P	response to biotic stimulus	124	618	9.40E-05	1.30E-03
GO:0009653	P	anatomical structure morphogenesis	117	584	1.70E-04	2.30E-03
GO:0022414	P	reproductive process	86	404	2.50E-04	3.10E-03
GO:0007165	P	signal transduction	107	534	3.60E-04	4.30E-03
GO:0050794	P	regulation of cellular process	107	534	3.60E-04	4.30E-03
GO:0023046	P	signaling process	108	545	5.00E-04	5.60E-03
GO:0023060	P	signal transmission	108	545	5.00E-04	5.60E-03
GO:0006810	P	transport	181	1010	6.20E-04	6.50E-03
GO:0051234	P	establishment of localization	181	1010	6.20E-04	6.50E-03
GO:0051179	P	localization	181	1010	6.20E-04	6.50E-03
GO:0090066	P	regulation of anatomical structure size	46	191	8.60E-04	8.20E-03
GO:0016049	P	cell growth	46	191	8.60E-04	8.20E-03
GO:0008361	P	regulation of cell size	46	191	8.60E-04	8.20E-03
GO:0032535	P	regulation of cellular component size	46	191	8.60E-04	8.20E-03
GO:0023052	P	signaling	121	636	9.70E-04	9.00E-03
GO:0042592	P	homeostatic process	40	161	1.10E-03	9.90E-03
GO:0019725	P	cellular homeostasis	40	161	1.10E-03	9.90E-03
GO:0016301	F	kinase activity	95	286	1.10E-14	6.10E-13

Table 10. Cont.

GO Term	Ontology	Description	Number in USEARCH 4040	Number in At Cp-targeted Proteome	p-value	FDR
GO:0016772	F	transferase activity, transferring phosphorus-containing groups	95	286	1.10E-14	6.10E-13
GO:0000166	F	nucleotide binding	215	934	1.50E-13	5.60E-12
GO:0003824	F	catalytic activity	406	2207	1.10E-11	3.10E-10
GO:0005488	F	binding	497	2878	4.20E-11	9.30E-10
GO:0016740	F	transferase activity	164	760	3.90E-08	7.20E-07
GO:0016818	F	hydrolase activity, acting on acid anhydrides, in phosphorus-containing anhydrides	17	31	2.70E-06	2.70E-05
GO:0016817	F	hydrolase activity, acting on acid anhydrides	17	31	2.70E-06	2.70E-05
GO:0003774	F	motor activity	17	31	2.70E-06	2.70E-05
GO:0016462	F	pyrophosphatase activity	17	31	2.70E-06	2.70E-05
GO:0017111	F	nucleoside-triphosphatase activity	17	31	2.70E-06	2.70E-05
GO:0009579	C	thylakoid	154	524	1.90E-18	3.70E-16
GO:0016020	C	membrane	321	1712	2.70E-09	2.60E-07
GO:0009536	C	plastid	460	2749	1.50E-07	1.00E-05

GO terms from one *Arabidopsis thaliana* proteins from each of the 737 shared protein sequences were analyzed by agriGO to identify enriched GO terms. Chi-square test was performed with a p-value cutoff of 0.01.

Comparing the USEARCH40/40 dataset along with that of the UCLUST50 dataset reveals that the two methods agree upon 439 shared plastid-targeted sequences, with 58 present only in UCLUST50 and 298 in USEARCH40/40. Both datasets have similar enriched GO terms with 49 enriched compared to the entire *Arabidopsis thaliana* plastid-targeted proteomes. However, USEARCH40/40 enriched GO terms contain an additional 10 biological process not represented in the UCLUST50 analysis; anatomical structure development (GO:0009653), regulation of biological quality (GO:0065008), anatomical structure morphogenesis (GO:0009653), reproductive process (GO:0022414), cell growth (GO:0016049), regulation of anatomical structure size (GO:0090066), regulation of cell size (GO:0008361), regulation of cellular component size (GO:0032535), homeostatic process (GO:0042592), and cellular homeostasis (GO:0019725).

The UCLUST50 enriched GO terms include five biological processes [catabolic process (GO:0009056), response to external stimulus (GO:0009719), nitrogen compound metabolic process (GO:0006807), nucleobase, nucleoside, nucleotide and nucleic acid metabolic process (GO:0006139), and response to endogenous stimulus (GO:0009719)] and two cellular components [cytoplasmic part (GO:0044444) and cytoplasm (GO:0005737)] not represented in the USEARCH40/40 analysis. A complete list of the 478 *Arabidopsis thaliana* loci from the UCLUST50 dataset and the 737 protein sequences from the USEARCH40/40 dataset along with their respective GO terms are provided in File S5.

Proteins conserved in this study were further compared with those from GreenCut2. GreenCut2 represents a collection of 597 nuclear-encoded proteins determined to be conserved across 20 photosynthetic eukaryotes, but absent in non-photosynthetic organisms [56]. A total of 677 unique loci from *Arabidopsis thaliana* (33 redundant of the original set of 710) were compared with those identified as shared between the seven species examined in this study both with USEARCH 40/40, as well as UCLUST

50% using Venny [57]. This comparison identified 70 proteins present in all three datasets, and a substantial set unique to each dataset. Sequence information from this comparison is provided in File S6.

In order to look at the overlap of homologous proteins between each species, the Jaccard similarity coefficient was determined for the USEARCH40/40 (Table 11). This comparison displays that the plastid-targeted proteomes of apple, strawberry, Arabidopsis and poplar are most similar to that of Peach, while the predicted plastid-targeted proteomes of peach, tomato and grape are most similar to *Arabidopsis*. An additional modified Jaccard similarity coefficient matrix was generated with the UCLUST50 analysis. In this matrix the predicted plastid-targeted of all species are most similar to that of peach, while that of peach is most similar to apple.

Discussion

This study reveals several interesting aspects about the constitution of predicted plastid-targeted proteomes for the species analyzed. A large portion of a plant's nuclear genome is dedicated to plastid-targeted proteins, many of which lack identity to plastid-targeted proteins in other species. Some plastid-targeted proteins have identity with potentially alternatively-targeted proteins in other systems. Of the predicted plastid-targeted proteins in *Arabidopsis*, 737 have significant identity to predicted plastid-targeted proteins in each of six investigated species, suggesting an evolutionarily core conserved set of plastid-targeted proteins. The caveat is that TargetP accuracy is not well defined, and has been shown to differ between experiments [58]. van Wijk and Baginsky determined that TargetP has a 35% false positive rate, suggesting that the predicted datasets established here will be greatly reduced upon future confirmatory experiments. Experimental approaches to characterize the proteome of plastids or some of their

Table 11. Jaccard's similarity coefficient matrix for seven species based on predicted plastid-targeted proteomes.

USEARCH 40/40

	Strawberry	Arabidopsis	Apple	Peach	Tomato	Poplar	Grape
Strawberry	1.000	0.409	0.365	0.449	0.369	0.322	0.335
Arabidopsis	0.409	1.000	0.391	0.466	0.447	0.378	0.397
Apple	0.365	0.391	1.000	0.423	0.342	0.310	0.303
Peach	0.449	0.466	0.423	1.000	0.427	0.389	0.393
Tomato	0.369	0.447	0.342	0.427	1.000	0.352	0.365
Poplar	0.322	0.378	0.310	0.389	0.352	1.000	0.330
Grape	0.335	0.397	0.303	0.393	0.365	0.330	1.000

UCLUST50

	Strawberry	Arabidopsis	Apple	Peach	Tomato	Poplar	Grape
Strawberry	1.000	0.194	0.199	0.267	0.171	0.158	0.173
Arabidopsis	0.194	1.000	0.192	0.253	0.222	0.193	0.210
Apple	0.199	0.192	1.000	0.268	0.162	0.156	0.164
Peach	0.267	0.253	0.268	1.000	0.229	0.224	0.240
Tomato	0.171	0.222	0.162	0.229	1.000	0.176	0.198
Poplar	0.158	0.193	0.156	0.224	0.176	1.000	0.184
Grape	0.173	0.210	0.164	0.240	0.198	0.184	1.000

Similarity between each species' plastid-targeted proteomes was represented by calculating the Jaccard coefficient from the UCLUST50 comparison and a USEARCH 4040 analysis.

constituents have relied largely upon mass spectrometry techniques [59,60,61,62]. A study in 2004 focused on the isolation and classification of the constituents of the *Arabidopsis thaliana* chloroplast proteome resulting in the identification of 604 nuclear-encoded proteins [60]. However, TargetP (at that time) was only able to correctly predict the plastid localization of 62.3% of these proteins, with 6.1% predicted to target the mitochondria, 8.1% secreted, and 23.5% predicted to have "any other location". When excluding the envelope proteins, TargetP chloroplast localization accuracy increased to 67.2%. An additional study which identified 241 stromal proteins from *Arabidopsis thaliana* chloroplasts identified through MALDITOF MS and nano-LC-ESI-MS/MS had a much higher predictability to be chloroplast targeted by TargetP with 88% accuracy [61]. Yet another study of 916 nuclear-encoded *Arabidopsis* plastid-targeted proteins revealed that 86% were correctly predicted using TargetP [63]. Such studies indicate that localization prediction methods need to be improved. Some reasons for this could include the complexities of the experimental system, sequence data, dual targeting, splice variants, presence of lesser characterized transport systems, or simply a lack of understanding of mechanisms of localization. As genomes become better characterized and targeting prediction improves, our ability to better understand the commonalities and diversities of plastid compositions and functions will also likely improve.

While it may be expected that plastid-targeted proteins would be highly conserved in this 7-species analysis, a previous study demonstrated a striking lack of similarity between the plastid proteomes of *Arabidopsis* and *Oryza sativa* [20]. In this study, a predicted *Arabidopsis* chloroplast proteome containing 2,100 proteins shared only 900 with the 4,800 plastid proteins in *Oryza sativa*. These 900 proteins were largely involved in transcription, energy, and metabolism. It would not be surprising to see this shared set of proteins shrink substantially as the number of species compared increases. A large focus has been put on the identification of proteins found only in photosynthetic organisms termed the GreenCut [64]. The first draft of this protein set contained 349 proteins conserved in photosynthetic eukaryotes, and absent in non-photosynthetic organisms. This was later updated generating GreenCut2 with 597 conserved proteins [56]. The comparison of the conserved plastid-targeted protein identified in this study by UCLUST50 and USEARCH40/40 methods reveals that there is only a minor overlap with GreenCut2 (Figure 6). Additionally, clusters do not consist of members from either a single species, or all seven. Instead, there are a significant number of clusters in our analysis with members from various combinations of species. These clusters could be helpful in the identification of proteins involved in plastidial pathways or traits conserved within a set of species. The comparison of more closely related angiosperms may in fact yield more similar plastid proteomes, while including more distantly related angiosperms likely reduces this similarity. It is worth noting that these differences could potentially occur due to the loss or gain of chloroplast transit peptides, rearrangement of protein domains or gene duplication, to name a few plausible mechanisms.

Differences in the outcome of comparative analysis projects can be attributed to the utilization of alignment or clustering methods. Previous studies in microbial comparative proteomic and genomic studies have utilized an alignments of 50% identity and 50% coverage to predictively resolve paralogs from orthologs [65,66] and 90% nucleic acid identity in algae (Bayer et al., 2012). A previous study which compared plastid proteomes between *Arabidopsis* and rice utilized BLASTP with e-value cut-off of $10xe^{-10}$ [20] while that of GreenCut utilized a BLASTP mutual

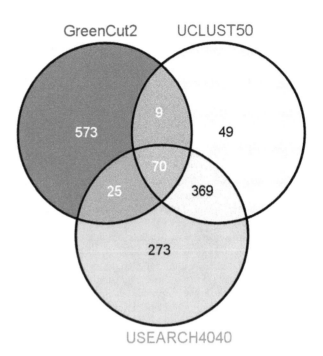

Figure 6. Comparison of conserved plastid-targeted protein datasets with GreenCut2. *Arabidopsis thaliana* loci associated with conserved proteins within the plastid-targeted proteomes identified in this study were compared with those of GreenCut2. A total of 70 sequences were found to be conserved between the three datasets.

best hit between *Chalydomonas* and Arabidopsis and human to identify paralogs, co-orthologs, and orthologs of *Chlamydomonas* [64]. GreenCut2 further expanded upon these parameters again utilizing mutual best hit analysis with BLASTP cut-off of $10xe^{-10}$ to identify orthologs and included sequences with over 50% amino acid identity as in-paralogs [56]. For this study, a predicted plastid-targeted protein was considered "unique" to a species if a global USEARCH alignment at 40% identity over 40% of the query sequence matched no sequences from the other 6 species. These parameters were chosen instead of 50% identity and 50% coverage as more matches were identified with a higher confidence with paralogous proteins removed. UCLUST, based on of global alignments, was then utilized to identify clusters with 50% amino acid identity, presumably including both orthologs and in-paralogs [33]. Of course the functionality of a protein cannot be ascertained through sequence identity alone. This study may not identify genes that have undergone rearrangement yet retain similar gene product functions, or examples of convergent evolution. The assumption in this study is that if such arrangements occur and functions are retained, or divergent plants adapt to create the same function in separate gene sequences, that an appropriate sequence from one of the remaining six other species would retain or have similar sequence identity for a 40/40 alignment to occur. However, the datasets that we present could be further investigated in the future through the use of a BLASTP-based comparison, and with other genomes as they become available.

The unique plastid-targeted proteins within each of the investigated species possess varied levels of functional information. Despite the lack of spatial and temporal transcriptome and proteome expression context, this information has a large referential value for future work. Blast2GO analysis and GO term enrichment analysis provide a glimpse as to what these proteins are likely participating in within the plastids. However, only a few GO terms were found to be enriched, most of which differed between species. This suggests that there are likely not specific classes of proteins which fall into this category of "unique" for the plastid proteomes of each species. As expected, due to substantial research performed on it, a large proportion of the proteins unique to the Arabidopsis plastid proteome (98%) possess associated GO terms (Table 8). However, tomato, which is used in the scientific community to understand and characterize the chloroplast to chromoplast transition, has the lowest percentage of unique proteins without GO terms, at 56.5%. Our analysis reinforces the lack of understanding about plastid biology especially in non-model systems, and the urgent need for further functional characterization of novel biological processes that these organelles harbor.

Plastids play an integral part in plant development, photosynthesis, and several other known biochemical processes. However in fruits their role has remained uncharacterized. Several important biochemical processes for synthesis and storage of pigments, nutraceutical and medically important compounds as well as aromatic compounds are resident in fruit plastids. These components are important both for consumer appeal as well as nutritional value. In addition, various physiological disorders, such as sunscald in apple, are associated with the inability of fruits to adequately quench excessive energy from sunlight. Through furthering our understanding of the plastid function in non-model plant systems and organs such as fruit, novel mechanisms for enhancing photosynthetic efficiency and crop productivity could be discovered.

The results presented in this work indicate that the current state of knowledge regarding plastid biology, mostly derived from model systems, is not comprehensive enough. In each plant species evaluated in this work, plastids are predicted to host a plethora of biological and metabolic processes necessitating subsequent wet-lab validation in non-model systems. New plant genomes are expected to enable the identification of potentially new plastid-targeted proteins that will aid in studying novel roles of plastids in plant development, metabolism and adaptation.

Conclusions

While previous studies have advocated the integration of multiple protein localization prediction techniques [20] it appears that no significant difference exists between a custom analysis utilizing multiple approaches and TargetP analysis for identifying plastid-targeted proteins in apple. Such results suggest that initial data mining with only TargetP may, in fact, be sufficient, depending upon the application. This is with the caveat that TargetP has an approximate 35% false positive rate in detecting plastid-targeted proteins [58]. As the understanding of protein localization improves, and complete genome sequences from larger number of plants become available, such predictive techniques will likely become a more reliable method to generate draft plastid proteomes.

The TargetP-based analysis indicates that a large subset of a plant's nuclear-encoded proteome is predicted to be localized to the plastid. However, the proportion of transcripts encoding plastid-targeted proteins varies, compromising 10–20% of the transcriptome depending upon the species investigated. Of the nuclear-encoded plastid proteome, there appears to be a significant subset that is species-specific. Many of these proteins have homology to proteins not predicted to be plastid-targeted in other systems, indicating that it may be common for proteins to gain or lose targeting peptides during evolution. If this is indeed the case, it would be interesting to investigate the evolutionary and mechanistic context of gain or loss of target peptides across species.

Through using two comparative methods, a USEARCH-based approach as well as a semi-global UCLUST-based approach, we displayed that very few plastid-targeted proteins are conserved between the predicted nuclear-encoded plastid proteomes. This value varied based upon the comparative technique and user parameters, but our predictions identify 497 and 737 Arabidopsis proteins which contain predicted plastid-targeted homologs in all other examined angiosperms. GO term enrichment analysis suggests that specific functions are significantly conserved in plastids, namely photosynthesis, many metabolic processes, transport, and cell death. Knowledge about these conserved proteins can be utilized in future studies to better understand and potentially predict those proteins which are plastid-targeted in other non-model systems and additionally identify novel plastid-targeted proteins.

The expression of genes encoding plastid-proteins appears to be very diverse within the fruit developmental continuum with 64 significant expression patterns detected (those containing five or more genes). However, most of the genes investigated can be clustered into nine expression patterns. These expression patterns can be overlapped with important milestones within the development of fruit to find plastid proteins which may be responsible for novel fruiting milestones or processes. While expression data is available for ~13,000 genes, subsequent developmental transcriptomics, metabolomics, and proteomics investigations are expected to provide a comprehensive understanding of the roles of plastids in apple fruit development.

Supporting Information

File S1 Expression data from Janssen et al. for *Malus × domestica* genes predicted to encode plastid-targeted proteins. Expression data for each cluster of co-expressed genes with sequence header and expression data in the form of Log2 of expression relative to the lowest value.

File S2 GO term information for *Malus × domestica* plastid-targeted protein clusters.

File S3 Header information for predicted plastid-targeted proteins from the investigated seven species. Excel file containing identifiers of predicted plastid-targeted protein sequences generated through with TargetP analysis.

File S4 Plastid-targeted proteins unique to each examined species with GO term information. Excel file containing the header information and associated GO terms.

File S5 Plastid-targeted proteins shared between all species. Excel file containing the header information for the UCLUST50 analysis separated by species. Information is provided for the 737 *Arabidopsis thaliana* sequences shared in the USEARCH comparative analysis.

File S6 Blast2GO annotation file containing header and GO term information for 70 protein sequences present in GreenCut2, UCLUST50, and USEARCH4040 analyses.

Acknowledgments

The authors would like to thank Ms. Vandhana Krishnan for her guidance in performing clustering analysis.

Author Contributions

Conceived and designed the experiments: SS AD. Performed the experiments: SS AH RR. Analyzed the data: SS AH RR PJ AD. Contributed reagents/materials/analysis tools: PJ AD. Wrote the paper: SS PJ AD.

References

1. Kutschera U, Niklas K (2005) Endosymbiosis, cell evolution, and speciation. Theory Biosci 124: 1–24.
2. Mettal U, Boland W, Beyer P, Kleinig H (1988) Biosynthesis of monoterpene hydrocarbons by isolated chromoplasts from daffodil flowers. Eur J Biochem 170: 613–616.
3. Kiss JZ, Hertel R, Sack FD (1989) Amyloplasts are necessary for full gravitropic sensitivity in roots of Arabidopsis thaliana. Planta 177: 198–206.
4. Fischer K, Weber A (2002) Transport of carbon in non-green plastids. Trends Plant Sci 7: 345–351.
5. Martin W, Herrmann RG (1998) Gene transfer from organelles to the nucleus: how much, what happens, and Why? Plant Physiol 118: 9–17.
6. Rochaix JD (1997) Chloroplast reverse genetics: new insights into the function of plastid genes. Trends Plant Sci 2: 419–425.
7. Martin W, Rujan T, Richly E, Hansen A, Cornelsen S, et al. (2002) Evolutionary analysis of Arabidopsis, cyanobacterial, and chloroplast genomes reveals plastid phylogeny and thousands of cyanobacterial genes in the nucleus. Proc Natl Acad Sci U S A 99: 12246–12251.
8. Jarvis P (2008) Targeting of nucleus-encoded proteins to chloroplasts in plants. New Phytol 179: 257–285.
9. Keegstra K, Olsen LJ, Theg SM (1989) Chloroplastic precursors and their transport across the envelope membranes. Annu Rev Plant Physiol Plant Mol Biol 40: 471–501.
10. Vonheijne G, Steppuhn J, Herrmann RG (1989) Domain-structure of mitochondrial and chloroplast targeting peptides. Eur J Biochem 180: 535–545.
11. Zhang XP, Glaser E (2002) Interaction of plant mitochondrial and chloroplast signal peptides with the Hsp70 molecular chaperone. Trends Plant Sci 7: 14–21.
12. Chen MH, Huang LF, Li HM, Chen YR, Yu SM (2004) Signal peptide-dependent targeting of a rice a-amylase and cargo proteins to plastids and extracellular compartments of plant cells. Plant Physiol 135: 1367–1377.
13. Miras S, Salvi D, Ferro M, Grunwald D, Garin J, et al. (2002) Non-canonical transit peptide for import into the chloroplast. J Biol Chem 277: 47770–47778.
14. Schemenewitz A, Pollmann S, Reinbothe C, Reinbothe S (2007) A substrate-independent, 14: 3: 3 protein-mediated plastid import pathway of NADPH: protochlorophyllide oxidoreductase A. Proc Natl Acad Sci U S A 104: 8538–8543.
15. Schein AI, Kissinger JC, Ungar LH (2001) Chloroplast transit peptide prediction: a peek inside the black box. Nucleic Acids Res 29.
16. Bannai H, Tamada Y, Maruyama O, Nakai K, Miyano S (2002) Extensive feature detection of N-terminal protein sorting signals. Bioinformatics 18: 298–305.
17. Emanuelsson O, Brunak S, von Heijne G, Nielsen H (2007) Locating proteins in the cell using TargetP, SignalP and related tools. Nat Protoc 2: 953–971.
18. Emanuelsson O, Nielsen H, Brunak S, von Heijne G (2000) Predicting subcellular localization of proteins based on their N-terminal amino acid sequence. J Mol Biol 300: 1005–1016.
19. Small I, Peeters N, Legeai F, Lurin C (2004) Predotar: A tool for rapidly screening proteomes for N-terminal targeting sequences. Proteomics 4: 1581–1590.
20. Richly E, Leister D (2004) An improved prediction of chloroplast proteins reveals diversities and commonalities in the chloroplast proteomes of Arabidopsis and rice. Gene 329: 11–16.
21. Verde I, Abbott A, Scalabrin S, Jung S, Shu S, et al. (2013) The high-quality draft genome of peach (Prunus persica) identifies unique patterns of genetic diversity, domestication and genome evolution. Nat Genet 45: 487–494.
22. Jaillon O, Aury JM, Noel B, Policriti A, Clepet C, et al. (2007) The grapevine genome sequence suggests ancestral hexaploidization in major angiosperm phyla. Nature 449: 463–U465.
23. Sato S, Tabata S, Hirakawa H, Asamizu E, Shirasawa K, et al. (2012) The tomato genome sequence provides insights into fleshy fruit evolution. Nature 485: 635–641.
24. Shulaev V, Sargent DJ, Crowhurst RN, Mockler TC, Folkerts O, et al. (2011) The genome of woodland strawberry (Fragaria vesca). Nat Genet 43: 109–U151.
25. Tuskan GA, DiFazio S, Jansson S, Bohlmann J, Grigoriev I, et al. (2006) The genome of black cottonwood, Populus trichocarpa (Torr. & Gray). Science 313: 1596–1604.
26. Velasco R, Zharkikh A, Affourtit J, Dhingra A, Cestaro A, et al. (2010) The genome of the domesticated apple (Malus × domestica Borkh.). Nat Genet 42: 833–+.
27. Kreuz K, Kleinig H (1984) Synthesis of prenyl lipids in cells of spinach leaf – compartmentation of enzymes for formation of isopentenyl diphosphate. Eur J Biochem 141: 531–535.
28. Camara B, Hugueney P, Bouvier F, Kuntz M, Moneger R (1995) Biochemistry and molecular biology of chromoplast development. Int Rev Cytol 163: 175–247.
29. Boyer J, Liu R (2004) Apple phytochemicals and their health benefits. Nutr J 3.
30. Liu YS, Roof S, Ye ZB, Barry C, van Tuinen A, et al. (2004) Manipulation of light signal transduction as a means of modifying fruit nutritional quality in tomato. Proc Natl Acad Sci U S A 101: 9897–9902.
31. Robinson N, Hewitt J, Bennett A (1988) Sink metabolism in tomato fruit. Plant Physiol 87.
32. Phan CT (1973) Chloroplasts of the peel and the internal tissues of apple-fruits. Experientia 29: 1555–1557.
33. Edgar RC (2010) Search and clustering orders of magnitude faster than BLAST. Bioinformatics 26: 2460–2461.
34. Hunter S, Apweiler R, Attwood TK, Bairoch A, Bateman A, et al. (2009) InterPro: the integrative protein signature database. Nucleic Acids Res 37: D211–D215.
35. Krogh A, Larsson B, von Heijne G, Sonnhammer ELL (2001) Predicting transmembrane protein topology with a hidden Markov model: Application to complete genomes. J Mol Biol 305: 567–580.
36. Ostlund G, Schmitt T, Forslund K, Kostler T, Messina DN, et al. (2010) InParanoid 7: new algorithms and tools for eukaryotic orthology analysis. Nucleic Acids Res 38: D196–D203.
37. Youens-Clark K, Buckler E, Casstevens T, Chen C, DeClerck G, et al. (2011) Gramene database in 2010: updates and extensions. Nucleic Acids Res 39: D1085–D1094.
38. Flicek P, Amode MR, Barrell D, Beal K, Brent S, et al. (2011) Ensembl 2011. Nucleic Acids Res 39: D800–D806.
39. Kinsella RJ, Kahari A, Haider S, Zamora J, Proctor G, et al. (2011) Ensembl BioMarts: a hub for data retrieval across taxonomic space. Database.
40. Lu Y, Savage LJ, Larson MD, Wilkerson CG, Last RL (2011) Chloroplast 2010: a database for large-scale phenotypic screening of Arabidopsis mutants. Plant Physiol 155: 1589–1600.
41. Lamesch P, Berardini TZ, Li DH, Swarbreck D, Wilks C, et al. (2012) The Arabidopsis Information Resource (TAIR): improved gene annotation and new tools. Nucleic Acids Res 40: D1202–D1210.
42. Conesa A, Gotz S, Garcia-Gomez JM, Terol J, Talon M, et al. (2005) Blast2GO: a universal tool for annotation, visualization and analysis in functional genomics research. Bioinformatics 21: 3674–3676.
43. Gotz S, Garcia-Gomez JM, Terol J, Williams TD, Nagaraj SH, et al. (2008) High-throughput functional annotation and data mining with the Blast2GO suite. Nucleic Acids Res 36: 3420–3435.
44. Quevillon E, Silventoinen V, Pillai S, Harte N, Mulder N, et al. (2005) InterProScan: protein domains identifier. Nucleic Acids Res 33: W116–W120.
45. Myhre S, Tveit H, Mollestad T, Laegreid A (2006) Additional Gene Ontology structure for improved biological reasoning. Bioinformatics 22: 2020–2027.
46. Kanehisa M (2002) The KEGG database. In silico simulation of biological processes 247: 91–103.
47. Kanehisa M, Goto S, Sato Y, Furumichi M, Tanabe M (2012) KEGG for integration and interpretation of large-scale molecular data sets. Nucleic Acids Res 40: D109–D114.
48. Du Z, Zhou X, Ling Y, Zhang Z, Su Z (2010) agriGO: a GO analysis toolkit for the agricultural community. Nucleic Acids Res 38: W64–70.
49. Janssen BJ, Thodey K, Schaffer RJ, Alba R, Balakrishnan L, et al. (2008) Global gene expression analysis of apple fruit development from the floral bud to ripe fruit. BMC Plant Biol 8.
50. Saeed AI, Hagabati NK, Braisted JC, Liang W, Sharov V, et al. (2006) TM4 microarray software suite. DNA Microarrays, Part B: Databases and Statistics 411: 134–+.
51. Saeed AI, Sharov V, White J, Li J, Liang W, et al. (2003) TM4: A free, open-source system for microarray data management and analysis. Biotechniques 34: 374–+.
52. Ben-Dor A, Shamir R, Yakhini Z (1999) Clustering gene expression patterns. J Comput Biol 6: 281–297.
53. Kaul S, Koo HL, Jenkins J, Rizzo M, Rooney T, et al. (2000) Analysis of the genome sequence of the flowering plant Arabidopsis thaliana. Nature 408: 796–815.
54. Clijsters H (1969) On the photosynthetic activity of developing apple fruits. Qualitas Plantarum et Materiae Vegetabiles 19: 129–140.
55. Meinke D, Muralla R, Sweeney C, Dickerman A (2008) Identifying essential genes in Arabidopsis thaliana. Trends Plant Sci 13: 483–491.
56. Karpowicz SJ, Prochnik SE, Grossman AR, Merchant SS (2011) The GreenCut2 resource, a phylogenomically derived inventory of proteins specific to the plant lineage. J Biol Chem 286: 21427–21439.

57. Oliveros JC (2007) VENNY. An interactive tool for comparing lists with Venn Diagrams.

58. van Wijk KJ, Baginsky S (2011) Plastid proteomics in higher plants: current state and future goals. Plant Physiol 155: 1578–1588.

59. Barsan C, Sanchez-Bel P, Rombaldi C, Egea I, Rossignol M, et al. (2010) Characteristics of the tomato chromoplast revealed by proteomic analysis. J Exp Bot 61: 2413–2431.

60. Kleffmann T, Russenberger D, von Zychlinski A, Christopher W, Sjolander K, et al. (2004) The *Arabidopsis thaliana* chloroplast proteome reveals pathway abundance and novel protein functions. Curr Biol 14: 354–362.

61. Peltier JB, Cai Y, Sun Q, Zabrouskov V, Giacomelli L, et al. (2006) The oligomeric stromal proteome of *Arabidopsis thaliana* chloroplasts. Mol Cell Proteomics 5: 114–133.

62. Wang YQ, Yang Y, Fei Z, Yuan H, Fish T, et al. (2013) Proteomic analysis of chromoplasts from six crop species reveals insights into chromoplast function and development. J Exp Bot.

63. Zybailov B, Rutschow H, Friso G, Rudella A, Emanuelsson O, et al. (2008) Sorting signals, N-terminal modifications and abundance of the chloroplast proteome. PLoS One 3.

64. Merchant SS, Prochnik SE, Vallon O, Harris EH, Karpowicz SJ, et al. (2007) The Chlamydomonas genome reveals the evolution of key animal and plant functions. Science 318: 245–251.

65. Friis C, Wassenaar TM, Javed MA, Snipen L, Lagesen K, et al. (2010) Genomic Characterization of *Campylobacter jejuni* Strain M1. PLoS One 5.

66. Lukjancenko O, Ussery DW, Wassenaar TM (2012) Comparative Genomics of Bifidobacterium, Lactobacillus and Related Probiotic Genera. Microb Ecol 63: 651–673.

7

De novo Assembly of the Grass Carp *Ctenopharyngodon idella* Transcriptome to Identify miRNA Targets Associated with Motile Aeromonad Septicemia

Xiaoyan Xu[1◐], Yubang Shen[1◐], Jianjun Fu[1], Liqun Lu[3], Jiale Li[1,2]*

1 Key Laboratory of Exploration and Utilization of Aquatic Genetic Resources, Shanghai Ocean University, Ministry of Education, Shanghai 201306, PR China, **2** E-Institute of Shanghai Universities, Shanghai Ocean University, 999 Huchenghuan Road, 201306 Shanghai, PR China, **3** National Pathogen Collection Center for Aquatic Animals, College of Fisheries and Life Science, Shanghai Ocean University, 999 Huchenghuan Road, 201306 Shanghai, PR China

Abstract

Background: *De novo* transcriptome sequencing is a robust method of predicting miRNA target genes, especially for organisms without reference genomes. Differentially expressed miRNAs had been identified previously in kidney samples collected from susceptible and resistant grass carp (*Ctenopharyngodon idella*) affected by *Aeromonas hydrophila*. Target identification for these differentially expressed miRNAs poses a major challenge in this non-model organism.

Results: Two cDNA libraries constructed from mRNAs of susceptible and resistant *C. idella* were sequenced by Illumina Hiseq 2000 technology. A total of more than 100 million reads were generated and *de novo* assembled into 199,593 transcripts which were further extensively annotated by comparing their sequences to different protein databases. Biochemical pathways were predicted from these transcript sequences. A BLASTx analysis against a non-redundant protein database revealed that 61,373 unigenes coded for 28,311 annotated proteins. Two cDNA libraries from susceptible and resistant samples showed that 721 unigenes were expressed at significantly different levels; 475 were significantly up-regulated and 246 were significantly down-regulated in the SG samples compared to the RG samples. The computational prediction of miRNA targets from these differentially expressed genes identified 188 unigenes as the targets of 5 conserved and 4 putative novel miRNA families.

Conclusion: This study demonstrates the feasibility of identifying miRNA targets by transcriptome analysis. The transcriptome assembly data represent a substantial increase in the genomic resources available for *C. idella* and will provide insights into the gene expression profile analysis and the miRNA function annotations in further studies.

Editor: Daniel Doucet, Natural Resources Canada, Canada

Funding: This work was supported by grants from National Key Technology R&D Program of China (2012BAD26B02), the China's Agricultural Research System (CARS-46-04), the Agricultural Seed Development Program of Shanghai City (2012NY10), Shanghai Ocean University Doctoral Research Foundation, Shanghai Universities First-class Disciplines Project of Fisheries and the Innovation Plan of Shanghai Graduate Education. The funders had no role in study design, data collection and analysis, decision to publish, or preparation of the manuscript.

Competing Interests: The authors have declared that no competing interests exist.

* Email: jlli2009@126.com

◐ These authors contributed equally to this work.

Background

Next-generation sequencing (NGS) -based RNA sequencing for transcriptome methods (RNA-seq) allow simultaneous acquisition of sequences for gene discovery as well as identification of transcripts involved in specific biological processes. This is especially suitable for non-model organisms whose genomic sequences are unknown [1,2]. In addition, the dynamic range, sensitivity and specificity of RNA-seq also make it ideal for quantitatively analyzing various aspects of gene regulation [3]. These techniques do not require prior knowledge of genomic sequence and are much advanced in terms of time, cost, labor, amount of data produced, data coverage, sensitivity, and accuracy compared to traditional sequencing methods [4,5].

The grass carp (*Ctenopharyngodon idella*) is one of the most important farmed fish species in China, with a cultural history dating back to the 7th century CE (Tang Dynasty) [6]. According to the FAO, the value of farmed *C. idella* reached more than 6.46 billion USD for a production of 5.03 billion tons in 2012, thus accounting for the highest production and third highest value of major cultured fish species worldwide at single species level [7]. Despite favorable growth traits, farmed *C. idella* are rather susceptible to various disease. Outbreaks of disease associated with bacteria such as *Aeromonas hydrophila* have caused high mortality, resulting in reduced production and considerable economic losses [8].

A. *hydrophila* is a causative agent of a wide spectrum of diseases in humans and animals [9]. While originally thought to be an opportunistic pathogen in immunocompromised humans, an increasing number of intestinal and extraintestinal disease cases suggest that it is an emerging human pathogen irrespective of the immune status of the host [10]. The pathogenesis, pathogenic mechanism, and virulence factors responsible for selected A. *hydrophila* infections in different species are not well understood [11]. A. *hydrophila* is a Gram-negative motile bacillus widely distributed in aquatic environments. It causes motile aeromonad septicemia (MAS), which results in great economic losses in worldwide freshwater fish farming [12]. Thus, more effective measures against A. *hydrophila* infection in fish are needed. Identification of differentially expressed genes (DGEs) following A. *hydrophila* infection is important for an improved understanding of fish MAS.

MicroRNAs (miRNAs) are 20–22 nt non-coding RNAs that play important roles in post-transcriptional gene regulation. In animal cells, miRNAs regulate their targets by translational inhibition and mRNA destabilization [13]. MicroRNAs (miRNAs) are key effectors in mediating host-pathogen interactions and constitute a family of small RNA species; they are considered a promising candidate for regulating the interaction between host and pathogen [14,15]. Therefore, dissecting the biological functions of miRNAs may help us understand the pathogenic mechanism of motile aeromonad septicemia in C. *idella*. Many studies have identified miRNAs and mRNA transcriptome in fish species, like common carp [16,17], nile tilapia [18,19] rainbow trout [20,21] channel catfish [22,23] and silver carp [24,25]. To thoroughly interpret the biological functions of these miRNAs, a first step is to predict their targets. Therefore, establishing a more powerful transcriptome data for target identification is preferred.

Although two parallel C. *idella* expressed sequence tag analyses have already been conducted using head kidney tissue [26,27], the data presented here represent the first effort to analyze the transcriptome of C. *idella* affected by A. *hydrophila*. Two cDNA libraries from SG and RG C. *idella* used for our miRNA analysis were constructed and sequenced with Illumina Hiseq 2000. The obtained reads were assembled into transcripts and annotated by BLAST analysis against various databases before screening the results for differentially expressed genes and the prediction of miRNA targets. Our work will provide an approach to identify the target genes of miRNAs and to characterize their functional/regulatory network to increase our understanding of hemorrhagic septicemia outbreaks in C. *idella*.

Materials and Methods

Ethics statement

All handling of fishes was conducted in accordance with the guidelines on the care and use of animals for scientific purposes set up by the Institutional Animal Care and Use Committee (IACUC) of the Shanghai Ocean University, Shanghai, China. The IACUC has specially approved this study within the project "Breeding of Grass Carp" (approval number is SHOU-09-007).

Sample collection

C. *idella* with an average weight of 50 g were cultured individually at the Wujiang National Farm of Chinese Four Family Carps (Jiangsu Province, China). Animals were raised at 28°C in 400-L aerated tanks for one week before the experiment and fed twice daily (in the morning and late in the afternoon) at a ratio of 5% of total biomass. Two groups (30 animals in each group) were maintained in two aquariums and intraperitoneally injected with A. *hydrophila* AH10 (Aquatic Pathogen Collection Center of Ministry of Agriculture, China) at a dose of 7.0×10^6 cells suspended in 100 μl PBS per fish. All fish were observed every 4 h for any mortality and samples were taken until the termination of the experiment at 240 h post-challenge. C. *idella* died in the first 72 h post-challenge were classified as susceptible group (SG) for their high sensitivity to A. *hydrophila*, while the animals that survived over 240 h post-challenge were considered as resistant group (RG) [28]. Spleen and kidney samples were immediately snap-frozen in liquid nitrogen and stored at −80°C until further use.

cDNA library construction and sequencing

Experimental protocols for the cDNA sequence were performed according to the manufacturer's technical instructions. The spleen and kidney tissues of randomly-selected three fish from both the susceptible and resistant groups were collected and, labeled as SG and RG, respectively. The RNA from the same tissue of three fishes of SG and RG C. *idella* was pooled with equal quantity for the construction of SG and RG cDNA libraries. The pooled total RNA was isolated from each spleen and kidney samples with TRIZOL reagent (Invitrogen, Grand Island, NY, USA). RNA integrity was confirmed using a 2100 Bioanalyzer (Agilent Technologies, Inc.) by RNA 6000 nano with a minimum RNA integrity number (RIN) value of 7.0. Poly (A) mRNA was purified from the total RNA using oligo (dT) magnetic beads. Equal amounts of the high-quality mRNA samples were obtained from each group for cDNA library preparation using the NEBNext Ultra RNA Library Prep Kit for Illumina (New England Biolabs, Ipswich, MA, USA) and purified using Agencount AMPure XP beads (Beckman Coulter, Krefeld, Germany) according to the manufacturer's recommendations. The concentration of the cDNA library was determined on an Agilent Technologies 2100 Bioanalyzer by Agilent DNA-1000. Libraries were sequenced, at the Novogene Bioinformatics Institute (Beijing, China) on an Illumina HiSeq 2000 instrument (Illumina, San Diego CA, USA) that generated paired-end reads of 101 nucleotides.

Data processing, assembly, and functional annotation

Raw reads generated by Illumina Hiseq 2000 were then cleaned by removing the adaptor containing sequences, any ambiguous base>10% reads and low quality reads (1/2 reads with Q-value≤ 5) to get clean reads. Then, all clean reads were assembled using the *de novo* assembly program Trinity [29]: first, short reads were assembled into high-coverage contigs that could not be extended farther in either direction in a k-mer-based approach for fast and efficient transcript assembly. Then, the related contigs were clustered and a de Bruijn graph for each cluster was constructed. Finally, in the context of the corresponding de Bruijn graph and all plausible transcripts, alternatively spliced isoforms and transcripts were derived.

All assembled transcripts were compared with publicly available databases including Nr (NCBI non-redundant protein sequences), Nt (NCBI non-redundant nucleotide sequences) [30], KOG/COG (Clusters of Orthologous Groups of proteins) [31], Swiss-Prot (a manually annotated and reviewed protein sequence database) [32], KO (KEGG ortholog database) [33], Pfam (protein family) [34] and GO (Gene Ontology) (http://www.geneontology.org/). Nr, Nt, KOG/COG, Swiss-Prot, and KO used the BLASTx analysis with a cut-off E-value of 10^{-5}, Pfam used Hmmerscan and GO used Blast2GO [35]. The best Blast hits from all Blast results were parsed for a homology-based functional

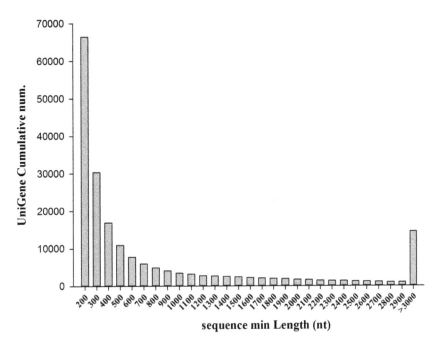

Figure 1. Length distribution of assembled unigenes in the sequenced cDNA library.

annotation. For the nr annotations, the Blast2GO program was used to obtain GO annotations of unique assembled transcripts to describe biological processes, molecular functions, and cellular components.

Differentially expressed genes between the SG/RG libraries

High-quality reads were mapped to reference sequences (unigenes from the transcriptome data of the cDNA library) using RSEM [36]. Gene expression levels were calculated using the fragments per kilo bases per million mapped reads (FPKM) method [37]. The calculation of unigene expression levels and the identification of unigenes that were differentially expressed between the libraries were performed by DEGseq [38] based on TMM normalized counts. The settings "q.value <0.005" and "|log2.Fold change.normalized|>2" were used as thresholds for judging significant differences in transcript expression. Differen-

tially expressed genes across the samples were further annotated by GO and KEGG pathway analysis

MiRNA target prediction

The SG and RG kidney miRNA-seq analysis were conducted in the same biological samples as mRNA-seq. Small RNA libraries were constructed using a Small RNA Cloning Kit (Takara). RNA was purified by polyacrylamide gel electrophoresis (PAGE) to enrich for the molecules in the range of 17–27 nt, then was ligated with 5' and 3' adapters. The resulting samples were used as templates for cDNA synthesis followed by PCR amplification. The obtained sequencing libraries were subjected to Solexa sequencing-by-synthesis method. After the run, image analysis, sequencing quality evaluation and data production summarization were performed with Illumina/Solexa pipeline. The sequencing data was pretreated to discard low quality reads, no 3'-adaptor reads, 5'-adaptor contaminants and sequences shorter than 18 nucleo-

Table 1. Summaries of sequencing cDNA library.

Sample name	SG	RG
Total reads	73,063,654	62,737,669
Clean reads	70,210,307	60,668,815
Total mapped to unigenes readcounts	61,396,052.97	51,785,549.97
Reads length (bp)	101	
GC content (%)	48.43	47.13
Number of unigenes	199,554	
Total length of unigenes (bp)	195,075,872	
Mean length of unigenes (bp)	977	
N50 of unigenes (bp)	2,117	
Maximal length of unigenes (bp)	27,185	

Table 2. Statistics of the annotation results for the C. idella unigenes.

	All	Nr	Nt	Pfam	KOG	Swiss-Prot	KO	GO
Number of unigenes	61373	28311	46653	33013	16980	22674	27775	34207
% of unigenes	100	46.1	76.0	53.8	27.7	36.9	45.3	55.7

Nr: NCBI non-redundant protein sequences, Nt: NCBI non-redundant nucleotide sequences, Pfam: Protein family, KOG: Clusters of Orthologous Groups of proteins, Swiss-Prot: A manually annotated and reviewed protein sequence database, KO: KEGG Ortholog database and GO: Gene Ontology.

tides. After trimming the 3′ adaptor sequence, sequence tags were mapped onto the transcriptome of *C. idella* using bowtie. Any small RNAs having exact matches to transcriptome of *C. idella* were used from further analysis. The mapped reads were compared to the miRBase (19.0) to annotate conserved miRNAs. To predict novel miRNAs, the miREvo [39] and mirdeep2 [40] were used.

Computational identification of differentially expressed miRNA targets was performed using the miRanda toolbox [41], using the complementary region between miRNAs and mRNAs and the thermodynamic stability of the miRNA-mRNA duplex. All mRNAs used for target prediction came from the differentially expressed unigenes obtained as described above. The miRanda toolbox employed a dynamic programming algorithm to search the complementary regions between the miRNA and the 3′-UTR of the mRNA, and the scores were based on sequence complementary as well as minimum free energy of RNA duplexes, and were calculated with the Vienna RNA package [42]. All detected targets with scores and energies less than the threshold parameters of S>90 (single-residue pair scores) and $\Delta G < -17$ kcal/mol (minimum free energy) were selected as potential targets.

Real time PCR validation

The sequencing results were validated by real time PCR using One Step PrimeScript miRNA cDNA Synthesis Kit for miRNA (TaKaRa) reversely transcribed, PrimeScript RT reagent Kit with gDNA Eraser (TaKaRa) for mRNA and SYBR *Premix Ex Taq* II (2x) (Takara) for qPCR according to the manufacturer protocols. Specific primer assays for *miR-21* [F: 5′-TAGCTTATCA-GACTGGTGTTGGC-3′, R: Uni-miR qPCR Primer (TaRaKa)], *JNK1* (F: 5′- TGGTCAGAGGTAGTGTGTTG-3′, R: 5′-AGTTTGTTGTGGTCCGAGTC-3′) and *ccr7* (F: 5′-CAAGCCAAGAACTTTGAGAGG-3′, R: 5′-GGCA-TAAAGGCGAATGTTGTC-3′) were purchased from sangon biotech and real time PCR quantification was carried out in CFX96 Real-Time PCR System (Bio-Rad, CA, USA). To normalize the expression values, *miR-22a* for miRNA and *18s* for mRNA were used as housekeeping control [43,44]. Expression levels were quantitatively analyzed using the $2^{-\Delta\Delta CT}$ method. One-way ANOVA tests were performed using SPSS 17.0 to determine significant differences. Each experiment was repeated in triplicates.

Results and Discussion

Antagonistic bacteria, such as *A. hydrophila*, enhance non-specifically immune-related enzyme activities and disease resistance in *C. idella* and provide a theoretical basis for disease prevention in aquaculture. However, the molecular mechanisms of this disease are still far from fully understood. The identification and characterization of candidate genes involved in MAS would represent the first step in understanding the genetic basis of this process in *C. idella*.

De novo assemblies and unigenes annotation

The sequencing generated 135.80 million raw reads. After trimming, 130.88 million clean reads remained, corresponding to 13.22 GB clean bases. The dataset of each sample, SG and RG, was represented by over 60 million clean reads, a read density sufficient for the subsequent quantitative analysis of genes. The raw sequencing reads have been submitted to the NCBI Short Read Archive under the accession number of SRR1124206 and SRR1125014. Then, all clean reads were assembled using a *de*

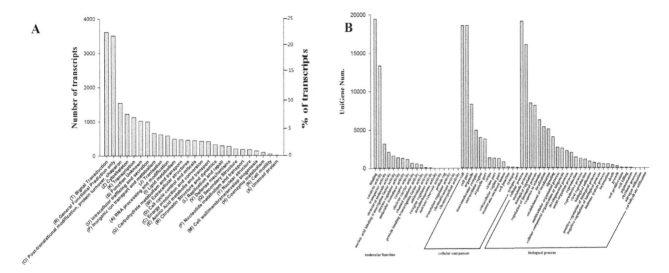

Figure 2. Functional annotations of the unigenes of _C. idella._ (A) KOG annotation. (B) Level 2 GO term distribution for the biological process, cellular component and molecular function categories.

novo assembly program Trinity [45]. These short reads were further assembled into 199,554 transcripts with an average length of 977 bp (Table 1). The size distribution of these transcripts ranged from 201 to 27,185 bp, of which 27,334 were larger than 2,000 bp (Figure 1). The assembled transcriptome data were deposited in NCBI's Transcriptome Shotgun Assembly (TSA) database under the accession numbers from SUB583458.

Annotation of predicted proteins

A total of 61,373 distinct sequences (30.75% of the transcripts) matched known genes corresponding to 28,311 of the annotated proteins (Table 2, Table S1). An additional functional annotation of the unigenes of _C. idella_ was performed searching for putative orthologs and paralogs within the KOG database [31]. A total of 16980 unigenes (27.7%) were assigned to 26 eukaryotic orthologous groups (Figure 2A). The category "signal transduction", which contained 3,611 unigenes (21.27% of 16980 unigenes), was the largest, followed by the categories "general functional prediction only" (3506, 20.65%), "post-translational modification, protein turnover, chaperone" (1539, 9.06%) and "cytoskeleton" (1226, 7.22%).

GO annotation and KEGG pathway analyses

After GO annotation, _C. idella_ transcripts could be assigned to three categories: biological processes, molecular functions and cellular components. Within the various biological processes, cellular processes (19,121 unigenes) metabolic processes (16,091) and biological regulation (8,463) were the most highly represented members (Figure 2B). Important functions, such as cell death (389) and immune system processes (540), were also identified in this category. Similarly, cell (18,586) as well as cell part (18,586) and binding (19,460) were the most represented sub-categories in the cellular component and molecular function categories, respectively.

Searching against the Kyoto Encyclopedia of Genes and Genomes Pathway database (KEGG) [33] revealed that 10,561 unigenes could be matched to 298 KEGG pathways. The most-represented pathways hierarchy 2 were the "infectious diseases" pathway (3210 unigenes) and the "signal transduction" pathway (2316) (Table 3). Some pathways related to immune system were also identified, such as the "Toll-like receptor signaling" pathway (110) [46] and the "chemokine signaling" pathway (224) [47] (Table 4).

Table 3. Top 10 list of the gene number of Pathway Hierarchy 2.

Pathway Hierarchy 2	Unigene Number
Infectious Diseases	3210
Signal Transduction	2316
Cancers	2235
Nervous System	1648
Immune System	1554
Neurodegenerative Diseases	932
Digestive System	875
Endocrine System	861
Cell Communication	859
Signaling Molecules and Interaction	763

Table 4. Top 10 list of pathways related to immune system.

Pathway Hierarchy 2	KEGG Pathway	Unigene Numbers
Immune System	Chemokine signaling pathway	224
Immune System	Leukocyte transendothelial migration	199
Immune System	T cell receptor signaling pathway	156
Immune System	Fc gamma R-mediated phagocytosis	155
Immune System	Natural killer cell mediated cytotoxicity	129
Immune System	B cell receptor signaling pathway	110
Immune System	Toll-like receptor signaling pathway	110
Immune System	Antigen processing and presentation	86
Immune System	Fc epsilon RI signaling pathway	84
Immune System	RIG-I-like receptor signaling pathway	67

Digital gene expression library sequencing

Based on the transcriptome sequence data, two DGE libraries were constructed to identify the differentially expressed unigenes between the SG and RG samples. After removing low-quality reads, 70,210,307 and 60,668,815 clean reads were generated from the SG and RG libraries, respectively (Table 1). Among these clean reads, 61,396,052.97 of the SG and 51,785,549.97 of the RG readcounts were mapped to unigenes.

Differential gene expression between the SG and RG libraries

The results suggest that the expression of 721 genes differed significantly between the SG and RG groups of *C. idella*. Of these genes, 475 were up-regulated and 246 were down-regulated in the SG samples compared to the RG samples (Figure 3 and Table S2). GO enrichment analysis of DEGs indicated that these genes were significantly enriched in oxidation-reduction processes (biological process), integral to membrane (cellular components), and protein binding (molecular function) (Table S3). Pathway enrichment analysis found the DEGs to be mainly enriched in complement and coagulation cascades, *Staphylococcus aureus* infection and porphyrin and chlorophy II metabolism (Table S4). Notably, several genes involved in the immune and inflammatory response were also identified, such as C-type lectin [48] and matrix metalloproteinase-9 [49].

MiRNA target prediction

The identification of miRNAs and their targets is important for understanding the physiological functions of miRNAs and the functional roles of differentially expressed miRNAs between healthy and diseased fish. We were thus interested in predicting miRNA target genes involved in the immune response or immune system, according to the KEGG analysis. In a previous study, small RNA deep-sequencing data were aligned with miRBase 18.0 to search for known miRNAs with complete matches, namely,

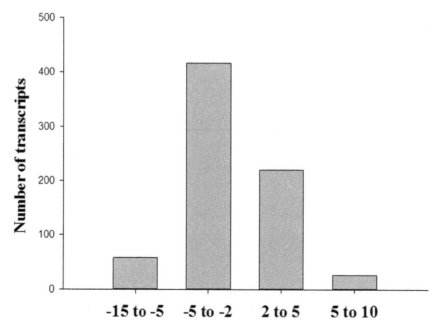

Figure 3. Number and Fold change distribution of differentially expressed genes between the SG/RG libraries.

Table 5. Differentially expressed miRNAs in *C. idella* kidney between SG and RG.

miRNA	Sequence (5′-3′)	SG (TMP)	RG (TMP)	log2 (Fold_change) normalized	p-value
let-7i	UGAGGUAGUAGUUUGUGCUGUU	3419.67473	1475.401006	1.212751983	4.62E-174
miR-142a-3p	UGUAGUGUUUCCUACUUUAUGGA	4770.81118	9798.344345	−1.038303405	0
miR-21	UAGCUUAUCAGACUGGUGUUGGC	3431.54242	1640.502276	1.064719594	4.12E-142
miR-217	UACUGCAUCAGGAACUGAUUGG	240.9140969	2983.647091	−3.630486183	0
miR-223	UGUCAGUUUGUCAAAUACCCC	1460.912578	6800.683353	−2.218809871	0
novel_115	UGAAGGCCGAAGUGGAGA	3.560306851	17.95531055	−2.334337113	0.001253505
novel_131	UGCCCGCAUUCUCCACCA	7.713998177	40.28996513	−2.384869847	9.85E-07
novel_154	CCCAGCCAUAUUUGUUUGAAC	16.02138083	0	4.768565529	4.44E-05
novel_3	UGUUUCUGGCUCUGAUAUUUGCU	32.04276166	71.82124218	−1.164412111	8.07E-05

conserved miRNAs (data unpublished). Meanwhile, miRNAs predicted by miRDeep 2.0 that could form stable secondary structures, were identified as novel miRNAs.

We have used a single algorithm, miRanda [50] to predict miRNA targets. As miRNA binding to target 3′ UTR generally results in mRNA destabilization and degradation [51], we chose to

Table 6. 26 pathways were related to immune and diseases in all pathways.

KEGG Pathway	Pathway Hierarchy 2	Target gene of different expression miRNA
Toll-like receptor signaling pathway	Immune system	JNK, TLR5, TBK1
Hepatitis C	Infectious diseases: Viral	JNK, TBK1, LDLR
Salmonella infection	Infectious diseases: Bacterial	JNK, TLR5
Fc epsilon RI signaling pathway	Immune system	JNK, FYN
Tuberculosis	Infectious diseases: Bacterial	JNK, PLK3, CTSS
Influenza A	Infectious diseases: Viral	JNK, DNAJB1, TBK1
Measles	Infectious diseases: Viral	FYN, TBK1
MAPK signaling pathway	Signal transduction	JNK, MKP
Toxoplasmosis	Infectious diseases: Parasitic	JNK, LDLR
Chemokine signaling pathway	Immune system	CCR7, ADCY7, DOCK2
HTLV-I infection	Infectious diseases: Viral	JNK, ADCY7, SLC2A1
Herpes simplex infection	Infectious diseases: Viral	JNK, TBK1
Fc gamma R-mediated phagocytosis	Immune system	DOCK2
Prion diseases	Neurodegenerative diseases	FYN
Viral myocarditis	Cardiovascular diseases	FYN
Pathways in cancer	Cancers: Overview	JNK, SLC2A1
Transcriptional misregulation in cancer	Cancers: Overview	CCR7
Type II diabetes mellitus	Endocrine and metabolic diseases	JNK
Cytokine-cytokine receptor interaction	Signaling molecules and interaction	CCR7, TNFSF12
Antigen processing and presentation	Immune system	CTSS
Pertussis	Infectious diseases: Bacterial	JNK
Natural killer cell mediated cytotoxicity	Immune system	FYN
T cell receptor signaling pathway	Immune system	FYN
Chagas disease	Infectious diseases: Parasitic	JNK
Staphylococcus aureus infection	Infectious diseases: Bacterial	CFB
Complement and coagulation cascades	Immune system	CFB

List of gene abbreviations: JNK: c-Jun N-terminal kinase, TLR5: toll-like receptor 5, TBK1: TANK-binding kinase 1, LDLR: low-density lipoprotein receptor, FYN: tyrosine-protein kinase Fyn, CTSS: cathepsin S, PLK3: polo-like kinase 3, DNAJB1: DnaJ homolog subfamily B member 1, MKP: dual specificity MAP kinase phosphatase, DOCK2: dedicator of cytokinesis protein 2, ADCY7: adenylate cyclase 7, SLC2A1: MFS transporter, SP family, solute carrier family 2 (facilitated glucose transporter), member 1, CCR7: C-C chemokine receptor type 7, TNFSF12: tumor necrosis factor ligand superfamily member 12, CFB: component factor B.

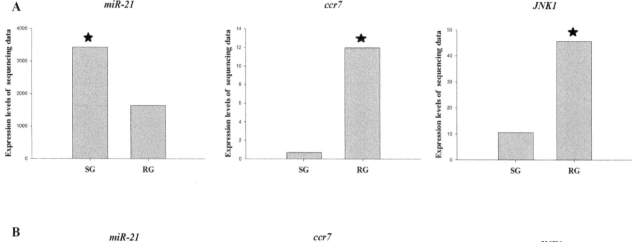

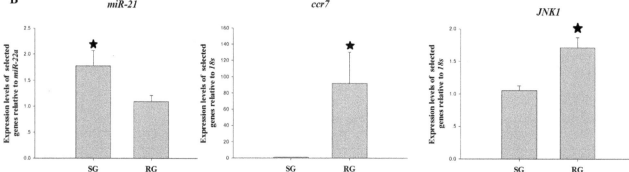

Figure 4. The expression analysis of selected genes from the expression profile by relative quantitative real-time PCR. A Transcriptome sequencing data, B Real-time PCR data. Increases and decreases in relative levels of transcripts with respect to the control 18s for mRNA and *miR-22a* for miRNA are shown. The settings "q.value <0.005" and "|log2.Fold change.normalized|>2" were used as thresholds for judging significant differences in transcript expression. One-way ANOVA tests were performed using SPSS 17.0 to determine significant differences for Real-time PCR data. Statistical significance of the relative expression ratio is indicated *.

narrow down potential targets to those showing differential expression in the opposite direction as the mRNA. This approach increases the strength to discover true target genes and functions affected by miRNA dysregulation. In total, 188 of the target genes predicted by miRanda were differentially expressed in the opposite direction in the target tissue. The identification of 188 unigenes (Table S5) as the predicted target genes of 5 conserved and 4 putative novel miRNA families (Table 5). The identified target genes involved in biological processes, molecular functions, and cellular components were defined using GO annotations. GO analysis demonstrated that these targets were involved in a broad range of physiological processes, including gene expression, transcription regulation, immune system processes, and responses to stress or stimuli (Table S6).

Searching against the KEGG indicated that 188 unigenes mapped to 48 KEGG pathways. 26 pathways were related to immune and diseases in all pathways (Table 6). These included the categories "Toll-like receptor signaling pathway", "Fc epsilon RI signaling pathway", "Chemokine signaling pathway", "Fc gamma R-mediated phagocytosis", "Antigen processing and presentation", "Natural killer cell mediated cytotoxicity", "T cell receptor signaling pathway" and "Complement and coagulation cascades" related immune functions. Toll-like receptor signaling pathway induce the expression of a variety of host defense genes. These include chemokine signaling pathway and other effectors necessary to arm the host cell against the invading pathogen [52]. Cytokine-cytokine receptor interaction play a pivotal role in the generation

of immunological responses during bacterial infection [53]. This implicated functions that are likely regulated by miRNAs and suggests regulation of different pathways during immune activation in susceptible and resistant *C. idella*.

Of particular interest to our study is the fact that several of the most highly expressed miRNAs in SG have been shown to have a role in immunity. let-7i and associated TLR4 expression are involved in cholangiocyte immune responses against *C. parvum* infection [54]. miR-21 targets multiple genes associated with the immunologically localized disease [55]. *miR-21* which is up-regulated in SG is predicted to target 28 differentially expressed genes (Table S7) in *C. idella*. Frequently represented in the top immune functions were protein kinase JNK1 (*JNK1*) and chemokine (C-C motif) receptor 7 (*ccr7*) (Table 6). *JNK1* and *ccr7* showed clearly down-regulated expression profiles in the SG samples compared to the RG samples (Figure 4A   http://www.plosone.org/article/info%3Adoi%2F10.1371%2Fjournal.pone.0073506-pone.0073506.s003   ). The decreased expression profile of these targets in the susceptible samples supported our previous finding that the expression of *miR-21* was significantly up-regulated, with TMP of 3,432 and 1,642 in the SG and RG groups, respectively (Table 5). We validated the *miR-21*, *JNK1* and *ccr7* which had expression change in the sequencing data by performing real time PCR in the same samples used for the sequencing (Figure 4B).

JNK1 is involved in apoptosis, neurodegeneration, cell differentiation and proliferation, inflammatory conditions and cytokine production mediated by AP-1 (activation protein 1), such as

Regulated upon activation normal T cell expressed and presumably secreted (*RANTES*), Interleukin 8 (*IL-8*), and Granulocyte-macrophage colony-stimulating factor (*GM-CSF*) [56]. It has been reported that JNK plays an important role in the innate immune response to microbial challenge [57,58]. Most importantly, JNK1 serves as a negative regulatory factor for MAP kinase phosphatase 5 (*MKP5*) that plays an essential role in innate immune responses [59]. The chemokine receptor CCR7 acts as an important organizer of the primary immune response [60]. A previous study demonstrated a discrete CCR7 requirement in the activation of different T cell subsets during bacterial infection [61]. CCR7 is differentially regulated by macrophages in exposure to bacteria, as it is triggered by exposure to both Gram-negative and Gram-positive bacteria [62].

The discovery of microRNAs dramatically changed our perspective on eukaryotic gene expression regulation [63]. MicroRNAs play important gene-regulatory roles in animals and plants by pairing to the mRNAs of protein-coding genes to direct their posttranscriptional repression [64,65]. The identification of miRNA target genes is an important step in understanding their role in gene regulatory networks. Most miRNA-associated computational methods comprise the prediction of miRNA genes and their targets, and an increasing number of computational algorithms and web-based resources are being developed to fulfill the needs of scientists performing miRNA research, like miRanda [42], TargetScan [66], RNAhybrid [67] and PicTar [68]. However, animal miRNA targets are difficult to predict since miRNA: mRNA duplexes often contain several mismatches, gaps, and G:U base pairs in many positions, thus limiting the maximum length of contiguous sequences of matched nucleotides [69]. The predicted interactions using these computational methods are inconsistent and the expected false positive rates are still high. Recently, several authors suggested integrating expression profiles from both miRNA and mRNA with *in silico* target predictions to reduce the number of false positives and increase the number of biologically relevant targets [70]. These methods have been shown to be effective in identifying the most prominent interactions from the databases of putative targets [71]. To minimize false positive rates in our study, the RNA-seq and miRNA-seq analysis were conducted in the same biological samples. Likewise, to reduce the number of putative target genes, the miRNA targets were predicted from differentially expressed genes. However, some false positive predictions proved inevitable. Further studies will focus on the experimental validation of the differentially expressed mRNA and miRNAs identified in this study. They are likely to be central regulators of the innate immune response to *A. hydrophila* and thus represent potential therapeutic targets or novel biomarkers of infection and inflammation.

Conclusions

In this study, we used high-throughput sequencing data to characterize the transcriptome of *C. idella*, a species for which little genomic data are available. Further, DGE tags were mapped to the assembled transcriptome for further gene expression analysis. A large number of candidate genes involved in MAS were identified. This represents a fully characterized transcriptome, and provides a valuable resource for genetic and genomic studies in *C. idella*. Additionally, DGE profiling provides new leads for functionally studies of genes involved in MAS.

Finally, comparison with our previous miRNA profiling, this study strongly indicates that miRNA is a critical factor in determining mRNA abundance and regulation during MAS. Our on-going effort using experimental approach such as knock-down or over-express candidate miRNAs and mRNAs in vitro is expected to provide new evidence in understanding these regulatory mechanisms of MAS in *C. idella*.

Supporting Information

Table S1   Summary of unigene annotation against Nr, Nt, Pfam, KOG, Swiss-Prot, KO and GO database.

Table S2   721 genes differed significantly between the SG and RG groups of *C. idella*.

Table S3   GO enrichment analysis of 721 genes differed significantly.

Table S4   Pathway enrichment analysis of 721 genes differed significantly.

Table S5   188 genes as the predicted target genes of 9 different expression miRNA between SG and RG groups of *C. idella*.

Table S6   GO enrichment analysis of 188 genes as the predicted target genes of different expression miRNA.

Table S7   28 target gene information of *miR-21*.

Acknowledgments

We thank Zhiwei Wang and Liang Zhang (Novogene Bioinformatics Technology Co. Ltd) for their help in sequencing and data analysis.

Author Contributions

Conceived and designed the experiments: JL LL. Performed the experiments: XX YS. Analyzed the data: XX. Contributed reagents/materials/analysis tools: XX YS JF. Wrote the paper: XX JL.

References

1. Huang da W, Sherman BT, Lempicki RA (2009) Bioinformatics enrichment tools: paths toward the comprehensive functional analysis of large gene lists. Nucleic Acids Res 37: 1–13.

2. Wang Y, Pan Y, Liu Z, Zhu X, Zhai L, et al. (2013) De novo transcriptome sequencing of radish (Raphanus sativus L.) and analysis of major genes involved in glucosinolate metabolism. BMC Genomics 14: 836.

3. Graveley BR (2008) Molecular biology: power sequencing. Nature 453: 1197–1198.

4. Ozsolak F, Kapranov P, Foissac S, Kim SW, Fishilevich E, et al. (2010) Comprehensive polyadenylation site maps in yeast and human reveal pervasive alternative polyadenylation. Cell 143: 1018–1029.

5. Bhardwaj J, Chauhan R, Swarnkar MK, Chahota RK, Singh AK, et al. (2013) Comprehensive transcriptomic study on horse gram (Macrotyloma uniflorum): De novo assembly, functional characterization and comparative analysis in relation to drought stress. BMC Genomics 14: 647.

6. Renkui C (1991) Development History of Freshwater Culture in China. China Press of Science & Technology, Beijing, China.

7. FAO (2014) FAO Yearbook of Fishery and Aquaculture Statistics Summary tables. Rome: FAO.

8. Huang Q, Tang S, Zhang J (1983) Ichthyopathology. Shanghai, China: Shanghai Press of Science & Technology.

9. Igbinosa I, Igumbor E, Aghdasi F, Tom M, Okoh A (2012) Emerging Aeromonas species infections and their significance in public health. The Scientific World Journal 2012: 625023.

10. Figueras MJ (2005) Clinical relevance of Aeromonas sM503. Reviews in Medical Microbiology 16: 145–153.

11. Rahman M, Colque-Navarro P, Kühn I, Huys G, Swings J, et al. (2002) Identification and characterization of pathogenic Aeromonas veronii biovar sobria associated with epizootic ulcerative syndrome in fish in Bangladesh. Applied and environmental microbiology 68: 650–655.

12. Xu X-Y, Shen Y-B, Fu J-J, Liu F, Guo S-Z, et al. (2012) Matrix metalloproteinase 2 of grass carp Ctenopharyngodon idella (CiMMP2) is involved in the immune response against bacterial infection. Fish & shellfish immunology 33: 251–257.

13. Bushati N, Cohen SM (2007) MicroRNA functions. Annual Review of Cell and Developmental Biology. pp.175–205.

14. Zhang P, Li C, Zhu L, Su X, Li Y, et al. (2013) De novo assembly of the sea cucumber Apostichopus japonicus hemocytes transcriptome to identify miRNA targets associated with skin ulceration syndrome. PLoS One 8: e73506.

15. Bartel DP (2004) MicroRNAs: genomics, biogenesis, mechanism, and function. Cell 116: 281–297.

16. Zhu YP, Xue W, Wang JT, Wan YM, Wang SL, et al. (2012) Identification of common carp (Cyprinus carpio) microRNAs and microRNA-related SNPs. BMC Genomics 13: 413.

17. Ji P, Liu G, Xu J, Wang X, Li J, et al. (2012) Characterization of common carp transcriptome: sequencing, de novo assembly, annotation and comparative genomics. PloS one 7: e35152.

18. Yan B, Guo J-T, Zhao L-H, Zhao J-L (2012) microRNA expression signature in skeletal muscle of Nile tilapia. Aquaculture 364–365: 240–246.

19. Tao W, Yuan J, Zhou L, Sun L, Sun Y, et al. (2013) Characterization of gonadal transcriptomes from Nile tilapia (Oreochromis niloticus) reveals differentially expressed genes. PLoS One 8: e63604.

20. Mennigen JA, Panserat S, Larquier M, Plagnes-Juan E, Medale F, et al. (2012) Postprandial regulation of hepatic microRNAs predicted to target the insulin pathway in rainbow trout. PLoS One 7: e38604.

21. Salem M, Rexroad CE 3rd, Wang J, Thorgaard GH, Yao J (2010) Characterization of the rainbow trout transcriptome using Sanger and 454-pyrosequencing approaches. BMC Genomics 11: 564.

22. Mu X, Pridgeon JW, Klesius PH (2011) Transcriptional profiles of multiple genes in the anterior kidney of channel catfish vaccinated with an attenuated Aeromonas hydrophila. Fish Shellfish Immunol 31: 1162–1172.

23. Barozai MY (2012) The MicroRNAs and their targets in the channel catfish (Ictalurus punctatus). Mol Biol Rep 39: 8867–8872.

24. Fu B, He S (2012) Transcriptome analysis of silver carp (Hypophthalmichthys molitrix) by paired-end RNA sequencing. DNA Res 19: 131–142.

25. Chi W, Tong C, Gan X, He S (2011) Characterization and comparative profiling of MiRNA transcriptomes in bighead carp and silver carp. PLoS One 6: e23549.

26. Chen J, Li C, Huang R, Du F, Liao L, et al. (2012) Transcriptome analysis of head kidney in grass carp and discovery of immune-related genes. BMC Vet Res 8: 108.

27. Liu F, Wang D, Fu J, Sun G, Shen Y, et al. (2010) Identification of immune-relevant genes by expressed sequence tag analysis of head kidney from grass carp (Ctenopharyngodon idella). Comp Biochem Physiol Part D Genomics Proteomics 5: 116–123.

28. Heng JF, Su JG, Huang T, Dong J, Chen LJ (2011) The polymorphism and haplotype of TLR3 gene in grass carp (Ctenopharyngodon idella) and their associations with susceptibility/resistance to grass carp reovirus. Fish & Shellfish Immunology 30: 45–50.

29. Grabherr MG, Haas BJ, Yassour M, Levin JZ, Thompson DA, et al. (2011) Full-length transcriptome assembly from RNA-Seq data without a reference genome. Nat Biotechnol 29: 644–652.

30. Pruitt KD, Tatusova T, Maglott DR (2007) NCBI reference sequences (RefSeq): a curated non-redundant sequence database of genomes, transcripts and proteins. Nucleic Acids Res 35: D61–65.

31. Koonin EV, Fedorova ND, Jackson JD, Jacobs AR, Krylov DM, et al. (2004) A comprehensive evolutionary classification of proteins encoded in complete eukaryotic genomes. Genome Biol 5: R7.

32. Boeckmann B, Bairoch A, Apweiler R, Blatter MC, Estreicher A, et al. (2003) The SWISS-PROT protein knowledgebase and its supplement TrEMBL in 2003. Nucleic Acids Res 31: 365–370.

33. Kanehisa M (2000) KEGG: Kyoto Encyclopedia of Genes and Genomes. Nucleic Acids Research 28: 27–30.

34. Bateman A, Birney E, Cerruti L, Durbin R, Etwiller L, et al. (2002) The Pfam protein families database. Nucleic Acids Res 30: 276–280.

35. Götz S, García-Gómez JM, Terol J, Williams TD, Nagaraj SH, et al. (2008) High-throughput functional annotation and data mining with the Blast2GO suite. Nucleic acids research 36: 3420–3435.

36. Li B, Dewey CN (2011) RSEM: accurate transcript quantification from RNA-Seq data with or without a reference genome. BMC Bioinformatics 12: 323.

37. Mortazavi A, Williams BA, McCue K, Schaeffer L, Wold B (2008) Mapping and quantifying mammalian transcriptomes by RNA-Seq. Nat Methods 5: 621–628.

38. Wang L, Feng Z, Wang X, Wang X, Zhang X (2010) DEGseq: an R package for identifying differentially expressed genes from RNA-seq data. Bioinformatics 26: 136–138.

39. Wen M, Shen Y, Shi S, Tang T (2012) miREvo: an integrative microRNA evolutionary analysis platform for next-generation sequencing experiments. BMC Bioinformatics 13: 140.

40. Friedländer MR, Mackowiak SD, Li N, Chen W, Rajewsky N (2012) miRDeep2 accurately identifies known and hundreds of novel microRNA genes in seven animal clades. Nucleic Acids Res 40: 37–52.

41. John B, Enright AJ, Aravin A, Tuschl T, Sander C, et al. (2004) Human MicroRNA targets. PLoS Biol 2: e363.

42. Enright AJ, John B, Gaul U, Tuschl T, Sander C, et al. (2003) MicroRNA targets in Drosophila. Genome Biol 5: R1.

43. Xu XY, Shen YB, Fu JJ, Lu LQ, Li JL (2014) Determination of reference microRNAs for relative quantification in grass carp (Ctenopharyngodon idella). Fish & Shellfish Immunology 36: 374–382.

44. Su J, Zhang R, Dong J, Yang C (2011) Evaluation of internal control genes for qRT-PCR normalization in tissues and cell culture for antiviral studies of grass carp (Ctenopharyngodon idella). Fish Shellfish Immunol 30: 830–835.

45. Grabherr MG, Haas BJ, Yassour M, Levin JZ, Thompson DA, et al. (2011) Full-length transcriptome assembly from RNA-Seq data without a reference genome. Nature Biotechnology 29: 644–U130.

46. Akira S, Takeda K (2004) Toll-like receptor signalling. Nature Reviews Immunology 4: 499–511.

47. Rot A, von Andrian UH (2004) Chemokines in innate and adaptive host defense: basic chemokine grammar for immune cells. Annu Rev Immunol 22: 891–928.

48. Liu F, Li J, Fu J, Shen Y, Xu X (2011) Two novel homologs of simple C-type lectin in grass carp (Ctenopharyngodon idellus): potential role in immune response to bacteria. Fish Shellfish Immunol 31: 765–773.

49. Xu XY, Shen YB, Fu JJ, Liu F, Guo SZ, et al. (2013) Characterization of MMP-9 gene from grass carp (Ctenopharyngodon idella): An Aeromonas hydrophila-inducible factor in grass carp immune system. Fish Shellfish Immunol 35: 801–807.

50. Enright AJ, John B, Gaul U, Tuschl T, Sander C, et al. (2004) MicroRNA targets in Drosophila. Genome Biology 5: R1–R1.

51. Guo H, Ingolia NT, Weissman JS, Bartel DP (2010) Mammalian microRNAs predominantly act to decrease target mRNA levels. Nature 466: 835–840.

52. Janeway CA, Medzhitov R (2002) Innate immune recognition. Annual Review of Immunology 20: 197–216.

53. Plata-Salamán C, Ilyin S, Gayle D, Flynn MC (1998) Gram-negative and gram-positive bacterial products induce differential cytokine profiles in the brain: Analysis using an integrative molecular-behavioral in vivo model. International journal of molecular medicine 1: 387–398.

54. Chen XM, Splinter PL, O'Hara SP, LaRusso NF (2007) A cellular micro-RNA, let-7i, regulates Toll-like receptor 4 expression and contributes to cholangiocyte immune responses against Cryptosporidium parvum infection. J Biol Chem 282: 28929–28938.

55. Liu PT, Wheelwright M, Teles R, Komisopoulou E, Edfeldt K, et al. (2012) MicroRNA-21 targets the vitamin D-dependent antimicrobial pathway in leprosy. Nat Med 18: 267–273.

56. Oltmanns U, Issa R, Sukkar MB, John M, Chung KF (2003) Role of c-jun N-terminal kinase in the induced release of GM-CSF, RANTES and IL-8 from human airway smooth muscle cells. Br J Pharmacol 139: 1228–1234.

57. Boutros M, Agaisse H, Perrimon N (2002) Sequential Activation of Signaling Pathways during Innate Immune Responses in Drosophila. Developmental Cell 3: 711–722.

58. Lee J, Mira-Arbibe L, Ulevitch RJ (2000) TAK1 regulates multiple protein kinase cascades activated by bacterial lipopolysaccharide. J Leukoc Biol 68: 909–915.

59. Zhang Y, Dong C (2005) MAP kinases in immune responses. Cell Mol Immunol 2: 20–27.

60. Förster R, Schubel A, Breitfeld D, Kremmer E, Renner-Müller I, et al. (1999) CCR7 Coordinates the Primary Immune Response by Establishing Functional Microenvironments in Secondary Lymphoid Organs. Cell 99: 23–33.

61. Kursar M, Hopken UE, Koch M, Kohler A, Lipp M, et al. (2005) Differential requirements for the chemokine receptor CCR7 in T cell activation during Listeria monocytogenes infection. J Exp Med 201: 1447–1457.

62. Nau GJ, Richmond JF, Schlesinger A, Jennings EG, Lander ES, et al. (2002) Human macrophage activation programs induced by bacterial pathogens. Proc Natl Acad Sci U S A 99: 1503–1508.

63. Mendes ND, Freitas AT, Sagot MF (2009) Current tools for the identification of miRNA genes and their targets. Nucleic Acids Res 37: 2419–2433.

64. Krol J, Loedige I, Filipowicz W (2010) The widespread regulation of microRNA biogenesis, function and decay. Nature Reviews Genetics 11: 597–610.

65. Bartel DP (2009) MicroRNAs: target recognition and regulatory functions. Cell 136: 215–233.

66. Lewis BP, Shih IH, Jones-Rhoades MW, Bartel DP, Burge CB (2003) Prediction of mammalian microRNA targets. Cell 115: 787–798.

67. Rehmsmeier M, Steffen P, Hochsmann M, Giegerich R (2004) Fast and effective prediction of microRNA/target duplexes. RNA 10: 1507–1517.

68. Krek A, Grun D, Poy MN, Wolf R, Rosenberg L, et al. (2005) Combinatorial microRNA target predictions. Nat Genet 37: 495–500.

69. Stark A, Brennecke J, Russell RB, Cohen SM (2003) Identification of Drosophila microRNA targets. PLoS biology 1: e60.

70. Nazarov PV, Reinsbach SE, Muller A, Nicot N, Philippidou D, et al. (2013) Interplay of microRNAs, transcription factors and target genes: linking dynamic expression changes to function. Nucleic Acids Res 41: 2817–2831.

71. Muniategui A, Pey J, Planes FJ, Rubio A (2013) Joint analysis of miRNA and mRNA expression data. Brief Bioinform 14: 263–278.

8

Uromodulin Retention in Thick Ascending Limb of Henle's Loop Affects SCD1 in Neighboring Proximal Tubule: Renal Transcriptome Studies in Mouse Models of Uromodulin-Associated Kidney Disease

Marion Horsch[1], Johannes Beckers[1,2,3], Helmut Fuchs[1], Valérie Gailus-Durner[1], Martin Hrabě de Angelis[1,2,3,4], Birgit Rathkolb[1,5], Eckhard Wolf[5], Bernhard Aigner[5], Elisabeth Kemter[5]*

1 German Mouse Clinic, Institute of Experimental Genetics, Helmholtz Zentrum München GmbH, German Research Center for Environmental Health, Neuherberg, Germany, 2 German Center for Diabetes Research (DZD), Neuherberg, Germany, 3 Experimental Genetics, Center of Life and Food Sciences Weihenstephan, Technische Universität München, Freising-Weihenstephan, Germany, 4 German Center for Vertigo and Balance Disorders, University Hospital Munich, Campus Grosshadern, Munich, Germany, 5 Molecular Animal Breeding and Biotechnology, and Laboratory for Functional Genome Analysis (LAFUGA), Gene Center, LMU München, Munich, Germany

Abstract

Uromodulin-associated kidney disease (UAKD) is a hereditary progressive renal disease which can lead to renal failure and requires renal replacement therapy. UAKD belongs to the endoplasmic reticulum storage diseases due to maturation defect of mutant uromodulin and its retention in the enlarged endoplasmic reticulum in the cells of the thick ascending limb of Henle's loop (TALH). Dysfunction of TALH represents the key pathogenic mechanism of UAKD causing the clinical symptoms of this disease. However, the molecular alterations underlying UAKD are not well understood. In this study, transcriptome profiling of whole kidneys of two mouse models of UAKD, $Umod^{A227T}$ and $Umod^{C93F}$, was performed. Genes differentially abundant in UAKD affected kidneys of both $Umod$ mutant lines at different disease stages were identified and verified by RT-qPCR. Additionally, differential protein abundances of SCD1 and ANGPTL7 were validated by immunohistochemistry and Western blot analysis. ANGPTL7 expression was down-regulated in TALH cells of $Umod$ mutant mice which is the site of the mutant uromodulin maturation defect. SCD1 was expressed selectively in the S3 segment of proximal tubule cells, and SCD1 abundance was increased in UAKD affected kidneys. This finding demonstrates that a cross talk between two functionally distinct tubular segments of the kidney, the TALH segment and the S3 segment of proximal tubule, exists.

Editor: Ines Armando, University of Maryland School of Medicine, United States of America

Funding: This work has been funded by the German Research Foundation (DFG) (grant number KE1673/1-1), by the German Federal Ministry of Education and Research (Infrafrontier grant 01KX1012) and by the German Center for Diabetes Research (DZD e.V.). The funders had no role in study design, data collection and analysis, decision to publish, or preparation of the manuscript.

Competing Interests: The authors have declared that no competing interests exist.

* Email: kemter@lmb.uni-muenchen.de

Introduction

Uromodulin-associated kidney disease (UAKD) is a rare dominant hereditary renal disease caused by amino acid-changing mutations in the uromodulin (*UMOD*) gene [1–3]. Patients with UAKD exhibit impaired urinary concentration ability, in most cases hyperuricemia, morphological kidney alterations like progressive tubulointerstitial damage and sometimes renal cysts and constantly develop disease progression up to renal failure. Dysfunction of thick ascending limb of Henle's loop (TALH) cells due to mutant UMOD maturation retardation and retention in the hyperplastic endoplasmic reticulum (ER) represents the key pathogenic mechanism of UAKD.

Uromodulin is selectively expressed in cells of the TALH and of the early distal convoluted tubules [4]. After UMOD synthesis in the ER and extensive glycosylation, the mature protein is translocated to the luminal cell membrane and released into urine by proteolytic cleavage. UMOD represents the most abundant protein in human urine. Although this glycoprotein was already discovered in the early fifties by Tamm and Horsfall (therefore initially named Tamm-Horsfall-glycoprotein) [5], the biological function of UMOD is still obscure. Studies on *Umod* knockout mice revealed a protective role of UMOD against ascending urinary tract infections e.g. of type 1-fimbriated *E. coli*, and a protective role against calcium oxalate crystal formation in the kidney [6–8]. Further, UMOD might have a role in innate immune system and might act as an endogenous danger signal after tubular damage that leads to exposition of UMOD protein to mononuclear cells like dendritic cells in the kidney interstitium [9]. In various genome-wide association studies, common allelic *UMOD* promoter variants were identified to be associated with increased risk for complex trait diseases like chronic kidney disease

(CKD), hypertension and kidney stones (reviewed in [3]). These variants of the *UMOD* promoter lead to increased UMOD expression and secretion which results, by influencing salt reabsorption in the kidney, to increased risk of developing hypertension and CKD [10].

UMOD maturation defect and retention in TALH cells in UAKD lead to unfolded protein response and activation of non-canonical NF-κB signaling in the TALH segment as demonstrated recently in two mouse models of UAKD, the mutant mouse lines $Umod^{C93F}$ and $Umod^{A227T}$ [11]. The focus of this study is to analyze transcriptional alterations in UAKD affected kidneys to obtain further insights into the pathogenesis of UAKD and/or the biological function of UMOD.

Materials and Methods

Animals

Both $Umod^{A227T}$ and $Umod^{C93F}$ mutant mouse lines, generated by ENU mutagenesis, exhibit the key features of UAKD like maturation defect and retention of UMOD in the hyperplastic ER of TALH cells and impaired kidney function with mild defect in urinary concentration ability, reduced fractional excretion of uric acid, and reduced UMOD excretion [12,13]. $Umod^{A227T}$ and $Umod^{C93F}$ mutant mice differ in onset, severity and speed of progression of disease symptoms as these are dependent on kind of *Umod* mutation and allelic status [13].

Homozygous $Slc12a1^{I299T}$ mutant mice exhibit key features of Type I Bartter syndrome with salt-wasting polyuria and reduced fractional excretion of uric acid due to impaired ion transport activity of the main ion transporter of the TALH segment, the $Na^+-K^+-2Cl^-$ ion transporter NKCC2, which is the protein derived from the *Slc12a1* gene [14].

All three mouse lines were maintained on the C3HeB/FeJ (C3H) genetic background. Mouse husbandry was done under a continuously controlled specific pathogen-free (SPF) hygiene standard according to the FELASA recommendations (http://www.felasa.eu) [15]. Mouse husbandry and all tests were carried out under the approval of the responsible animal welfare authority (Regierung von Oberbayern, Germany).

Genome-wide transcriptome analysis of UAKD-affected kidneys

The genome-wide transcriptome analyses of whole kidney lysates of *Umod* mutant mice were part of the systemic, comprehensive phenotypic analysis carried out in the German Mouse Clinic at the Helmholtz Zentrum München using standardized examination protocols (http://www.mouseclinic.de) [16,17]. First, the analysis of kidneys of four male homozygous $Umod^{A227T}$ mutant mice and four male wild-type littermates was carried out at an age of 17 weeks, followed by the analysis of kidneys of four male heterozygous $Umod^{C93F}$ mutants and their wild-type littermates at an age of 38 weeks.

Total RNA of whole kidney was isolated using Trizol (Invitrogen, Karlsruhe, Germany) in combination with RNeasy Midi Kits according to manufacturer's protocol (Qiagen, Hilden, Germany). cDNA microarrays covering 17,346 mouse genes were in-house produced, dual color hybridization was performed and arrays were scanned as previously described [18]. As a first step of the expression profiling data from kidney low-quality array elements were eliminated applying several filter methods (background checking for both channels with a signal/noise threshold of 2.0, one bad tolerance policy parameter and flip dye consistency checking). In total, 9,237 useable expression values remained and were used for the identification of genes with significant differential

gene regulation using SAM (Significant analysis of microarrays [18–20]). Genes were ranked according to the mean fold change (mean mutant divided by mean wild-type signal intensity) and selected as significantly differentially expressed with ratio >1.4 in combination with false discovery rate (FDR) <10%. The FDR is a value which estimates the percentage of nonsense genes by calculating 1000 permutations.

Over-represented functional annotations within the data sets were provided as GO (Gene Ontology terms of the category 'functions and diseases' (Ingenuity Pathway Analysis, IPA)). IPA was also used to identify biological associations based on co-citations and co-expression between the regulated genes of each of the two mutant lines and the NF-κB signaling complex. In the GEO database [21], the complete microarray dataset is available under accession number GSE58513.

Quantitative real time PCR analysis

After isolation of total RNA from whole kidneys of four-month-old male homozygous *Umod* mutant mice, three-month-old male homozygous $Slc12a1^{I299T}$ and corresponding age-matched male wild-type littermates with the Trizol method, cDNA synthesis was performed using cDNA EcoDry Premix Double Primed (Clontech). NCBI/Primer-BLAST (http://www.ncbi.nlm.nih.gov/tools/primer-blast/index.cgi?LINK_LOC=BlastHome) was used to select cDNA specific primers from different exons (except for *Hba-a1*) and each primer pair was tested for comparability of amplification efficiencies by performance of real-time PCR-based standard curve analyses. Primer sequences are listed in Table 1. Transcript abundance was quantified using FastStart Universal SYBR Green Master (ROX) (Roche) on a StepOne Real-Time PCR system (Applied Biosystems) and using LinRegPCR software [22]. All real-time PCR measurements were performed in duplicates and included no template controls. Transcript abundances of *Angptl7*, *Hba-a1*, *Odc1*, *Scd1*, and *Wfdc15b* were calculated in relation to the expression of the housekeeping genes *Sdha*, *Tbp*, and *Rpl13a* using the $2^{-\Delta CT}$ method.

Immunohistochemical analyses

Histological analyses of kidneys of homozygous *Umod* mutant mice and wild-type mice were performed as described previously [12]. Immunohistochemistry was performed using the following primary antibodies: rat monoclonal antibody against mouse ANGPTL7 (clone 538401; R&D Systems), rabbit monoclonal antibody against mouse SCD1 (#2794, Cell Signaling), and rat monoclonal antibody against mouse UMOD (clone 774056; R&D Systems). Immunoreactivity was visualized using 3,3-diaminobenzidine tetrahydrochloride dihydrate (DAB) (brown color) or using Vector Red Alkaline Phosphatase Substrate Kit I (red color). Nuclear counterstaining was done with hemalum. Cells of TALH segment were identified by UMOD immunostaining. Cells of proximal tubule segment were identified by morphological criteria of presence of luminal microvilli.

Western blot analyses

Renal tissue (whole kidney or outer medullary region) of homozygous *Umod* mutant mice and wild-type mice was homogenized in Laemmli extraction buffer (20 mM Tris, 2% Triton-X100, 20% 5× Laemmli buffer). Outer medullary region contained a higher fraction of TALH segments and was prepared as described previously [11]. Protein concentration was determined by BCA assay. Equal amounts of denatured proteins per lane were separated on 12% SDS-polyacrylamide minigels and blotted on PVDF membranes. Equal loading was controlled by Ponceau staining.

Table 1. List of primer sequences used for RT-qPCR.

Gene	Sequence (5'-3')		Length	Accession No.
Angptl7	forward: TCCGAAAAGGTGGCTACTGG	reverse: ATGCCATCCATGTGCTTTCG	97 nt	NM_001039554.3
Hba-a1	forward: GTGCATGCCTCTCTGGACA	reverse: GGTACAGGTGCAAGGGAGAG	127 nt	NM_008218.2
Odc1	forward: CCGGCTCTGACGATGAAGAT	reverse: CTTCTCGTCTGGCTTGGGTC	146 nt	NM_013614.2
Rpl13a	forward: GACCTCCTCCTTTCCCAGGC	reverse: GCCTCGGCCATCCAATACC	70 nt	NM_009438.5
Scd1	forward: AGGCCTGTACGGGATCATACT	reverse: AGAGCGCTGGTCATGTAGTAG	84 nt	NM_009127.4
Sdha	forward: AACACTGGAGGAAGCACACC	reverse: AGTAGGAGCGGATAGCAGGA	135 nt	NM_023281.1
Tbp	forward: TCTGGAATTGTACCGCAGCTT	reverse: ATGATGACTGCAGCAAATCGC	131 nt	NM_013684.3
Wfdc15b	forward: TTTCGCATACGGAGGACAGT	reverse: GGGCCAGGTGTGGTTATGTC	79 nt	NM_001045554.1

cDNA-specific primers for amplification of mouse angiopoietin-like 7 (*Angptl7*), hemoglobin alpha adult chain 1 (*Hba-a1*), ornithine decarboxylase structural 1 (*Odc1*), stearoyl-Coenzyme A desaturase 1 (*Scd1*), WAP four-disulfide core domain 15B (*Wfdc15b*), and the housekeeping genes ribosomal protein L13A (*Rpl13a*), succinate dehydrogenase complex subunit A flavoprotein (*Sdha*) and TATA box binding protein (*Tbp*).

The following primary antibodies were used: rat monoclonal antibody against mouse ANGPTL7 (clone 538401; R&D Systems), rabbit monoclonal antibody against GAPDH (#2118, Cell Signaling), rabbit polyclonal antibody against mouse SCD1 (#2438, Cell Signaling). Bound antibodies were visualized using ECL reagent (GE Healthcare Amersham Biosciences). Signal intensities were quantified using ImageQuant (GE Healthcare). Standardization of equal loading was referred to the signal intensities of GAPDH of the corresponding PVDF membrane.

Statistical analysis

Statistical analyses of genome-wide transcriptome analyses were described above. Statistical analyses of data derived by RT-qPCR analyses and Western blot analyses were carried out by One-way ANOVA with Tukey's Multiple Comparison Post hoc Test if comparing three groups, and by unpaired Student's *t*-test if comparing two groups. Statistical analyses of plasma urea data were performed by Two-way ANOVA with Bonferroni Multiple Comparison Post hoc Test. Data are shown as mean ± standard deviations. Significant differences are indicated for $P<0.05$, 0.01, and 0.001.

Results

Transcriptome profiling of UAKD-affected kidneys

The primary phenotype reports of both mutant lines *Umod*^{A227T} (see line "HST012") and *Umod*^{C93F} (see line "HST001") of the German Mouse Clinic are accessible online (http://146.107.35.38/phenomap/jsp/annotation/public/phenomap.jsf) [17]. For the following analysis, the kidney transcriptome data of both *Umod* mutant lines were analyzed using enhanced settings in the software programs (see Materials and Methods). For genome-wide transcriptome profiling of whole kidneys, one group of UAKD-affected mice with initial mild disease phenotype, the young-adult homozygous *Umod*^{A227T} mutants, and another group with progressed disease state of UAKD including morphological kidney alterations, the aged heterozygous *Umod*^{C93F} mutant mice, were used (Figure S1).

First, genome-wide transcriptome profiling analysis of kidneys from homozygous *Umod*^{A227T} mutant mice, carried out at an age of 17 weeks, revealed 104 differentially expressed genes (DEGs) compared to age-matched wild-type littermate controls (Figure S2). In order to identify over-represented functional annotations among the set of regulated genes, Ingenuity Pathway Analysis (IPA) was employed. In kidney of *Umod*^{A227T} mutants, the DEGs

were associated with proliferation of cells, necrosis, inflammation, metabolism of lipid and proteins as well as renal cancer (Table 2, Figure S4). Several up-regulated genes are known to be expressed in proximal and/or distal tubules like *Abcc4*, *Atp1a1*, *Atoh8*, *Clcnkb*, *Col18a1*, *Car15* and *Mt1* and are required for normal reabsorption and urine concentration. *Fabp4*, *Gsn*, *Il6st*, and *Ly6e*, were associated with nephritis and *Col18a1*, *Prom1*, and *Scd1* with renal cell carcinoma. Further, several differential abundant genes were associated with repression or activation of NF-κB signaling complex (*Atp1a1*, *Cxcl12*, *Fabp4*, *Hbb*, *Itgav*, *Mt1*, *Scd1*, *Serpina1*, and *Tacstd2*). The other way round, the NF-κB signaling complex plays roles in the expression of *Cxcl12*, *Fabp4*, *Fgf1*, *Gclc*, *Odc1*, and *Tap*.

Second, 54 DEGs were detected in whole kidney lysate of 38-week-old heterozygous *Umod*^{C93F} mutants when compared to age-matched wild-type littermate controls (Figure S3). Necrosis, inflammation, lipid metabolism, hypertension and distinct renal and urological functions/disease were over-represented GO terms in this dataset (Table 3, Figure S5). Expression in the proximal and/or distal renal tubules was described for *Kap*, *Lgmn*, *Pdzk1*, *Slc13a1*, and *Slco4c1*, which were down-regulated in kidneys of *Umod*^{C93F} mutants. Further genes were annotated with nephritis (*Cndp2*, *Dnase1* and *Scd1*). While *Timp1* and *Odc1* expression depends on the activity of the NF-κB signaling complex, *Akr1b1*, *Casp9*, *Mt2* and *Scd1* diminish the activity of the transcription factor complex.

To identify DEGs irrespective of secondary morphological kidney alterations due to different UAKD disease progression stage, transcriptome datasets of the *Umod*^{A227T} mutant mouse line were compared with that of the *Umod*^{C93F} mutant mouse line. Hierarchical cluster analysis of all differentially expressed genes in both mutant lines displayed a similar tendency of transcriptional alteration for about 43% of the genes identified as DEGs in one of both *Umod* mutant mouse lines but which were only significantly regulated in one of the two lines (data not shown). An overlap of 5 DEGs (*Angptl7*, *Hba-a1*, *Odc1*, *Scd1* and *Wfdc15b*) was identified between the datasets of regulated genes of both *Umod* mutant lines. *Scd1* was overexpressed and *Angptl7*, *Hba-a1*, *Odc1* and *Wfdc15b* displayed significant down-regulation in *Umod*^{A227T} and *Umod*^{C93F}.

Validation of microarray datasets by quantitative RT-real time PCR

Genes identified as equally differentially abundant in both *Umod* mutant mouse lines by genome-wide transcriptome analyses

Table 2. Functional classification of differentially expressed genes in kidney of $Umod^{A227T}$ mutant line.

Categories	Diseases or Functions Annotation	p-Value	Genes	# Genes
Cancer	abdominal neoplasm	6.37E-03	Abcc4, Acsl5, Actg1, Ahnak, Aldoa, Anxa5, Atp1a1, Col18a1, Cpa1, Cxcl12, Dnajc10, Dpp10, Egr1, Eno1, Fabp4, Fgf1, Grin2a, Gsn, Hba1/Hba2, Hbb, Il6st, Itgav, Ivns1abp, Ly6e, Mapre1, Mark3, Mt1, Nup155, Odc1, Pcolce, Prom1, Rbm3, Rimbp2, Rps4y1, S100a11, Sat1, Scd, Serpina1, Tacstd2, Tap1, Wfdc2	41
Cellular Growth and Proliferation	proliferation of cells	1.47E-05	Abcc4, Acsl5, Actg1, Ahnak, Aldoa, Cd24a, Col18a1, Cxcl12, Dnph1, Eef1a1, Egr1, Eno1, Fabp4, Fgf1, Gclc, Grin2a, Gsn, Hba1/Hba2, Igkv1-117, Il6st, Itgav, Ivns1abp, Mapre1, Morf4l1, Mt1, Odc1, Rbm3, Rps14, S100a11, Sat1, Scd, Serpina1, Tacstd2, Tap1	34
Cell Death and Survival	necrosis	3.36E-06	Abcc4, Aldoa, Ap2b1, Atp1a1, Cd24a, Col18a1, Cxcl12, Eef1a1, Egr1, Eno1, Fabp4, Fbxo32, Fgf1, Gclc, Grin2a, Gsn, Gsta1, Igkv1-117, Il6st, Itgav, Ivns1abp, Mt1, Odc1, Rbm3, S100a11, Sat1, Scd, Serpina1, Ucp1	29
Inflammatory Response	inflammation of organ	1.11E-07	Abcc4, Ahnak, Aldoa, Anxa5, Atp1a1, Cxcl12, Eef1a1, Egr1, Eno1, Fabp4, Fbxo32, Gsn, H3f3a/H3f3b, Hbb, Il6st, Mt1, Odc1, Prom1, S100a11, Scd	20
Molecular Transport	transport of molecule	6.05E-05	Abcc4, Acsl5, Atp1a1, Clcnka, Cxcl12, Egr1, Fabp4, Fgf1, Grin2a, Gsn, Hba1/Hba2, Hbb, Il6st, Itgav, Ly6e, Nup155, Scd, Tap1, Ucp1	19
Lipid Metabolism	synthesis of lipid	4.20E-05	Abcc4, Acsl5, Atp1a1, Cxcl12, Eef1a1, Egr1, Fgf1, Gsta3, Hbb, Odc1, Scd, Serpina1, Ucp1	13
Protein Synthesis	metabolism of protein	7.61E-04	Cpa1, Eef1a1, Fbxo32, Grin2a, Gsn, Pcolce, Rbm3, Rps14, Rps4y1, Sat1, Serpina1, Wfdc2	12
Metabolic Disease	diabetes mellitus	6.97E-03	Anxa5, Fabp4, Fgf1, Grin2a, Hba1/Hba2, Hbb, Itgav, Mt1, Nup155, Scd, Tap1	11
Carbohydrate Metabolism	quantity of carbohydrate	6.48E-04	Anxa5, Eef1a1, Fabp4, Fgf1, Il6st, Mark3, Mt1, Scd, Ucp1	9
Cardiovascular Disease	hypertension	1.97E-03	Atp1a1, Clcnka, Cpa1, Fbxo32, Fgf1, Gsn, Gsta3, Hba1/Hba2, Il6st	9
Renal Disease	renal cancer	4.83E-03	Ahnak, Anxa5, Gsn, Mt1, Nup155, S100a11, Scd, Tacstd2	8
Cell-To-Cell Signaling	binding of cells	7.10E-03	Anxa5, Cd24a, Col18a1, Cxcl12, Fgf1, Gsn, Itgav	7

Table 3. Functional classification of differentially expressed genes in kidney of $Umod^{C93F}$ mutant line.

Categories	Diseases or Functions Annotation	p-Value	Genes	# genes
Cancer	cancer	5.85E-03	Acox1, Acsm2a, Agps, Akr1b1, Ank1, C11orf54, Casp9, Cchcr1, Cndp2, Cyp51a1, Dnase1, Gpm6a, Grm2, Hba1/Hba2, Hgd, Hsd17b11, Il5ra, Inmt, Lgmn, Macrod1, Mt2, Odc1, Pah, Pank1, Pank3, Pdzk1, Rtn3, Scd, Slc13a3, Snx27, Timp1, Tmem174, Ttr, Ube2a, Ugt2b28, Ugt2b7	36
Cell Death and Survival	necrosis	5.17E-03	Akr1b1, Casp9, Cchcr1, Dnase1, Gpm6a, Grm2, Il5ra, Kap, Lgmn, Mt2, Odc1, Pdzk1, Scd, Timp1, Ttr	15
Molecular Transport	transport of molecule	1.00E-02	Ank1, Casp9, Cchcr1, Grm2, Hba1/Hba2, Pdzk1, Scd, Slc13a3, Snx27, Ttr	10
Inflammatory Response	inflammation	3.68E-03	Aco2, Acox1, Dnase1, Il5ra, Lgmn, Mt2, Odc1, Scd, Timp1	9
Lipid Metabolism	concentration of lipid	5.51E-04	Agps, Akr1b1, Casp9, Kap, Pank1, Pdzk1, Scd, Timp1, Ttr	9
Cardiovascular Disease	hypertension	6.97E-03	Acox1, Cyb5b, Hba1/Hba2, Inmt, Kap, Pah	6
Protein Synthesis	quantity of protein in blood	1.94E-03	Hba1/Hba2, Lgmn, Mt2, Pdzk1, Scd, Ttr	6
Renal Disease	renal cancer	1.99E-02	Cyp51a1, Gpm6a, Mt2, Scd, Timp1	5
	failure of kidney	1.98E-04	Dnase1, Hba1/Hba2, Lgmn, Mt2, Ttr	5
	cell death of renal tubule	4.90E-04	Dnase1, Kap, Mt2	3
	proliferation of kidney cell lines	6.54E-03	Cchcr1, Mt2, Ttr	3
	nephritis	2.41E-02	Dnase1, Il5ra, Mt2	3
	abnormal morphology of renal tubule	1.57E-03	Akr1b1, Lgmn, Mt2	3
	end stage disease	1.96E-03	Lgmn, Mt2, Ttr	3
	glomerulosclerosis	3.08E-03	Kap, Lgmn, Mt2	3
	urination disorder	3.84E-02	Akr1b1, Kap, Lgmn	3
Post-Translational Modification	hydrolysis of protein fragment	2.69E-02	Casp9, Cndp2, Lgmn	3

were selected for quantitative RT-real time PCR (RT-qPCR) analyses of whole kidney lysates of four-month-old male mice (n = 5 per group) (Figure 1). *Scd1* transcripts were significantly higher abundant in kidneys of both homozygous *Umod*^C93F and *Umod*^A227T mutant mice compared to wild-type mice by RT-qPCR analyses, whereas *Wfdc15b* exhibited a significantly decreased transcript abundance in both *Umod* mutant mouse lines. *Angptl7* and *Odc1* exhibited significantly decreased transcript abundances in whole kidney lysates of homozygous *Umod*^C93F mutant mice and showed the tendency of decreased transcript abundances also in homozygous mutants of the *Umod*^A227T mouse line which exhibited a less severe UAKD phenotype and TALH dysfunction compared to age-matched mutant mice of the *Umod*^C93F mouse line [13]. RT-qPCR of *Hba-a1* revealed similar transcript abundances irrespective of genotype. In summary, RT-qPCR analyses of independent groups of mice confirmed the results for *Angptl7*, *Odc1*, *Wfdc15b* and *Scd1*, obtained by array hybridization, and differed only for *Hba-a1*.

Localization and quantification of ANGPTL7 and SCD1 in the kidney by immunohistochemistry and Western blot analysis

In wild-type mice, ANGPTL7 protein was localized in the cytoplasm of all tubular cells, with more intense staining in TALH cells than in cells of other tubular segments (Figure 2A). In contrast to wild-type mice, the staining intensity of ANGPTL7 in TALH cells were less intense in *Umod* mutant mice, exhibiting a nearly similar staining intensity of ANGPTL7 in TALH cells like in cells of other tubular segments. Staining intensity of ANGPTL7 in non-TALH cells was similar irrespective of genotype. Identification of TALH segment was enabled by detection of UMOD. Western blot analyses of ANGPTL7 revealed a significantly decreased abundance of ANGPTL7 in the outer medulla of

kidneys of *Umod*^A227T and *Umod*^C93F mutant mice compared to wild-type controls (Figure 2B).

SCD1 protein was localized in the cytoplasm selectively of proximal tubular cells of the kidney, concretely in the straight (S3) segment of proximal tubule (Figure 3A). Other tubule segments like TALH cells were negative for SCD1. In homozygous *Umod* mutant mice of both lines, SCD1 appeared to be abundant in a larger fraction of proximal tubular cells compared to wild-type mice, and the average staining intensity of SCD1 positive cells appeared to be more prominent. Western blot analyses of SCD1 revealed a 2.7-fold and 4.6-fold higher abundance in whole kidneys of *Umod*^A227T and *Umod*^C93F homozygotes compared to wild-type controls (Figure 3B).

Evaluation of the role of TALH-dysfunction derived salt wasting and volume depletion state on Scd1 transcript abundance

The bumetanide-sensitive ion transporter NKCC2 is mainly expressed in the kidney in the apical membrane of the cells of TALH and macula densa, and impaired function of NKCC2 due to inactivating mutations or due to the action of loop diuretics leads to salt wasting polyuria with reduced fractional excretion of uric acid [23]. Homozygous *Slc12a1*^I229T mutant mice, suffering on TALH-dysfunction derived salt wasting and volume depletion state [14], exhibited a significantly increased *Scd1* transcript abundance in their kidneys compared to their littermate wild-type controls (Figure 4).

Discussion

UAKD is a progressive hereditary disease and belongs to the endoplasmic reticulum (ER) storage diseases due to maturation defect of mutant UMOD and its retention in the enlarged ER of

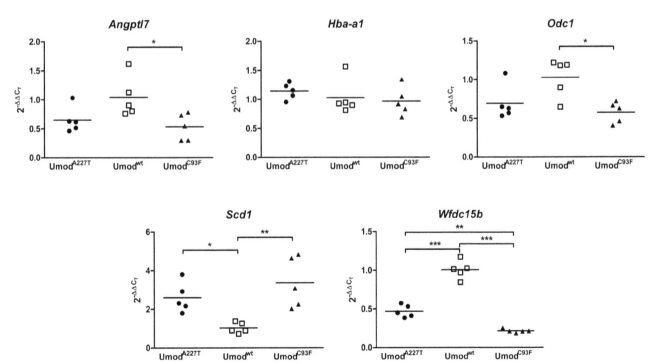

Figure 1. Verification of DEGs, identified by transcriptome profiling of whole kidneys, by RT-qPCR. Data are shown as scatter dot plot with mean (n = 5 per group). Age of mice analyzed: four months. One-way-ANOVA with Tukey's Multiple Comparison Post hoc Test: p vs. wild-type, *, p<0.05; **, p<0.01; ***, p<0.001.

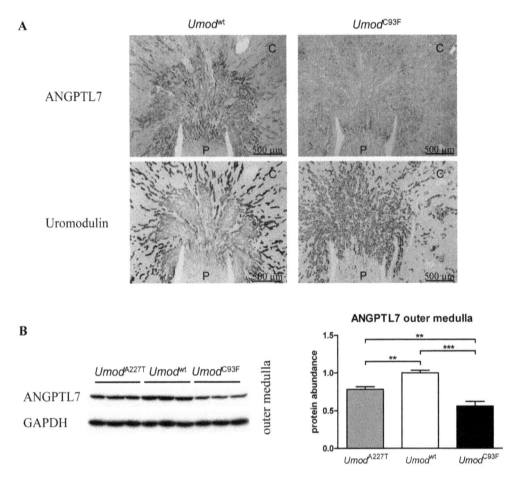

Figure 2. Analysis of localization and protein abundance of ANGPTL7 in healthy and UAKD-affected kidneys. (A) ANGPTL7 was predominantly detected in the cytoplasmic compartment of tubular cells, predominantly in TALH cells, of wild-type mice. Compared to the staining intensity of ANGPTL7 in TALH cells of wild-type mice, TALH cells of *Umod* mutant mice exhibited a lower cytoplasmic staining intensity of ANGPTL7. Age of mice analyzed: four months. *Umod*^{wt}: wild-type mouse; *Umod*^{C93F}: homozygous *Umod*^{C93F} mutant mouse. UMOD immunohistochemistry enabled identification of TALH segments. Serial kidney sections were used for ANGPTL7 and uromodulin immunohistochemistry and corresponding kidney regions are shown. C: renal cortex; P: renal papilla. Chromogen: DAB; nuclear staining: hemalum. **(B)** Protein abundance of ANGPTL7 in the outer medulla of kidneys of homozygous *Umod* mutant mice of both lines was decreased compared to wild-type mice. Signal intensities of ANGPTL7 were corrected for GAPDH signal intensities of the same PVDF-membrane. Mean of protein abundance of wild-type mice was set on a value of 1 [mean (wild-type) = 1]. One-way-ANOVA with Tukey's Multiple Comparison Post hoc Test: p vs. wild-type, **, $p<0.01$; ***, $p<0.001$. Age of mice analyzed: four months.

TALH cells [3]. Until know, little is known about the molecular alterations in UAKD affected kidneys, except of the induction of unfolded protein response and the recently identified activation of non-canonical NF-κB signaling in cells of the TALH segment [11]. In this study, we performed genome-wide transcriptome profiling of whole kidneys from young-adult homozygous *Umod*^{A227T} and from aged heterozygous *Umod*^{C93F} mutant mice exhibiting different severities of UAKD. With these analyses, we identified numerous DEGs in both mouse models of UAKD, which were functionally annotated with renal diseases and renal dysfunction. Further, numerous of the identified DEGs were found to be associated with the NF-κB signaling pathway. However, only 5 DEGs were identified being commonly differentially abundant in both *Umod* mutant mouse lines, but about 43% of the genes identified as DEGs in one of both *Umod* mutant mouse lines displayed a similar tendency of transcriptional alterations but being only significantly differentially abundant in one of the two mutant lines. This finding might be related to two facts of our study design:

First, in UAKD, primary functional and morphological alterations occur selectively in TALH cells, and TALH segments represent approximately 20% of whole kidney fraction. Using whole kidney for transcriptome profiling, this might be not suitable for detecting genes with altered abundances selectively in a small fraction of the kidney like in the TALH segment but being normally expressed also in other tubular segments as these molecular alterations in a small segment will be highly likely masked by the unaffected transcript abundance in the majority of kidney cells. However, our approach is highly suitable for identifying DEGs in UAKD expressed in a specific tubular fraction or whose transcript abundance is excessively altered in UAKD.

Second, the two groups of UAKD affected mice used for genome-wide transcriptome profiling exhibited different disease stages of UAKD. As recently shown, severity of UMOD maturation defect and speed of progression of UAKD *in vivo* strongly depends on the particular *Umod* mutation itself and the zygosity status [13]. The A227T mutation of *Umod* causes a milder UMOD maturation defect than the C93F mutation, and

A

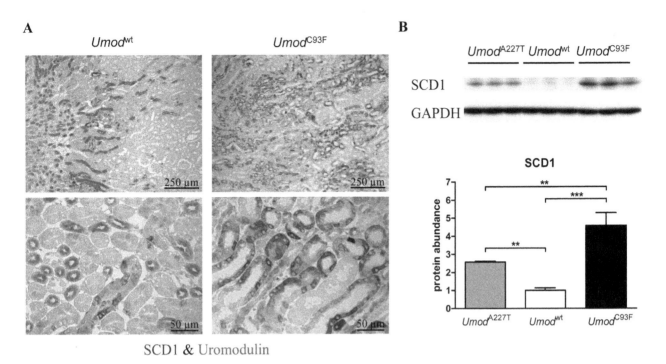

SCD1 & Uromodulin

Figure 3. Analysis of localization and protein abundance of SCD1 in healthy and UAKD-affected kidneys. (A) SCD1 was detected in the cytoplasmic compartment selectively of proximal tubular cells (in the straight S3 segment). In the kidney of the homozygous $Umod^{C93F}$ mutant mouse, SCD1 appeared to be abundant in a larger fraction of proximal tubular cells compared to the kidney of the wild-type mouse, and the average staining intensity of SCD1 positive cells appeared to be more prominent. Age of mice analyzed: four months. $Umod^{wt}$: wild-type mouse; $Umod^{C93F}$: homozygous $Umod^{C93F}$ mutant mouse. Uromodulin immunohistochemistry enabled identification of TALH segments. Proximal tubule segment are morphologically characterized by luminal microvilli. Chromogen: DAB for SCD1, Vector RED for UMOD; nuclear staining: hemalum. **(B)** Protein abundance of SCD1 in whole kidney lysate of homozygous $Umod$ mutant mice of both lines was increased compared to wild-type mice. Signal intensities of SCD1 were corrected for GAPDH signal intensities of the same PVDF-membrane. Mean of protein abundance of wild-type mice was set on a value of 1 [mean (wild-type) = 1]. One-way-ANOVA with Tukey's Multiple Comparison Post hoc Test: p vs. wild-type, **, $p<0.01$; ***, $p<0.001$. Age of mice analyzed: four months.

homozygosity of *Umod* mutation causes a stronger UMOD maturation defect and TALH dysfunction compared to heterozygous *Umod* mutant mice. Severity of UMOD maturation defect and severity of UAKD phenotype are similar between homozygous $Umod^{A227T}$ mutant mice and age-matched heterozygous $Umod^{C93F}$ mutant mice. UAKD is more progressed in 9-month-old heterozygous $Umod^{C93F}$ mutant mice than in 3-month-old homozygous $Umod^{A227T}$ mutant mice analyzed. Thus, the latter mice showed only an initial UAKD phenotype whereas the first group already exhibited progressed secondary morphological kidney lesions like interstitial fibrosis and inflammatory cell infiltrations. Consequently, on the one side, this sample selection of kidneys for genome-wide transcriptome profiling exhibiting different disease stages of UAKD might be responsible for the low number of identified genes commonly differentially abundant in both *Umod* mutant mouse lines. On the other side, these DEGs which were further positive validated by RT-qPCR of additional groups of UAKD-affected mice (*Angptl7*, *Odc1*, *Wfdc15b* and *Scd1*) might be the genes most constantly affected by UAKD irrespective of disease stage and severity. We selected two of the overlapping DEGs (*Angptl7* and *Scd1*) for more detailed analyses.

Angiopoietin-like protein 7 (ANGPTL7) was identified to be less abundant in TALH cells of UAKD affected kidneys. ANGPTL7 is a member of the ANGPTL protein family exhibiting structural homology to the angiopoietins, which have important functions in angiogenesis [24]. However the function of ANGPTLs might differ from angiopoietins as they do not bind to the receptors classically targeted by angiopoietins. So far, biological function of ANGPTL7, which is a secreted protein, is poorly understood. Due to its role in extracellular matrix formation of trabecular meshwork of the eye, a role of ANGPTL7 in glaucoma was assumed [25]. In this context, ANGPTL7 was postulated as

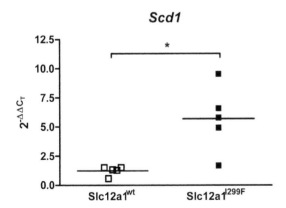

Figure 4. Evaluation of the role of TALH-dysfunction derived salt wasting state on renal *Scd1* transcript abundance. Increased *Scd1* transcript abundance in kidneys of homozygous $Slc12a1^{I299F}$ mutant mice compared to the kidneys of littermate controls were detected by RT-qPCR analysis. Data are shown as scatter dot plot with mean (n = 5 per group). Age of mice analyzed: three months. Student's t test: p vs. wild-type, *, $p<0.05$.

potential target gene of the WNT/β-catenin signaling pathway [26]. Recently, ANGPTL7 was identified as a biomarker upregulated selectively in the early stage of acute kidney injury, with higher transcript abundances 4 to 10 hours after ischemia reperfusion injury (according to the data in patent #EP25828 40A1, http://www.google.com/patents/EP2582840A1?cl=en). Further, strong upregulation of *Angptl7* transcription was detected in kidneys of hypertensive mice [27]. In UAKD, TALH cells are the site of primary pathogenesis due to mutant UMOD maturation retardation causing ER hyperplasia and TALH dysfunction. Decreased ANGPTL7 abundance in UAKD-affected TALH cells could be on the one side related to impaired synthesis due to impaired ER capacity and dysfunction. On the other side, as the biological function of ANGPTL7 in TALH cells is so far unknown, decreased ANGPTL7 synthesis might also contribute to TALH cell dysfunction leading to impaired kidney function in UAKD.

Increased abundance of stearoyl-coenzyme A desaturase 1 (SCD1) was present in proximal tubule segment of UAKD affected kidneys. SCD1 is a lipogenic enzyme catalyzing the critical step in the biosynthesis of monounsaturated fatty acids of cellular lipids [28]. The degree of unsaturation of cellular lipids is known to influence cell signaling and membrane fluidity and, thus, cell function. Due to its critical role in cell function, SCD1 expression is highly regulated. For instance, SCD1 expression can be induced by glucose, saturated fatty acids, insulin, and by the actions of lipogenic transcription factor sterol regulatory element binding protein-1c (SREBP-1c) and the nuclear receptor LXR. In the kidney, SCD1 was reported to be selectively expressed in the proximal straight tubular cells [29], which is in line with our results. It is an astonishing finding that disturbed TALH function as it is present in UAKD influenced transcript abundance of a gene selectively expressed in the tubule segment upstream of TALH. One key feature of UAKD represents a reduced renal fractional excretion of uric acid [3], and uric acid reabsorption and excretion is mainly regulated in the proximal tubule segment [30]. As a compensatory mechanism due to volume depletion caused by TALH dysfunction in UAKD, increased sodium absorption in proximal tubule, which is coupled with uric acid cotransport, was assumed in UAKD, leading to increased reabsorption and decreased excretion of uric acid [12]. Thus, higher SCD1 expression in proximal tubule of UAKD affected kidneys could be due to functional adaption of proximal tubule function to compensate TALH dysfunction for maintaining fluid and electrolyte homeostasis of the body. This hypothesis was underlined by our finding of increased *Scd1* abundance in kidneys of homozygous *Slc12a1*^{I299T} mutant mice. These mice exhibited Type I Bartter syndrome with salt-wasting polyuria and reduced uric acid clearance due to an inactivating mutation of the Na$^+$-K$^+$-2Cl$^-$ ion transporter, which represents the main ion transporter of the TALH segment [14].

A direct cross talk between the two functionally distinct tubular segments, TALH and the morphologically contiguous S3 segment of proximal tubule, was assumed due to protective properties of UMOD, located in TALH segment, to acute injury damage in proximal tubules [31]. So, recovery of proximal tubule injury by ischemia was delayed in *Umod* knockout mice, which could be drawn back to increased inflammatory response in injured proximal tubule and increased neutrophil infiltration in *Umod* knockout mice compared to the ischemic injury kidney response in wild-type mice. A conceptual model of the UMOD dependent protective cross-talk of TALH segment on S3 segment of proximal tubule in acute kidney injury was proposed, postulating three putative paths of cross-talk between TALH and proximal tubule

segment: (1) direct role of basolaterally translocated and released UMOD in down-regulation of inflammatory signaling in neighboring proximal tubule, (2) mediation of cross-talk by a putative secondary paracrine mediator released by TALH segment, or (3) involvement of interstitial cells in mediating cross-talk between the two functionally distinct tubular segments [32]. The increased abundance of SCD1 in straight segment of proximal tubule of UAKD-affected kidneys, as shown in our study, might be a further indication of the impact of TALH segment on the proximal tubule segment. The first proposed path of cross-talk due to reported translocated UMOD after ischemic injury might be of minor importance in UAKD-affected kidneys as we did not observe extracellular UMOD immunopositivity in renal interstitium. However, NF-κB signaling is activated in UAKD-affected TALH segment [11] what could be a putative mediator pathway of nephron segment cross-talk but this hypothesis has to be investigated in further analyses. Increased abundance of SCD1 itself might influence function of proximal tubule cells, by influencing their lipid metabolism and thus may alter the composition of phospholipids, triglycerides and cholesterol ester content in cell membrane [29]. Further, protective effects against lipotoxicity of free cholesterol and free saturated fatty acids, improvement of cell membrane fluidity and enhancement of cell function and integrity are assumed due to increased SCD1 abundance. Thus, the assumed increased sodium- and uric acid-cotransport in proximal tubule might be caused not only to compensate volume depletion in UAKD but also might be in part caused by altered cell function of proximal tubule cells itself.

In conclusion, transcriptome profiling analyses in whole kidneys of two *Umod* mutant mouse models for UAKD at two different stages of the disease resulted in the description of differentially regulated genes which were further classified according to disease and/or functions annotations. Localization and quantification analyses of ANGPTL7 and SCD1 gave novel hints for the function of these proteins in healthy and UAKD affected kidneys. A cross talk between two functionally distinct tubular segments, the TALH segment and the S3 segment of proximal tubule, was demonstrated, which might occur by direct contact or due to functional compensatory properties.

Supporting Information

Figure S1 Clinical and morphological phenotype of young adult and aged *Umod*^{A227T} and *Umod*^{C93F} mutant mice. (A) Plasma urea concentrations of young adult and aged *Umod*^{A227T} and *Umod*^{C93F} mutant mice [13]. Data points show means ± SD. Age of clinical-chemical analysis is indicated. n = 6–16 per genotype and *Umod* mutant line. Two-way-ANOVA with Bonferroni Multiple Comparison Post hoc Test: Homozygous *Umod* mutants vs. wild type: ***, $p < 0.001$; Heterozygous *Umod* mutants vs. wild type: ###, $p < 0.001$. (B) Multifocal tubulointerstitial fibrosis and moderate inflammatory cell infiltration predominantly in the corticomedullary region were found in UAKD-affected kidneys of 12-month-old heterozygous *Umod*^{C93F} mutant mice. These histological alterations were not found in three-month-old young adult homozygous *Umod*^{A227T} mutant mice (not shown, [12]). Histological staining, age of mice and genotype are indicated. Het: heterozygous mutant, homo: homozygous mutant of the indicated *Umod* mutant mouse line.

Figure S2 Heat map of DEGs in kidney between homozygous *Umod*^{A227T} mutant mice compared to wild-type controls. The genes are ranked according to their fold change from largest positive to largest negative value. The first column shows the mean linear fold changes calculated as ratio of the mean mutant intensities (n = 4) and the arithmetic mean signal of the corresponding wild-type mice (n = 4). Columns represent the regulated genes for single mutant mice and the rows the differential expression of a gene across the animals. The color code displays the fold changes according to the scale bar at the bottom of the heat map: Blue represents down- and yellow up-regulation in mutant animals compared to wild-types.

Figure S3 Heat map of DEGs in kidney between heterozygous $Umod^{C93F}$ mutant mice compared to wild-type controls. The genes are ranked according to their fold change from largest positive to largest negative value. The first column shows the mean linear fold changes calculated as ratio of the mean mutant intensities (n = 4) and the arithmetic mean signal of the corresponding wild-type mice (n = 4). Columns represent the regulated genes for single mutant mice and rows represent the differential expression of a gene across the animals. The color code displays the fold changes according to the scale bar at the bottom of the heat map: Blue represents down- and yellow up-regulation in mutant animals compared to wild-types.

Figure S4 Functional classification of DEGs in kidney of homozygous $Umod^{A227T}$ mutant mice. The first three columns list the categories of the diseases or functions annotations and the respective p-value for each over-represented term. The following columns show those DEGs associated with each term including a heat map and the mean fold change. The genes of each term are ranked according to the mean fold changes. A column represents single mutant mice and rows the differential expression across the

animals. The color code displays the fold changes according to the scale bar at the bottom of the heat map: Blue represents down- and yellow up-regulation in mutant animals compared to wild-types.

Figure S5 Functional classification of DEGs in kidney of heterozygous $Umod^{C93F}$ mutant mice. The first three columns list the categories of the diseases or functions annotations and the respective p-value for each over-represented term. The following columns show those DEGs associated with each term including a heat map and the mean fold change. The genes of each term are ranked according to the mean fold changes. A column represents single mutant mice and rows the differential expression across the animals. The color code displays the fold changes according to the scale bar at the bottom of the heat map: Blue represents down- and yellow up-regulation in mutant animals compared to wild-types.

Acknowledgments

We would like to thank Sandra Geißler for excellent technical support on cDNA microarray performance, and Stephanie Schmid for excellent technical support on immunohistological analyses.

Author Contributions

Conceived and designed the experiments: MH JB BR EW BA EK. Performed the experiments: MH JB EK. Analyzed the data: MH JB EK. Contributed reagents/materials/analysis tools: HF VGD MHdA EW. Wrote the paper: MH EK.

References

1. Hart TC, Gorry MC, Hart PS, Woodard AS, Shihabi Z, et al. (2002) Mutations of the UMOD gene are responsible for medullary cystic kidney disease 2 and familial juvenile hyperuricaemic nephropathy. J Med Genet 39: 882–892.

2. Rampoldi L, Caridi G, Santon D, Boaretto F, Bernascone I, et al. (2003) Allelism of MCKD, FJHN and GCKD caused by impairment of uromodulin export dynamics. Hum Mol Genet 12: 3369–3384.

3. Rampoldi L, Scolari F, Amoroso A, Ghiggeri G, Devuyst O (2011) The rediscovery of uromodulin (Tamm-Horsfall protein): from tubulointerstitial nephropathy to chronic kidney disease. Kidney Int 80: 338–347.

4. Serafini-Cessi F, Malagolini N, Cavallone D (2003) Tamm-Horsfall glycoprotein: biology and clinical relevance. Am J Kidney Dis 42: 658–676.

5. Tamm I, Horsfall FL Jr (1950) Characterization and separation of an inhibitor of viral hemagglutination present in urine. Proc Soc Exp Biol Med 74: 106–108.

6. Bates JM, Raffi HM, Prasadan K, Mascarenhas R, Laszik Z, et al. (2004) Tamm-Horsfall protein knockout mice are more prone to urinary tract infection: rapid communication. Kidney Int 65: 791–797.

7. Mo L, Zhu XH, Huang HY, Shapiro E, Hasty DL, et al. (2004) Ablation of the Tamm-Horsfall protein gene increases susceptibility of mice to bladder colonization by type 1-fimbriated Escherichia coli. Am J Physiol Renal Physiol 286: F795–F802.

8. Mo L, Huang HY, Zhu XH, Shapiro E, Hasty DL, et al. (2004) Tamm-Horsfall protein is a critical renal defense factor protecting against calcium oxalate crystal formation. Kidney Int 66: 1159–1166.

9. Darisipudi MN, Thomasova D, Mulay SR, Brech D, Noessner E, et al. (2012) Uromodulin triggers IL-1beta-dependent innate immunity via the NLRP3 inflammasome. J Am Soc Nephrol 23: 1783–1789.

10. Trudu M, Janas S, Lanzani C, Debaix H, Schaeffer C, et al. (2013) Common noncoding UMOD gene variants induce salt-sensitive hypertension and kidney damage by increasing uromodulin expression. Nat Med 19: 1655–1660.

11. Kemter E, Sklenak S, Rathkolb B, Hrabe de Angelis M, Wolf E, et al. (2014) No amelioration of uromodulin maturation and trafficking defect by sodium 4-phenylbutyrate in vivo: studies in mouse models of uromodulin-associated kidney disease. J Biol Chem 289: 10715–10726.

12. Kemter E, Rathkolb B, Rozman J, Hans W, Schrewe A, et al. (2009) Novel missense mutation of uromodulin in mice causes renal dysfunction with alterations in urea handling, energy, and bone metabolism. Am J Physiol Renal Physiol 297: F1391–1398.

13. Kemter E, Prueckl P, Sklenak S, Rathkolb B, Habermann FA, et al. (2013) Type of uromodulin mutation and allelic status influence onset and severity of uromodulin-associated kidney disease in mice. Hum Mol Genet 22: 4148–4163.

14. Kemter E, Rathkolb B, Bankir L, Schrewe A, Hans W, et al. (2010) Mutation of the Na+-K+-2Cl- cotransporter NKCC2 in mice is associated with severe polyuria and a urea-selective concentrating defect without hyperreninemia. Am J Physiol Renal Physiol 298: F1405–F1415.

15. Nicklas W, Baneux P, Boot R, Decelle T, Deeny AA, et al. (2002) Recommendations for the health monitoring of rodent and rabbit colonies in breeding and experimental units. Lab Anim 36: 20–42.

16. Fuchs H, Gailus-Durner V, Adler T, Aguilar-Pimentel JA, Becker L, et al. (2011) Mouse phenotyping. Methods 53: 120–135.

17. Kemter E, Pruckl P, Rathkolb B, Micklich K, Adler T, et al. (2013) Standardized, systemic phenotypic analysis of Umod(C93F) and Umod(A227T) mutant mice. PLoS One 8: e78337.

18. Horsch M, Schadler S, Gailus-Durner V, Fuchs H, Meyer H, et al. (2008) Systematic gene expression profiling of mouse model series reveals coexpressed genes. Proteomics 8: 1248–1256.

19. Tusher VG, Tibshirani R, Chu G (2001) Significance analysis of microarrays applied to the ionizing radiation response. Proc Natl Acad Sci U S A 98: 5116–5121.

20. Saeed AI, Sharov V, White J, Li J, Liang W, et al. (2003) TM4: a free, open-source system for microarray data management and analysis. Biotechniques 34: 374–378.

21. Edgar R, Domrachev M, Lash AE (2002) Gene Expression Omnibus: NCBI gene expression and hybridization array data repository. Nucleic Acids Res 30: 207–210.

22. Ruijter JM, Ramakers C, Hoogaars WM, Karlen Y, Bakker O, et al. (2009) Amplification efficiency: linking baseline and bias in the analysis of quantitative PCR data. Nucleic Acids Res 37: e45.

23. Gamba G (2005) Molecular physiology and pathophysiology of electroneutral cation-chloride cotransporters. Physiol Rev 85: 423–493.

24. Santulli G (2014) Angiopoietin-like proteins: a comprehensive look. Front Endocrinol (Lausanne) 5: 4.

25. Kuchtey J, Kallberg ME, Gelatt KN, Rinkoski T, Komaromy AM, et al. (2008) Angiopoietin-like 7 secretion is induced by glaucoma stimuli and its concentration is elevated in glaucomatous aqueous humor. Invest Ophthalmol Vis Sci 49: 3438–3448.

26. Comes N, Buie LK, Borras T (2011) Evidence for a role of angiopoietin-like 7 (ANGPTL7) in extracellular matrix formation of the human trabecular meshwork: implications for glaucoma. Genes Cells 16: 243–259.

27. Puig O, Wang IM, Cheng P, Zhou P, Roy S, et al. (2010) Transcriptome profiling and network analysis of genetically hypertensive mice identifies

potential pharmacological targets of hypertension. Physiol Genomics 42A: 24–32.

28. Sampath H, Ntambi JM (2011) The role of stearoyl-CoA desaturase in obesity, insulin resistance, and inflammation. Ann N Y Acad Sci 1243: 47–53.

29. Zhang Y, Zhang X, Chen L, Wu J, Su D, et al. (2006) Liver X receptor agonist TO-901317 upregulates SCD1 expression in renal proximal straight tubule. Am J Physiol Renal Physiol 290: F1065–1073.

30. Choi HK, Mount DB, Reginato AM (2005) Pathogenesis of gout. Ann Intern Med 143: 499–516.

31. El-Achkar TM, McCracken R, Rauchman M, Heitmeier MR, Al-Aly Z, et al. (2011) Tamm-Horsfall protein-deficient thick ascending limbs promote injury to neighboring S3 segments in an MIP-2-dependent mechanism. Am J Physiol Renal Physiol 300: F999–1007.

32. El-Achkar TM, Wu XR (2012) Uromodulin in kidney injury: an instigator, bystander, or protector? Am J Kidney Dis 59: 452–461.

9

A Bumpy Ride on the Diagnostic Bench of Massive Parallel Sequencing, the Case of the Mitochondrial Genome

Kim Vancampenhout[1], Ben Caljon[2], Claudia Spits[1], Katrien Stouffs[1,2], An Jonckheere[3], Linda De Meirleir[1,3], Willy Lissens[1,2], Arnaud Vanlander[4], Joël Smet[4], Boel De Paepe[4], Rudy Van Coster[4], Sara Seneca[1,2]*

1 Research Group Reproduction and Genetics (REGE), Vrije Universiteit Brussel (VUB), Brussels, Belgium, 2 Center for Medical Genetics, UZ Brussel, Vrije Universiteit Brussel (VUB), Brussels, Belgium, 3 Department of Pediatric Neurology, UZ Brussel, Vrije Universiteit Brussel (VUB), Brussels, Belgium, 4 Department of Pediatrics, Division of Pediatric Neurology and Metabolism, University Hospital Ghent, Ghent University, Ghent, Belgium

Abstract

The advent of massive parallel sequencing (MPS) has revolutionized the field of human molecular genetics, including the diagnostic study of mitochondrial (mt) DNA dysfunction. The analysis of the complete mitochondrial genome using MPS platforms is now common and will soon outrun conventional sequencing. However, the development of a robust and reliable protocol is rather challenging. A previous pilot study for the re-sequencing of human mtDNA revealed an uneven coverage, affecting predominantly part of the plus strand. In an attempt to address this problem, we undertook a comparative study of standard and modified protocols for the Ion Torrent PGM system. We could not improve strand representation by altering the recommended shearing methodology of the standard workflow or omitting the DNA polymerase amplification step from the library construction process. However, we were able to associate coverage bias of the plus strand with a specific sequence motif. Additionally, we compared coverage and variant calling across technologies. The same samples were also sequenced on a MiSeq device which showed that coverage and heteroplasmic variant calling were much improved.

Editor: Robert Lightowlers, Newcastle University, United Kingdom

Funding: This work was supported by Fonds voor Wetenschappelijk Onderzoek Vlaanderen (FWO; www.FWO.be) G.0.200; The 'Association Belge contre les Maladies Neuro-Musculaires (ABMM)' (http://www.hospichild.be/fr/associations/maladies-neuro-musculaires/association-belge-contre-les-maladies-neuro-musculaires-asbl-abmm-maladies-genetiques), and Vrije Universiteit Brussel (with reference OZR1928 and OZRMETH3). Authors who received funding: CS LDM SS. The funders had no role in study design, data collection and analysis, decision to publish, or preparation of the manuscript.

Competing Interests: The authors have declared that no competing interests exist.

* Email: sara.seneca@uzbrussel.be

Introduction

The human mitochondrial DNA (mtDNA) is a small circular double stranded molecule that comprises 16569 bp and codes for 13 protein genes, 22 tRNAs and 2 rRNAs. All these are essential elements to the correct function of the oxidative phosphorylation (OXPHOS) system, a fundamental process of the cellular role of mitochondria. For over 25 years, the pathogenicity of certain alterations of the mitochondrial genome has been clearly established in mtDNA disease. Despite the existence of mutation hotspot genes and regions, and the occurrence of recurrent mutations, these pathogenic aberrations are scattered over the entire mitochondrial genome. This makes it necessary to completely analyze this small genome to confirm or exclude pathogenic mtDNA changes. Molecular analysis often requires different and complementary methods, e.g. Southern blot, long range (LR)-PCR, Denaturing Gradient Gel Electrophoresis (DGGE), High Resolution Melting (HRM), quantitative (q)PCR and Sanger sequencing for the detection and quantification of

mtDNA. The emergence of MPS technologies has provided the diagnostic bench with a new and highly valuable tool for the evaluation of human mtDNA integrity. However, these new sequencing platforms have pitfalls, and crucial biases might be created [1] such as the loss of coverage in regions with GC-extreme (high or low) content, or the limited ability to analyze homopolymeric stretches [2] [3]. As a result, heteroplasmic variant calling might be severely complicated or even erroneous, as the nucleotide representation can be too weak or unreliable in some of these regions. In a recent study by Seneca et al. [4], the mitochondrial genomes of 32 DNA samples were analyzed using an Ion Torrent PGM system after enrichment with LR-PCR amplification of the mtDNA. A major bias in read depth between the positive and negative strand was seen for almost 10% of the mitochondrial genome, despite the fact that the sequencing was carried out at an average coverage of 6000. Moreover, in some regions the data for the positive strand dropped severely, reaching a critically low coverage. This difference in read depth between both strands made it challenging to distinguish true low-level heteroplasmic variants from sequencing errors. Therefore, we tried to develop an improved MPS-based protocol for the analysis of the human mitochondrial genome. Several library preparation methods and sequencing technologies were tested in order to ameliorate the present sequencing protocol, and their outputs were compared. We were also able to identify the specific nature of the systematically undercovered nucleotide motifs. We are convinced that our findings are of interest to all laboratories working on MPS for the mtDNA, both in a research or clinical setting.

Materials and Methods

Ethics Statement

This study was approved by the ethics committee of the Institutional Review Board (IRB) of the University Hospital (UZ Brussel, Vrije Universiteit Brussel). For all control samples a written informed consent was obtained. The informed consent form was also reviewed and approved by the local ethics committee of the IRB. For the patient samples, during clinical consultation oral consent was given to study their genetic material by any methods relevant to diagnostically confirm or rule out mutations in their mtDNA. This procedure does not require a written consent by the patient, and oral consent is recorded in a protected medical patient file. This is a standard procedure that is approved within the Center for Medical Genetics and accepted by the ethics committee of the IRB of the hospital.

Sample collection and DNA

Six DNA samples, corresponding to three controls (samples 1, 2, 4 in [4]) and three patients (samples 9, 14, 21 in [4]), were randomly selected from the previous sample cohort [4]. Total DNA had been extracted from leukocytes using standard DNA isolation techniques (Chemagen, Perkin Elmer, Zaventem, Belgium). An overview of the samples and techniques used is given in Supporting Information S1.

Long range PCR

MPS data files, obtained from a previous study, were mainly generated by the sequencing of three overlapping LR-PCR fragments covering the whole mitochondrial genome (all six samples were amplified using the 'three overlapping' fragment approach, two were additionally generated with a 'single fragment' method) [4]. However, as was demonstrated in a previous study, one large single LR-PCR product allowed the detection of variants, indels and large deletions simultaneously, a situation that is advantageous due to time and cost constrains for clinical genetic testing. For this single LR-PCR a 16.2 kb fragment [5] was generated using the LongAmp *Taq* PCR kit (New England Biolabs, Bioke, Leiden, The Netherlands). The mitochondrial genome was amplified from 200 ng gDNA as template in a 50 µL PCR assay according to manufacturer's recommendations. The PCR protocol was adapted to an initial 30 s denaturation at 94°C, followed by 15 cycles with first a denaturation of 10 s at 92°C, annealing at 67°C for 30 s and an extension of 10 min at 68°C. This was followed by 18 cycles with a denaturation of 10 s at 92°C and an extension of 10 min +20 s every cycle at 68°C. A final extension step was performed at 68°C for 7 min. Successful PCR amplification was assessed using 0.8% agarose gel electrophoresis, and products were purified with AMpure beads (Analis, Champion, Belgium).

Ion Torrent PGM sequencing

Ion Torrent semi-conductor sequencing technology detects the incorporation of each of the four nucleotides as small changes in pH that are provoked by the release of a proton. Library and template preparation include an amplification step. The latter is known as an emulsion PCR which takes place in aqueous droplets suspended in oil.

The data files of six samples, previously sequenced using the Ion Torrent PGM assay according to the manufacturer's instructions [4], were regarded as benchmark material for a comparative study of the new protocols described in the present study. We evaluated the following modifications to the standard protocol: different shearing methodologies and avoiding the amplification step in the library preparation of the Ion Torrent PGM protocol. To test the fragmentation methods, LR-PCR products were sheared using the Covaris M220 sonicator (Life Technologies Europe, Gent, Belgium) and the NEBNext dsDNA Fragmentase (Bioke). For the first fragmentation method, a dilution to 100 ng in 50 µL of LR-PCR products were subjected to sonication for 130 s with a duty factor of 20%, a peak incident power of 50W, a temperature of 20°C and 200 cycles per burst, to tailor the DNA molecules into fragments with a median size of 200 bp (Ion Xpress Plus gDNA Fragment Library Preparation, Appendix B). A standard procedure was followed for the NEBNext dsDNA Fragmentase assay. Briefly, 1 µg of PCR product was added to 2 µL 10x Fragmentase reaction buffer and 0.2 µL of 100x BSA. This mixture was placed on ice for 5 min prior to the addition of 2 µL of NEBNext dsDNA Fragmentase and an incubation at 37°C for 30 min. The reaction was stopped by adding 5 µL of 0.5 M EDTA solution to the DNA fragments. Sheared samples were purified using AMPure beads. The size distribution of the fragmented DNA was assessed on the Bioanalyzer (Agilent, Diegem, Belgium), using the High Sensitivity Assay (Agilent, Diegem, Belgium). All further downstream manipulations were performed according to the Ion Torrent PGM protocol's instructions (Ion Xpress Plus gDNA Fragment Library preparation, Life Technologies, Gent, Belgium). Briefly, samples were end repaired, ligated with adaptors, nick repaired and bead purified prior to amplification of size selected (E-gel system, Life Technologies) fragments around 330 bp long. Fragment sizes were assessed using the Bioanalyzer system and quantified with the Qubit 2.0 fluorimeter (Life Technologies, Gent, Belgium). Pooled libraries were used for emulsion PCR amplification. Sequencing reactions were run on the Ion Torrent PGM using Ion 316 version 2 chips and the Ion PGM 200 sequencing kit (Life Technologies, Gent, Belgium).

Illumina MiSeq sequencing

To obtain 350 bp fragments LR-PCR products were sheared with the Covaris M220 sonicator (Life Technologies Europe, Gent, Belgium) and the NEBNext dsDNA Fragmentase enzyme (Bioke, Leiden, The Netherlands), both starting with 1 µg LR-PCR product. Covaris sheared LR-PCR products were fragmented using custom instrument specifications (TruSeq DNA PCR-Free Sample Preparation Guide). The protocol described, before concerning the NEBNext dsDNA Fragmentase, was the same except for the incubation time that was adapted to 15 min to obtain 350 bp fragments. Next, samples were further processed using the TruSeq DNA PCR-Free Sample Preparation protocol as instructed by the supplier (Illumina, Eindhoven, The Netherlands). After fragmentation, end repair, adenylation, and indexed paired end adapter ligation, samples were pooled and processed on the MiSeq sequencer with the MiSeq Reagent Micro Kit, v2 (Illumina). Conversely, all six samples were also processed using the Nextera XT kit (Illumina). A single Nextera tagmentation enzymatic reaction was used where LR-PCR products were simultaneously fragmented and tagged with adaptors. Finally, a limited cycle PCR protocol (12 cycles) was applied, adding simultaneously sequencing indexes (Nextera XT DNA Sample Preparation Guide, Illumina).

Detection threshold determination for the MiSeq

The technical error rate of the MiSeq platform was determined with the methodology used for the Ion Torrent PGM system [4]. For the latter device, which unlike PhiX for the Illumina MiSeq lacks an endogenous control sample, a well typed pUC19 plasmid was used. The use of the same pUC19 DNA sample also allowed a comparison of sequencing results across platforms. One µg of pUC19 plasmid DNA (Thermo Fisher, Erembodegem-Aalst, Belgium) was sheared by the Covaris or NEBNext dsDNA Fragmentase. Subsequently, samples were processed using the TruSeq DNA PCR-Free Sample Preparation protocol, and sequenced on the MiSeq. The error rate of the sequencing process was computed by calculating the ratio of non-reference versus total bases per position. Taking the average of all ratios per position resulted in the average error rate of the pUC19 plasmid DNA.

Data analysis

FastQ files from all datasets, generated by either the Ion Torrent PGM or MiSeq platforms, were mapped to the mitochondrial revised Cambridge Reference Sequence (rCRS, NC 012920.1) using BWA-MEM (version 0.7.5) [6]. As a metric for coverage bias, the relative coverage was used. Applying the SAMtools software (version 0.1.18) [7] the number of reads mapping to each reference base was counted. The mean coverage was calculated by averaging this value across each base in the sequence. By computing the ratio of the coverage of a given reference base and the mean coverage of all reference bases, the relative coverage was obtained. This was calculated for the plus and minus strand separately, for the total coverage of both strands together, and was presented in graphical illustrations. To visualize the relative coverage resulting from all different protocols and methods tested, circular plots were generated with the freeware Circos-0.64 software [8]. The Circos plots demonstrated in this article are restricted to sample 1, as the coverage profiles were consistent across all samples. To compare different methodologies, datasets were down sampled to an average coverage of 3000 using Picard (http://picard.sourceforge.net). The average relative coverage was collected for all samples processed with the same protocol resulting in seven datasets (Ion Torrent standard, Ion Torrent without amplification step, Ion Torrent Covaris, Ion Torrent NEBNext dsDNA Fragmentase, TruSeq Covaris, TruSeq NEBNext dsDNA Fragmentase and Nextera XT). For each dataset the fraction with a relative coverage <0.50; <0.25; <0.10; <0.05 and <0.01 was determined. To identify the nucleotide composition of undercovered regions GC, AT along with CT, AG, AC and GT dinucleotide motif plots were created and correlated to the total relative coverage, as well as the relative coverage from each strand separately. Both the incidence (in percentages) of the dinucleotide motifs in the mtDNA molecule, and the relative coverage were calculated in bins of 150 nucleotides and illustrated as bias plots.

For variant calling, three different strategies were employed and compared. First, all data were analyzed using an in-house pipeline based on GATK. FastQ files were aligned to the rCRS using BWA-MEM and sorted. Next, GATK realignment around indels and recalibration was performed. The GATK Unified Genotyper was used for variant calling, without at random down sampling of reads to reduce coverage. Subsequently, all variants with a quality score <400 were filtered from the vcf data. Second, all data were also analyzed using the CLC Genomics Workbench (version 6.0.5)

against the rCRS. Only variants with an average quality score > 25 were selected. A third and last strategy was only implemented on the Ion Torrent data. PGM files were mapped and variants were called using the Torrent Suite 4.2.

For each sample analyzed with the Ion Torrent PGM or MiSeq device, the sequencing error was determined for each position of the genome sequence, with exception of the true variants (versus rCRS) detected in each sample. The average sequencing error and their standard deviations were determined for these six samples. Potential low heteroplasmic variant levels were compared to these values and utilized as a reliable baseline (index) to reduce the false positive rate of the data [4].

Results and Discussion

Assessment of different PGM protocols

We have recently studied the use of the Ion Torrent PGM sequencer system in a diagnostic setting for the nucleotide analysis of human mitochondrial genomes of patient and control samples. The results uncovered a rather poor performance for some of the mtDNA regions [4]. Although it is well known that the PGM sequencing technology has problems handling homopolymeric stretches, an additional limitation was revealed, as a major difference in read depth between both strands was exposed for about 10% of the mitochondrial genome regions. For these sequences, the relative coverage of the positive strand dropped below 0.1. These particular patterns were reproduced in replicates of the same and between different samples, but never observed for pUC19 plasmid samples (Figure 1A). The causes of this remained unknown. Previous experiments had already excluded primer, LR-PCR or sample dependence, and it was assumed that the discrepancy originated from the enzymatic shearing step included in the Ion Torrent assay [4]. Altering fragmentation in the original Ion Torrent PGM assay could thus promote a change of the coverage profile. Hence, the standard enzymatic shearing step was omitted and substituted with an enzymatic treatment with NEBNext dsDNA Fragmentase or with physical shearing with a Covaris M220 sonicator device, leaving all further downstream process steps unchanged. Nonetheless, MPS data demonstrated that none of the altered protocols induced an equilibrated strand representation. Neither did they show an improvement of the under-representation of the plus strand. Uneven coverage was still produced (Figure 1B). Both shearing methods resulted still in 7 to 7.8% of the 16.2 kb fragment to have a relative coverage of the plus strand <0.1. Moreover, 2% of the 16.2 kb region showed a relative coverage of the plus strand <0.01 (Table 1). Further experiments, such as omission of the first PCR amplification step in the PGM library preparation protocol were carried out and subjected to MPS. But also this intervention did not lead to a reduced bias (Figure 1C, Table 1). By exchanging the Platinum Taq DNA polymerase for Kapa HiFi in the nick translation and amplification step during library preparation, Quail et al. [9] had demonstrated a reduced bias in PGM data. Therefore, it was proposed that the DNA amplification treatment during the library preparation and/or the emulsion PCR mediated a bias interfering with all further analysis of the mitochondrial genome. In order to further characterize the underlying mechanisms of poor PGM results across parts of the mitochondrial genome, the depth of the relative coverage seen at each position was tabulated for both strands separately. Hence, a possible association with its nucleotide composition was investigated systematically. In-house Perl scripts were used to calculate the content of GC or AT rich motifs, as well as any other dinucleotide rich combination. This analysis did not disclose any relationship between GC or AT rich regions and poor strand representation (Figure 2). The findings of Quail et al. [9] about very low coverage from GC or AT rich motifs for *P.falciparum* were not confirmed by the analysis of mtDNA. In contrast, reduced coverage was detected for AC and CT rich motifs. Particularly, coverage of the plus strand was negatively influenced by these two motifs. Moreover, the relative coverage of the plus strand dropped almost to zero for 80% (and more) AC rich motifs (Figure 2). The sequencing bias is seen for a high AC-content (range 70–80%) which corresponds to the figures of 80% and more for the GC and AT motifs presented by Ross et al. [1]. It is already known for a long time that the nucleotide composition of both mtDNA strands is different. The plus strand or *light* strand is C-rich, while the minus strand or *heavy* strand is G-rich. The rCRS is based on the L-strand and corresponds to the underrepresented plus strand in our sequencing results. The analyses were also performed for the pUC19 plasmid DNA. As expected, no correlation between its nucleotide composition and coverage data was observed (Supporting Information S2). We therefore hypothesize that the troughs generated by the Ion Torrent PGM system rather originate from the proliferation of the sheared mtDNA sequences and not from the fragmentation method *per se*. In fact, it might be inherent to the combination of the DNA polymerases used in the PCR amplification steps included in the standard protocols, and the nature of the mitochondrial genome sequence.

Comparison PGM-MiSeq

We proceeded to study mitochondrial genome resequencing on a MiSeq platform, using two different strategies. The results of the PCR amplification free protocol of TruSeq were compared with those of the Nextera XT kit, a method including one PCR amplification step in the library preparation step. Experiments were carried out according to the manufacturer's instructions. The average read depth for the different datasets generated with the MiSeq were 3723, 4701 and 19418 for the TruSeq Covaris, TruSeq NEBNext dsDNA Fragmentase and the Nextera XT methods, respectively. The reads generated by MiSeq (paired end reads), had a 150 bp fixed length, while reads generated by Ion Torrent PGM showed a variable single-end read length with an average of 145 bp. To compare different methodologies, datasets were down sampled to an average coverage of 3000. Relative coverage analysis showed a major improvement in strand equilibration for the TruSeq data. Data from the TruSeq sheared with the Covaris protocol, and the TruSeq enzymatically digested with NEBNext dsDNA Fragmentase achieved an impressive relative coverage, with few areas (only 1.6% and 1.6%, respectively) of the plus strand <0.5. The Nextera XT data did not show strand bias as seen with the PGM data. However, a general unevenness of coverage of both strands was seen. Indeed, regions of both strands (9.2% of the plus strand and 9.6% of min strand) showed a relative coverage <0.5. (Figure 1D, Table 1) which were associated with CT rich motifs. Unlike for the PGM, where mainly the positive strand was involved, both strands were affected, however not as severe as for the Ion Torrent data (Figure 2).

Detection limit of the MiSeq

The detection threshold for the identification of base variants was set on 5% for the Ion Torrent chemistry. This value was based on the determination of the sequencing error and the sensitivity and specificity experiments previously performed [4]. To set the detection threshold for the MiSeq, the same pUC19 plasmid DNA sample was sheared with two different methods, once using the Covaris M220 sonicator and secondly using the NEBNext dsDNA

Fragmentase. Both differentially sheared samples were sequenced on the MiSeq following TruSeq PCR free library preparation and a 100% coverage was obtained with an average read depth of 30 440 and 30 966 respectively. Similar average sequencing error results were obtained with 0.27% and 0.19% for the Covaris sheared sample and the enzymatic sheared sample respectively. These values are in concordance with the error rate obtained by the PhiX, which presented with an error rate of 0.35%. These error rates in turn correspond to previously reported data for the MiSeq platform [9]. By applying these results to determine the variant threshold for the mitochondrial resequencing, a detection threshold level of 2% is possible. However as the PGM data were previously investigated with a detection threshold of 5%, these settings were also used for the MiSeq data.

Variant calling

Last, we assessed variant detection in all samples using the data panel of nucleotide alterations reflecting the Sanger sequencing previously performed. The majority of these variants were identified on both platforms (Table 2; Supporting Information S3). Results were collected for a PGM, TruSeq or Nextera XT dataset. Two variant calling pipelines, an in-house pipeline based on GATK and the Quality-based variant detection method (CLC Genomics Workbench) were applied to MiSeq datasets, and subsequently compared to the results of our previous study. The TS4.2 was only used with the PGM data. The first pipeline resulted in 99.5% of the variants detected in the TruSeq and Nextera XT dataset, while the PGM dataset showed a 92.4% concordance with the Sanger sequencing results. The CLC Genomics Workbench pipeline requires the variant to be present on both strands. 93.4%, 97.7% and 97.2% of the Sanger sequencing variants were called in the TruSeq Covaris, TruSeq NEBNext dsDNA Fragmentase and the Nextera XT dataset, respectively. Applying these terms to the PGM data resulted in 84.7% concordance with Sanger sequencing. However, omitting the strand parameter identified 95.2% of the variants for PGM data. These figures demonstrated clearly the effect of strand bias on variant calling for the PGM data. Indeed, 67 out of 98 false negative results were present on one strand only. An additional analysis with the TS4.2 software identified 96.6% of the variants. Three positions, m.294T>C, m.16183A>C and the polymorphic 302_316 region, presented as false negative results in the PGM data sets. An additional false negative variant, at position m.5899_5900insC escaped variant calling. All of these variants are situated near a homopolymeric stretch and, with the exception of m.5899_5900insC, are also located in regions with significant AC contents and its associated strand bias (relative coverage of the plus strand <0.2). It must be pointed out that, despite the well documented shortcoming in homopolymer calling, the propriety software is clearly well fitted for the PGM needs in variant calling. Comparing the various algorithms applied in this present and the previous study, the TS4.2 software was noticeably the better performer. Compared to the former TS3.6 version, a remarkable improvement was noticed for the false positive rate. Reanalyzing all PGM samples with the TS4.2 release showed a reduction in false positives from 13,4% to 8,9%, with a detection threshold level of 5%. The highest sensitivity for the MiSeq results (TruSeq and Nextera XT data) was obtained by our in-house pipeline based on GATK. Indeed, the only false negative result for these data was one specific variation in the polymorphic 302_316 region in sample 21. Two single nucleotide insertions were detected in this region with Sanger sequencing (m.309_310insC and m.315_316insC), but MiSeq identified them incorrectly as a heteroplasmic sequence mixture of molecules with an insertion of

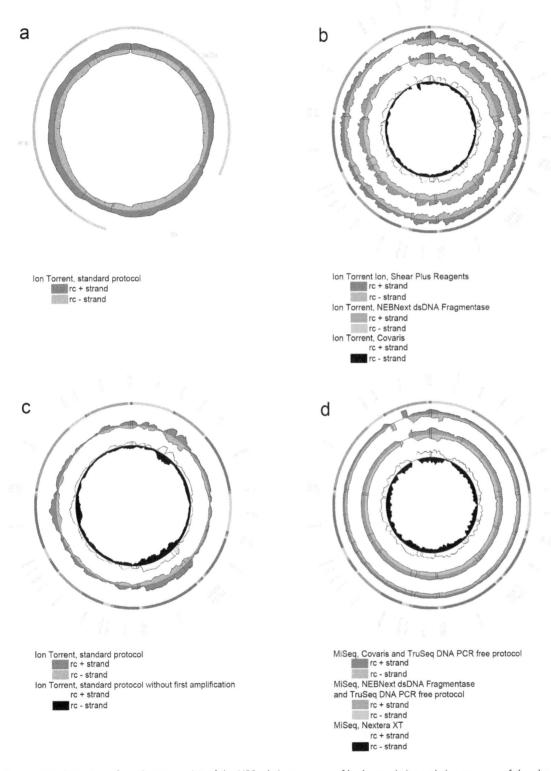

Figure 1. Genome Coverage plots. Representation of the MPS relative coverage of both strands (rc+: relative coverage of the plus strand, rc-: relative coverage of the negative strand) of the pUC19 plasmid, or mtDNA molecules obtained from the Ion Torrent PGM or MiSeq sequencing system. The outer circle symbolizes the pUC19 (A) or mtDNA (B, C, D) gene structure, respectively. **1A:** Use of the Ion Torrent PGM standard protocol on the pUC19 plasmid. **1B:** Use of three different fragmentation methods in combination with the Ion Torrent sequencing protocol on the mtDNA: Ion Shear Plus Reagents (enzymatic), NEBNext dsDNA Fragmentase (enzymatic) and Covaris (physical). **1C:** Use of an Ion Torrent PGM protocol without PCR amplification in the library construction on the mtDNA. **1D:** LR-PCR products of the mtDNA were Covaris (physical) or NEBNext dsDNA Fragmentase (enzymatic) sheared, followed by a TruSeq DNA PCR free protocol on a MiSeq instrument. The same six samples were processed with a Nextera XT kit (enzymatic shearing and PCR amplification in library preparation) prior to MiSeq analysis.

Table 1. Comparison between different methods and technologies based on relative coverage (RC) analysis of the data.

Ion Torrent PGM

RC	Standard			no library amplification			Covaris			NEBNext ds Fragmentase		
	Total	Plus	Min	Total	Plus	Min	Total	Plus	Min	Total	Plus	Min
<0.5	15.14	23.43	3.96	16.10	24.32	4.64	13.15	22.33	2.58	11.17	19.25	1.60
<0.25	1.36	13.12	0.02	1.10	13.73	0.09	0.88	12.70	0.35	0.27	11.52	0.29
<0.10	0.01	7.66	0.01	0.04	8.38	0.04	0.01	7.83	0.01	0.00	7.05	0.01
<0.05	0.00	5.47	0.01	0.00	6.02	0.01	0.00	5.81	0.01	0.00	4.95	0.00
<0.01	0.00	2.04	0.00	0.00	2.49	0.00	0.00	2.73	0.00	0.00	1.96	0.00

Illumina MiSeq

RC	TruSeq-Covaris			TruSeq-NEBNext ds Fragmentase			Nextera XT		
	Total	Plus	Min	Total	Plus	Min	Total	Plus	Min
<0.5	0.27	1.56	1.47	0.39	1.64	1.14	7.03	9.20	9.57
<0.25	0.01	0.81	0.83	0.01	0.73	0.42	1.64	2.61	2.48
<0.10	0.01	0.38	0.35	0.01	0.12	0.02	0.01	0.26	0.01
<0.05	0.00	0.23	0.20	0.00	0.01	0.00	0.01	0.06	0.01
<0.01	0.00	0.00	0.00	0.00	0.00	0.00	0.00	0.00	0.00

For all samples processed with a same protocol the average relative coverage was calculated and resulted in 7 different datasets. For each dataset, the fraction with a relative coverage <0.50, <0.25, <0.10, <0.05, <0.01 was determined. These analyses were performed for each strand separately (Plus, Min) and the total relative coverage (Total).

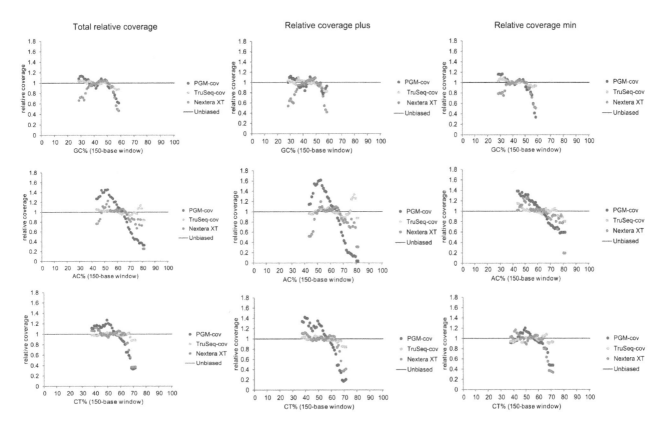

Figure 2. Nucleotide GC, AC and CT bias plots for the human mtDNA. The relative coverage as seen in this illustration is based on the average of the relative coverage of the six samples processed with the different protocols: Covaris shearing followed by the Ion Torrent protocol, Covaris shearing followed by the TruSeq procedure and the Nextera XT method. The average relative coverage was calculated for the total relative coverage and for both strand separately.

two or four C's at position 309. Analysis of the Ion Torrent PGM data previously had revealed four variants (m.7989T>C, m.9769T>C, m.10866T>C, m.12071T>C) hitherto not identified by Sanger sequencing. These same variants were identified by the Illumina system with analogous allele frequencies (Table 3). In this project, a detection threshold of 5% was used for all data analysis. However, it must be pointed out that a more stringent detection threshold of 2% is possible for both the PGM and MiSeq data. From a diagnostic perspective these low detection limits are not always relevant. Most pathogenic mutations have a disease threshold well >60%. Nonetheless, in the context of genetic counseling of asymptomatic female relatives for family planning, low detection limits might be indicated. Adjusting the detection limit to 2% in our sample cohort identified two additional heteroplasmic variants on the MiSeq platform. A novel heteroplasmic variant m.8207C>T (p.(Pro208Ser)) in the *MT-CO2* gene was revealed in the mtDNA of leukocytes of patient 9. Another heteroplasmic variant, m.5609T>C, was identified in leukocytes of patient 14 in the *MT-TA* gene. Both allele frequencies, 2% and 4% respectively, were below the applied detection limit of the PGM sequencer. Both nucleotide variants, however, were acknowledged by the PGM data as well, as was indicated by review of the BAM files in IGV and reanalysis of the data using a detection threshold of 0.8% (corresponding to the sequencing error rate of the PGM device). It must be pointed out that, although the accuracy of low level heteroplasmy determination is heavily dependent on the depth of coverage, it is also defined by the sequencing error of the system. The latter being related to

PCR, platform technologies, and the various algorithms implemented at the different steps of data processing.

Conclusion

MPS analysis is a powerful tool able to simultaneously detect and quantify sequencing variants. However, diagnostic settings have high demands regarding accuracy of test results. A high sensitivity is crucial to avoid a misdiagnosis, while a low false positive rate is necessary to minimize additional Sanger sequencing work for confirmation of pathogenic discoveries. Our current findings have illustrated that MPS protocols demand a thorough evaluation of their data, and validation of the result files before a possible implementation as a diagnostic test should be considered. In many laboratories MPS analysis is now part of daily diagnostic work. Selecting an appropriate methodology for MPS projects envisioned deserves the necessary attention. Assessment of the nucleotide content of DNA samples to be analyzed proved here to be an essential parameter, among others, for evaluation of the performance of a sequencing methodology or technology. In our hands, the current Ion Torrent PGM standard assay, even with modifications, suffered from lack of coverage consistency of the L-strand of the human mitochondrial genome, making an evaluation of heteroplasmy in these underrepresented regions cumbersome. Comparison of the PGM and MiSeq Nextera XT data results with the MiSeq PCR free sequencing method suggest that coverage bias might be generated by the enzymes involved in the amplification rounds of the MPS processes. Indeed, the Nextera XT method, which included a PCR amplification step, produced also more variation in coverage than the samples processed with

Table 2. Comparison between the number of variants detected with the MPS and Sanger sequencing technologies.

Sample	Sanger sequencing	PGM[a] Ion shear enzymes vs Sanger	FN	extra	Covaris vs Sanger	FN	extra	NEBNext dsDNA Fragmentase vs Sanger	FN	extra	MiSeq[b] Covaris vs Sanger	FN	extra	NEBNext dsDNA Fragmentase vs Sanger	FN	extra	Nextera XT vs Sanger	FN	extra
1	33	32	1	1	32	1	1	32	1	1	33	–	1	33	–	1	33	–	1
2	13	11	2	1	12	1	1	12	1	1	13	–	1	13	–	1	13	–	1
4	33	32	1	–	31	2	–	31	2	–	33	–	–	33	–	–	33	–	0
9	36	35	1	2	35	1	2	35	1	2	36	–	3	36	–	2	36	–	2
14	57	56	1	–	56	1	–	56	1	–	57	–	1	57	–	1	57	–	–
21	42	40	2	–	41	1	–	41	1	–	41	1	–	41	1	–	41	1	–
Total	**214**	**206**	**8**	**4**	**207**	**7**	**4**	**207**	**7**	**4**	**213**	**1**	**6**	**213**	**1**	**6**	**213**		

FN: false negative result, extra: additional low allele frequency variants identified compared to Sanger sequencing.
[a] Ion Torrent PGM results obtained with the TS4.2 software.
[b] MiSeq data obtained with the in-house GATK pipeline.

Table 3. Overview of the minor allele frequency (in %) of heteroplasmic variants detected in this study.

Sample	Variant	PGM Ion shear	Covaris	NEBNext	average	stddev	MiSeq Covaris	NEBNext	Nextera XT	average	stddev
1	m.12071T>C	12.2	12.2	12.1	12.2	0.1	12	12	23	15.7	6.4
2	m.7989T>C	17.1	14.7	14.1	15.3	1.6	19	19	13	17.0	3.5
9	m.9769T>C	9.7	9	8.5	9.1	0.6	8	8	8	8.0	0.0
9	m.10866T>C	7.3	6	6.4	6.6	0.7	7	9	7	7.7	1.2
9	m.8207C>T	1.5	1.3	1.5	1.4	0.1	2	1	2	1.7	0.6
14	m.5609T>C	4	8	4.8	5.6	2.1	4	4	4	4.0	0.0
14	m.7453G>A	52	55	53	53.3	1.5	53	54	56	54.3	1.5

the TruSeq DNA PCR-Free Sample Preparation protocol. Nextera XT certainly reduced, but did not resolve the coverage inconsistency. Further modifications, such as the use of another DNA polymerase in both amplification steps of the standard PGM workflow (one in the library preparation, and another in the emulsion PCR) may lead to further improvements. However, this might be a complex process and beyond the time management and financial scope of this project. Consequently, at this very moment the TruSeq DNA PCR-Free Sample Preparation protocol on the MiSeq system might be the most appropriate technology to address low copy number mtDNA heteroplasmy adequately.

Supporting Information

Supporting Information S1   Overview experiments.

Supporting Information S2   Nucleotide GC, AC and CT bias plots for the pUC19 plasmid. pUC19 DNA was processed with the Standard Ion Torrent protocol, Covaris

sheared followed by the TruSeq procedure and the Nextera XT method.

Supporting Information S3 Overview of the variant calling results obtained by the TS4.2 software for the Ion Torrent data and our in-house pipeline based on GATK for all MiSeq data and Ion Torrent PGM data. P: Ion Torrent PGM sequencing; T: TruSeq DNA PCR-Free Sample Preparation protocol; NXT: Nextera XT method; S: sample; Ion: Ion shear enzymes; C: Covaris; N: NEBNext dsDNA Fragmentase; TS4.2: analyzed with the Torrent Suite 4.2 software; GATK: analyzed with our in-house pipeline based on GATK.

Author Contributions

Conceived and designed the experiments: KV BC KS WL SS. Performed the experiments: KV BC. Analyzed the data: KV BC CS SS. Contributed reagents/materials/analysis tools: AJ LDM AV JS BDP RVC. Contributed to the writing of the manuscript: KV BC CS WL KS SS.

References

1. Ross MG, Russ C, Costello M, Hollinger A, Lennon NJ, et al. (2013) Characterizing and measuring bias in sequence data. Genome Biol 14: R51.
2. Dohm JC, Lottaz C, Borodina T, Himmelbauer H (2008) Substantial biases in ultra-short read data sets from high-throughput DNA sequencing. Nucleic Acids Res 36: e105.
3. Oyola SO, Otto TD, Gu Y, Maslen G, Manske M, et al. (2012) Optimizing illumina next-generation sequencing library preparation for extremely at-biased genomes. BMC Genomics 13: 1.
4. Seneca S, Vancampenhout K, Van Coster R, Smet J, Lissens W, et al. (2014) Analysis of the whole mitochondrial genome: translation of the Ion Torrent Personal Genome Machine system to the diagnostic bench? Eur. J. Hum. Genet. doi:10.1038/ejhg.2014.49.
5. Cheng S, Higuchi R, Stoneking M (1994) Complete mitochondrial genome amplification. Nat. Genet 7: 350–351.
6. Li H (2013) Aligning sequence reads, clone sequences and assembly contigs with BWA-MEM. Available: http://www.arxiv.org/abs/1303.3997. Accessed 29 April 2014.
7. Li H, Handsaker B, Wysoker A, Fennell T, Ruan J, et al. (2009) The Sequence Alignment/Map format and SAMtools. Bioinformatics 25: 2078–2079.
8. Krzywinski M, Schein J, Birol I, Connors J, Gascoyne R, et al. (2009) Circos: An information aesthetic for comparative genomics. Genome Res 19: 1639–1645.
9. Quail M, Smith ME, Coupland P, Otto TD, Harris SR, et al. (2012) A tale of three next generation sequencing platforms: comparison of Ion Torrent, pacific biosciences and illumina MiSeq sequencers. BMC Genomics 13: 341.

10

Zinc-Finger Nuclease Knockout of Dual-Specificity Protein Phosphatase-5 Enhances the Myogenic Response and Autoregulation of Cerebral Blood Flow in FHH.1[BN] Rats

Fan Fan[1], Aron M. Geurts[2], Mallikarjuna R. Pabbidi[1], Stanley V. Smith[1], David R. Harder[3], Howard Jacob[2], Richard J. Roman[1]*

1 Department of Pharmacology and Toxicology, University of Mississippi Medical Center, Jackson, Mississippi, United States of America, 2 Human and Molecular Genetics Center, Medical College of Wisconsin, Milwaukee, Wisconsin, United States of America, 3 Department of Physiology and Cardiovascular Research Center, Medical College of Wisconsin, Milwaukee, Wisconsin, United States of America

Abstract

We recently reported that the myogenic responses of the renal afferent arteriole (Af-Art) and middle cerebral artery (MCA) and autoregulation of renal and cerebral blood flow (RBF and CBF) were impaired in Fawn Hooded hypertensive (FHH) rats and were restored in a FHH.1[BN] congenic strain in which a small segment of chromosome 1 from the Brown Norway (BN) containing 15 genes including dual-specificity protein phosphatase-5 (Dusp5) were transferred into the FHH genetic background. We identified 4 single nucleotide polymorphisms in the Dusp5 gene in FHH as compared with BN rats, two of which altered CpG sites and another that caused a G155R mutation. To determine whether Dusp5 contributes to the impaired myogenic response in FHH rats, we created a Dusp5 knockout (KO) rat in the FHH.1[BN] genetic background using a zinc-finger nuclease that introduced an 11 bp frame-shift deletion and a premature stop codon at AA121. The expression of Dusp5 was decreased and the levels of its substrates, phosphorylated ERK1/2 (p-ERK1/2), were enhanced in the KO rats. The diameter of the MCA decreased to a greater extent in Dusp5 KO rats than in FHH.1[BN] and FHH rats when the perfusion pressure was increased from 40 to 140 mmHg. CBF increased markedly in FHH rats when MAP was increased from 100 to 160 mmHg, and CBF was better autoregulated in the Dusp5 KO and FHH.1[BN] rats. The expression of Dusp5 was higher at the mRNA level but not at the protein level and the levels of p-ERK1/2 and p-PKC were lower in cerebral microvessels and brain tissue isolated from FHH than in FHH.1[BN] rats. These results indicate that Dusp5 modulates myogenic reactivity in the cerebral circulation and support the view that a mutation in Dusp5 may enhance Dusp5 activity and contribute to the impaired myogenic response in FHH rats.

Editor: Jaap A. Joles, University Medical Center Utrecht, Netherlands

Funding: This work was funded in part by National Institutes of Health R01 HL36279 and DK104184 (RJR), H105997 (Harder), GO grant HL-101681 (Jacob) and New innovator award OD-8396 (Geurts), VA Research Career Scientist Award (Harder) and Scientist Development Grant 13SDG14000006 (Pabbidi) from American Heart Association. The funders had no role in study design, data collection and analysis, decision to publish, or preparation of the manuscript.

Competing Interests: The authors have declared that no competing interests exist.

* Email: rroman@umc.edu

Introduction

The myogenic response is an intrinsic property of vascular smooth muscle cells (VSMC) that initiates contraction of arterioles in response to elevations in transmural pressure [1,2] and contributes to autoregulation of renal and cerebral blood flow (RBF, CBF). [3–6] We recently reported that the myogenic responses of the renal afferent arteriole (Af-Art) and middle cerebral artery (MCA) and autoregulation of RBF and CBF were impaired in Fawn Hooded hypertensive (FHH) rats and were restored in a FHH.1[BN] congenic strain in which a small segment of chromosome 1 from the Brown Norway (BN) containing 15 genes, including dual-specificity protein phosphatase-5 (Dusp5) were transferred into FHH genetic background. [7–9] However, the genes that contribute to the impaired myogenic response and the mechanisms involved remain to be determined.

Dusp5 is a serine-threonine phosphatase that inactivates MAPK activity[10–14] by dephosphorylating ERK1/2 MAP kinases [15] which modulate the activities of the large conductance Ca^{2+}-activated K^+ channel (BK) and transient receptor potential (TRP) channels. Both of these channels influence vascular reactivity and the myogenic response. [1,16–19] In the present study, we found that there were 17 SNPs in the Dusp5 gene in FHH relative to BN rats. One SNP was in the 5′-UTR and three were in the coding region. Of these, two altered potential CpG methylation sites and one introduced a G155R mutation. To determine whether Dusp5 regulates vascular tone and reactivity and if the sequence variants in this gene contribute to the impaired myogenic response in FHH rats, we created and characterized a Dusp5 Zinc-finger nuclease

(ZFN) knockout (KO) rat in the FHH.1[BN] genetic background since transfer of this region of chromosome 1 containing the Dusp5 gene was shown to restore the myogenic response in cerebral arteries. We first compared the myogenic response of the MCA and autoregulation of CBF in Dusp5 KO, FHH.1[BN] and FHH rats. We then compared the expression of Dusp5 in multiple tissues isolated from Dusp5 ZFN KO, FHH.1[BN] and FHH rats. We also investigated whether there are differences in the expression of p-ERK1/2 in cerebral microvessels isolated from these strains as they are the primary substrates normally dephosphorylated and inactivated by Dusp5 [15,20].

Materials and Methods

General

Experiments were performed on 33 FHH, 68 FHH.1[BN] and 92 Dusp5 KO male rats bred in our in house colonies and 16 age-matched Sprague-Dawley (SD) male rats purchased from Charles River Laboratories (Wilmington, MA). The animal care facility at the University of Mississippi Medical Center is approved by the American Association for the Accreditation of Laboratory Animal Care. The rats had free access to food and water throughout the study and all protocols received prior approval by the Institutional Animal Care and Use Committees (IACUC) of the University of Mississippi Medical Center.

Identification and confirmation of SNPs in Dusp5 in FHH versus FHH.1[BN] rats

We first performed an *in silico* analysis of the sequence of the Dusp5 gene in FHH versus BN rats, which is publically available from the Rat Genome database (RGD, http://rgd.mcw.edu/rgdweb/report/gene/main.html?id=620854). To confirm that the SNPs identified in the database are present in our FHH and FHH.1[BN] colonies at both the DNA and mRNA levels, we isolated genomic DNA from tail biopsies using PureLink Genomic DNA Kits (Life Technologies, Grand Island, NY) and RNA from cerebral arteries using TRIzol solution (Life Technologies, Grand Island, NY) and sequenced across the regions of interest. RNA (1 μg) was reverse transcribed using an iScript cDNA Synthesis Kit (Bio-Rad) to produce cDNA. The regions of interest were amplified in a 25 μl PCR reaction containing 25 ng of genomic DNA or 4 ng of cDNA, 25 ng of each primer, 20 mM Tris-HCl buffer (pH 8.4), 50 mM KCl, 1.5 mM MgCl₂, 200 μM of each dNTP and 0.5 U Taq DNA polymerase (QIAGEN) using several primer pairs to cover the full length sequence. For amplification of exon 1, the following forward and reverse primers were used: 5′-AGCTTTCCGGGGCAGCGAGTG-3′ and 5′-TCAGGA-TACTGTGAGTAGAAG-3′. Exons 2, 3 and a portion of exon 4 that is in the codon region were amplified using the following forward and reverse primers: 5′-CGTGCTGGACCAGGG-CAGCCG-3′ and 5′-GACAGAGAGAGGTCTTCAGTATTG-3′. Intronic primers were used to amplify the regions of interest from genomic DNA. Amplification of exon 1 or 2 required an additional 5 μl of Q solution since these regions were GC-rich. The PCR products were purified using a PureLink PCR Purification Kit (Life Technologies, Grand Island, NY) and then ligated into a pCR4-TOPO TA vector (Life Technologies, Grand Island, NY) by incubating at room temperature for 20 min. One Shot MAX Efficiency DH5α-T1[R] Competent cells (Life Technologies, Grand Island, NY) were transformed using the ligated vectors according to manufacturer's instructions. The colonies were incubated at 37°C overnight in LB media with 100 μg/ml of Ampicillin. The plasmids were extracted using a QIAprep Spin Miniprep Kit (QIAGEN, Valencia, CA) and sequenced using M13

primers. The data were analyzed using ABI software (Applied Biosystems, Grand Island, NY) and compared to the BN reference sequence available on the NCBI GeneBank and RGD databases.

Comparison of the expressions of Dusp5, p-PKC and p-ERK1/2 in FHH and FHH.1[BN] rats

RT-qPCR. RNA was isolated from microdissected cerebral arteries in FHH and FHH.1[BN] rats that was reverse transcribed as described above. Fast SYBR Green Real-Time PCR Master Mixes (Life Technologies, Grand Island, NY) which contain a blend of dTTP/dUTP that is compatible with Uracil N-Glycosylase (UNG) to eliminate DNA contamination from PCR products synthesized in the presence of dUTP were mixed with 4 ng of cDNA and 25 ng of forward (5′-CTT AAA GGT GGG TAC GAG ACC TTC TAC -3′) and reverse (5′-GAG AAT GGG CTT TCC GCA CTG -3′) primers and amplified using a real-time PCR system (Mx3000P, Stratagene, La Jolla, CA). The data were analyzed with Mxpro qPCR software (Stratagene, La Jolla, CA) using the $2^{-\Delta\Delta CT}$ Method [21]. The PCR products were also separated on a 1% agarose gel using a Tris-borate-EDTA (TBE) buffer visualized the intensity of the bands with 100 mg/ml of ethidium bromide (Sigma, St. Louis, MO) and analyzed using ChemiDoc MP Imaging System (Bio-Rad, Hercules, CA).

Western Blot. Cerebral microvessels were isolated using the Evans blue sieving procedure as previously described [22,23] . Cerebral microvessels and brain tissue obtained from FHH and FHH.1[BN] rats were homogenized in ice-cold RIPA buffer (R0278, Sigma-Aldrich, St. Louis, MO) in the presence of protease and phosphatase inhibitors (Cat# 88663, Thermo Scientific, Pittsburgh, PA) and the proteasome inhibitor MG 132 (Sigma-Aldrich, St. Louis, MO) [12] using a ground glass homogenizer followed by a FastPrep-24 homogenizer (MP Biomedicals, Santa Ana, CA). The homogenate was centrifuged at 1,000 g for 10 minutes at 4°C. Aliquots of supernatant protein (40 μg for cerebral microvessels and 100 μg for brain tissues) were separated on a 10% SDS-PAGE gel, transferred to nitrocellulose membranes and probed with a pan p-PKC antibody (Cat 9371, Cell signaling, Danvers, MA) at 1:1,000 dilution. Antibodies that against p-ERK1/2 and total ERK1/2 (Cat# 4377 and Cat# 4695, Santa Cruz, Santa Cruz, CA) were used at a 1:1,000 dilution followed by a 1:4.000 dilution of a horseradish peroxidase (HRP)-coupled anti-rabbit secondary antibody. The membranes were then stripped and re-probed with a 1: 8,000 dilution of an anti-beta Actin antibody (ab6276, Abcam, Cambridge, MA) followed by a 1: 20,000 dilution of anti-mouse HRP-coupled secondary antibody as a loading control.

Generation of the Dusp5 ZFN KO rats in the FHH.1[BN] genetic background

ZFNs targeting the following sequence CAGGGCAGCCGC-CACtggcaGAAGCTGCGGGAGGA in exon 1 of the rat Dusp5 gene (NM_133578) were obtained from Sigma-Aldrich (St. Louis, MO) and were used to generate a Dusp5 KO rat in the FHH.1[BN] genetic background as previously described [24–26]. The ZFN mRNA was injected into the pronucleus [25,27] of fertilized FHH.1[BN] embryos and transferred to the oviduct of pseudopregnant females to generate Dusp5 ZFN KO founders. Tail biopsies were obtained and digested with 0.2 mg/ml proteinase K in a direct PCR lysis reagent (102-T, Viagen Biotech) at 85°C with rotation for 45 minutes. Founders were identified using the CEL-1 assay [28] and the mutations were confirmed by Sanger DNA sequencing [29]. Positive founders were backcrossed to parental strain to generate heterozygous F1 rats and the siblings were

SNP	Chr	RGSC Genome Assembly v3.4 start	stop	position on genomic DNA	position on mRNA	reference nuc	variant nuc	AA	reference AA	variant AA	variant ID
1	1	259754340	259754341	107	107 (5'UTR)	C(G)	T(G)				296415523
2	1	259754563	259754564	330	330 (exon 1)	G	T	52	L	L	296415524
3	1	259756067	259756068	1834	(intron 1)	T	C				293523829
4	1	259756482	259756483	2249		G	A				296415525
5	1	259757182	259757183	2949		G	A				296415526
6	1	259758392	259758393	4159		T	C				293523830
7	1	259758588	259758589	4355		A	G				296415527
8	1	259759032	259759033	4799		G	A				296415528
9	1	259759139	259759140	4906	627 (exon 2)	C(G)	T(G)	151	L	L	296415529
10	1	259759148	259759149	4915	637 (exon 2)	G	A	155	G	R	296415530
11	1	259759596	259759597	5363	(intron 2)	G	C				261240240
12	1	259759660	259759661	5427		T	C				296415531
13	1	259762138	259762139	7905		C	T				261240242
14	1	259763857	259763858	9624	(intron 3)	A	G				296415532
15	1	259766107	259766108	11874		T	C				293523831
16	1	259766563	259766564	12330	(intron 4)	G	C				293523832
17	1	259767470	259767471	13237		A	G				296415533

Figure 1. Comparison of sequence variants in the Dusp5 gene in FHH and Brown Norway (BN) rats. Analysis of the NeXT Generation sequence data available on the Rat Genome database (RGD, http://rgd.mcw.edu/rgdweb/report/gene/main.html?id=620854). These results indicate that there are 17 SNPs in the Dusp5 (NM -133578) gene in FHH/EurMcwi(variant nuc) rats as compared to Brown Norway (reference nuc) rats. Most of the SNPs are located in introns. There is a C107T SNP is located in the 5'-UTR, and three are found in the coding region including a G330T SNP in exon 1, a C627T and a G637A SNP in exon 2. The C107T SNP alters a CpG site and the C627T SNP alters one of the six 5 CpG's methylation sites previously identified in exon 2. The G637A SNP causes G155R mutation in the Dusp5 protein in FHH rats.

intercrossed to produce homozygous animals. Thereafter, the rats were genotyped using the following primers: Dusp5-F: 5'-GCT GCA GGA GGG CGG CGG CG -3', Dusp5 R: 5'-CTT TAA GGA AGT AGA CCC G -3'. These primers amplified a 155 bp band for the wild type allele and a 144 bp band for the knockout allele.

Characterization of the Dusp5 ZFN KO rats

Western Blot. Cerebral microvessels, liver, brain and spleen tissues were isolated from Dusp5 KO and FHH.1[BN] rats as described above. White blood cells were also harvested using Ficoll-Paque Premium 1.084 (GE Healthcare) according to the manufacturer's protocol. A 100 μg aliquot of protein isolated from brain, liver, spleen and white blood cells (WBCs) and a 40 μg aliquot of protein isolated from cerebral microvessels was separated on a 10% SDS-PAGE gel, transferred to nitrocellulose membranes which were probed with antibodies to Dusp5 (H00001847-M04, Abnova, Taiwan; 1:1,000) targeting AA286-384 in the C-terminus of the Dusp5 protein and antibodies raised against p-ERK1/2, total ERK1/2 and beta-Actin as described above.

Myogenic response on MCA. MCAs were microdissected from the brains of 9–12 week old Dusp5 KO, FHH.1[BN], FHH and SD rats and mounted on glass micropipettes in a myograph. The bath solution was equilibrated with O_2 (95%) and CO_2 (5%) to provide adequate oxygenation and to maintain at pH 7.4. The diameters of the vessels were measured using a videomicrometer (VIA-100, Boeckeler Instruments) at intraluminal pressures ranging from 40 to 140 mmHg in steps of 20 mmHg as previously described. [8,9,30]

Autoregulation of CBF. CBF autoregulation were determined on 9–12 week old male Dusp5 KO, FHH.1[BN], FHH and SD rats. The rats were anesthetized with ketamine (30 mg/kg, *i.m.*) and Inactin (50 mg/kg, *i.p.*) and were mechanically ventilated throughout experiment to maintain pO_2 and pCO_2 at 100 and 35 Torr, respectively. Body temperature was maintained

at 37°C during experiment. Catheters (PE-50) were placed in the femoral artery and vein and the rats received an intravenous infusion of 0.9% NaCl solution at a rate of 100 μl/min to replace surgical fluid losses. The scalp was exposed and the cranial bone was thinned 3 mm lateral and 6 mm posterior to the Bregma using a handheld drill until the pial vessels became visible through the thinned window. CBF was monitored bilaterally with a laser-Doppler flow meter (PF5001, Perimed Corp, Jarfalla, Sweden). After surgery and a 30-min equilibration period, mean arterial pressure (MAP) was lowered to 90–100 mmHg by increasing the depth of pentobarbital anesthesia (1–5 mg/kg, *i.v.*). Baseline regional CBF was measured, then systemic pressure was elevated in steps of 10–20 mmHg by graded *i.v.* infusion of phenylephrine (0.5–5 μg/min). [31–33] MAP was maintained for 3–5 min until a new steady-state level of CBF was obtained. CBF was expressed as a percentage of the baseline laser-Doppler flow signal.

Statistics

Mean values ± SEM are presented. The significance of the differences in the expression of various proteins and mRNA in FHH, FHH.1[BN] and Dusp5 KO rats was determined using one-way ANOVA. The significance of the differences in mean values between and within groups in the myogenic responses and autoregulation of CBF was determined using a two-way ANOVA for repeated measures and Holm-Sidak test for preplanned comparisons. A P value <0.05 was considered to be statistically significant.

Results

Sequence analysis and expression of Dusp5, p-ERK1/2 and p-PKC in FHH and FHH.1[BN] rats

The results of the comparative sequence analysis are presented in **Figure 1**. We identified 17 SNPs in the Dusp5 gene in FHH versus the BN reference sequence. Most of the SNPs were in introns, however, there were four SNPs in the Dusp5 mRNA

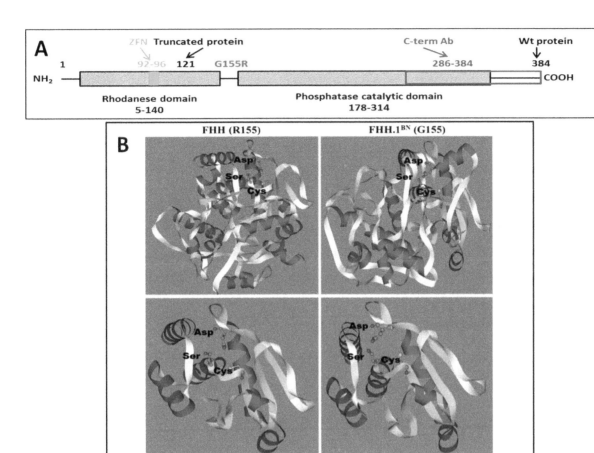

Figure 2. Identification of the Zn-finger target site and deletion in the Dusp5 KO strain. Panel A presents a schematic model of the Dusp5 protein. The Dusp5 Zinc-finger construct targets amino acids (AA) 92–96 in the N-terminal regulatory rhodanese domain (AA5-140) resulting in the introduction of a premature stop codon at AA121 that is predicted to produce a truncated protein. The Dusp5 antibody used in these studies targets AA286-384 in the C-terminal phosphatase catalytic domain (AA178-314). **Panel B** presents a comparison of I-TASSER predicted structure and the folding of the Dusp5 protein in FHH (155R) and FHH.1^{BN} (155G) rats. The upper panels show the predicted structure of the Dusp5 protein in both strains based on the complete AA sequence. The putative catalytic triad (Asp232/Ser268/Cys263) is shown in a "stick figure" form and the 3-letter AA codes are labeled in black. The rest of the protein is represented as ribbon running along the backbone. Secondary structural elements are depicted by color with helices, beta sheets and coils represented in red, cyan and white, respectively. The putative catalytic triad is magnified and shown in "stick figure" form in the lower panel and the 3-letter AA codes are labeled with in black. Only residues 174–320 of Dusp5 protein are presented in order to enhance the view of the putative catalytic triad. There are significant structural differences both in the overall folding of the Dusp5 protein that impact on the structure of the active site/catalytic triad region between the strains. This may account for the observed differences in the activity of the Dusp5 protein in FHH versus FHH.1^{BN} rats.

including a C107T SNP in the 5'-UTR, a G330T SNP in exon 1, a C627T and a G637A in exon 2. All of these SNPs were verified in our FHH and FHH.1^{BN} strains by sequencing cDNAs derived from mRNA extracted from the isolated cerebral vessels. The C107T SNP altered a CpG site and the C627T SNP altered one of six CpGs in exon 2 that were previously identified as methylation sites by bisulfite modification. [34]. Moreover, the G637A SNP caused a G155R mutation (**Figure 2A**) that is predicted using I-TASSER modeling package [35–37] may alter the folding of the protein and the active site conformation of the Dusp5 protein in FHH versus FHH.1^{BN} rats as shown in **Figure 2B.**

To determine if the C107T and C627T SNPs that altered CpG sites and the G637A SNP that caused a G155R mutation might alter the expression of Dusp5 in FHH versus FHH.1^{BN} rats, RT-qPCR and western blot experiments were performed. The results presented in **Figure 3A** indicate that the expression of Dusp5 mRNA is 2-fold higher in cerebral arteries of FHH than in the

FHH.1^{BN} control rats but the expression of protein is not different between these strains (**Figure 3C**). Moreover, as presented in **Figure 3B and 3C**, the expression of p-ERK1/2, the primary substrates for dephosphorylation by Dusp5, is significantly reduced in cerebral microvessels and the brains of FHH relative to FHH.1^{BN} rats, while total ERK1/2 levels are not significantly altered. The expression of p-PKC protein is also significantly elevated in FHH.1^{BN} as compared to FHH rats.

Generation and characterization of the Dusp5 ZFN KO rats

The homozygous Dusp5 knockout (KO) and wild type FHH.1^{BN} control strain were derived from an intercross of the heterozygous Dusp5 founders. Genotyping and sequencing of the Dusp5 KO strain indicated that there is a 14 bp deletion and a 3 bp insertion between nucleotides 449–464 in Dusp5 mRNA that creates a frame shift mutation which is predicted to introduce a premature stop codon at amino acid (AA) 121 (**Figure 2A,**

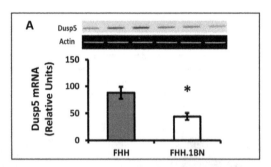

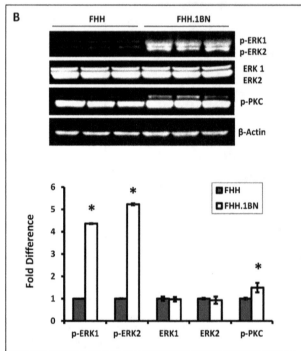

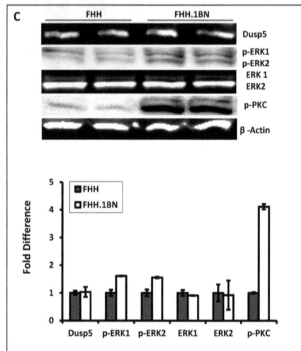

Figure 3. Comparison of the expression and activity of Dusp5 in FHH versus FHH.1^BN rats. Panel A presents a comparison of the expression of Dusp5 mRNA in cerebral arteries of FHH versus FHH.1^BN rats. The upper portion of the figure presents the representative images of gels showing the qPCR products and the bar graph below compares the expression levels. **Panel B** presents a comparison of the expression of phosphorylated-ERK1/2, total ERK1/2, phosphorylated-PKC and beta-Actin in the brain of FHH and FHH.1^BN rats. The upper panel presents the representative images and the lower panel presents the relative quantitation. **Panel C** presents a comparison of expression of these proteins in cerebral microvessels of FHH as compared with FHH.1^BN rats. All of the vessels isolated from one strain were pooled into a single sample. The upper panel presents the representative images and the lower panel presents the quantitation of the images. Mean values ± SE from 3 rats per strain are presented in Panel A and Panel B. Panel C represents the results from duplicate aliquots run from a single pooled microvessel sample isolated from 8–10 rats per strain. * indicates a significant difference from the corresponding value in FHH rats.

4A, 4B). The genotypes of the animals were verified using PCR as shown in **Figure 4C**.

Dusp5 is a ubiquitous protein that is abundantly expressed in the brain, spleen and WBCs [11,12,38]. The expression of Dusp5 protein in WBCs isolated from Dusp5 KO versus FHH.1^BN control rats was compared using an anti-Dusp5 antibody that targeted AA286-384 (**Figure 2A**) in the C-terminus of the protein to confirm that the introduction of the new stop codon produces a truncated Dusp5 protein in the KO animals. The results presented in **Figure 5** indicate that the expression of Dusp5 protein is markedly reduced in WBCs (**Figure 5A**) and cerebral microvessels (**Figure 5B**) isolated from Dusp5 KO rats relative to the FHH.1^BN control strain and in other tissues (brain, liver and spleen, data not shown). We also compared the expression of the primary substrates of Dusp5, p-ERK1/2, in these strains. The results presented in **Figure 5** indicate that the levels of p-ERK1/2

are enhanced, while the expression of total ERK1/2 is not significantly different in WBCs (**Figure 5A**) and cerebral microvessels (**Figure 5B**) in Dusp5 KO rats compared to FHH.1^BN control rats.

Comparison of the myogenic response of the MCA of Dusp5 KO and FHH.1^BN rats

The luminal diameter of the MCA in FHH.1^BN rats (n = 12) decreased by 20±2% when the perfusion pressure was increased from 40 to 140 mmHg. The myogenic response of MCA isolated from Dusp5 KO rats was significantly greater, and the diameter of these vessels decreased by 34±7% when the perfusion pressure was increased over the same range. In contrast, the myogenic response of the MCA of FHH rats was markedly impaired as the diameter of these vessels only increased by 10±4% when pressure was increased from 40 to 140 mmHg (**Figure 6A**). The passive

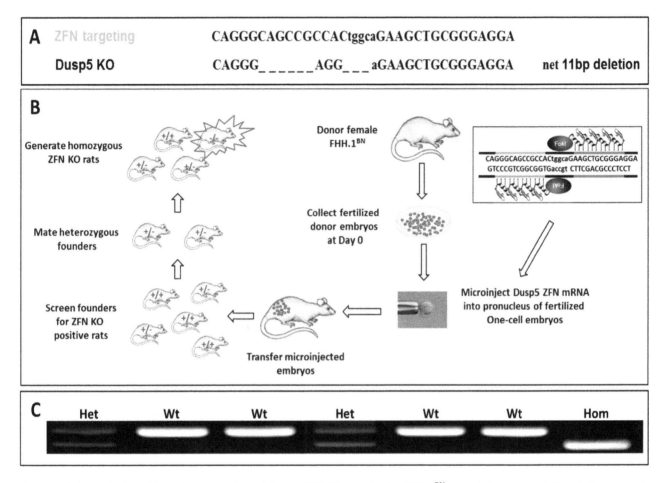

Figure 4. Schematic describing the generation of Dusp5 ZFN KO rats in the FHH.1^{BN} genetic background. Panel A presents the sequence of the Dusp5 ZFN. The Dusp5 specific ZFN introduced a 14 bp deletion and a 3 bp insertion resulting in a net 11 bp deletion between nucleotides 449–464 in the Dusp5 sequence in the KO animals that introduced a frame shift mutation. **Panel B** presents the strategy for the generation of Dusp5 ZFN KO rats in the FHH.1^{BN} genetic background. Fertilized donor embryos from female FHH.1^{BN} rats were collected and the Dusp5 ZFN mRNA was microinjected into pronuclei of the fertilized one-cell embryos. These embryos were transferred back to a foster mother. The heterozygous founders were brother-sister mated to generate the homozygous ZFN KO founders and the FHH.1^{BN} wild type control rats. **Panel C** presents an example of PCR genotyping of the region of interest in FHH.1^{BN} and Dusp5 KO rats. The rats were genotyped using the following primers: Dusp5-F: 5′-GCT GCA GGA GGG CGG CGG CG -3′, Dusp5 R: 5′-CTT TAA GGA AGT AGA CCC G-3′. These primers amplify a 155 bp band for the wild type allele and a 144 bp band for the knockout allele. Both bands are observed in heterozygous rats.

diameter curves generated in Ca^{2+} free solution in all strains were not significantly different (**Figure 6B**).

Comparison of the autoregulation of CBF of Dusp5 KO and FHH.1^{BN} rats

Autoregulation was markedly impaired and CBF increased by $54\pm6\%$ in FHH rats when MAP was increased from 100 to 160 mmHg. CBF was autoregulated to a greater extent in the FHH.1^{BN} and Dusp5 KO rats were not significant different and only increased by $26\pm3\%$ and $12\pm3\%$, respectively, when MAP was increased over the same range. The range of the autoregulation of CBF since CBF rose by $33\pm4\%$ when pressure was increased from 100 to 190 mmHg in Dusp5 KO rats versus an increase of $65\pm5\%$ in the FHH.1^{BN} rats and $99\pm3\%$ in the FHH animals (**Figure 7**).

Discussion

We recently reported that the myogenic response of the MCA and autoregulation of CBF were markedly impaired in FHH rats

and were restored in a FHH.1^{BN} congenic strain in which Chromosome 1 from the BN rats containing 15 genes was transferred into the FHH genetic background. [7–9] However, the gene or genes that contribute to the impairment of vascular function and the mechanisms involved still remain obscure. In the present study, we identified 17 SNPs in the Dusp5 gene in FHH versus FHH.1^{BN} rats. Most were in the intronic region, but four were in exons including a C107T SNP in the 5′-UTR, a G330T SNP in exon 1, a C627T and a G637A in exon 2. Both of the SNPs, C107T and C627T, altered CpG sites and the C637A SNP caused a G155R mutation. To determine if the altered CpG sites and/or the G155R mutation might underlie the loss of the myogenic response in FHH rats, we created Dusp5 KO rats in the FHH.1^{BN} genetic background using ZFN KO technology [25–27,39]. Site specific Zn-fingers fused to *Fok I* nuclease were introduced into the pronucleus of one cell embryos to induce double-strand DNA breaks at the target site, followed by error-prone non-homologous DNA repair which resulted in a frameshift mutation and formation of a premature stop codon leading to a truncated or non-functional protein. [25–27,40] The ZFNs

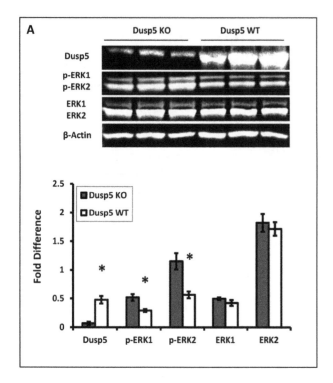

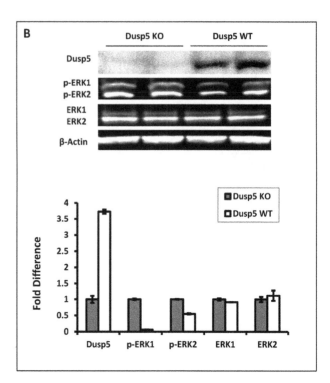

Figure 5. Comparison of the expression and activity of Dusp5 in WBCs isolated from Dusp5 ZFN KO versus FHH.1^{BN} rats. Panel A: The expression of Dusp5 protein in WBCs is nearly absent in Dusp5 KO versus FHH.1^{BN} rats. The levels of phosphorylated-ERK1/2 protein are significantly increased in Dusp5 KO compared to FHH.1^{BN} rats, but there is no change in the expression of total ERK or beta-Actin. **Panel B** presents the results of the expression of Dusp5, p-ERK1/2, total ERK1/2 and β-Actin protein from duplicate aliquots of a single pooled microvessel sample isolated from 8–10 Dusp5 KO and FHH.1^{BN} rats. The upper panel presents a representative image of the gels and the lower panel presents the quantitation of the images above. Mean values ± SE from 3 rats per strain are presented in Panel A. * indicates a significant difference from the corresponding value in Dusp5 KO rats.

targeted AA92-96 in the N-terminal regulatory rhodanese domain of Dusp5 protein that is the ERK1/2 binding site (**Figure 2A**). [41,42] The Dusp5 ZFN KO strain was successfully generated and we confirmed that this strain had a net 11 bp deletion between nucleotides 449–464 in Dusp5 mRNA that introduced a frame shift mutation and a premature stop codon at AA121 (**Figure 4A, 2A**). We also confirmed that the expression of Dusp5 at the protein level was nearly absent in multiple tissues including cerebral microvessels using an antibody directly targeting AA286-384 in the C-terminus that are beyond the predicted site of the newly introduced stop codon. The loss of Dusp5 protein in the KO animals was associated with an expected increase in p-ERK1/2 levels in various tissues and cerebral microvessels compared to FHH.1^{BN} wild type animals because Dusp5 is a serine-threonine phosphatase that inactivates MAPK activity [10–14] by specifically dephosphorylating p-ERK1/2 MAP kinases. [15,38]

An increase in the dephosphorylation of p-ERK1/2 by Dusp5 phosphatase would be expected to modulate BK and TRP channel activities [1,16–19] and downregulate PKC, Rho/ROCK[20,43] and STAT pathways[11] which are regulated by the MAP kinase system. Activation of BK and TRP channel activities alter the myogenic response of small blood vessels by modulating calcium entry in VSMCs. [1,19,22] Discovered over 100 years ago by Bayliss, the myogenic response is an intrinsic property of VSMC that initiates contraction of arterioles in response to elevations in transmural pressure. [1,2] It is impaired following cerebral vasospasm, stroke or traumatic brain injury and the autoregulatory range is shifted to higher pressures in hypertension [3,4,44–46]. Autoregulation of CBF is one of the major mechanisms to protect

the brain from elevations in perfusion pressure that promote vascular leakage and swelling of the brain [9,19,46]_ENREF_64. The myogenic response of MCA plays a major role in autoregulation of CBF and contributes about 50% to overall compensation to elevations in perfusion pressure [47].

In the present study, we found that the myogenic response of the MCA was greater in Dusp5 KO animals than in wild type controls and FHH rats. There was no difference in the passive pressure diameter relationships measured in Ca^{2+} free solution between these strains. The increased myogenic response was associated with enhanced autoregulation of CBF in response to elevations in systemic pressure from 100–160 mmHg in the Dusp5 KO compared to FHH.1^{BN} and FHH strains. Moreover, the range of effective autoregulation of CBF was extended to higher pressures in the Dusp5 KO rats versus FHH.1^{BN} and FHH animals. Our findings are consistent with the results of a recent study by Wickramasekera, *et al* demonstrating that downregulation of the expression of Dusp5 by siRNA in cultured cerebral arteries enhanced pressure-dependent myogenic constriction [20]. Together, these findings confirm that alteration in the expression or activity of Dusp5 modulates the myogenic response of the MCA *in vitro* and autoregulation of CBF *in vivo*.

We also examined whether the sequence variants we identified in the Dusp5 gene in FHH versus FHH.1^{BN} rats might contribute to the impaired myogenic response and autoregulation of CBF in FHH rats by altering the expression of Dusp5 and its phosphatase activity. Our RT-qPCR results indicated that the expression of Dusp5 at the message level in microdissected cerebral arteries was 2-fold higher in FHH relative to FHH.1^{BN} rats, but the expression

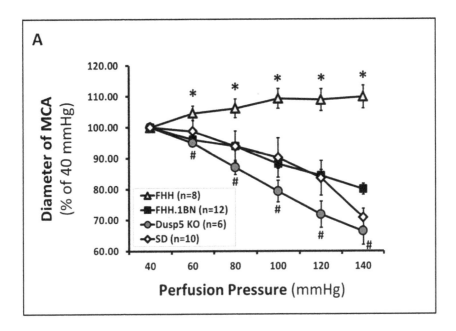

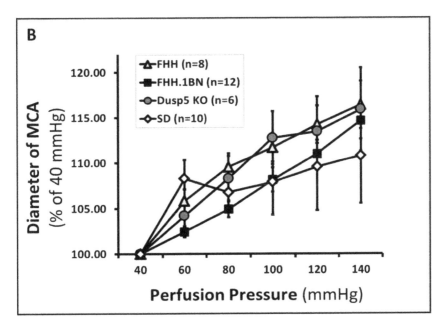

Figure 6. Comparison of the myogenic response in middle cerebral artery (MCA) isolated from Dusp5 ZFN KO versus FHH.1^{BN} and FHH rats. Panel A presents the passive pressure-diameter curves in Ca^{2+} free solution at each pressure in all strains. **Panel B**: The luminal diameter of the MCA decreased from 100 to $66 \pm 4\%$ in Dusp5 KO rats and from 100 to $80 \pm 2\%$ in FHH.1^{BN} rats when the perfusion pressure was increased from 40 to 140 mmHg, whereas it was dilated in FHH rats (from 100 to $110 \pm 4\%$). The MCA also constricted in Sprague Dawley rats that is widely used as a control strain for the myogenic response. Mean values $\pm$ SE are presented. Numbers in parentheses indicate the number of vessels studied per group. * indicates a significant difference in the corresponding value in FHH versus all the other strains. # indicates there is a significant difference between Dusp5 KO and FHH.1^{BN} rats.

at protein level was not significantly different. This suggests that the difference in the myogenic response of MCA between FHH and FHH.1^{BN} rats is not due to changes in the expression of Dusp5 secondary to the two SNPs (C107T and C627T) in the Dusp5 gene in FHH rats that is predicted to alter CpG sites that may possibly cause DNA demethylation and alter transcriptional activity. [34] However, we found that p-ERK1/2 levels were significantly decreased in FHH compared to FHH.1^{BN} rats. Although more work will be needed to rigorously test this

hypothesis, a decrease in p-ERK1/2 levels is entirely consistent with the observed reduction in the myogenic response in the MCA and autoregulation of CBF observed in FHH rats. Moreover, we also found that a G637A SNP causes a G155R mutation in Dusp5 protein in FHH rats. This G155R mutation localized between the N-terminal regulatory rhodanese domain and the C-terminal phosphatase catalytic domain (**Figure 2A**) converts a nonpolar amino acid Glycine (G) to a basic polar Arginine (R) and is predicted by the I-TASSER program [35–37] to affect the folding

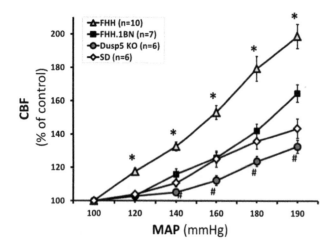

Figure 7. Comparison of autoregulation of CBF in Dusp5 ZFN KO versus FHH.1^{BN} and FHH rats. The relationships between cerebral blood flow and mean arterial pressures in 9–12 week old Dusp5 ZFN KO, FHH.1[BN], FHH and Sprague Dawley rats are compared. Mean values ± SE are presented. * indicates significantly difference in the corresponding value in FHH rats versus all the other strains. # indicates a significant difference in the corresponding values in Dusp5 KO and FHH.1[BN] rats. Numbers in parentheses indicate numbers of animal studied per strain.

of the Dusp5 protein and the conformation of the active site. **Figure 2B** illustrates the changes in the folding based on theoretical structural models (upper panel) and the expanded view presents the confirmation of the active/catalytic site (lower panel). These differences in global folding and active site conformation in FHH compared to FHH.1[BN] rats might lead to differences in protein stability, interactions with binding partners, catalytic efficiency or catalytic activity.

An impaired myogenic response and autoregulation of CBF has been reported in various pathological conditions in patients and

experimental animals including: in subarachnoid hemorrhage (SAH), [48–53] ischemic stroke [3,54–56] and traumatic brain injury. [57–60] In hypertensive patients, impaired autoregulation of CBF accelerates the development of a cognitive decline. [61] However, the mechanisms involved have been difficult to directly study due to lack of an animal model in which autoregulation of CBF is altered. The present findings indicating that the myogenic response in the cerebral arteries and autoregulation of CBF are impaired in FHH rats and restored in FHH.1[BN] congenic strain and are enhanced in our newly generated Dusp5 ZFN KO rats now fill this knowledge gap and provide an important new model system to study the mechanisms by which genetic defects in myogenic mechanisms contribute to the development of small vessel disease and brain damage.

Perspectives and Significance

The present study reports on the creation of a Dusp5 KO rat and provides the first *in vivo* evidence that Dusp5 plays an important role in modulating the myogenic response of cerebral arteries and autoregulation of CBF. We identified a G155R mutation that might contribute to an increase in Dusp5 phosphatase activity and reduced the phosphorylation of ERK1/2 that is consistent with the impaired myogenic response and autoregulation of CBF in FHH rats. Our newly generated Dusp5 KO rat model also provides the scientific community a new model to investigate the mechanisms of impaired myogenic response in FHH rats, and the essential role of Dusp5 in the regulation of MAP kinase activity in vascular reactivity, immune response, cell proliferation and apoptosis and cancer.

Author Contributions

Conceived and designed the experiments: FF DRH HJ RJR. Performed the experiments: FF AMG MRP. Analyzed the data: FF AMG MRP SVS RJR. Contributed reagents/materials/analysis tools: AMG MRP HJ RJR. Wrote the paper: FF SVS RJR.

References

1. Davis MJ, Hill MA (1999) Signaling mechanisms underlying the vascular myogenic response. Physiol Rev 79: 387–423.
2. Bayliss WM (1902) On the local reactions of the arterial wall to changes of internal pressure. J Physiol 28: 220–231.
3. Paulson OB, Strandgaard S, Edvinsson L (1990) Cerebral autoregulation. Cerebrovasc Brain Metab Rev 2: 161–192.
4. Strandgaard S (1991) Cerebral blood flow in the elderly: impact of hypertension and antihypertensive treatment. Cardiovasc Drugs Ther 4 Suppl 6: 1217–1221.
5. Johansson B (1989) Myogenic tone and reactivity: definitions based on muscle physiology. J Hypertens Suppl 7: S5–8; discussion S9.
6. Mellander S (1989) Functional aspects of myogenic vascular control. J Hypertens Suppl 7: S21–30; discussion S31.
7. Burke M, Pabbidi M, Fan F, Ge Y, Liu R, et al. (2013) Genetic basis of the impaired renal myogenic response in FHH rats. American Journal of Physiology-Renal Physiology 304: F565–F577.
8. Pabbidi MR, Mazur O, Fan F, Farley JM, Gebremedhinm D, et al. (2014) Enhanced large conductance K+ channel (BK) activity contributes to the impaired myogenic response in the cerebral vasculature of Fawn Hooded Hypertensive rats. Am J Physio heart and circulatory phys 2014 Apr 1;306(7): H989–H1000.
9. Pabbidi MR, Juncos J, Juncos L, Renic M, Tullos HJ, et al. (2013) Identification of a region of rat chromosome 1 that impairs the myogenic response and autoregulation of cerebral blood flow in fawn-hooded hypertensive rats. Am J Physiol Heart Circ Physiol 304: H311–317.
10. Owens DM, Keyse SM (2007) Differential regulation of MAP kinase signalling by dual-specificity protein phosphatases. Oncogene 26: 3203–3213.
11. Kovanen PE, Bernard J, Al-Shami A, Liu C, Bollenbacher-Reilley J, et al. (2008) T-cell development and function are modulated by dual specificity phosphatase DUSP5. J Biol Chem 283: 17362–17369.

12. Kucharska A, Rushworth LK, Staples C, Morrice NA, Keyse SM (2009) Regulation of the inducible nuclear dual-specificity phosphatase DUSP5 by ERK MAPK. Cell Signal 21: 1794–1805.
13. Zassadowski F, Rochette-Egly C, Chomienne C, Cassinat B (2012) Regulation of the transcriptional activity of nuclear receptors by the MEK/ERK1/2 pathway. Cell Signal 24: 2369–2377.
14. Patterson KI, Brummer T, O'Brien PM, Daly RJ (2009) Dual-specificity phosphatases: critical regulators with diverse cellular targets. Biochem J 418: 475–489.
15. Mandl M, Slack DN, Keyse SM (2005) Specific inactivation and nuclear anchoring of extracellular signal-regulated kinase 2 by the inducible dual-specificity protein phosphatase DUSP5. Mol Cell Biol 25: 1830–1845.
16. Sun CW, Falck JR, Harder DR, Roman RJ (1999) Role of tyrosine kinase and PKC in the vasoconstrictor response to 20-HETE in renal arterioles. Hypertens 33: 414–418.
17. Murphy TV, Spurrell BE, Hill MA (2002) Cellular signalling in arteriolar myogenic constriction: involvement of tyrosine phosphorylation pathways. Clin Exp Pharmacol Physiol 29: 612–619.
18. Kamkin A, Kiseleva I, Isenberg G (2000) Stretch-activated currents in ventricular myocytes: amplitude and arrhythmogenic effects increase with hypertrophy. Cardiovasc Res 48: 409–420.
19. Toth P, Csiszar A, Tucsek Z, Sosnowska D, Gautam T, et al. (2013) Role of 20-HETE, TRPC channels, and BKCa in dysregulation of pressure-induced Ca2+ signaling and myogenic constriction of cerebral arteries in aged hypertensive mice. Am J Physiol Heart Circ Physiol 305: H1698–1708.
20. Wickramasekera NT, Gebremedhin D, Carver KA, Vakeel P, Ramchandran R, et al. (2013) Role of dual-specificity protein phosphatase-5 in modulating the myogenic response in rat cerebral arteries. J Appl Physiol 114: 252–261.
21. Livak KJ, Schmittgen TD (2001) Analysis of relative gene expression data using real-time quantitative PCR and the 2(-Delta Delta C(T)) Method. Methods 25: 402–408.

22. Fan F, Sun CW, Maier KG, Williams JM, Pabbidi MR, et al. (2013) 20-Hydroxyeicosatetraenoic Acid Contributes to the Inhibition of K+ Channel Activity and Vasoconstrictor Response to Angiotensin II in Rat Renal Microvessels. PLoS One 8: e82482.

23. Dunn KM, Renic M, Flasch AK, Harder DR, Falck J, et al. (2008) Elevated production of 20-HETE in the cerebral vasculature contributes to severity of ischemic stroke and oxidative stress in spontaneously hypertensive rats. Am J Physiol Heart Circ Physiol 295: H2455–2465.

24. Chen CC, Geurts AM, Jacob HJ, Fan F, Roman RJ (2013) Heterozygous knockout of transforming growth factor-beta1 protects Dahl S rats against high salt-induced renal injury. Physiol Genomics 45: 110–118.

25. Geurts AM, Cost GJ, Freyvert Y, Zeitler B, Miller JC, et al. (2009) Knockout rats via embryo microinjection of zinc-finger nucleases. Science 325: 433.

26. Geurts AM, Cost GJ, Remy S, Cui X, Tesson L, et al. (2010) Generation of gene-specific mutated rats using zinc-finger nucleases. Methods Mol Biol 597: 211–225.

27. Rangel-Filho A, Lazar J, Moreno C, Geurts A, Jacob HJ (2013) Rab38 modulates proteinuria in model of hypertension-associated renal disease. J Am Soc Nephrol 24: 283–292.

28. Miller JC, Holmes MC, Wang J, Guschin DY, Lee YL, et al. (2007) An improved zinc-finger nuclease architecture for highly specific genome editing. Nat Biotechnol 25: 778–785.

29. Sanger F, Coulson AR (1975) A rapid method for determining sequences in DNA by primed synthesis with DNA polymerase. J Mol Biol 94: 441–448.

30. Burke M, Pabbidi M, Fan F, Ge Y, Liu R, et al. (2013) Genetic basis of the impaired renal myogenic response in FHH rats. Am J Physiol Renal Physiol 304: F565–577.

31. Bellapart J, Fraser JF (2009) Transcranial Doppler assessment of cerebral autoregulation. Ultrasound Med Biol 35: 883–893.

32. Purkayastha S, Raven PB (2011) The functional role of the alpha-1 adrenergic receptors in cerebral blood flow regulation. Indian J Pharmacol 43: 502–506.

33. Wagner BP, Ammann RA, Bachmann DC, Born S, Schibler A (2011) Rapid assessment of cerebral autoregulation by near-infrared spectroscopy and a single dose of phenylephrine. Pediatr Res 69: 436–441.

34. Fu Q, McKnight RA, Yu X, Callaway CW, Lane RH (2006) Growth retardation alters the epigenetic characteristics of hepatic dual specificity phosphatase 5. FASEB J 20: 2127–2129.

35. Roy A, Kucukural A, Zhang Y (2010) I-TASSER: a unified platform for automated protein structure and function prediction. Nat Protoc 5: 725–738.

36. Zhang Y (2008) I-TASSER server for protein 3D structure prediction. BMC Bioinformatics 9: 40.

37. Roy A, Yang J, Zhang Y (2012) COFACTOR: an accurate comparative algorithm for structure-based protein function annotation. Nucleic Acids Res 40: W471–477.

38. Huang CY, Tan TH (2012) DUSPs, to MAP kinases and beyond. Cell Biosci 2: 24.

39. Rangel-Filho A, Sharma M, Datta YH, Moreno C, Roman RJ, et al. (2005) RF-2 gene modulates proteinuria and albuminuria independently of changes in glomerular permeability in the fawn-hooded hypertensive rat. Journal of the American Society of Nephrology 16: 852–856.

40. Geurts AM, Moreno C (2010) Zinc-finger nucleases: new strategies to target the rat genome. Clin Sci (Lond) 119: 303–311.

41. Caunt CJ, Keyse SM (2013) Dual-specificity MAP kinase phosphatases (MKPs): shaping the outcome of MAP kinase signalling. FEBS J 280: 489–504.

42. (2003) USRDS: the United States Renal Data System. Am J Kidney Dis 42: 1–230.

43. Zhao Y, Zhang L, Longo LD (2005) PKC-induced ERK1/2 interactions and downstream effectors in ovine cerebral arteries. Am J Physiol Regul Integr Comp Physiol 289: R164–171.

44. Faraci FM, Baumbach GL, Heistad DD (1990) Cerebral circulation: humoral regulation and effects of chronic hypertension. J Am Soc Nephrol 1: 53–57.

45. Faraci FM, Mayhan WG, Heistad DD (1987) Segmental vascular responses to acute hypertension in cerebrum and brain stem. Am J Physiol 252: H738–742.

46. Walsh MP, Cole WC (2013) The role of actin filament dynamics in the myogenic response of cerebral resistance arteries. J Cereb Blood Flow Metab 33: 1–12.

47. Faraci FM, Heistad DD (1998) Regulation of the cerebral circulation: role of endothelium and potassium channels. Physiol Rev 78: 53–97.

48. Ishii R (1979) Regional cerebral blood flow in patients with ruptured intracranial aneurysms. J Neurosurg 50: 587–594.

49. Voldby B, Enevoldsen EM, Jensen FT (1985) Regional CBF, intraventricular pressure, and cerebral metabolism in patients with ruptured intracranial aneurysms. J Neurosurg 62: 48–58.

50. Voldby B, Enevoldsen EM, Jensen FT (1985) Cerebrovascular reactivity in patients with ruptured intracranial aneurysms. J Neurosurg 62: 59–67.

51. Dernbach PD, Little JR, Jones SC, Ebrahim ZY (1988) Altered cerebral autoregulation and CO2 reactivity after aneurysmal subarachnoid hemorrhage. Neurosurgery 22: 822–826.

52. Lang EW, Diehl RR, Mehdorn HM (2001) Cerebral autoregulation testing after aneurysmal subarachnoid hemorrhage: the phase relationship between arterial blood pressure and cerebral blood flow velocity. Crit Care Med 29: 158–163.

53. Roman RJ, Renic M, Dunn KM, Takeuchi K, Hacein-Bey L (2006) Evidence that 20-HETE contributes to the development of acute and delayed cerebral vasospasm. Neurological research 28: 738–749.

54. Agnoli A, Fieschi C, Bozzao L, Battistini N, Prencipe M (1968) Autoregulation of cerebral blood flow. Studies during drug-induced hypertension in normal subjects and in patients with cerebral vascular diseases. Circulation 38: 800–812.

55. Hoedt-Rasmussen K, Skinhoj E, Paulson O, Ewald J, Bjerrum JK, et al. (1967) Regional cerebral blood flow in acute apoplexy. The "luxury perfusion syndrome" of brain tissue. Arch Neurol 17: 271–281.

56. Olsen TS, Larsen B, Herning M, Skriver EB, Lassen NA (1983) Blood flow and vascular reactivity in collaterally perfused brain tissue. Evidence of an ischemic penumbra in patients with acute stroke. Stroke 14: 332–341.

57. Cold GE (1981) Cerebral blood flow in the acute phase after head injury. Part 2: Correlation to intraventricular pressure (IVP), cerebral perfusion pressure (CPP), PaCO2, ventricular fluid lactate, lactate/pyruvate ratio and pH. Acta Anaesthesiol Scand 25: 332–335.

58. Cold GE, Christensen MS, Schmidt K (1981) Effect of two levels of induced hypocapnia on cerebral autoregulation in the acute phase of head injury coma. Acta Anaesthesiol Scand 25: 397–401.

59. Cold GE, Jensen FT (1978) Cerebral autoregulation in unconscious patients with brain injury. Acta Anaesthesiol Scand 22: 270–280.

60. Overgaard J, Tweed WA (1974) Cerebral circulation after head injury. 1. Cerebral blood flow and its regulation after closed head injury with emphasis on clinical correlations. J Neurosurg 41: 531–541.

61. Lammie GA (2002) Hypertensive cerebral small vessel disease and stroke. Brain Pathol 12: 358–370.

11

Disclosure of Genetic Information and Change in Dietary Intake: A Randomized Controlled Trial

Daiva E. Nielsen, Ahmed El-Sohemy*

Department of Nutritional Sciences, University of Toronto, 150 College St, Toronto, ON, M5S 3E2, Canada

Abstract

Background: Proponents of consumer genetic tests claim that the information can positively impact health behaviors and aid in chronic disease prevention. However, the effects of disclosing genetic information on dietary intake behavior are not clear.

Methods: A double-blinded, parallel group, 2:1 online randomized controlled trial was conducted to determine the short- and long-term effects of disclosing nutrition-related genetic information for personalized nutrition on dietary intakes of caffeine, vitamin C, added sugars, and sodium. Participants were healthy men and women aged 20–35 years (n = 138). The intervention group (n = 92) received personalized DNA-based dietary advice for 12-months and the control group (n = 46) received general dietary recommendations with no genetic information for 12-months. Food frequency questionnaires were collected at baseline and 3- and 12-months after the intervention to assess dietary intakes. General linear models were used to compare changes in intakes between those receiving general dietary advice and those receiving DNA-based dietary advice.

Results: Compared to the control group, no significant changes to dietary intakes of the nutrients were observed at 3-months. At 12-months, participants in the intervention group who possessed a risk version of the *ACE* gene, and were advised to limit their sodium intake, significantly reduced their sodium intake (mg/day) compared to the control group (−287.3±114.1 vs. 129.8±118.2, p = 0.008). Those who had the non-risk version of *ACE* did not significantly change their sodium intake compared to the control group (12-months: −244.2±150.2, p = 0.11). Among those with the risk version of the *ACE* gene, the proportion who met the targeted recommendation of 1500 mg/day increased from 19% at baseline to 34% after 12 months (p = 0.06).

Conclusions: These findings demonstrate that disclosing genetic information for personalized nutrition results in greater changes in intake for some dietary components compared to general population-based dietary advice.

Trial Registration: ClinicalTrials.gov NCT01353014

Editor: Margaret M. DeAngelis, University of Utah, United States of America

Funding: This study was supported by a grant from the Advanced Foods and Materials Network (AFMNet) and the Canadian Institutes of Health Research (CIHR; MOP-89829). AE-S holds a Canada Research Chair in Nutrigenomics. DEN is a recipient of an Ontario Graduate Scholarship and a Banting & Best Diabetes Centre Graduate Studentship. The funders had no role in study design, data collection and analysis, decision to publish, or preparation of the manuscript.

Competing Interests: The authors have read the journal's policy and the authors of this manuscript have the following competing interests: AE-S holds shares in Nutrigenomix Inc., a genetic testing company for personalized nutrition.

* Email: a.el.sohemy@utoronto.ca

Introduction

Personal genetic information has become easily obtainable, in large part due to the advancement of the consumer genetic testing industry. As a result of the decreasing costs to carry out genotyping, individuals can now receive personalized feedback regarding their susceptibility to a number of different health conditions at a relatively low cost [1]. The impact that this information may have on health behaviors is of particular interest [2,3], since chronic diseases such as cardiovascular disease and type 2 diabetes have become major public health concerns. There is considerable evidence that these conditions are associated with a number of modifiable health behaviors such as diet, physical activity and smoking, but lifestyle interventions aimed at achieving positive health behavior changes are often ineffective at producing the long-term changes necessary to mitigate disease risk [4]. As a result, proponents of personalized medicine claim that health recommendations tailored to an individual's genetic profile may be more effective at producing behavior change than generic population-based recommendations. A growing body of qualitative research shows strong public interest in genomics and personalized medicine for disease prevention [5–9], but there is limited quantitative evidence to support the claim that personalized genomics can be employed as a useful prevention tool.

The study of how human genetic variations modify an individual's response to diet on various health outcomes, often referred to as nutrigenomics or nutrigenetics, is a key part of personalized medicine [10] because nutrition is arguably one of the most important modifiers of chronic disease risk [11]. Many direct-to-consumer genetic tests provide single nucleotide polymorphism (SNP)-based estimates of disease susceptibility that do not take into account environmental factors. For complex diseases, including diet-related chronic diseases, risk estimates based solely on genetic variation without consideration of environmental interactions can be inaccurate [12]. Therefore, genetic testing for personalized nutrition using *modifier* or *metabolic* genes may have the potential to be more useful than genetic testing for disease risk using disease susceptibility genes because the advice that is given from a personalized nutrition test is more specific and actionable than advice from a disease susceptibility test. Indeed, a previous study demonstrated that individuals consider DNA-based dietary advice to be more useful and understandable than general population-based dietary recommendations, and individuals report that they would be more motivated to change their diet if provided with personalized nutrition information based on their genetics [13]. Individuals who have had their genomes analyzed report that the genetic information impacted their dietary behaviors, although the genetic information they received was not necessarily linked to any specific dietary modification [14,15]. Despite this evidence, no previous study has examined the effect of disclosing personalized genetic information based on nutrigenomics testing on dietary intake behavior. In addition, previous studies investigating the impact of personal genomic information related to disease susceptibility on health behaviors have lacked long-term follow-up data. As a result, the short- and long-term effects of personal genomic information on health behaviors are largely unknown. Therefore, the objective of the present study was to determine the short- and long-term effects of disclosing genetic information for personalized nutrition on dietary intake in a population of young adults using a randomized controlled trial (Table 1).

Materials and Methods

Ethics Statement

Ethics approval was obtained from the University of Toronto Institutional Review Board and the study is registered with http://clinicaltrials.gov (NCT 01353014). All subjects provided written informed consent. The protocol for this trial and supporting CONSORT checklist are available as supporting information; see Checklist S1 and Protocol S1.

Study Design and Subjects

The present study was intended to mimic the nature of a direct-to-consumer genetic test such that all study materials were distributed and completed in the mail or electronically and no in-person contact was made with subjects for the present study. Details on the study design have been published elsewhere [13]. Briefly, subjects (n = 157) who had previously participated in a nutrigenomics research study and had provided a blood sample were invited to complete a 196-item, semi-quantitative Toronto-modified Willet food frequency questionnaire (FFQ) and were then randomized to an intervention or control group (Figure 1). Subject recruitment occurred from May 2011 to August 2011, the 3-month follow-up assessment occurred from September 2011 to January 2012 and the 12-month follow-up assessment took place from June 2012-October 2012. Since the recommendations in this study were based on caffeine, vitamin C, sugar, and sodium,

eligible participants were those who consumed at least 100 mg of caffeine per day, 10% of energy from total sugars per day, and 1,500 mg of sodium per day and did not take vitamin C-containing supplements. Eligible women who were pregnant or breast-feeding at the time of recruitment were excluded from the study Subjects were given information on portion sizes, which were indicated for most FFQ items, and were asked to select how frequently they consumed the items over the past month from a list of frequency responses. The FFQ was used to collect detailed information on intake of fruits and vegetables, dairy products, meats and alternatives, grain products, sweets and baked goods, processed and prepared foods, and caffeinated and non-caffeinated beverages. Nutrient analyses were carried out at the Harvard School of Public Health Channing Laboratory using the USDA National Nutrient Database for Standard Reference.

Randomization was done by study personnel who used a computer software program (Random Allocation Software) that generated a random list of assignments. A 2:1 ratio of participants in the intervention group compared to the control group was applied since the intervention group consisted of those who would have either the "risk" or "non-risk" genotype for each of the four genes (Table 1). Participants were informed that they would receive DNA-based dietary advice at some point during the study and those who were randomized to the control group were given the DNA-based advice after the final follow-up assessment was completed. Dietary intakes were self-reported by participants on the FFQ with no assistance from study personnel, and the nutrient analyses were made without knowledge of study group assignment. DNA was isolated from whole blood with the GenomicPrep Blood DNA Isolation kit (Amersham Pharmacia Biotech Inc, Piscataway, NJ) and genotyping was completed using either real-time polymerase chain reaction on an ABI 7000 Sequence Detection System (Applied Biosystems) or a multiplex restriction fragment length polymorphism (RFLP) polymerase chain reaction method, as described previously [16,17]. Genotyping was verified by using positive control subjects in each 96-well plate as well as a second genotyping of ~5% of a random selection of samples with 100% concordance.

Dietary Reports and Recommendations

Subjects in the intervention group were genotyped for variants that affect caffeine metabolism (*CYP1A2*) [18,19], vitamin C utilization (*GSTT1* and *GSTM1*) [17], sweet taste perception (*TAS1R2*) [20], and sodium-sensitivity (*ACE*) [21,22] (Table 1). These genes were selected as representative sample tests from consumer genetic testing companies, which are based on studies of gene-diet interactions using a candidate gene approach. Although newer experimental approaches such as genome-wide association studies (GWAS) are being used to identify genetic variants associated with disease susceptibility, to date only a few GWAS studies have investigated gene-diet interactions on health outcomes [23,24]. Thus, the purpose of the present study was to evaluate the behavioral response to disclosing genetic information, not to validate the clinical efficacy of the genetic variants provided in the dietary advice reports. Dietary reports were sent to all subjects by e-mail. The dietary report for subjects in the intervention group informed subjects of their genotypes and included a corresponding DNA-based dietary recommendation for daily intake of caffeine, vitamin C, added sugars and sodium (Nutrigenomix Inc., Toronto, Canada). Those who possessed the genotype that has been associated with increased risk of a health outcome when consuming above or below a certain daily amount were given a "targeted" dietary recommendation. For caffeine and sodium, this recommendation was more stringent than the current general

Table 1. Prevalence of risk alleles in intervention group (n = 92) and associated risk.

Dietary Component	Gene	Risk Allele n (%)	Non-Risk Allele	Associated Risk
Caffeine	CYP1A2	48 (52)	44 (48)	Increased risk of myocardial infarction and hypertension when consuming above 200 mg of caffeine/day
				General recommendation: ≤300 mg/day for women of child-bearing age
				≤400 mg/day for other adults
				Targeted Recommendation: ≤200 mg/day for those with risk version of CYP1A2
Vitamin C	GSTM1 + GSTT1	52 (57)	40 (43)	Increased risk of serum ascorbic acid deficiency when consuming below the RDA for vitamin C
				General recommendation: RDA[a] for women: ≥75 mg/day
				RDA for men: ≥90 mg/day
				Targeted Recommendation: Same as general recommendation
Added Sugars	TAS1R2	41 (45)	51 (55)	Increased risk of over-consuming sugars
				General recommendation: ≤10% energy/day
				Targeted Recommendation: Same as general recommendation
Sodium	ACE	64 (70)	28 (30)	Increased risk of sodium-sensitive hypertension when consuming above the AI for sodium
				General recommendation: UL[b]: ≤2300 mg/day
				Targeted Recommendation: AI[c]: ≤1500 mg/day for those with risk version of ACE

[a]RDA: Recommended dietary allowance.
[b]UL: Tolerable upper intake level.
[c]AI: Adequate intake.

recommendation for daily intake and was based on previous work that evaluated health outcomes according to genotype at different levels of intake [18,19,21,22]. For added sugars and vitamin C, subjects were informed to be particularly mindful of meeting the current general recommendation for daily intake, since no previous study has examined how individuals respond to consuming various levels of these nutrients according to genotype and, therefore, a different intake level could not be recommended. Subjects who possessed the genotype that has not been associated with increased risk received the current general recommendation for daily intake [25–27]. The control group was e-mailed a report of current general recommendations for the same nutrients without genetic information. Subjects were e-mailed a monthly reminder of their dietary report and additional FFQs were collected at 3- and 12-month follow-up assessments.

Study outcomes

The primary study outcome was change in dietary intakes of caffeine, vitamin C, added sugars, and sodium between the control and intervention groups from baseline to the follow-up assessments. Changes in dietary intakes were examined between baseline and 3-months to determine the short-term effects of the intervention, while changes in intakes between baseline and 12-months were examined to determine the long-term effects. The secondary study outcome was to compare the proportion of participants who met the recommendations for intake before and after the intervention for dietary components that significantly changed between the control and intervention groups.

Statistical Analyses

Statistical analyses were performed using the Statistical Analysis Software (version 9.2; SAS Institute Inc., Cary, NC). The α error was set at 0.05 and all reported p-values are two-sided. Subject

characteristics between the intervention and control group were compared using a Chi-square test for categorical variables and a Student's t-test for continuous variables. The distributions of nutrient intakes were examined and a log transformation was applied to those that deviated from normality. In these cases, the p-values from models using transformed values are reported, but untransformed means and measures of spread are reported to facilitate interpretation. Subjects who were likely under-reporters (consuming less than 800 kcal/day) were excluded from the analyses, since dietary intake data from these individuals may not have been reliable. Baseline mean intakes of vitamin C, sugar, sodium and caffeine were compared between ethnocultural groups using general linear models to determine if any significant dietary differences were present between groups at the start of the study. General linear models were also conducted to test for changes in dietary intakes between baseline and 3-months, and baseline and 12-months, in order to determine the effect of the dietary advice over a short- and long-term period. The Tukey-Kramer test for multiple comparisons was applied to determine whether any changes in intake of the intervention groups differed from the change in intake of the control group. The Chi-square test was used to compare the proportion of subjects meeting the recommendations for intake between baseline and the follow-up assessments. Fisher's Exact Test was used if a proportion category consisted of fewer than 5 subjects.

Results

Subject Characteristics

Of the 157 subjects who were sent the baseline FFQ, 125 completed the 12-month study giving an overall retention rate of 80% (Figure 1). In relation to those who were randomized (n = 138), this represents 91% of subjects who completed the 12-month study.

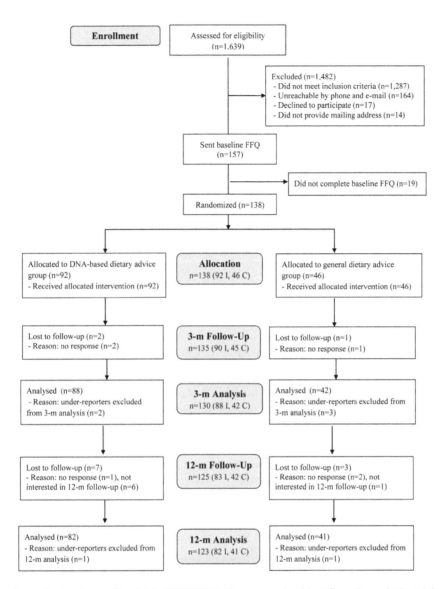

Figure 1. Consolidated standards of reporting trials (CONSORT) diagram and subject flow through the trial.

The mean age of the participants was 26.5±3.0 years and 78% were female. The study population was multi-ethnic with Caucasian, East Asian, and South Asian groups representing the majority of ethnic backgrounds. Over half of the population possessed at least an undergraduate degree. There were no significant differences between the characteristics of participants in the intervention group when compared to the control group (Table 2). However, a significant difference in baseline sodium intake (mg/day) was observed between the East Asian and Caucasian groups (1837±147 vs. 2319±88, p = 0.03). As a result, the general linear models examining changes in dietary intakes are adjusted for ethnocultural group. At baseline, the proportion of subjects who did not meet the general recommendation for caffeine, vitamin C, added sugars and sodium were 9%, 14%, 24% and 39%, respectively. Thirty eight percent of subjects did not meet the targeted recommendation (for those with elevated risk) for caffeine intake at baseline, while 80% did not meet the targeted recommendation for sodium intake. The targeted recommendation for vitamin C and added sugars was the same as the general recommendations.

Changes in Dietary Intakes

Of the 138 subjects who were randomized, 135 completed the 3-month follow-up and 130 were included in the 3-month analyses (n = 5 under-reporters excluded). At 12 months, 125 subjects completed the follow-up and 123 were included in the analyses (n = 2 under-reporters excluded). There were no differences in the baseline characteristics (e.g. age, proportion of males/females) between those who were included in the final analysis and those who were not. Compared to the control group, no significant changes from baseline were observed for intakes of caffeine, vitamin C, added sugars, or sodium at the 3-month follow-up among subjects in the intervention group who carried a risk version of the corresponding gene (intervention risk group) or among subjects who carried the non-risk version (intervention non-risk group). At the 12-month follow-up, subjects in the intervention group who were informed that they possessed the risk version of the *ACE* gene, and who were given the targeted advice to consume below the Adequate Intake (AI) of 1500 mg/day of sodium, significantly reduced their mean sodium intake (mg/day) from baseline when compared to the control group

Table 2. Participant characteristics.

	Intervention (n = 92)	Control (n = 46)	p-value
	n (%)		
Age (years)[*]	27±3	26±3	0.82
Female	69 (75)	37 (80)	0.48
Ethnicity			
Caucasian	59 (64)	24 (52)	0.18
East Asian	19 (21)	12 (26)	0.47
South Asian	9 (10)	6 (13)	0.56
Other	5 (5)	4 (9)	0.46
Education			
Some college or undergraduate training	9 (10)	8 (17)	0.20
College or undergraduate degree	50 (54)	22 (48)	0.47
Graduate degree	33 (36)	16 (35)	0.90

[*]Values shown are mean ± standard deviation.

(-287.3 ± 114.1 vs. 129.8 ± 118.2, $p = 0.008$), which did not receive genetic information and was given the general recommendation for sodium intake (Tolerable Upper Intake Level (UL): $\leq$ 2300 mg/day). The mean change in sodium intake (mg/day) among subjects who were informed that they possessed the non-risk version of the *ACE* gene, and who were advised to follow the general recommendation for sodium intake, did not differ from the change in intake of the control group at 12-months (-244.2 ± 150.2 vs. 129.8 ± 118.2, $p = 0.11$). The mean changes in intakes from baseline for caffeine, vitamin C and added sugars did not differ from the control group at the 12-month follow-up among either the intervention-risk or intervention non-risk groups (Table 3).

At the 12-month follow-up assessment 66% of subjects in the intervention risk group, 65% of subjects in the intervention non-risk group and 68% of subjects in the control group met the general recommendation for sodium intake of ≤ 2300 mg/day. In addition, 34% of subjects in the intervention risk group, 19% of subjects in the intervention non-risk group and 24% of subjects in the control group met the targeted recommendation for sodium intake of ≤ 1500 mg/day. These proportions were not significantly different between the control and intervention groups. Among those in the intervention group who had the risk version of the *ACE* gene, 19% met the targeted recommendation of ≤ 1500 mg/day at baseline compared to 34% after 12 months ($p = 0.06$), and 59% met the general recommendation of ≤ 2300 mg/day at baseline compared to 66% after 12 months ($p = 0.41$).

Discussion

The present study is the first to evaluate the effects of disclosing genetic information related to personalized nutrition on dietary intake and the findings show that DNA-based dietary advice results in greater changes in intake for some dietary components compared to population-based dietary advice. Dietary modification is an important health behavior for chronic disease prevention. Changes in health behaviors have not been frequently reported in previous studies that have investigated the effect of disclosing genetic information related to disease risk [28–31] and a 2010 Cochrane review concluded that disclosing genetic risk information for disease has little impact on actual behavior,

although it has a small effect on one's intention to change [32]. However, the genomic information provided in those studies was related to disease susceptibility, not personalized nutrition, and the studies lacked a long-term follow-up assessment. Moreover, participants in previous studies were not provided with personalized recommendations on what behavioral strategies should be followed to mitigate disease risk. Results from the present study provide evidence that genetic testing for personalized nutrition may be more clinically useful for motivating favorable dietary changes than testing for disease susceptibility, since a change in sodium intake was observed after 12-months among the intervention risk group. In line with this finding, a previous study comparing a personalized, DNA-based weight loss diet with a traditional weight loss diet reported that subjects on the personalized diet had greater dietary adherence, longer-term maintenance of weight loss and greater improvements in fasting blood glucose levels [33]. In addition, a study investigating health behavior changes after revealing genetic risk for Alzheimer's disease reported that the addition of a vitamin E supplement was the most common change to vitamin or medication use among subjects who were informed that they were at greater genetic risk [34].

Although changes were observed in dietary intakes of sodium among the intervention-risk group of subjects at the 12-month follow-up, no changes in intakes of caffeine, vitamin C or added sugars were observed at either follow-up assessment. This may be due to the baseline intakes of these nutrients that were already mostly in line with the recommendations that were given to the subjects who possessed a risk allele, which is a limitation of the present study. Nevertheless, variants in other genes involved in reward pathways may play a role in one's ability to reduce consumption of some of these dietary components [16,35]. Indeed, the National Human Genome Research Institute has recommended investigation into the potential for genomic information to improve behavior change interventions by customizing interventions to individuals based on genetic markers of adherence [36,37]. Despite the lack of an intervention effect on intake of these three dietary components in the present study, it is worth noting that subjects who possessed a non-risk allele for the corresponding genes did not shift to a less desirable level of intake by increasing

Table 3. Changes in dietary intake after 3-months and 12-months.

	Baseline (n = 133)			3-months (n = 130)			12-months (n = 123)		
	n	Mean ± SEM	p-value‡ compared to control group	n	Mean change ± SEM	p-value‡ compared to control group	n	Mean change ± SEM	p-value‡ compared to control group
Caffeine (mg/day)									
Intervention risk	46	181.4±16.8	0.82	45	-3.0±14.8	0.61	41	-18.9±18.8	0.92
Intervention non-risk	44	194.8±17.8	0.92	43	-24.7±15.3	0.66	41	1.5±19.4	0.99
Control	43	183.5±16.3		42	-7.3±14.8		41	-0.3±17.8	
Vitamin C (mg/day)									
Intervention risk	50	197.3±33.6	0.85	49	49.5±37.6	0.99	45	36.6±43.1	0.73
Intervention non-risk	40	226.2±35.1	0.96	39	-13.9±38.4	0.22	37	-58.4±43.5	0.42
Control	43	220.0±31.9		42	44.1±37.1		41	-21.4±40.1	
Added sugars (%e/day)									
Intervention risk	38	8.9±0.8	0.99	37	-0.9±0.9	0.18	33	0.4±0.9	0.98
Intervention non-risk	52	8.3±0.7	0.54	51	0.5±0.8	0.99	49	-0.4±0.8	0.85
Control	43	9.3±0.7		42	0.6±0.8		41	-0.4±0.8	
Sodium (mg/day)									
Intervention risk	63	2144.5±124.4	0.51	62	-143.0±109.0	0.20	56	-287.3±114.1	0.008
Intervention non-risk	27	2224.9±171.0	0.31	26	97.8±145.6	0.97	26	-244.2±150.2	0.11
Control	43	2000.8±131.2		42	82.2±119.2		41	129.8±118.2	

‡p-values are for log-transformed values.
Results are adjusted for ethnicity.

their consumption of caffeine, sodium or added sugars, or decreasing their intake of vitamin C. Although we did not report a detrimental impact on dietary intake behavior as a result of disclosing genetic information indicating no increased risk, proper communication of genetic test results is needed to prevent individuals from misunderstanding or misinterpreting the information and to guide them toward making appropriate lifestyle changes where necessary [38]. As such, providing this type of information through a qualified healthcare professional might be more appropriate than providing such information direct-to-consumer. If the results of a genetic test require dietary modification then a dietitian might be best suited to guide the consumer whereas a genetic counselor would be better suited for communicating results of tests for high penetrance genes that may require a more severe intervention.

Another limitation of the present study is the use of a FFQ to assess dietary intake, which is more useful in larger, population-based studies, as it provides a measure of relative intake rather than actual intake. However, the objective of the present study was to assess change in dietary intakes, which is a relative measure of intake. In addition, the sample size was small, yet comparable to previous studies examining the impact of disclosing genetic information on particular health behaviors [30,33,34], and subjects were highly educated and recruited from a previous nutrigenomics study. The reported reduction of nearly 300 mg of sodium per day in the intervention risk group was not sufficient to reduce the average sodium intake to the AI of 1500 mg/day, which was the targeted recommendation provided in the dietary report. Nevertheless, a recent Institute of Medicine report concluded that there is no benefit to sharply restricting sodium intake to the level of the AI [39] and a 2010 computer-simulated model examining the effect of dietary salt reduction on future cardiovascular disease projected that a 1 g/day reduction in average population salt intake, which is equivalent to about 400 mg/day of sodium, would prevent up to 28,000 deaths from any cause and would be more cost-effective than using medications to manage hypertension [40]. Therefore, the approximate 300 mg/day reduction in sodium intake reported in the present study would be considered clinically relevant.

Strengths of the present study are the inclusion of a control group, which provided a method of comparing the utility of DNA-based dietary advice to population-based recommendations, and

the randomized design, which minimizes the potential for confounding effects. Including a 3- and 12-month follow-up assessment enabled us to examine the short- and long-term effects of the intervention. The finding that sodium intake was significantly reduced compared to the control group after 12-months among subjects in the intervention group with the risk version of the *ACE* gene suggests that longer-term studies are required to fully determine the impact of disclosing genetic information. Moreover, conducting the present study so that it closely resembles a consumer genetic test increases the validity of the findings to reflect the real world effects among consumer genetic test users. Early adopters of consumer genetic testing are more likely to be highly educated and Caucasian, with a substantial proportion of users between the ages of 18–49 years [15,28,41]. One study has reported a larger proportion of female consumers [41]. As a result, the subjects in the present study are representative of the early adopters of consumer genetic testing.

The present study was the first to empirically test the effect of DNA-based personalized nutritional advice on dietary intake behavior compared to population-based dietary advice. The findings show that DNA-based dietary advice can impact dietary intake to a greater extent than general population-based recommendations and provide supportive evidence for the clinical utility of personalized nutrition to assist in chronic disease prevention.

Acknowledgments

The authors thank Alyssa Katzikowski, Maria Tassone, and Sarah Shih for their assistance with subject recruitment and data collection.

Author Contributions

Conceived and designed the experiments: AE-S DEN. Performed the experiments: DEN. Analyzed the data: DEN. Contributed reagents/materials/analysis tools: AE-S. Wrote the paper: DEN AE-S.

References

1. Caulfield T, McGuire AL (2012) Direct-to-consumer genetic testing: perceptions, problems, and policy responses. Annu Rev Med 63: 23–33.

2. McBride CM, Koehly LM, Sanderson SC, Kaphingst KA (2010) The behavioral response to personalized genetic information: will genetic risk profiles motivate individuals and families to choose more healthful behaviors? Annu Rev Public Health 31: 89–103.

3. Christensen KD, Green RC (2013) How could disclosing incidental information from whole-genome sequencing affect patient behavior? Per Med 10: 377–386.

4. Desroches S, Lapointe A, Ratte S, Gravel K, Legare F, et al. (2011) Interventions to enhance adherence to dietary advice for preventing and managing chronic diseases in adults: a study protocol. BMC Public Health 11: 111.

5. Leighton JW, Valverde K, Bernhardt BA (2011) The general public's understanding and perception of direct-to-consumer genetic test results. Public Health Genomics 15: 11–21.

6. Cherkas LF, Harris JM, Levinson E, Spector TD, Prainsack B (2010) A survey of UK public interest in internet-based personal genome testing. PLoS One 5: e13473.

7. Goddard KA, Duquette D, Zlot A, Johnson J, Annis-Emeott A, et al. (2009) Public awareness and use of direct-to-consumer genetic tests: results from 3 state population-based surveys, 2006. Am J Public Health 99: 442–445.

8. Stewart-Knox BJ, Bunting BP, Gilpin S, Parr HJ, Pinhao S, et al. (2009) Attitudes toward genetic testing and personalised nutrition in a representative sample of European consumers. Br J Nutr 101: 982–989.

9. Kolor K, Duquette D, Zlot A, Foland J, Anderson B, et al. (2012) Public awareness and use of direct-to-consumer personal genomic tests from four state population-based surveys, and implications for clinical and public health practice. Genet Med 14: 860-867.

10. Kaput J (2008) Nutrigenomics research for personalized nutrition and medicine. Curr Opin Biotechnol 19: 110–120.

11. Nielsen DE, El-Sohemy A (2012) Applying genomics to nutrition and lifestyle modification. Per Med 9: 739–749.

12. Wesselius A, Zeegers MP (2013) Direct-to-consumer genetic testing. OA Epidemiology. 1:4.

13. Nielsen DE, El-Sohemy A (2012) A randomized trial of genetic information for personalized nutrition. Genes Nutr 7: 559-566.

14. Maher B (2011) Nature readers flirt with personal genomics. Nature 478: 19.

15. Kaufman DJ, Bollinger JM, Dvoskin RL, Scott JA (2012) Risky business: risk perception and the use of medical services among customers of DTC personal genetic testing. J Genet Couns 21: 413–422.

16. Eny KM, Corey PN, El-Sohemy A (2009) Dopamine D2 receptor genotype (C957T) and habitual consumption of sugars in a free-living population of men and women. J Nutrigenet Nutrigenomics 2: 235–242.

17. Cahill LE, Fontaine-Bisson B, El-Sohemy A (2009) Functional genetic variants of glutathione S-transferase protect against serum ascorbic acid deficiency. Am J Clin Nutr 90: 1411–1417.

18. Cornelis MC, El-Sohemy A, Kabagambe EK, Campos H (2006) Coffee, CYP1A2 genotype, and risk of myocardial infarction. JAMA 295: 1135–1141.

19. Palatini P, Ceolotto G, Ragazzo F, Dorigatti F, Saladini F, et al. (2009) CYP1A2 genotype modifies the association between coffee intake and the risk of hypertension. J Hypertens 27: 1594–1601.

20. Eny KM, Wolever TM, Corey PN, El-Sohemy A (2010) Genetic variation in TAS1R2 (Ile191Val) is associated with consumption of sugars in overweight and obese individuals in 2 distinct populations. Am J Clin Nutr 92: 1501–1510.

21. Giner V, Poch E, Bragulat E, Oriola J, Gonzalez D, et al. (2000) Renin-angiotensin system genetic polymorphisms and salt sensitivity in essential hypertension. Hypertension 35: 512–517.

22. Poch E, Gonzalez D, Giner V, Bragulat E, Coca A, et al. (2001) Molecular basis of salt sensitivity in human hypertension. Evaluation of renin-angiotensin-aldosterone system gene polymorphisms. Hypertension 38: 1204–1209.

23. He J, Kelly TN, Zhao Q, Li H, Huang J, et al. (2013) Genome-wide association study identifies 8 novel loci associated with blood pressure responses to interventions in Han Chinese. Circ Cardiovasc Genet 6: 598–607.

24. Hamza TH, Chen H, Hill-Burns EM, Rhodes SL, Montimurro J, et al. (2011) Genome-wide gene-environment study identifies glutamate receptor GRIN2A as a Parkinson's disease modifier gene via interaction with coffee. PLoS Genet 7: e1002237.

25. Nawrot P, Jordan S, Eastwood J, Rotstein J, Hugenholtz A, et al. (2003) Effects of caffeine on human health. Food Addit Contam 20: 1–30.

26. National Research Council (2006) Dietary Reference Intakes: The Essential Guide to Nutrient Requirements. The National Academies Press: Washington, DC.

27. Nishida C, Uauy R, Kumanyika S, Shetty P (2004) The joint WHO/FAO expert consultation on diet, nutrition and the prevention of chronic diseases: process, product and policy implications. Public Health Nutr 7: 245–250.

28. Bloss CS, Schork NJ, Topol EJ (2011) Effect of direct-to-consumer genomewide profiling to assess disease risk. N Engl J Med 364: 524–534.

29. Collier R (2012) Predisposed to risk but not change. CMAJ 184: E407–408.

30. Grant RW, O'Brien KE, Waxler JL, Vassy JL, Delahanty LM, et al. (2013) Personalized genetic risk counseling to motivate diabetes prevention: a randomized trial. Diabetes Care 36: 13–19.

31. Bloss CS, Madlensky L, Schork NJ, Topol EJ (2011) Genomic information as a behavioral health intervention: can it work? Per Med 8: 659–667.

32. Marteau TM, French DP, Griffin SJ, Prevost AT, Sutton S, et al. (2010) Effects of communicating DNA-based disease risk estimates on risk-reducing behaviours. Cochrane Database Syst Rev: CD007275.

33. Arkadianos I, Valdes AM, Marinos E, Florou A, Gill RD, et al. (2007) Improved weight management using genetic information to personalize a calorie controlled diet. Nutr J 6: 29.

34. Chao S, Roberts JS, Marteau TM, Silliman R, Cupples LA, et al. (2008) Health behavior changes after genetic risk assessment for Alzheimer disease: The REVEAL Study. Alzheimer Dis Assoc Disord 22: 94–97.

35. Cornelis MC, Monda KL, Yu K, Paynter N, Azzato EM, et al. (2011) Genome-wide meta-analysis identifies regions on 7p21 (AHR) and 15q24 (CYP1A2) as determinants of habitual caffeine consumption. PLoS Genet 7: e1002033.

36. Green ED, Guyer MS (2011) Charting a course for genomic medicine from base pairs to bedside. Nature 470: 204–213.

37. McBride CM, Bryan AD, Bray MS, Swan GE, Green ED (2012) Health behavior change: can genomics improve behavioral adherence? Am J Public Health 102: 401–405.

38. Ferguson LR, Barnett MPG (2012) Research in nutrigenomics and potential applications to practice. Nutrition Diet 69: 198–202.

39. Institute of Medicine (2013) Sodium Intake in Populations: Assessment of Evidence. Washington, DC: The National Academies Press.

40. Bibbins-Domingo K, Chertow GM, Coxson PG, Moran A, Lightwood JM, et al. (2010) Projected effect of dietary salt reductions on future cardiovascular disease. N Engl J Med 362: 590–599.

41. Gollust SE, Gordon ES, Zayac C, Griffin G, Christman MF, et al. (2012) Motivations and perceptions of early adopters of personalized genomics: perspectives from research participants. Public Health Genomics 15: 22–30.

12

A Simplified and Versatile System for the Simultaneous Expression of Multiple siRNAs in Mammalian Cells Using Gibson DNA Assembly

Fang Deng[1,2], Xiang Chen[2], Zhan Liao[2,3], Zhengjian Yan[2,4], Zhongliang Wang[2,4], Youlin Deng[2,4], Qian Zhang[2,4], Zhonglin Zhang[2,5], Jixing Ye[2,6], Min Qiao[2,4,3], Ruifang Li[2,5], Sahitya Denduluri[2], Jing Wang[2,4], Qiang Wei[2,4], Melissa Li[2], Nisha Geng[2], Lianggong Zhao[2,7], Guolin Zhou[2], Penghui Zhang[2,4], Hue H. Luu[2], Rex C. Haydon[2], Russell R. Reid[2,8], Tian Yang[1]*, Tong-Chuan He[2,3]*

1 Department of Cell Biology, Third Military Medical University, Chongqing, 400038, China, 2 Molecular Oncology Laboratory, Department of Orthopaedic Surgery and Rehabilitation Medicine, The University of Chicago Medical Center, Chicago, IL, 60637, United States of America, 3 Department of Orthopaedic Surgery, the Affiliated Xiang-Ya Hospital of Central South University, Changsha, 410008, China, 4 Ministry of Education Key Laboratory of Diagnostic Medicine, and the Affiliated Hospitals of Chongqing Medical University, Chongqing, 400016, China, 5 Department of Surgery, the Affiliated Zhongnan Hospital of Wuhan University, Wuhan, 430071, China, 6 School of Bioengineering, Chongqing University, Chongqing, 400044, China, 7 Department of Orthopaedic Surgery, the Second Affiliated Hospital of Lanzhou University, Lanzhou, Gansu, 730000, China, 8 The Laboratory of Craniofacial Biology, Department of Surgery, The University of Chicago Medical Center, Chicago, IL, 60637, United States of America

Abstract

RNA interference (RNAi) denotes sequence-specific mRNA degradation induced by short interfering double-stranded RNA (siRNA) and has become a revolutionary tool for functional annotation of mammalian genes, as well as for development of novel therapeutics. The practical applications of RNAi are usually achieved by expressing short hairpin RNAs (shRNAs) or siRNAs in cells. However, a major technical challenge is to simultaneously express multiple siRNAs to silence one or more genes. We previously developed pSOS system, in which siRNA duplexes are made from oligo templates driven by opposing U6 and H1 promoters. While effective, it is not equipped to express multiple siRNAs in a single vector. Gibson DNA Assembly (GDA) is an *in vitro* recombination system that has the capacity to assemble multiple overlapping DNA molecules in a single isothermal step. Here, we developed a GDA-based pSOK assembly system for constructing single vectors that express multiple siRNA sites. The assembly fragments were generated by PCR amplifications from the U6-H1 template vector pB2B. GDA assembly specificity was conferred by the overlapping unique siRNA sequences of insert fragments. To prove the technical feasibility, we constructed pSOK vectors that contain four siRNA sites and three siRNA sites targeting human and mouse β-catenin, respectively. The assembly reactions were efficient, and candidate clones were readily identified by PCR screening. Multiple β-catenin siRNAs effectively silenced endogenous β-catenin expression, inhibited Wnt3A-induced β-catenin/Tcf4 reporter activity and expression of Wnt/β-catenin downstream genes. Silencing β-catenin in mesenchymal stem cells inhibited Wnt3A-induced early osteogenic differentiation and significantly diminished synergistic osteogenic activity between BMP9 and Wnt3A *in vitro* and *in vivo*. These findings demonstrate that the GDA-based pSOK system has been proven simplistic, effective and versatile for simultaneous expression of multiple siRNAs. Thus, the reported pSOK system should be a valuable tool for gene function studies and development of novel therapeutics.

Editor: Jun Sun, Rush University Medical Center, United States of America

Funding: The reported work was supported in part by research grants from the National Institutes of Health (AT004418, AR50142, AR054381 to TCH, RCH and HHL), and the National Natural Science Foundation of China (NSFC grant #81271770 to TY). DF was a recipient of a doctorate fellowship from the China Scholarship Council. This work was also supported in part by The University of Chicago Core Facility Subsidy grant from the National Center for Advancing Translational Sciences (NCATS) of the National Institutes of Health through grant UL1 TR000430. The funders had no role in study design, data collection and analysis, decision to publish, or preparation of the manuscript.

Competing Interests: The authors have declared that no competing interests exist.

* Email: tche@uchicago.edu (TCH); tiany@163.net (TY)

Introduction

RNA interference (RNAi) was first discovered in *C. elegans* as a protecting mechanism against invasion by foreign genes and has subsequently been demonstrated in diverse eukaryotes, such as insects, plants, fungi and vertebrates [1–7]. RNAi is a cellular process of sequence-specific, post-transcriptional gene silencing initiated by double-stranded RNAs (dsRNA) homologous to the gene being suppressed. The dsRNAs are processed by Dicer to generate duplexes of approximately 21nt, so-called short interfering RNAs (siRNAs), which cause sequence-specific mRNA degradation. Dicer-produced siRNA duplexes comprise two 21 nucleotide strands, each bearing a 5′ phosphate and 3′ hydroxyl group, paired in a way that leaves two-nucleotide overhangs at the 3′ ends. Target regulation by siRNAs is mediated by the RNA-induced silencing complex (RISC). Since its discovery, RNAi has become a valuable and powerful tool to analyze loss-of-function phenotypes *in vitro* and *in vivo* [2–7]. Given its gene-specific targeting nature, RNAi also offers unprecedented opportunities for developing novel and effective therapeutics for human diseases [8–13].

The practical applications of siRNA duplexes to interfere with the expression of a given gene require target accessibility and effective delivery of siRNAs into target cells and for certain applications long-term siRNA expression [8–14]. While RNAi can be achieved by delivering synthetic short double-stranded RNA duplexes into cells, a more commonly-used approach is to express short hairpin RNAs (shRNAs) or siRNAs in cells [11,12,14]. In this case, the endogenous expression of siRNAs is achieved by using various Pol III promoter expression cassettes that allow transcription of functional siRNAs or their precursors [14]. However, one of the formidable technical challenges is to effectively construct these RNAi expression vectors, especially when gene silencing necessitates the use of multiple siRNA target sites for a gene of interest. We previously developed the pSOS system, in which the siRNA duplexes are made from an oligo template driven by opposing U6 and H1 promoters [15]. While effective, it usually requires to make multiple vectors and multiple-round infections to achieve effective knockdown when multiple siRNA sites are used. On the other hand, there are clear needs to simultaneously deliver multiple siRNAs that target more than one genes.

Gibson DNA Assembly (GDA), so named after the developer of the method [16], is one of commonly-used synthetic biology techniques that offer restriction enzyme-free, scarless, largely sequence-independent, and multi-fragment DNA assembly [17,18]. GDA is an *in vitro* recombination system that has the capacity to assemble and repair multiple overlapping DNA molecules in a single isothermal step [16,17]. The optimized GDA contains three essential components: an exonuclease (e.g., 5′-T5 exonuclease) that removes nucleotides from the ends of double-stranded (ds) DNA molecules so exposing complementary single-stranded (ss) DNA overhangs that are specifically annealed; a DNA polymerase (e.g., Phusion DNA polymerase) that fills in the ssDNA gaps of the joined molecules; and a DNA ligase (e.g., Taq ligase) that covalently seals the nicks [17]. Thus, this assembly method can be used to seamlessly construct synthetic and natural genes, genetic pathways, and entire genomes as useful molecular engineering tools [16–18].

Here, we sought to use the GDA technique to establish a simplified one-step assembly system for constructing a single vector that expresses multiple siRNA target sites. To achieve this, we have engineered the GDA destination retroviral vector pSOK, based on our previously reported pSOS vector [15], which can be linearized with SwaI for assembly reactions. The assembly fragments containing multiple siRNA sites are generated by PCR amplifications using the back-to-back U6-H1 promoter vector pB2B as a template. The first fragment overlaps with the 3′-end of U6 promoter while the last fragment overlaps with the 3′-end of H1 promoter. The ends of the middle fragments overlap the specific siRNA target sequences, which confers assembly specificity. After the GDA reactions, single vectors expressing multiple siRNA target sites are generated. To prove the feasibility of this pSOK system, we have developed the vectors that contain four siRNA sites and three siRNA sites that target human and mouse β-catenin, respectively. We demonstrate that the assembly reactions are

efficient, and that candidate clones are readily identified by PCR screening, although vectors containing three siRNAs are seemingly more favorably assembled under our assembly condition. Functional analyses demonstrate that the multiple β-catenin siRNA constructs can effectively silence endogenous β-catenin expression, inhibit Wnt3A-induced β-catenin/Tcf4 reporter activity and the expression of Wnt/β-catenin downstream target genes. In mesenchymal stem cells, silencing β-catenin inhibits Wnt3A-induced early osteogenic differentiation and significantly diminishes the synergistic osteogenic activity between BMP9 and Wnt3A both *in vitro* and *in vivo*. Taken together, our results have demonstrated that the GDA-based pSOK system is proven simplistic, effective and versatile for simultaneous expression of multiple siRNA target sites. Thus, the pSOK system should be a valuable tool for gene function studies and the development of therapeutics.

Materials and Methods

Cell culture and chemicals

HEK-293 and human colon cancer SW480 lines were purchased from ATCC (Manassas, VA) and maintained in complete Dulbecco's Modified Eagle's Medium (DMEM) containing 10% fetal bovine serum (FBS, Invitrogen, Carlsbad, CA), 100 units of penicillin and 100 μg of streptomycin at 37°C in 5% CO_2 [19–24]. The reversibly immortalized mouse embryonic fibroblasts (iMEFs) were previously characterized [25,26]. The recently engineered 293pTP line was used for adenovirus amplification [27]. Both 293pTP and iMEFs were maintained in complete DMEM. Unless indicated otherwise, all chemicals were purchased from Sigma-Aldrich (St. Louis, MO) or Fisher Scientific (Pittsburgh, PA).

Construction of the retroviral vector pSOK and PCR template vector pB2B for Gibson DNA Assembly reactions

As illustrated in **Figure 1A** and **Figure S1A**, the MSCV retroviral vector pSOK was constructed on the base of our previously reported pSOS vector [15], which contains the opposing U6 and H1 promoters to drive siRNA duplex expression. The linker sites of the pSOS vector were modified and a SwaI site was engineered for linearizing the vector for Gibson Assembly (**Figure 1A, panel a**). This vector also confers Blasticidin S resistance for generating stable mammalian cell lines. The pB2B vector was constructed on the base of our previously reported pMOLuc vector [28]. Briefly, the high-fidelity PCR amplified U6 and H1 promoter fragments were subcloned into the EcoRI/HindIII sites of pMOLuc in a back-to-back orientation, and ligated at MluI site (**Figure S1B**). Both U6 and H1 promoters contain a string of "AAAAA" immediately preceding their transcription start sites, which serves as transcription termination signal for the reverse strand. The full-length vector sequences and maps are available at: http://www.boneandcancer.org/MOLab%20Vectors%20after%20Nov%201%202005/pSOK.pdf and http://www.boneandcancer.org/MOLab%20Vectors%20after%20Nov%201%202005/pBOK%20vector%20map%20and%20sequence%202013-12-02.pdf.

Gibson DNA Assembly (GDA) reactions for generating pSOK vectors expressing siRNA sites targeting human and mouse β-catenin and the generation of stable cell lines

The GDA reactions were carried out by using the Gibson Assembly Master Mix from New England Biolabs (Ipswich, MA)

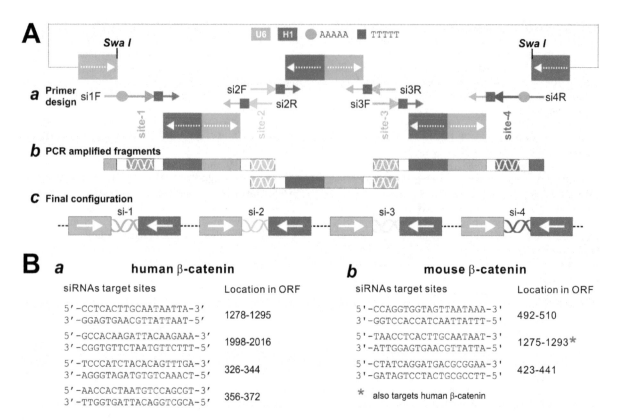

Figure 1. Schematic depiction of the one-step system pSOK for expressing multiple siRNAs. (A) Schematic representation of a tandem siRNA targeting configuration (4 sites listed as an example). The pSOK vector was constructed based on the previously reported pSOS vector, which contains opposing U6 and H1 promoters to drive siRNA duplex expression (**Figure S1A**) [15]. The linker sites of the pSOS vector were modified and a SwaI site was created for linearizing the vector for Gibson Assembly (*a*). The primers were designed according to the guidelines outlined in **Figure S2A**. Using the pB2B as a template vector (**Figure S1B**), the back-to-back U6-H1 promoter fragments with different siRNA target sites were generated. The first fragment overlaps with the 3'-end of the U6 promoter while the last fragment overlaps with the 3'-end of the H1 promoter (*b*). The ends of the middle fragments overlaps the specific siRNA target sequences (*b*). After the Gibson Assembly reaction, a single vector expressing 4 siRNA target sites is constructed (*c*). It is noteworthy that the siRNA sites may target the same or different genes. **(B)** The targeting sequences and locations of the designed siRNA sites on human (*a*) and mouse (*b*) β-catenin open reading frame (ORF). All of these sites have been validated in previous studies [15,40]. Note that one of the mouse siRNAs also targets human β-catenin coding sequence.

following the manufacturer's instructions. The overlapping inserts were prepared by PCR amplifications using the Phusion High-Fidelity PCR kit (New England Biolabs). Each assembly reaction contained approximately 100 ng of each insert and 50 ng of the SwaI-linearized pSOK vector and incubated at 50°C for 45 min. After the assembly reactions, the reaction mix was briefly digested with SwaI and transformed into electro-competent DH10B cells. Colony PCR screening was carried out using a forward and reverse primer pair of the two neighboring siRNA sites. Positive clones were sequencing verified. Regardless the compositions of the obtained clones, vectors containing one, two, three or four siRNA sites targeting human β-catenin were designated as pSOK-siBC1, pSOK-siBC2, pSOK-siBC3, and pSOK-siBC4, respectively. For the mouse β-catenin siRNAs, we only chose the vector that contains all three siRNA sites, namely pSOK-simBC3 for this study. A control vector containing three scrambled sites that do not target any human and mouse genes (5'-GCAAAGACGCAA-TAATACA-3'; 5'-GCACAAAGAACGACTATAA-3'; 5'-GAAACACGATTAACAGACA-3') was also constructed, designated as pSOK-siControl.

The stable knockdown lines were generated using a retrovirus system as previously reported [15,29–31]. Briefly, the siRNA-containing pSOK vectors were co-transfected with the retrovirus packaging plasmid pCL-Ampho into HEK-293 cells. The packaged retrovirus supernatants were used to infect 293, SW480 (for siBC vectors) and iMEFs (for simBC3 vector). The infected cells were selected in Blasticidin S (4 μg/ml) for 5–7 days. The stable pools of cells were kept in LN2 for long-term storage. The resultant stable lines were designated such as 293-siBC4, 293-siControl, SW480-siBC4, SW480-siControl, iMEF-simBC3, and iMEF-siControl, to name a few.

Generation and amplification of recombinant adenoviruses expressing BMP9, Wnt3A, and GFP

Recombinant adenoviruses were generated using the AdEasy technology as described [30,32–34]. The coding regions of human BMP9 and mouse Wnt3A were PCR amplified and cloned into an adenoviral shuttle vector, and subsequently used to generate and amplify recombinant adenoviruses in HEK-293 or 293pTP cells [27]. The resulting adenoviruses were designated as AdBMP9 and AdWnt3A, both of which also express GFP [35–38]. Analogous adenovirus expressing only GFP (AdGFP) was used as controls [39–42]. For all adenoviral infections, polybrene (4–8 μg/ml) was added to enhance infection efficiency as previously reported [23].

Cell transfection and firefly luciferase reporter assay

Subconfluent cells were transfected with the Tcf/Lef reporter pTOP-Luc using Lipofectamine Reagent (Invitrogen) by following the manufacturer's instructions. For 293 and iMEF cells, the cells were co-transfected with pCMV-Wnt3A. At 48 h post transfection, cells were lysed for luciferase assays using Luciferase Assay System (Promega, Madison, WI) by following the manufacturer's instructions. Easy conditions were done in triplicate.

RNA isolation and quantitative real-time PCR (qPCR)

Total RNA was isolated by using TRIZOL Reagents (Invitrogen) and used to generate cDNA templates by reverse transcription reactions with hexamer and M-MuLV reverse transcriptase (New England Biolabs, Ipswich, MA). The cDNA products were used as PCR templates. The sqPCR were carried out as described [43–47]. PCR primers (**Table S1**) were designed by using the Primer3 program and used to amplify the genes of interest (approximately 150–250 bp). For qPCR analysis, SYBR Green-based qPCR analysis was carried out by using the thermocycler Opticon II DNA Engine (Bio-Rad, CA) with a standard pUC19 plasmid as described elsewhere [21,48–50]. The qPCR reactions were done in triplicate. The sqPCR was also carried out as described [15,24,25,27,29,39,47,51,52]. Briefly, sqPCR reactions were carried out by using a touchdown protocol: $94°C \times 20''$, $68°C \times 30''$, $70°C \times 20''$ for 12 cycles, with $1°C$ decrease per cycle, followed by 25–30 cycles at $94°C \times 20''$, $56°C \times 30''$, $70°C \times 20''$. PCR products were resolved on 1.5% agarose gels. All samples were normalized by the expression level of GAPDH.

Immunofluorescence staining

Immunofluorescence staining was performed as described [30,40,43,50,53,54]. Briefly, cells were infected with AdWnt3A or AdGFP for 48 h, fixed with methanol, permeabilized with 1% NP-40, and blocked with 10% BSA, followed by incubating with β-catenin antibody (Santa Cruz Biotechnology). After being washed, cells were incubated with Texas Red-labeled secondary antibody (Santa Cruz Biotechnology). Stains were examined under a fluorescence microscope. Stains without primary antibodies, or with control IgG, were used as negative controls.

Qualitative and quantitative assays of alkaline phosphatase (ALP) activity

ALP activity was assessed quantitatively with a modified assay using the Great Escape SEAP Chemiluminescence assay kit (BD Clontech, Mountain View, CA) and qualitatively with histochemical staining assay (using a mixture of 0.1 mg/ml napthol AS-MX phosphate and 0.6 mg/ml Fast Blue BB salt), as previously described [29,30,32,33,39,40,44,53]. Each assay condition was performed in triplicate and the results were repeated in at least three independent experiments.

iMEF cell implantation and ectopic bone formation

All animal studies were conducted by following the guidelines approved by the Institutional Animal Care and Use Committee (IACUC) of The University of Chicago (protocol #71108). Stem cell-mediated ectopic bone formation was performed as described [20,25,29,30,33,47,55–57]. Briefly, subconfluent iMEFsimBC3 and iMEF-siControl cells were infected with AdBMP9 and/or AdWnt3A, or AdGFP for 16 h, collected and resuspended in PBS for subcutaneous injection (5×10^6/injection) into the flanks of athymic nude (nu/nu) mice (5 animals per group, 4–6 wk old, female, Harlan Laboratories, Indianapolis, IN). At 4 weeks after implantation, animals were sacrificed, and the implantation sites were retrieved for histologic evaluation and Trichrome staining as described below.

Histological evaluation and Trichrome staining

Retrieved tissues were fixed, decalcified in 10% buffered formalin, and embedded in paraffin. Serial sections of the embedded specimens were stained with hematoxylin and eosin (H & E). Trichrome staining was carried out as previously described [20,25,26,44,47,52,55,56].

Statistical analysis

The quantitative assays were performed in triplicate and/or repeated three times. Data were expressed as mean ± SD. Statistical significances were determined by one-way analysis of variance and the student's t test. A value of $p < 0.05$ was considered statistically significant.

Results

Construction of the GDA vector pSOK for expressing multiple siRNA target sites in mammalian cells

We previously developed the pSOS system, in which the siRNA duplexes are made from an oligo template driven by opposing U6 and H1 promoters [15]. While effective, it usually requires to make multiple vectors and multiple-round infections to achieve effective knockdown if multiple siRNA target sites are used. In other cases, there are clear needs to deliver multiple siRNAs that target more than one genes. Here, we sought to establish a simplified one-step approach, based on the GDA technology, which will allow us to make a single vector that express multiple siRNA target sites against one gene or multiple genes.

As depicted in **Figure 1A**, the pSOK vector was constructed based on the previously reported pSOS vector [15], which contains the opposing U6 and H1 promoters to drive siRNA duplex expression (**Figure S1A**). The linker sites of the pSOS vector were modified and a SwaI site was engineered for linearizing the vector for Gibson Assembly (**Figure 1A, panel a**). This vector confers Blasticidin S resistance for generating stable mammalian cell lines. For examples, four siRNA target sites are exemplified to illustrate the primer design and construction process, the primers were designed according to the guidelines outlined in **Figure S2A**. Using the pB2B as a template vector (**Figures S1B, and S2B**), the back-to-back U6-H1 promoter fragments with different siRNA target sites were generated. The first fragment overlaps with the 3′-end of the U6 promoter while the last fragment overlaps with the 3′-end of the H1 promoter (**Figure 1A, panel b**). Thus, the ends of the middle fragments overlap the specific siRNA target sequences (**Figure 1A, panel b**). After the GDA reaction, a single vector expressing four siRNA target sites is constructed (**Figure 1A, panel c**).

To prove the principle and feasibility of the pSOK system, we designed four siRNA sites and three siRNA sites that target human and mouse β-catenin, respectively (**Figure 1B**). The targeting sequences and locations of the designed siRNA sites on human (**Figure 1B, panel a**) and mouse (**Figure 1B, panel b**) β-catenin open reading frame (ORF). These siRNA sites were previously demonstrated to effectively silence β-catenin expression [15,40]. As indicated, these siRNA sites target a broad region of β-catenin coding regions, and one of the mouse siRNAs also target human β-catenin (**Figure 1A, panel b**).

Construction and characterization of pSOK vectors that express multiple siRNAs targeting human β-catenin

We first chose to construct the pSOK vector expressing the four siRNAs that target human β-catenin. After performing PCR amplifications of the three inserts as depicted in **Figure 1A**, we carried out the GDA reactions using the three inserts and the SwaI-linearized pSOK vector. The potential recombinants were screened by colony PCR using the forward and reverse primer pairs of the neighboring siRNA sites (**Figure 2A**). Since there were multiple repetitive U6-H1 promoter units in the construct, we found the most robust and specific amplifications were obtained when the forward and reverse primer pairs of the neighboring siRNA sites were used. To further demonstrate this phenomenon, we used a representative pSOK-siBC4 clone as the template and tested PCR amplifications with different combinations of primer pairs. We found that the primer pairs 1F/2R, 2F/3R, and 3F/4R yielded robust and relatively specific products while primer pairs covering two or more siRNA sites produced multiple bands (**Figure 2A**).

Since the overlapping sequences for the inserts are only 19 bp, it is conceivable that a high exonuclease activity in the assembly reaction may over digest the overlapping sequences and cause mis-pairing of the siRNA sites. In fact, we found clones containing one to four siRNA sites as shown by the restriction digestion to release the inserts (in roughly 500 bp U6-H1 cassettes) (**Figure 2B**). More than candidate clones were verified by DNA sequencing and

blasting against the query sequence outlined in **Figure S2C**. Sequencing analysis of these clones revealed that different clones may contain different siRNA sites (**Figure 2C**). Statistically, clones containing three siRNA sites (i.e, siBC3) are the most abundant (accounted for about 50% of the clones), followed by clones containing two or four siRNA sites (i.e., siBC2 or siBC4) at about 20% each (**Figure 2D**). Surprisingly, clones containing one siRNA site only accounted for about 10% of the screened clones (**Figure 2D**). These results indicate that the assembly reactions are efficient although the assembly of three siRNA sites (e.g., two inserts) may be a more favorable event, at least under our reaction conditions. After sequencing verification, the clones containing one, two, three and four siRNA sites were pooled and designated as pSOK-siBC1, pSOK-siBC2, pSOK-siBC3 and pSOK-siBC4, respectively. An analogous vector containing three scrambled sites was also constructed as a control (e.g., pSOK-siControl). While we chose to construct multiple siRNAs to target the same gene (i.e., β-catenin), it is conceivable that one can assemble multiple siRNAs that target more than one genes.

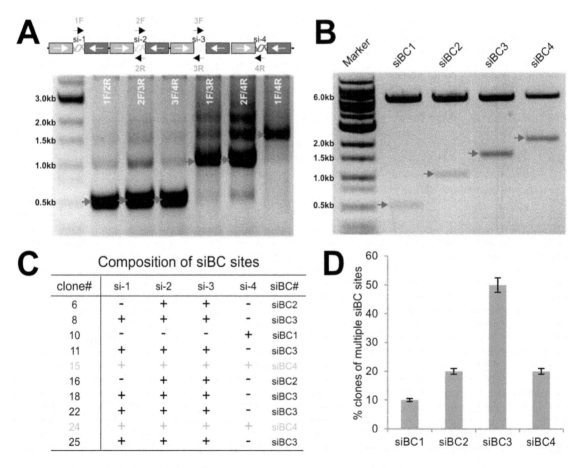

Figure 2. Construction and characterization of pSOK vectors that express multiple siRNAs targeting human β-catenin. (A) PCR screening strategy for candidate clones. A representative pSOK-siBC4 clone was used as a template and PCR amplified with different combinations of primer pairs, as depicted above the gel image. The PCR products were resolved on a 0.8% agarose gel. The arrows indicate the expected products. **(B)** Restriction digestion confirmation of the obtained clones. Representative clones containing 1 to 4 siRNA sites were digested with PmeI/HindIII to release the inserts. The digested products were resolved on a 0.8% agarose gel. The arrows indicate the expected products. **(C)** The frequency of the clones containing multiple copies of siRNA target expression units of human β-catenin. Plasmid DNA was isolated from approximately 80 individual SwaI-resistant clones and subjected to DNA sequencing. The presence of different copy numbers of siRNA sites was tabulated. **(D)** siRNA target site composition of 10 representative clones of human β-catenin. "+", site present, "−", site absent.

Functional validation of the pSOK-siBC4 vector that contains four siRNA sites targeting human β-catenin

Although most of these siRNA target sites have been tested for their silencing efficiency and effect on Wnt/β-catenin signaling activity, the knockdown efficiency may be compromised due to promoter competitions in the pSOK system because multiple U6-H1 expression cassettes are engineered in a single vector. To test this possibility, we established stable lines of HEK-293 and SW480 cells expressing the siBC sites or siControl using a retroviral system. We first assessed the knockdown efficiency of endogenous β-catenin in 293 cells and SW480 cells. Using qPCR analysis, we found that the endogenous β-catenin expression in 293-siBC4 was significantly lower than that of the 293-siControl's (**Figure 3A, panel a**). Similarly, using the human colon cancer line SW480 we found the β-catenin expression was drastically reduced in SW480-siBC4 cells compared with that in the SW480-siControl cells (**Figure 3A, panel b**). Overall, the siBC4-expressing 293 and SW480 cells exhibited marked decreases in the β-catenin expression, only about 32% and 4% of the control cells' (**Figure 3A, panel c**). It is noteworthy that we also found that endogenous β-catenin expression was significantly reduced in the 293 and SW480 cells that express siBC1, siBC2, and siBC3 (data not shown). When the 293-siBC4 (co-transfected with Wnt3A) and SW480-siBC4 cells were transfected with the β-catenin/Tcf reporter pTOP-Luc, the reporter activities were marked reduced at the tested time points in both 293 cells ($p<0.001$) (**Figure 3B, panel a**) and SW480 cells ($p<0.001$) (**Figure 3B, panel b**). Moreover, the Wnt3A-stimulated reporter activities in 293 cells stably expressing siBC1, siBC2, and siBC3 were also remarkably inhibited (**Figure S3A**), and similar results were obtained in SW480 cells, in which the Wnt/β-catenin signaling is constitutively active (**Figure S3B**). Furthermore, we examined the β-catenin knockdown efficiency in SW480 cells by immunofluorescence staining. We found that cytoplasmic/nuclear accumulation of β-catenin in SW480-siBC4 cells was significantly diminished, compared with that in the SW480-siControl cells (**Figure 3C**). Taken the above results together, the pSOK-siBC4 vector that expresses four human β-catenin siRNA sites can effectively silence β-catenin expression in human cells.

pSOK-simBC3 contains multiple siRNAs targeting mouse β-catenin and effectively inhibits canonical Wnt signaling activity in iMEFs

Our results in **Figure 2** indicate that three siRNA sites (two inserts) are seemingly more favorably assembled into pSOK vector. It is conceivable that in most cases three siRNA sites should be sufficiently effective in silencing a given gene. Here, we tested this possibility by constructing a vector, designated as pSOK-simBC, that expressed three siRNA sites targeting mouse β-catenin (**Figure 1B, panel b**). The construction and screening process were very efficient. After sequencing verification, the pSOK-simBC was packaged as retrovirus and used to generate the stable line iMEF-simBC3, along with a control line iMEF-siControl. The iMEFs were previously characterized multi-potent mesenchymal stem cells (MSCs) [20,25]. When the iMEF stable lines were infected with AdWnt3A or AdGFP and analyzed for β-catenin expression, we found that β-catenin expression was significantly reduced in iMEF-simBC3 cells, compared with that in iMEF-siControl cells ($p<0.001$) (**Figure 4A**).

Using the β-catenin/Tcf luciferase reporter, we found that iMEF-simBC3 cells exhibited significantly lower β-catenin/Tcf reporter activity upon Wnt3A stimulation (p<0.001) (**Figure 4B**). Accordingly, when the expression of two well-characterized Wnt/

β-catenin downstream target genes, Axin2 [58] and c-Myc [59], was examined, we found that Wnt3A was shown to significantly induce the expression of Axin2 and c-Myc in iMEF-siControl cells; but silencing β-catenin in iMEFs significantly diminished Wnt3A-induced expression of c-Myc (**Figure 4C, panel a**) and Axin2 (**Figure 4C, panel b**). Furthermore, immunofluorescence staining indicate that Wnt3A-induced cytoplasmic/nuclear accumulation of β-catenin protein was effectively reduced in iMEF-simBC3 cells, compared with that in iMEF-siControl cells (**Figure 4D**). Therefore, the above data demonstrate that the three-siRNA site-containing pSOK-simBC3 can effectively blunt the functional activities of Wnt3A/β-catenin in iMEF cells.

Silencing β-catenin diminishes the synergistic osteogenic activity between BMP9 and Wnt3A in iMEF cells

We further analyzed the functional consequences of β-catenin knockdown on MSC differentiation. We and others demonstrated that canonical Wnt signaling can induce osteogenic differentiation of mesenchymal stem cells [39,40,48,60]. We sought to determine if Wnt3A can induce early osteogenic marker alkaline phosphatase (ALP) activity in iMEFs, and if the induced ALP activity would be reduced when β-catenin is silenced in iMEFs. We found that Wnt3A effectively induced ALP activity in iMEF-siControl cells, which was significantly blunted in iMEF-simBC3 cells (**Figure 5A**). We previously demonstrated that BMP9 is one of the most potent osteogenic BMPs in mesenchymal cell stems [26,30,32,33,61]. We found that BMP9 stimulated robust ALP activity in iMEF-siControl cells while the BMP9-induced ALP activity was remarkably reduced in iMEF-simBC3 cells (**Figure 5A**).

We previously showed that Wnt3A and BMP9 act synergistically in regulating osteogenic differentiation of MSCs [40]. We found that the iMEF-siControl cells co-transduced with Wnt3A and BMP9 exhibited higher ALP activity than that of the cells transduced with either Wnt3A or BMP9 alone, which was remarkably blunted by β-catenin knockdown (**Figure 5A**). Quantitative ALP activity analysis revealed a similar trend, BMP9 and Wnt3A+BMP9 stimulated ALP activities were significantly inhibited by β-catenin knockdown $p<0.001$ (iMEF-simBC3 vs. iMEF-siControl) (**Figure 5B**). Thus, these results suggest that β-catenin may play an important role in this synergistic action between BMP9 and Wnt3A in osteogenic differentiation of MSCs.

BMP9-induced ectopic bone formation from iMEFs is potentiated by Wnt3A but attenuated by β-catenin knockdown

Using our previously established stem cell implantation assays [20,25,26,30,38,47,52,57], we tested the *in vivo* effect of β-catenin knockdown on BMP9 and Wnt3A-induced ectopic bone formation. Subconfluent iMEF-simBC3 and iMEF-siControl cells were transduced with AdBMP9, AdWnt3A, AdBMP9+AdWnt3A, or AdGFP, and injected subcutaneously into the flanks of athymic nude mice for 4 weeks. No recoverable masses were detected in the GFP or Wnt3A group. Robust bony masses were retrieved from both BMP9 and BMP9+Wnt3A transduced iMEF-siControl groups, while significantly smaller masses were recovered from the iMEF-simBC3 group (**Figure 6A, panels a & b vs. c**). BMP9+Wnt3A group formed a slightly larger bone masses (**Figure 6A, panels a vs. b**).

When the retrieved samples were subjected to H & E staining, we found that BMP9-transduced iMEF-siControl cells formed evident trabecular bone, which was even more robust in the

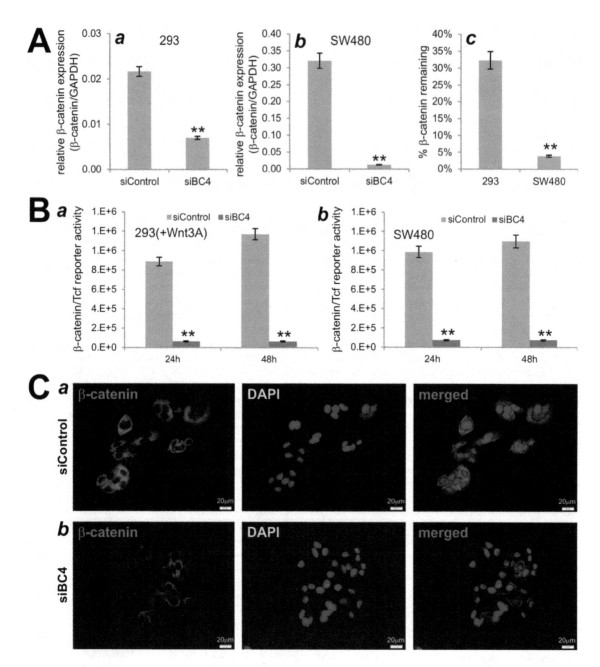

Figure 3. Functional validation of siRNAs targeting human β-catenin. (**A**) Efficient knockdown of endogenous β-catenin in HEK-293 and SW480 cells. Total RNA was isolated from subconfluent 293-siBC4, 293-siControl, SW480-siBC4, and SW480-siControl cells, and subjected to qPCR analysis using primers specific for human β-catenin. All samples were normalized with GAPDH. Each reaction was done in triplicate. Relative β-catenin expression was calculated by dividing β-catenin expression levels with respective GAPDH levels in 293 (*a*) and SW480 (*b*) cells. The % of remaining β-catenin expression was calculated by dividing the relative β-catenin expression in siBC4 with that of the respective siControl's (*c*). "******", $p<0.001$. (**B**) β-Catenin/Tcf transcription activity is significantly reduced in siBC4 cells. Subconfluent 293-siBC4 and 293-siControl cells were co-transfected with TOP-Luc reporter and pCMV-Wnt3A plasmids using Lipofectamine reagent (*a*), while SW480-siBC4 and SW480-siControl cells were transfected with TOP-Luc reporter plasmid using Lipofectamine reagent (*b*). At 24 h and 48 h after transfection, the cells were lysed and subjected to firefly luciferase assay using the Luciferase Reporter Assay System (Promega). Each assay condition was done in triplicate. "******", $p<0.001$. (**C**) siBCs can effectively block Wnt3a-induced β-catenin accumulation. Subconfluent SW480-siBC4 and SW480-siControl cells fixed and subjected to immunofluorescence staining with an anti-β-catenin antibody. The cell nuclei were counter stained with DAPI. Control IgG and minus primary antibody were used as negative controls (data not shown).

presence of both BMP9 and Wnt3A (**Figure 6B**). However, silencing β-catenin expression in iMEF-simBC3 cells significantly reduced trabecular bone formation induced by BMP9 or BMP9+Wnt3A, and formed cartilage-like small masses (**Figure 6B**). Trichrome staining confirmed that iMEF-siControl cells

transduced with BMP9 formed apparently mature and mineralized bone matrices, while a combination of BMP9 and Wnt3A induced more mature and highly mineralized bone matrices (**Figure 6C**). However, the maturity and mineralization were significantly diminished in iMEF-simBC3 cells transduced with

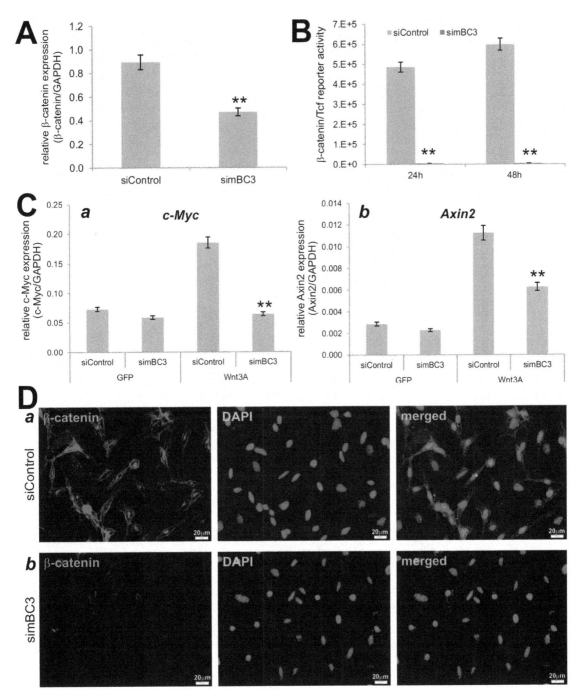

Figure 4. Multiple siRNAs targeting mouse β-catenin simBC3 effectively inhibit canonical Wnt signaling activity in iMEFs. (A) Reduced β-catenin expression in iMEF-simBC3 cells. Subconfluent iMEF-simBC3 and iMEF-siControl cells were infected with AdWnt3A or AdGFP. At 36 h after infection, total RNA was isolated and subjected to qPCR analysis using primers for mouse β-catenin and GAPDH. Relative expression was calculated by dividing the β-catenin expression levels with respective GAPDH expression. All samples were subjected to the subtraction of baseline (i.e., AdGFP infected cells) expression. Each assay was done in triplicate. "**", $p<0.001$. **(B)** iMEF-simBC3 cells exhibit significantly lower β-catenin/Tcf reporter activity upon Wnt3A stimulation. Subconfluent iMEF-simBC3 and iMEF-siControl cells were transfected with TOP-Luc reporter plasmid and infected with AdWnt3A or AdGFP. At 24 h and 48 h post transfection/infection, cells were lysed for luciferase assays. Relative β-catenin/Tcf reporter activity was subjected to subtractions of basal activity (i.e., AdGFP groups). Easy conditions were done in triplicate. "**", $p<0.001$. **(C)**. Wnt3A-induced expression of Wnt/β-catenin target genes was significantly decreased in iMEF-simBC3 cells. Subconfluent iMEF-simBC3 and iMEF-siControl cells were infected with AdWnt3A or AdGFP for 36 h. Total RNA was isolated and subjected to reverse transcription. The resultant cDNAs were used as templates for qPCR analysis using primers specific for mouse Axin2 and c-Myc transcripts. All samples were normalized by GAPDH levels. Each assay condition was done in triplicate. "**", $p<0.001$. **(D)** simBC3 can effectively block Wnt3a-induced β-catenin accumulation. Subconfluent iMEF-siControl (a) cells fixed and subjected to immunofluorescence staining with an anti-β-catenin antibody. The cell nuclei were counter stained with DAPI. Control IgG and minus primary antibody were used as negative controls (data not shown). Representative images are shown.

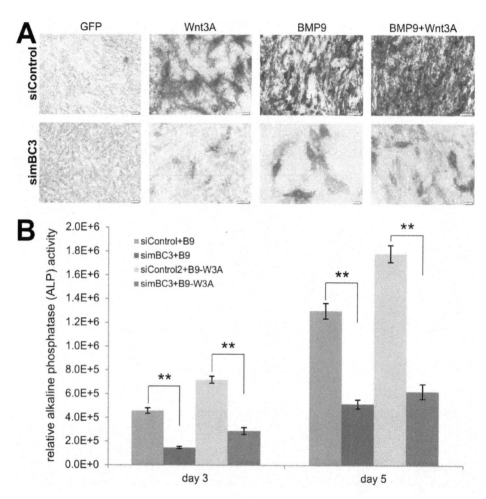

Figure 5. Silencing β-catenin diminishes the synergistic osteogenic activity between BMP9 and Wnt3A in iMEF cells. (A) Wnt3A and/or BMP9-induced early osteogenic marker alkaline phosphatase (ALP) activity is reduced in iMEF-simBC3 cells. Subconfluent iMEF-simBC3 and iMEF-siControl cells were infected with AdWnt3A, AdBMP9, AdGFP, or AdWnt3A+AdBMP9. At day 5 post infection, cells were fixed for ALP histochemical staining assay. Each assay conditions were done in triplicate. Representative results are shown. **(B)** Wnt3A and/or BMP9-induced ALP activity is decreased in the β-catenin silenced iMEFs. The experiments were set up in a similar fashion to that described in **(A)**. At days 3 and 5, cells were lysed for quantitative ALP activity assays. Basal ALP activities (e.g., GFP groups) were subtracted from all BMP9, Wnt3A, and Wnt3A+BMP9 groups. Each assay conditions were done in triplicate. "**", $p<0.001$ (iMEF-simBC3 vs. iMEF-siControl).

either BMP9 or BMP9+Wnt3A (**Figure 6C**). Taken together, these in vivo results strongly suggest that β-catenin may play an important role in mediating BMP9-induced bone formation, and the BMP9-Wnt3A may crosstalk in inducing osteoblastic differentiation of MSCs.

Discussion

To overcome the technical challenges in simultaneously expressing multiple siRNAs that silence one specific gene or different genes, here we sought to develop a simple, efficient and versatile method to express multiple siRNAs in a single vector by exploring the possible utility of the Gibson DNA Assembly. We take advantages of our previously established pSOS system, in which the siRNA duplexes are generated from oligo templates driven by opposing U6 and H1 promoters [15]. Since there are only a few Pol III promoters that are well characterized, we choose to use the same U6-H1 promoter cassette to drive the expression of multiple siRNA sites. However, the use of these repetitive U6-H1 expression units poses a technical challenge for choosing overlapping sequences for Gibson DNA Assembly. To overcome

this hurdle we design an assembly scheme that takes advantages of the unique sequences of different siRNA sites (e.g., stretches of 19 nucleotides). The assembly fragments containing multiple siRNA sites are generated by PCR amplifications using the back-to-back U6-H1 promoter vector pB2B as a template while the vector is SwaI-linearized pSOK.

We carried out the proof-of-principle studies using multiple siRNAs targeting human and mouse β-catenin. We demonstrate that the assembly reactions are efficient, and that candidate clones are readily identified by PCR screening. Functional analyses demonstrate that multiple β-catenin siRNA constructs can effectively silence endogenous β-catenin expression, inhibit Wnt3A-induced β-catenin/Tcf reporter activity and the expression of Wnt/β-catenin downstream target genes. Furthermore, in mesenchymal stem cells we found that silencing β-catenin inhibits Wnt3A-induced early osteogenic differentiation and significantly diminishes the synergistic osteogenic activity between BMP9 and Wnt3A both in vitro and in vivo. Therefore, our results have demonstrated that the Gibson Assembly-based pSOK system is proven simplistic, effective and versatile for simultaneous expression of multiple siRNA target sites.

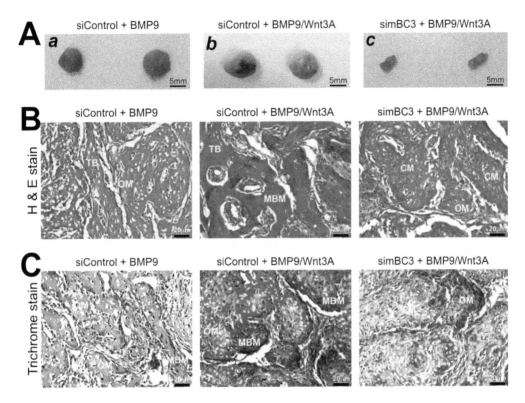

Figure 6. BMP9-induced ectopic bone formation from iMEFs is potentiated by Wnt3A but attenuated by β-catenin knockdown. (A) Gross images. Subconfluent iMEF-simBC3 and iMEF-siControl cells were infected with AdBMP9, AdWnt3A, AdBMP9+AdWnt3A, or AdGFP for 16 h. Cells were collected for subcutaneous injections (3×10^6 cells/site in 100 μl PBS) into the flanks of athymic nude mice (n = 5 each group). At 4 weeks after injection, the animals were sacrificed. Masses formed at the injection sites were retrieved from the groups injected with the iMEF-siControl cells transduced with BMP9 (*a*) or BMP9+Wnt3A (*b*), while very small masses were retrieved from the animals injected with the iMEFs transduced BMP9+Wnt3A (c) or BMP9 (not shown). No masses were retrieved from the animals injected with Wnt3A- or GFP-transduced iMEF cells. Representative results are shown. **(B)** and **(C)** Histologic evaluation and Trichrome staining. The retrieved samples were decalcified and subjected to paraffin-embedded sectioning for histologic evaluation, including H & E staining (*B*) and Trichrome staining (*C*). Representative results are shown. TB, trabecular bone; MBM, mineralized bone matrix; OM, osteoid matrix; CM, chondroid matrix.

Our findings have addressed at least two technical and functional concerns over the pSOK system. First, our design for the overlapping ends of the inserts is only 19 nucleotides. It's conceivable the overlapping sequences are too short and may comprise the assembly reactions. Our results indicate that the assembly efficiency is reasonably high although, in our attempt to assemble four siRNA sites, we do obtain clones that contain one, two or three sites. In fact, vectors containing three siRNAs are seemingly more favorably assembled under our assembly condition. Second, the repeated U6-H1 promoter cassettes may compromise the expression of multiple siRNA sites due to possible promoter competition [62]. Our results using the clones containing different numbers of siRNA sites strongly suggest that the use of repetitive U6-H1 expression cassettes may pose little or insignificant impact on the efficient expression of multiple siRNA sites although we did not analyze the precise expression levels of these siRNA duplexes. Given the nature of the short 19-nt overlapping sequences, we have found two critical technical parameters should be taken into consideration for an efficient assembly: 1) using 3~5 times more inserts than conventional ligation reactions; and 2) using shorter assembly reaction time (e.g., 30–45 min at 50°C). Furthermore, it is conceivable that the same assembly system can be introduced into recombinant adenovirus, adenovirus-associated virus, and other gene delivery vector systems.

In this study, we examined the functional consequences of β-catenin knockdown on Wnt3A and/or BMP9-induced osteogenic differentiation of mesenchymal stem cells. Wnts are a family of secreted glycoproteins that regulate many developmental processes [63]. Wnt signaling plays an important role in skeletal development [60,64]. Wnt proteins bind to their cognate receptor frizzled (Fz) and LRP-5/6 co-receptors, and activate distinct signaling pathways, including the canonical β-catenin pathway. In the absence of Wnt signaling, β-catenin is degraded by the proteasome system after GSK3β dependent phosphorylation. In the presence of Wnt signaling, unphosphorylated β-catenin accumulates in the cytoplasm and translocates into the nucleus where it associates with Tcf/LEF transcription factors to regulate the expression of target genes [59,65–67]. However, the precise function of Wnt/β-catenin in osteoblastic differentiation remains to be fully elucidated. We previously found that BMP9 (aka, GDF2) is one of the most potent osteogenic BMPs [26,30,32,33,61,68]. Through gene expression profiling, we found that Wnt3A and BMP9 regulated the expression of overlapping but distinct sets of downstream target genes in MSCs [39,48], suggesting that there may be an important crosstalk between BMP and Wnt-induced osteogenic signaling. In this study, we used iMEFs and demonstrated that Wnt3A and BMP9 can potentiate each other's ability to induce osteogenic differentiation *in vitro* and *in vivo*. Furthermore, β-catenin knockdown significantly diminishes BMP9-induced osteogenic differentiation of iMEFs, indicating that BMP9-induced osteoblastic differentiation requires functional β-catenin signaling.

In summary, we provide a conceptual design of a simplified and versatile system for the simultaneous expression of multiple siRNAs that silence one or different genes. A series of proof-of-concept studies have validated the technical feasibility and functional efficiency of the pSOK system by silencing human and mouse β-catenin expression. Thus, our results have demonstrated that the GDA-based pSOK system should be a valuable tool for gene function studies and the development of therapeutics.

Supporting Information

Figure S1 Schematic representations of the pSOK and pB2B vectors developed in this study. (**A**) The pSOK vector is a Murine Stem Cell Virus (MSCV) retroviral vector. It was derived from the previously developed pSOS vector [15]. The pSOK is a destination vector used for the one-step Gibson Assembly after SwaI linearization. This vector confers Blasticidin S resistance for generating stable mammalian cell lines. (**B**) The pB2B vector is a common template for PCR amplifications to generate the fragments with distinct siRNA target sites, which are subsequently used for Gibson Assembly with the SwaI-linearized pSOK vector. The full-length sequences and maps of these vectors are available at: http://www.boneandcancer.org/MOLab%20Vectors%20after %20Nov%201%202005/pSOK.pdf and http://www.boneandcan cer.org/MOLab%20Vectors%20after%20Nov%201%202005/p BOK%20vector%20map%20and%20sequence%202013-12-02. pdf.

Figure S2 A Guide for primer design and essential sequences for assembly analysis. (**A**) Primer design guide. To make a construct containing four siRNA target sites driven by opposing U6-H1

promoters, three PCR fragments will be made for the assembly reaction. Please note the sense-strand (upper case; driven by U6 promoter) and reverse-complement strand (lower case) of the chosen siRNA sites. (**B**) The DNA sequence of the H1-U6 back-to-back promoters in pB2B is used to amplify the different siRNA fragments. Please note the template sequence contains the "TTTTT" and "AAAAA" sequences to terminate siRNA transcripts. (**C**) The assembled query sequence for BLAST analysis of sequenced candidadte clones. One can simply replace the designed "X", "Y" and "Z" target site sequences (red and underlined) and use the modified sequence as a template to perform BLAST2 analysis and verify colony authenticity.

Figure S3 Function validation of the silencing efficiency of four siRNA sites targeting human β-catenin. 293 and SW480 cells stably expressing one, two, three, four siRNA sites, or siControl were generated as described in Methods. Subconfluent 293 lines were co-transfected with TOP-Luc and pCMV-Wnt3A plasmids (**A**) while the SW480 lines were just transfected with TOP-Luc reporter plasmid (**B**). At 24 h and 48 h after transfection, cells were lysed and subjected to firefly luciferase activity assays as described in Methods. Each assay condition was done in triplicate.

Table S1 Primers used for PCR analysis.

Author Contributions

Conceived and designed the experiments: TCH TY RRR RCH HHL FD. Performed the experiments: FD XC ZL ZY ZW. Analyzed the data: FD XC ZL ZY ZW. Contributed reagents/materials/analysis tools: YD QZ ZZ JY MQ RL SD JW QW ML NG LZ GZ PZ. Wrote the paper: TCH TY RCH HHL FD.

References

1. Hammond SM, Bernstein E, Beach D, Hannon GJ (2000) An RNA-directed nuclease mediates post-transcriptional gene silencing in Drosophila cells. Nature 404: 293–296.
2. Castel SE, Martienssen RA (2013) RNA interference in the nucleus: roles for small RNAs in transcription, epigenetics and beyond. Nat Rev Genet 14: 100–112.
3. Dykxhoorn DM, Novina CD, Sharp PA (2003) Killing the messenger: short RNAs that silence gene expression. Nat Rev Mol Cell Biol 4: 457–467.
4. Ghildiyal M, Zamore PD (2009) Small silencing RNAs: an expanding universe. Nat Rev Genet 10: 94–108.
5. Hammond SM, Caudy AA, Hannon GJ (2001) Post-transcriptional gene silencing by double-stranded RNA. Nat Rev Genet 2: 110–119.
6. Okamura K, Lai EC (2008) Endogenous small interfering RNAs in animals. Nat Rev Mol Cell Biol 9: 673–678.
7. Sarkies P, Miska EA (2014) Small RNAs break out: the molecular cell biology of mobile small RNAs. Nat Rev Mol Cell Biol 15: 525–535.
8. Bumcrot D, Manoharan M, Koteliansky V, Sah DW (2006) RNAi therapeutics: a potential new class of pharmaceutical drugs. Nat Chem Biol 2: 711–719.
9. Czech MP, Aouadi M, Tesz GJ (2011) RNAi-based therapeutic strategies for metabolic disease. Nat Rev Endocrinol 7: 473–484.
10. de Fougerolles A, Vornlocher HP, Maraganore J, Lieberman J (2007) Interfering with disease: a progress report on siRNA-based therapeutics. Nat Rev Drug Discov 6: 443–453.
11. Iorns E, Lord CJ, Turner N, Ashworth A (2007) Utilizing RNA interference to enhance cancer drug discovery. Nat Rev Drug Discov 6: 556–568.
12. Kim DH, Rossi JJ (2007) Strategies for silencing human disease using RNA interference. Nat Rev Genet 8: 173–184.
13. Pecot CV, Calin GA, Coleman RL, Lopez-Berestein G, Sood AK (2011) RNA interference in the clinic: challenges and future directions. Nat Rev Cancer 11: 59–67.
14. Fellmann C, Lowe SW (2014) Stable RNA interference rules for silencing. Nat Cell Biol 16: 10–18.
15. Luo Q, Kang Q, Song WX, Luu HH, Luo X, et al. (2007) Selection and validation of optimal siRNA target sites for RNAi-mediated gene silencing. Gene 395: 160–169.
16. Gibson DG, Young L, Chuang RY, Venter JC, Hutchison CA 3rd, et al. (2009) Enzymatic assembly of DNA molecules up to several hundred kilobases. Nat Methods 6: 343–345.
17. Gibson DG (2011) Enzymatic assembly of overlapping DNA fragments. Methods Enzymol 498: 349–361.
18. Lienert F, Lohmueller JJ, Garg A, Silver PA (2014) Synthetic biology in mammalian cells: next generation research tools and therapeutics. Nat Rev Mol Cell Biol 15: 95–107.
19. Wang N, Zhang H, Zhang BQ, Liu W, Zhang Z, et al. (2014) Adenovirus-mediated efficient gene transfer into cultured three-dimensional organoids. PLoS One 9: e93608.
20. Wang N, Zhang W, Cui J, Zhang H, Chen X, et al. (2014) The piggyBac Transposon-Mediated Expression of SV40 T Antigen Efficiently Immortalizes Mouse Embryonic Fibroblasts (MEFs). PLoS One 9: e97316.
21. Lamplot JD, Liu B, Yin L, Zhang W, Wang Z, et al. (2014) Reversibly Immortalized Mouse Articular Chondrocytes Acquire Long-Term Proliferative Capability while Retaining Chondrogenic Phenotype. Cell Transplant.
22. Wen S, Zhang H, Li Y, Wang N, Zhang W, et al. (2014) Characterization of constitutive promoters for piggyBac transposon-mediated stable transgene expression in mesenchymal stem cells (MSCs). PLoS One 9: e94397.
23. Zhao C, Wu N, Deng F, Zhang H, Wang N, et al. (2014) Adenovirus-mediated gene transfer in mesenchymal stem cells can be significantly enhanced by the cationic polymer polybrene. PLoS One 9: e92908.
24. Li R, Zhang W, Cui J, Shui W, Yin L, et al. (2014) Targeting BMP9-Promoted Human Osteosarcoma Growth by Inactivation of Notch Signaling. Curr Cancer Drug Targets.
25. Huang E, Bi Y, Jiang W, Luo X, Yang K, et al. (2012) Conditionally Immortalized Mouse Embryonic Fibroblasts Retain Proliferative Activity without Compromising Multipotent Differentiation Potential. PLoS One 7: e32428.
26. Wang J, Zhang H, Zhang W, Huang E, Wang N, et al. (2014) Bone Morphogenetic Protein-9 (BMP9) Effectively Induces Osteo/Odontoblastic Differentiation of the Reversibly Immortalized Stem Cells of Dental Apical Papilla. Stem Cells Dev 23: 1405–1416.
27. Wu N, Zhang H, Deng F, Li R, Zhang W, et al. (2014) Overexpression of Ad5 precursor terminal protein accelerates recombinant adenovirus packaging and amplification in HEK-293 packaging cells. Gene Ther 21: 629–637.

28. Feng T, Li Z, Jiang W, Breyer B, Zhou L, et al. (2002) Increased efficiency of cloning large DNA fragments using a lower copy number plasmid. Biotechniques 32: 992, 994, 996 passim.

29. Sharff KA, Song WX, Luo X, Tang N, Luo J, et al. (2009) Hey1 Basic Helix-Loop-Helix Protein Plays an Important Role in Mediating BMP9-induced Osteogenic Differentiation of Mesenchymal Progenitor Cells. J Biol Chem 284: 649–659.

30. Kang Q, Song WX, Luo Q, Tang N, Luo J, et al. (2009) A comprehensive analysis of the dual roles of BMPs in regulating adipogenic and osteogenic differentiation of mesenchymal progenitor cells. Stem Cells Dev 18: 545–559.

31. Yang R, Jiang M, Kumar SM, Xu T, Wang F, et al. (2011) Generation of melanocytes from induced pluripotent stem cells. J Invest Dermatol 131: 2458–2466.

32. Cheng H, Jiang W, Phillips FM, Haydon RC, Peng Y, et al. (2003) Osteogenic activity of the fourteen types of human bone morphogenetic proteins (BMPs). J Bone Joint Surg Am 85-A: 1544–1552.

33. Kang Q, Sun MH, Cheng H, Peng Y, Montag AG, et al. (2004) Characterization of the distinct orthotopic bone-forming activity of 14 BMPs using recombinant adenovirus-mediated gene delivery. Gene Ther 11: 1312–1320.

34. Luo J, Deng ZL, Luo X, Tang N, Song WX, et al. (2007) A protocol for rapid generation of recombinant adenoviruses using the AdEasy system. Nat Protoc 2: 1236–1247.

35. Kong Y, Zhang H, Chen X, Zhang W, Zhao C, et al. (2013) Destabilization of Heterologous Proteins Mediated by the GSK3beta Phosphorylation Domain of the beta-Catenin Protein. Cell Physiol Biochem 32: 1187–1199.

36. Liu X, Qin J, Luo Q, Bi Y, Zhu G, et al. (2013) Cross-talk between EGF and BMP9 signalling pathways regulates the osteogenic differentiation of mesenchymal stem cells. J Cell Mol Med.

37. Wang Y, Hong S, Li M, Zhang J, Bi Y, et al. (2013) Noggin resistance contributes to the potent osteogenic capability of BMP9 in mesenchymal stem cells. J Orthop Res 31: 1796–1803.

38. Gao Y, Huang E, Zhang H, Wang J, Wu N, et al. (2013) Crosstalk between Wnt/beta-Catenin and Estrogen Receptor Signaling Synergistically Promotes Osteogenic Differentiation of Mesenchymal Progenitor Cells. PLoS One 8: e82436.

39. Luo Q, Kang Q, Si W, Jiang W, Park JK, et al. (2004) Connective Tissue Growth Factor (CTGF) Is Regulated by Wnt and Bone Morphogenetic Proteins Signaling in Osteoblast Differentiation of Mesenchymal Stem Cells. J Biol Chem 279: 55958–55968.

40. Tang N, Song WX, Luo J, Luo X, Chen J, et al. (2009) BMP9-induced osteogenic differentiation of mesenchymal progenitors requires functional canonical Wnt/beta-catenin signaling. J Cell Mol Med 13: 2448–2464.

41. Zhang Y, Chen X, Qiao M, Zhang BQ, Wang N, et al. (2014) Bone morphogenetic protein 2 inhibits the proliferation and growth of human colorectal cancer cells. Oncol Rep.

42. Chen X, Luther G, Zhang W, Nan G, Wagner ER, et al. (2013) The E-F Hand Calcium-Binding Protein S100A4 Regulates the Proliferation, Survival and Differentiation Potential of Human Osteosarcoma Cells. Cell Physiol Biochem 32: 1083–1096.

43. Huang J, Bi Y, Zhu GH, He Y, Su Y, et al. (2009) Retinoic acid signalling induces the differentiation of mouse fetal liver-derived hepatic progenitor cells. Liver Int 29: 1569–1581.

44. Zhang W, Deng ZL, Chen L, Zuo GW, Luo Q, et al. (2010) Retinoic acids potentiate BMP9-induced osteogenic differentiation of mesenchymal progenitor cells. PLoS One 5: e11917.

45. Rastegar F, Gao JL, Shenaq D, Luo Q, Shi Q, et al. (2010) Lysophosphatidic acid acyltransferase beta (LPAATbeta) promotes the tumor growth of human osteosarcoma. PLoS One 5: e14182.

46. Su Y, Wagner ER, Luo Q, Huang J, Chen L, et al. (2011) Insulin-like growth factor binding protein 5 suppresses tumor growth and metastasis of human osteosarcoma. Oncogene 30: 3907–3917.

47. Huang E, Zhu G, Jiang W, Yang K, Gao Y, et al. (2012) Growth hormone synergizes with BMP9 in osteogenic differentiation by activating the JAK/STAT/IGF1 pathway in murine multilineage cells. J Bone Miner Res 27: 1566–1575.

48. Si W, Kang Q, Luu HH, Park JK, Luo Q, et al. (2006) CCN1/Cyr61 Is Regulated by the Canonical Wnt Signal and Plays an Important Role in Wnt3A-Induced Osteoblast Differentiation of Mesenchymal Stem Cells. Mol Cell Biol 26: 2955–2964.

49. Peng Y, Kang Q, Cheng H, Li X, Sun MH, et al. (2003) Transcriptional characterization of bone morphogenetic proteins (BMPs)-mediated osteogenic signaling. J Cell Biochem 90: 1149–1165.

50. Zhu GH, Huang J, Bi Y, Su Y, Tang Y, et al. (2009) Activation of RXR and RAR signaling promotes myogenic differentiation of myoblastic C2C12 cells. Differentiation 78 195–204.

51. Luo X, Sharff KA, Chen J, He TC, Luu HH (2008) S100A6 expression and function in human osteosarcoma. Clin Orthop Relat Res 466: 2060–2070.

52. Hu N, Jiang D, Huang E, Liu X, Li R, et al. (2013) BMP9-regulated angiogenic signaling plays an important role in the osteogenic differentiation of mesenchymal progenitor cells. J Cell Sci 126: 532–541.

53. Luo X, Chen J, Song WX, Tang N, Luo J, et al. (2008) Osteogenic BMPs promote tumor growth of human osteosarcomas that harbor differentiation defects. Lab Invest 88: 1264–1277.

54. Bi Y, Huang J, He Y, Zhu GH, Su Y, et al. (2009) Wnt antagonist SFRP3 inhibits the differentiation of mouse hepatic progenitor cells. J Cell Biochem 108: 295–303.

55. Luo J, Tang M, Huang J, He BC, Gao JL, et al. (2010) TGFbeta/BMP type I receptors ALK1 and ALK2 are essential for BMP9-induced osteogenic signaling in mesenchymal stem cells. J Biol Chem 285: 29588–29598.

56. Chen L, Jiang W, Huang J, He BC, Zuo GW, et al. (2010) Insulin-like growth factor 2 (IGF-2) potentiates BMP-9-induced osteogenic differentiation and bone formation. J Bone Miner Res 25: 2447–2459.

57. Zhang J, Weng Y, Liu X, Wang J, Zhang W, et al. (2013) Endoplasmic reticulum (ER) stress inducible factor cysteine-rich with EGF-like domains 2 (Creld2) is an important mediator of BMP9-regulated osteogenic differentiation of mesenchymal stem cells. PLoS One 8: e73086.

58. Yan D, Wiesmann M, Rohan M, Chan V, Jefferson AB, et al. (2001) Elevated expression of axin2 and hnkd mRNA provides evidence that Wnt/beta -catenin signaling is activated in human colon tumors. Proc Natl Acad Sci U S A 98: 14973–14978.

59. He TC, Sparks AB, Rago C, Hermeking H, Zawel L, et al. (1998) Identification of c-MYC as a target of the APC pathway [see comments]. Science 281: 1509–1512.

60. Kim JH, Liu X, Wang J, Chen X, Zhang H, et al. (2013) Wnt signaling in bone formation and its therapeutic potential for bone diseases. Ther Adv Musculoskelet Dis 5: 13–31.

61. Luu HH, Song WX, Luo X, Manning D, Luo J, et al. (2007) Distinct roles of bone morphogenetic proteins in osteogenic differentiation of mesenchymal stem cells. J Orthop Res 25: 665–677.

62. Conte C, Dastugue B, Vaury C (2002) Promoter competition as a mechanism of transcriptional interference mediated by retrotransposons. Embo J 21: 3908–3916.

63. Wodarz A, Nusse R (1998) Mechanisms of Wnt signaling in development. Annu Rev Cell Dev Biol 14: 59–88.

64. Deng ZL, Sharff KA, Tang N, Song WX, Luo J, et al. (2008) Regulation of osteogenic differentiation during skeletal development. Front Biosci 13: 2001–2021.

65. Tetsu O, McCormick F (1999) Beta-catenin regulates expression of cyclin D1 in colon carcinoma cells. Nature 398: 422–426.

66. He TC, Chan TA, Vogelstein B, Kinzler KW (1999) PPARdelta is an APC-regulated target of nonsteroidal anti-inflammatory drugs. Cell 99: 335–345.

67. Luo J, Chen J, Deng ZL, Luo X, Song WX, et al. (2007) Wnt signaling and human diseases: what are the therapeutic implications? Lab Invest 87: 97–103.

68. Lamplot JD, Qin J, Nan G, Wang J, Liu X, et al. (2013) BMP9 signaling in stem cell differentiation and osteogenesis. Am J Stem Cells 2: 1–21.

13

Comparison of the Transcriptome between Two Cotton Lines of Different Fiber Color and Quality

Wenfang Gong[1ϑ], Shoupu He[1ϑ], Jiahuan Tian[1], Junling Sun[1], Zhaoe Pan[1], Yinhua Jia[1], Gaofei Sun[1,2], Xiongming Du[1]*

1 State Key Laboratory of Cotton Biology, Institute of Cotton Research, Chinese Academy of Agricultural Sciences, Anyang, China, 2 Department of Computer Science and Information Engineering, Anyang Institute of Technology, Anyang, China

Abstract

To understand the mechanism of fiber development and pigmentation formation, the mRNAs of two cotton lines were sequenced: line Z128 (light brown fiber) was a selected mutant from line Z263 (dark brown fiber). The primary walls of the fiber cell in both Z263 and Z128 contain pigments; more pigments were laid in the lumen of the fiber cell in Z263 compared with that in Z128. However, Z263 contained less cellulose than Z128. A total of 71,895 unigenes were generated: 13,278 (20.26%) unigenes were defined as differentially expressed genes (DEGs) by comparing the library of Z128 with that of Z263; 5,345 (8.16%) unigenes were up-regulated and 7,933 (12.10%) unigenes were down-regulated. qRT-PCR and comparative transcriptional analysis demonstrated that the pigmentation formation in brown cotton fiber was possibly the consequence of an interaction between oxidized tannins and glycosylated anthocyanins. Furthermore, our results showed the pigmentation related genes not only regulated the fiber color but also influenced the fiber quality at the fiber elongation stage (10 DPA). The highly expressed flavonoid gene in the fiber elongation stage could be related to the fiber quality. DEGs analyses also revealed that transcript levels of some fiber development genes (Ca^{2+}/CaM, reactive oxygen, ethylene and sucrose phosphate synthase) varied dramatically between these two cotton lines.

Editor: Xianlong Zhang, National Key Laboratory of Crop Genetic Improvement, China

Funding: This study was supported by the "Twelfth Five-Year Plan" of the National Science and Technology Support Project (2011BAD35B05 and 2013BAD01B03) and Basic Scientific Research funds in National Nonprofit Institutes (SJA0608). The funders had no role in study design, data collection and analysis, decision to publish, or preparation of the manuscript.

Competing Interests: The authors have declared that no competing interests exist.

* Email: dujeffrey8848@hotmail.com

ϑ These authors contributed equally to this work.

Introduction

Upland cotton (*Gossypium hirsutum* L.) is the largest natural fiber producer of the plants. In recent years, interest in naturally colored cotton has grown because it may reduce pollution, making it preferable to white fiber which requires a dyeing process [1–3]. However, its commercial application is very limited due to the lack of fiber color diversity and low fiber quality [2]. Limited brown (different color depth) and green fiber lines, among other varieties, have been used in the textile industry. A previous study demonstrated that there was a significant negative correlation between the degree of fiber color and lint percentage and fiber quality traits in cotton [4]. Therefore, subsequent studies should focus on improving the fiber quality and revealing the underlying mechanisms for pigmentation formation in naturally colored cotton.

Early genetic analysis suggested that the brown color of cotton fiber was controlled by one incompletely dominant major gene [5]. Furthermore, gene expression analysis and dimethylaminocinnaldehyde staining showed that tannins could be the key chemical responsible for the brown color in cotton fiber [6]. Subsequent chemical research indicated that the brown pigment in cotton fiber

might be the chinone compound oxidated from condensed tannins, and the accumulation period of condensed tannins was from 10 DPA-25 DPA [7,8]. However, the molecular mechanism that underlies pigmentation in colored cotton fiber is still unknown.

With the development of next generation sequencing technology, RNA-seq provides a powerful tool to rebuild our knowledge of transcriptomics. By directly sequencing and assembling the mRNA, the whole transcriptome could be de novo reconstructed precisely and efficiently [9], aligned with public databases for function annotation and the critical genes could be assessed using pathway classification. In addition, gene expression can be measured and the number of transcripts can be obtained if the appropriate level of sequencing was performed. The application of RNA-seq technology has been used successfully for various species [10–14].

Lines Z263 and Z128, two brown fiber inbred lines with dark and light brown fiber respectively, both derived from a cross between white and brown fiber cotton. To reveal the whole transcriptome landscape of the natural colored cotton fiber, and to understand the molecular mechanism of pigment formation, the transcriptomes of these two lines at the whole early developmental stage (0 dpa–20 dpa) were sequenced using RNA-seq technology and analyzed.

Materials and Methods

1. Plant material

Line Z263 (*Gossypium hirsutum* L.), with dark brown fiber, was selected from a cross between white cotton (Zhong 6331) and brown fiber cotton (Crd). Line Z128 (*G. hirsutum* L.) was a selected mutant from Z263. They were planted in an experimental field at the Institute of Cotton Research, Chinese Academy of Agricultural Sciences (ICR, CAAS) under normal agronomic management conditions. The bolls at the day of anthesis (0 day post anthesis/DPA), 5 DPA, 15 DPA and 20 DPA were harvested and stored in an ice box, then the ovules of 0 DPA and fibers of other stages were dissected and separated on ice as fast as possible, and then stored at −80°C immediately.

2. Fiber quality measurement, fiber microstructure detection and cellulose test

To test the fiber quality of Z263 and Z128, the following measurements were included: upper half mean length (mm), uniformity index (%), micronaire, fiber elongation (%), fiber strength, 15 g lint samples of each line were analyzed using USTER HVI 1000 (USTER Technologies, Inc., Uster, Switzerland). The fiber microstructure detection and cellulose test were performed according to Ru et al [15].

3. RNA extraction and cDNA library construction

Total RNAs were extracted from each sample using the CTAB method described by Wan and Wilkins [16], with minor modifications to increase the yield. All RNA quality and quantity were measured by 1.0% agarose gel and an ultraviolet spectrophotometer. Four stages of RNAs (0 DPA, 5 DPA, 15 DPA and 20

DPA) with the same concentration and quality from each line were mixed together as one mixed library. The two libraries of Z263 and Z128 were constructed using the method described by Xia et al. [17].

4. RNA-seq and sequence de novo assembly

The sequencing of two libraries of Z263 and Z128 were performed on HiSeq 2000 (illumina) by the Beijing Genomics Institute (BGI) (Shenzhen, Guangdong, China). The raw reads, transformed from images, were first processed by removing adaptors and redundant fragments to generate clean reads. The clean reads de novo assembly was carried out using the short reads assembling program-SOAPdenovo [18]. Clean reads with a certain length of overlap were first combined to form "contigs". Then the clean reads were mapped back to the contigs; it was possible to detect contigs from the same transcript, as well as the distances between them, by paired-end reads. Next, SOAPdenovo connected the contigs into "scaffolds"; "N" was used to present unknown sequences. Paired-end reads were used again to fill the gaps between scaffolds, longer sequences which could not extend at either end were assembled and defined as "unigenes". In this study, the unigenes from the two libraries were further processed by sequence splicing and redundancy removal with the sequence cluster program-TGICL [19] to acquire non-redundant unigenes that were as long as possible.

5. Unigene functional annotation and functional categorization

All-unigenes were first searched using the blastx tool against public protein databases such as Non-redundant protein sequences (nr, http://www.ncbi.nlm.nih.gov), Swiss-Prot (http://www.expasy.org/sprot/), Kyoto Encyclopedia of Genes and Genomes (KEGG, http://www.genome.jp/kegg/pathway.html) and Cluster of orthologous groups for eukaryotic complete genomes (KOG,

Table 1. The primers used in this study.

Gene name	NCBI accession no.	Direction	Primer Sequence (5′→3′)	Product length
CHS	EF643507	F	GGTGTGGACATGCCTGGGGC	265
		R	CAGCTGCGGCACCATCACCA	
CHI	EF187439	F	ATCCGTTGAGTTTTTCAGAG	127
		R	CCAAATAGCAACGCAAT	
F3′H		F	CGAGGAGATGGATAAGGTGATTG	128
		R	GCAAGTTCAAGGGAGTAGATGGA	
F3H	EF187440	F	GCTTCTTGAGGTGTTGTCAGAGG	116
		R	CAGGTTGAGGGCATTTAGGATAG	
DFR	FJ713480	F	CGCGACCCTGGCAACTCGAA	417
		R	CCAGGCTGCTTGCTCTGCCA	
ANS	EU921264	F	AAGAGAAGTATGCCAACGAC	102
		R	AGAAGTAGTCCTCCCACTCA	
ANR	FJ713479	F	TCCTCAACAAAAGATACCCTGACTT	147
		R	CGGTTTGGTCGTAGATTTCCTC	
LAR		F	AAAGTAGCCAAAGCCCTTCA	260
		R	TAACAGTGCCGACAGAGTGAA	
Gh18S	L24145	F	TGACGGAGAATTAGGGTTCGA	100
		R	CCGTGTCAGGATTGGGTAATTT	

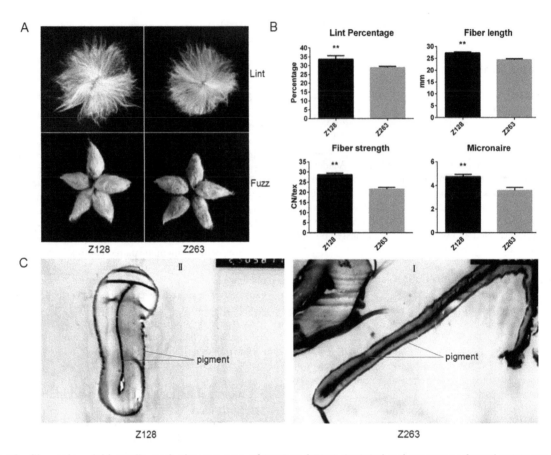

Figure 1. The fiber color, yield, quality and microstructure of Z128 and Z263. Statistical analyses were performed at 95% condence with IBM SPSS Statistics 11.0 (SPSS Inc., Chicago, USA). Values with an asterisk represented a significant differe- nce at P<0.05. The fiber microstructure of Z263 (C I) and Z128 (C II) was observed under a microscope (×3000 and ×2500, respectively).

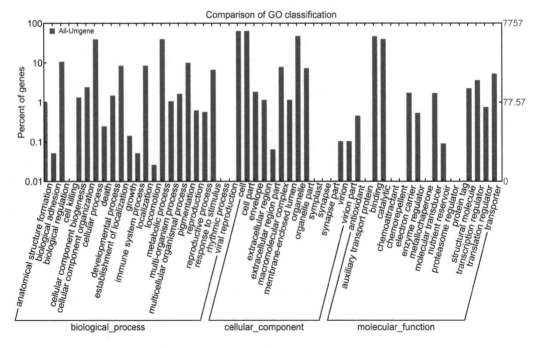

Figure 2. All-unigenes classified by GO analysis.

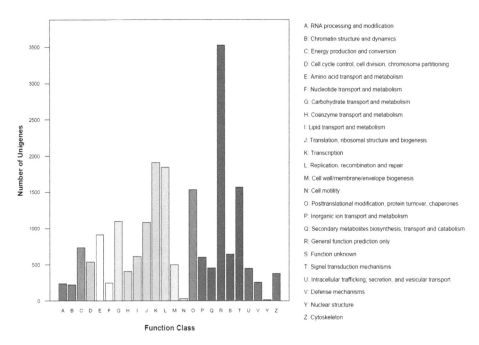

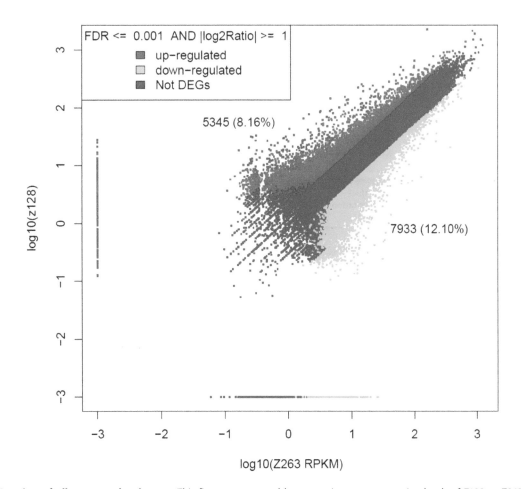

Figure 3. COG classification of all-unigenes. A–Z represented different functions classified by GO analysis, respectively.

Figure 4. Mapping of all expressed unigenes. This figure was created by comparing gene expression levels of Z128 to Z263. FDR≤0.001, |log₂Ratio| ≥1 was used as threshold, red dots represented up-regulated genes, green dots represented the down-regulated genes, and blue dots represented gene expression with no significant difference.

Table 2. The DEGs enriched terms in GO analysis (P-value<1).

Ontology	Gene Ontology term	Cluster frequency	Genome Frequency of use	Corrected P-value
Biological process	RNA-dependent DNA replication	15/597, 2.5%	47/4110, 1.1%	0.82163
Cellular component	cytoplasmic vesicle	137/696, 19.7%	750/4911, 15.3%	0.06344
	cytoplasmic membrane-bounded vesicle	135/696, 19.4%	745/4911, 15.2%	0.10093
	spliceosome	7/696, 1.0%	14/4911, 0.3%	0.23912
Molecular function	oxidoreductase activity, acting on paired donors, with incorporation or reduction of molecular oxygen, 2-oxoglutarate as one donor, and incorporation of one atom each of oxygen into both donors	7/769, 0.9%	13/5273, 0.2%	0.32203
	RNA-directed DNA polymerase activity	15/769, 2.0%	48/5273, 0.9%	0.76482
	DNA polymerase activity	15/769, 2.0%	49/5273, 0.9%	0.95719

http://genome.jgi.d oe.gov/Tutorial/tutorial/kog.html). An e-value<10^{-5} was used as the threshold. To further understand the distribution of gene function, the protein functional classification and pathway were annotated by Gene Ontology (http://www.geneontology.org/), KOGs and KEGG. With NR annotation, GO functional annotation and classification were obtained using the Blast2GO program [20] and WEGO software [21], respectively.

6. Differential expressed genes (DEGs) identification and enrichment analysis

Referring to the method described by Audic and Claverie [22], the Beijing Genomics Institute (BGI) developed a rigorous

Table 3. The top 10 DEGs enriched pathways in KEGG analysis.

No.	Pathway	DEGs with pathway annotation (3,213)	All genes with pathway annotation (20,242)	P-value	Pathway ID
1	Zeatin biosynthesis	16 (0.5%)	41 (0.2%)	0.000299	ko00908
2	Anthocyanin biosynthesis	4 (0.12%)	5 (0.02%)	0.002767	ko00942
3	ABC transporters	40 (1.24%)	164 (0.81%)	0.003	ko02010
4	Diterpenoid biosynthesis	15 (0.47%)	47 (0.23%)	0.004717	ko00904
5	Phenylpropanoid biosynthesis	56 (1.74%)	255 (1.26%)	0.006239	ko00940
6	3-Chloroacrylic acid degradation	18 (0.56%)	63 (0.31%)	0.007641	ko00641
7	Flavone and flavonol biosynthesis	14 (0.44%)	45 (0.22%)	0.007987	ko00944
8	Ubiquitin mediated proteolysis	108 (3.36%)	550 (2.72%)	0.00975	ko04120
9	Taurine and hypotaurine metabolism	5 (0.16%)	10 (0.05%)	0.012555	ko00430
10	Base excision repair	36 (1.12%)	158 (0.78%)	0.014168	ko03410

Table 4. The involved unigenes in Flavone and flavonol biosynthesis and Anthocyanin biosynthesis pathways.

	Orthology	Entry of enzyme	Unigenes	Ratio[1]	Status[2]	Encoded protein
Flavonoid biosynthesis	K00660	2.3.1.74	Unigene56280_All	−1.9	D	chalcone synthase (CHS)
	K00588	2.1.1.104	Unigene34142_All	−1.4	D	caffeoyl-CoA O-methyltransferase
			Unigene42690_All	−3.6	D	
			Unigene10308_All	−1.1	D	
	K01859	5.5.1.6	Unigene69406_All	−1.6	D	chalcone isomerase (CHI)
	K05277	1.14.11.19	Unigene56780_All	−2.3	D	anthocyanidin synthase (ANS)
			Unigene57093_All	−3.0	D	
			Unigene5167_All	−1.6	D	
	K08695	1.3.1.77	Unigene43073_All	−1.1	D	anthocyanidin reductase (ANR)
	K05280	1.14.13.21	Unigene71334_All	2.8	U	flavonoid 3'-hydroxylase (F3'H)
	K00475	1.14.11.9	Unigene70267_All	3.9	U	flavanone 3-hydroxylase (F3H)
			Unigene71550_All	4.2	U	
	K05278	1.14.11.23	Unigene70581_All	4.1	U	flavonol synthase (FLS)
			Unigene6472_All	1.0	U	
Flavone and flavonol biosynthesis	K05280	1.14.13.21	Unigene71334_All	2.8	U	flavonoid 3'-hydroxylase (F3'H)
	K05279	2.1.1.76	Unigene13368_All	−2.2	D	flavonol 3-O-methyltransferase (FOMT)
			Unigene21288_All	−1.4	D	
			Unigene38018_All	−1.2	D	
			Unigene47446_All	−2.4	D	
			Unigene47563_All	−3.7	D	
			Unigene50636_All	−2.7	D	
			Unigene57809_All	−1.7	D	
			Unigene58594_All	−1.9	D	
			Unigene70908_All	−1.7	D	
			Unigene9272_All	−1.1	D	
	K10757	2.4.1.91	Unigene50839_All	1.0	U	flavonol 3-O-glucosyltransferase (FOGT)
			Unigene8998_All	2.1	U	
			Unigene54635_All	−4.3	D	
Anthocyanin biosynthesis	K12338	2.4.1.298	Unigene18122_All	2.2	U	anthocyanin 5-O-glucosyltransferase (5GT)
			Unigene59143_All	12.0	U	
			Unigene67316_All	12.5	U	
			Unigene71660_All	12.9	U	

1: ratio indicated log2(z128_RPKM/Z263_RPKM).
2: "U" indicated that this unigene was up-regulated; "D" indicated down-regulated.

algorithm to identify DEGs from two samples of RNA-seq data. Because the expression of each gene occupies only a small part of the library, we denote the number of unambiguous clean tags from gene A as x, and the p(x) is in the Poisson distribution. The formula is as follows:

$$p(x) = \frac{e^{-\lambda}\lambda^x}{x!} \quad (\lambda \text{ is the real transcripts of the gene})$$

The total clean tag number of sample 1 is N^1, and total clean tag number of sample 2 is N^2; gene A holds x tags in sample 1 and y tags in sample 2. The probability of gene A expressed equally between two samples can be calculated with the following formula:

$$2\sum_{i=0}^{i=y} p(i|x) \text{ or } 2 \times (1 - \sum_{i=0}^{i=y} p(i|x)) \text{ (if } \sum_{i=0}^{i=y} p(i|x) > 0.5)$$

$$p(y|x) = \left(\frac{N^2}{N^1}\right)^y \frac{(x+y)!}{x!y!\left(1+\frac{N^2}{N^1}\right)^{(x+y+1)}}$$

The P value corresponds to the differential gene expression test.

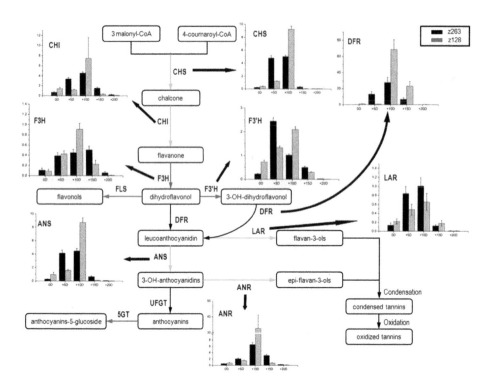

Figure 5. The schematic of pigment formation in cotton fiber. The red and green arrows indicated the up- and down-regulated status detected by comparing gene expression in Z128 with that in Z263. chalcone synthase (CHS); chalcone isomerase (CHI); flavanone 3-hydroxylase (F3H); flavonol synthase (FLS); dihydroflavonol 4-reductase (DFR); leucoanthocyanidin reductase (LAR); anthocyanidin synthase (ANS); UDP-favonoid glucosyl transferase (UFGT); anthocyanidin reductase (ANR); anthocyanin 5-O-glucosyltransferase (5GT).

The false discovery rate (FDR) is a method used to determine the threshold of the P value in multiple tests. In our study, the reads per kilobase of exon model per million mapped reads (RPKM) value, was calculated referring to the formula described by Mortazavi et al. [23] and was used to quantify the transcript level of Z128 versus Z263. FDR$\leq 10^{-3}$ and the absolute value of the log2Ratio (Z128_RPKM/Z263_RPKM) ≥ 1 as the threshold, were used to judge significant differences in gene expression. DEGs were then subjected to GO functional enrichment analysis and KEGG pathway enrichment analysis.

GO functional enrichment analysis provides GO terms, which significantly enrich in DEGs compared to the genome background, indicating that the DEGs are connected to interesting biological functions. All DEGs are firstly mapped to GO terms in the database, calculating gene numbers for every term, then the ultra-geometric test is used to find significantly enriched GO terms in DEGs compared to the genome background. The formula is:

$$P = 1 - \sum_{i=0}^{m-1} \frac{\binom{M}{i}\binom{N-M}{n-i}}{\binom{N}{n}}$$

Where N is the number of genes with GO annotation; n is the number of DEGs in N; M is the number of genes that are annotated according to the GO terms; m is the number of DEGs in M. The calculated P value≤ 0.05 was taken as a threshold. GO terms fulfilling this condition are defined as significantly enriched GO terms in DEGs. This analysis recognizes the main biological functions that DEGs participate in.

Pathway enrichment analysis identifies significantly enriched metabolic pathways or signal transduction pathways in DEGs compared with the whole genome background. The formula is the same as that in GO analysis. Here, N is the number of genes with KEGG annotation, n is the number of DEGs in N, M is the number of genes related to specific pathways, and m is number of DEGs in M (Qvalue≤ 0.05). All DEGs are further mapped on to each pathway; up-regulated genes are marked with red borders while down-regulated genes are marked with green borders.

7. qRT-PCR analysis for genes related to flavonoid synthesis

The genes and primers used for the gene expression analysis related to flavonoid synthesis were listed in Table 1. qRT-PCR was performed in a total volume of 20 μL with 10 μL SYBR Premix Ex Taq(2×) (Takara, Japan), PCR forward primer (10 μM) 0.4 μL,PCR reversed primer 0.4 μL (10 μM),ROX Reference Dye II (50×) 0.4 μL,cDNA template 2.0 μL and ddH$_2$O 6.8 μl on a 7500 real-time PCR machine (Applied Biosystems) according to the manufacturer's instructions. PCR amplification employed a 10 s denaturing step at 95°C, followed by 5 s at 95°C and 40 s at 60°C with 40 cycles. Relative mRNA levels were calculated by the $2^{-\Delta\Delta CT}$ method with *Gh18S* (accession number: L24145) as an internal control.

8. Statistical analysis

All of the experiments concerning data comparisons were performed three times. Statistical analyses were performed using the S-N-K method of independent-samples t-test at 95% confidence with IBM SPSS Statistics 11.0 (SPSS Inc., Chicago, USA). Values with different lowercases represent a significant difference at P<0.05.

Table 5. Fiber development related DEGs of brown and white fibers.

genes	Z263 (RPKM)	Z128 (RPKM)	p-value	FDR	Ratio[1]	Status[2]	Homologous proteins
Unigene31293_All	40.4792	19.2683	5.06E-15	2.40E-16	-1.1	D	extracellular Cu/Zn superoxide dismutase
Unigene15009_All	36.5823	3.0274	1.54E-52	2.92E-54	-3.6	D	class III peroxidase
Unigene22859_All	1.8516	5.7002	4.97E-04	1.53E-04	1.6	U	glutathione peroxidase
Unigene25776_All	2.1838	9.8675	1.78E-09	2.04E-10	2.2	U	peroxisomal membrane ABC transporter family
Unigene40537_All	80.13	230.84	2.93E-14	1.51E-15	1.5	U	fiber quinone-oxidoreductase
Unigene50168_All	2.78	1.20	3.83E-02	2.08E-02	-1.2	D	calcium-transporting ATPase(ACA9)
Unigene1176_All	16.85	45.17	2.03E-13	1.24E-14	1.4	U	calcium ion transmembrane transporter(ACA2)
Unigene55005_All	3.59	0.50	1.55E-03	5.46E-04	-2.8	D	calcium ion binding/transporter(ATNRT1:2)
Unigene15884_All	3.32	14.7	2.71E-14	1.38E-15	2.1	U	autoinhibited calcium ATPase
Unigene61435_All	1.1851	5.9315	3.37E-04	1.00E-04	2.3	U	ethylene receptor
Unigene40770_All	5.06	14.98	2.02E-07	3.10E-08	1.5	U	ethylene responsive element binding protein
Unigene26099_All	4.86	61.24	0.00	0.00	3.7	U	ethylene responsive transcription factor 2b
Unigene37372_All	22.708	90.2862	0.00	0.00	1.9	U	ethylene-responsive element binding protein ERF2
Unigene62869_All	0.86	9.38	2.65E-05	6.01E-06	3.4	U	Ethylene-responsive transcription factor 1B
Unigene30365_All	6.59	21.61	6.11E-14	3.33E-15	1.7	U	sucrose phosphate synthase

1: ratio indicated log2(z128_RPKM/Z263_RPKM);
2: "U" indicated that this unigene was up-regulated; "D" indicated down-regulated.

Results

1. The differences in fiber color and quality between Z263 and Z128

Line Z128 was a selected mutant from Z263 and both lines have similar genetic backgrounds. However, their fibers were different. As shown in Fig. 1, the fiber of Z263 was dark brown while that of Z128 was light. Though the primary walls of the fiber cell in both Z263 and Z128 contain pigments (Fig. 1), more pigments were laid in the lumen of the fiber cell in Z263 compared with that Z128. This resulted in a darker color in the Z263 fiber. However, the fiber yield and quality of Z128 was better than that of Z263. The lint percentages (%), upper half mean length (mm), micronaire and fiber strength in Z128 were also better than those in Z263 (Fig. 1). Furthermore, Z128 contained more cellulose (98.5%) than Z263 (94.5%).

2. The de novo assembled transcriptome of the fibers in Z263 and Z128

By removing useless sequences, a total of 38,114,054 (2,858,554,050 nucleotides) and 39,355,642 (2,951,673,150 nucleotides) 75 bp-length clean reads were obtained from the Z128 and Z263 mRNA libraries, respectively. A total of 170,201 and 182,404 contigs, 143,588 and 151,478 scaffolds and 59,926 and 71,895 unigenes were assembled in the Z128 and Z263 library, respectively. Because both of the samples for library construction were collected from the same tissue (ovule and fiber), two unigene libraries were taken forward for sequence clustering and redundancy removal to generate a new unigene library (All-unigene library) to make the non-redundant unigenes as long as possible. The All-unigene library contained 71,895 unigenes with an average length of 533 bp, which was obviously greater and longer than the other two libraries. The length of most of the unigenes in the three libraries was in the range of 100–1,000 bp, accounting for 93.08%, 92.44% and 88.21% in Z128-, Z263- and all-unigene library, respectively.

Since the all-unigene library contained the most complete and longest sequences, it was used to run batch alignment with a cut-off E-value of 10^{-5} on online public databases. A total of 49,941 (69.46%) and 31,714 (44.11%) unigenes received annotations from the NCBI non-redundant (nr) and Swiss-Prot databases, respectively. KEGG, KOGs and GO similarity analyses indicated that 20,241 (28.15%), 14,333 (19.94%) and 7,757 (10.79%) unigenes matched these databases, respectively. To investigate the genomic similarity between *Gossypium* and other species, we estimated all annotations of unigenes from nr. The result showed that the most abundant unigenes were annotated as "*Vitis vinifera*", "*Ricinus communis*" and "*Populus trichocarpa*", which accounted for 29.71%, 29.59% and 24.63%, respectively. Furthermore, we annotated 68.2% and 80.2% unigenes on the recently r- eleased A and D genome of diploid cotton (which were considered to be two donor g- enomes of the tetroploid cotton subgenome), respectively.

GO classification analysis showed that 7,757 all-unigenes were categorized into three main ontologies: biological process, cellular component and molecular function, which were further categorized as 54 terms, and all-unigenes were classified into different terms. One unigene might be repeatedly classified in different terms, therefore, a total of 32,935 all-unigenes (including 75 unigenes repeated in different categories) were distributed over 42 terms, 14,867 (45.14%) of them were categorized in cellular component ontology, 10,229 (31.06%) in biological process and 7,839 (23.80%) in molecular function (Fig. 2). The detailed classification demonstrated that the term "cell" (4,876) and "cell part" (4,876) in "cellular component" ontology, "metabolic process" (3,026) and "cellular process" (3,002) in "biological process" ontology, "binding" (3,538) and "catalytic" (3,030) in "molecular function" contained the most unigenes, respectively. Furthermore, 764 unigenes under the term "pigmentation" indicated that an abundance of pigment-related biological processes were involved in the development of colored cotton fiber.

The all-unigenes library was aligned to the KOG database to predict and classify possible functions. A total of 14,333 unigenes were matched and categorized into 24 classes (Fig. 3). The function class defined as "general function prediction" (code: R) had the most unigenes (24.61%), followed by "transcription" (K: 13.33%), "replication, recombination and repair" (L: 12.86%), "signal transduction mechanisms" (T: 10.95%) and "posttranslational modification, protein turnover, chaperones" (O: 10.70%).

The all-unigene library was aligned with the KEGG pathway database, and the result showed a total of 38,645 unigenes which were classified into six categories, mostly concentrated in three of them: metabolism (11,371; 29.42%), protein families (10,638; 27.53%) and cellular processes (6,959; 18.01%). Furthermore, 1,315 unigenes were clustered in the "biosynthesis of secondary metabolites" category (Table S1).

3. Differentially expressed genes (DEGs) analysis

The unigene expression level in the Z128 and Z263 libraries were compared. A total of 13,278 (20.26%) unigenes were significantly differentially expressed when these two non-redundant libraries were compared (FDR≤0.001, $|\log_2 \text{Ratio}| \geq 1$). A total of 5,345 (8.16%) of them were up-regulated, while 7,933 (12.10%) of them were down-regulated; the others were not DEGs (Fig. 4).

The GO analysis results showed that when corrected (P-value≤ 1), seven enriched terms belonged to three ontologies, one of them was categorized in "biological process", three were in "cellular component" and three were in "molecular function" (Table 2).

Table 3 showed the top 10 DEGs enriched pathways, seven of which were categorized as a "metabolism" pathway, two were categorized as "genetic information processing" pathways and the other one was categorized as "environmental information processing". Further identification indicated that five of the "metabolism" terms belonged to the "biosynthesis of secondary metabolites" sub-category.

4. Expression of related genes for color and fiber development in two cotton lines of different fiber color and quality

The pigmentation deposits in the brown colored cotton fiber were closely related to the flavonoid and proanthocyanidins biosynthesis. In this study, the "anthocyanin biosynthesis" and "flavone and flavonol biosynthesis" appeared on the top 10 list of the KEGG enrichment analysis (Table 3). Further analysis using the KEGG database indicated that a total of 14 unigenes were involved in the "flavone and flavonol biosynthesis" pathway (Table 4), which could be further classified as three orthology categories. "K05280" contained one up-regulated unigene which encoded flavonoid 3'-hydroxylase (F3'H). In addition, 10 down-regulated unigenes that encoded flavonol 3-O-methyltransferase (OMT) belonged to "K05279"; in "K10757", all unigenes encoded flavonol 3-O-glucosyltransferase (FOGT), two of them were up-regulated, and one was down-regulated. A total of four unigenes were mapped in the "anthocyanin biosynthesis" pathway, all of them were up-regulated. All four unigenes were involved in anthocyanin 5-O-glucosyltransferase (5GT) encoding.

Another important pigmentation pathway in plants is the "flavonoid biosynthesis pathway", which is also considered to be a key pathway for pigment formation in brown fiber cotton. Although it was not shown in Table 3, eight important genes were involved in this pathway. Chalcone synthase (*CHS*), chalcone isomerase (*CHI*), leucoanthocyanidin reductase (*LAR*), anthocyanidin reductase (*ANR*) and anthocyanidin synthase (*ANS*) were down-regulated in this pathway, while flavonoid 3'-hydroxylase (*F3'H*), flavanone 3-hydroxylase (*F3H*) and flavonol synthase (*FLS*) were all up-regulated. According to the distribution of DEGs in the entire pathway, most of the down-regulated DEGs (*CHS, CHI, LAR, ANR, ANS*) were enriched in upstream and downstream pathways, and the up-regulated DEGs were in the middle of the flavonoid biosynthetic pathway (*F3H, F3'H, FLS*). However, another gene in the middle of the pathway, the dihydroflavonol 4-reductase (*DFR*), was unchanged (Fig. 5). Furthermore, a gene that encodes 5-O-glucosyltransferase in the anthocyanin biosynthesis pathway was up-regulated. The down-regulated genes in the flavonoid biosynthesis pathway suggested that there were less pigments in Z128 compared with that in Z263. This was confirmed by the lighter brown fiber color in Z128 compared to the dark brown fiber in Z263.

To test the reliability of comparative transcriptional data, qRT-PCR analysis was performed for *CHS, CHI, LAR, DFR, F3H, F3'H, ANR* and *ANS*. Samples were selected from the flavonoid biosynthesis pathway across five developmental time points from 0 DPA to 20 DPA. Overall, the results of the qRT-PCR analysis were consistent with the results from the transcriptome for the mixed mRNAs of five developmental time points of the eight selected genes. However, when Z128 was compared to Z263 at 10 DPA, the selected eight genes had significantly higher transcript levels in Z128 (Fig. 5). This suggests that the genes involved in the flavonoid biosynthesis pathway at 10 DPA are related to better fiber quality formation. Moreover, the genes involved in flavonoid biosynthesis affected the fiber color and fiber quality of the brown fiber cotton. Some other genes such as the ethylene related factors also regulated fiber development [24]. According to Table 5, all of the ethylene related factors had higher expression levels in Z128 than in Z263. The DEGs analysis also revealed that transcript levels of different fiber development related factors varied dramatically in cotton fibers, such as reactive oxygen [25], ethylene [24], Ca^{2+}/CaM [26], and sucrose phosphate synthase [27] (Table 5).

Discussion

The RNA-seq approach based on next generation sequencing technology provided us with a new method to study the transcriptome of developing cotton fibers. It was not dependent on existing genome information and was an efficient way to quantify the expression level of a single gene without high background noise [28]. In recent years, this technology has been successfully applied in transcriptome studies for many non-model organisms [13,17,29–30].

In this study, we mixed the RNA from four important fibers at various developmental stages (5, 10, 15 and 20 DPA) to de novo assemble the transcriptome of developmental colored cotton fiber. A total of 125,014 unigenes were generated in two sequencing libraries, which were further assembled into 71,895 all-unigenes with an average length of 533 bp, 69.46% of which could be matched in the NCBI database (E-value≤10^{-5}). This volume of data was greater than that reported in previous studies on other species [17,30]. Approximately 20,000 unigenes identified from *G. hirsutum* were recorded in the NCBI database (Nov 2011). In this study, approximately 50,000 unigenes (EST) were matched with nr database records. Therefore, we believe that our unigene library contained almost all of the known unigenes from *G. hirsutum*. In conclusion, we acquired a high quality and well-assembled transcriptome library for developing colored cotton fibers.

In all nr-annotated unigenes, only 1,537 (3.08%) were directly annotated with the field of "*Gossypium hirsutum*"; most other unigenes were assigned to other species, which mainly included "*Vitis vinifera*", "*Ricinus communis*" and "*Populus trichocarpa*". This implies that the genome of cotton may be very similar to these species and this could be a reference for a prospective cotton sequencing project.

Z263 was the offspring which derived from a cross between white fiber cotton and dark brown cotton, and Z128 was an inbred line selected from Z263 with a lighter color and better fiber quality. Therefore, the similar genetic backgrounds provided a fine model with which to study the mechanism underlying the formation of brown color in cotton fiber. We compared the whole transcriptome for each and, unexpectedly, an abundance of unigenes (20.26%) revealed significant differential expression between Z263 and Z128 (FDR≤0.001, |log2Ratio|≥1). This evidence demonstrates that all DEGs are relatively evenly distributed in most of the relevant metabolism pathways. Namely, the divergence of cotton fiber color and quality was the result of complex processes generated by multiple metabolic processes.

There is limited information in the literature on the molecular mechanism that underlies the formation of fiber color in cotton. Xiao et al [6] cloned five flavonoid structure genes from brown cotton fiber and found that all the cloned genes could be involved in pigmentation metabolism for brown fiber. Several studies focused on chemicals also suggested that proanthocyanidins (condensed tannins) might be the precursor of pigmentation in natural colored cotton fiber [7–8]. Here, almost of all the unigenes which encoded the key enzymes (CHS, CHI, LAR, ANS and ANR) of the flavonoid biosynthesis pathway were down-regulated in Z128 compared with that in Z263, implying that accumulation of proanthocyanidins in Z263 might be more than that in Z128. Zhan et al. [7–8] suggested that the depth of brown color might be closely related to the accumulated quantity of condensed tannins. Another unexpected finding in this study was related to the "anthocyanin biosynthesis" pathway. As a downstream metabolic pathway, it is one of the most important elements of pigment biosynthesis in plants [31–33]. We found that all unigenes involved in this pathway were significantly up-regulated (read in Z263 = 0) in Z128 and homologous with anthocyanin 5-O-glucosyltransferase (5GT), which could make anthocyanin more stable by modification [34]. Another recent study demonstrated that the lack of glucose at the 5 position of anthocyanin could lead to color variation in carnations [35]. This result implied that the depth of brown cotton fiber color variation might be the consequence of an interaction between oxidized tannins and glycosylated anthocyanin.

F3H has been thoroughly studied for decades and it is predominantly expressed during the fiber elongation stage in *G. barbadense*, a process that has no relation to pigment formation [36]. Furthermore, when the *F3H–RNAi* segment was transferred into brown fiber plants, they yielded more stunted fibers. Transgenic analysis showed that the suppression of *F3H* not only inhibited fiber elongation but also retarded fiber maturation [37]. *F3H* was suppressed in Z263 but up-regulated in Z128. This evidence suggests that *F3H* is important in fiber development. Like *F3H*, *F3'H* and *FLS* were very important in the pigments synthesis. Therefore, it is possible that the genes in the middle were up-regulated. According to Xiao et al [6] and Zhan et al [7], tannins could be the key chemical responsible for the brown color in cotton fiber. As shown in Fig. 5, *LAR*, *ANR*, and *ANS* were the key enzymes to accumulate the tannins. The down-regulated *CHS*, *CHI*, *LAR*, *ANR*, and *ANS* genes in the main pathway of pigment formation should be related to lighter fiber color of Z128 than that of Z263.

The pigmentation related genes not only regulated the fiber color but also influenced the fiber quality. The flavonoids are abundant and widely distributed plant secondary metabolites, and known to be an active participant in fiber development [38]. Previous studies showed that in fiber cells, the flavonoid genes were dominantly expressed in the fiber elongation stage [39–41]. In our study, *CHS*, *CHI*, *LAR*, *DFR*, *F3H*, *F3'H*, *ANR* and *ANS* showed higher expression levels at 10 DPA in Z128, thus highlighting that flavonoid metabolism represents a novel pathway with the potential for cotton fiber improvement. Our GWAS analysis of SNP in cotton germplasm indicated that the genes involved in flavonoid biosynthesis were also associated with fiber quality traits (unpublished data). Therefore, the highly expressed flavonoid gene in the fiber elongation stage in Z128 should be related to better fiber quality.

Cotton fibers are single-celled trichomes that differentiate from the ovule epidermis, including fiber initiation, elongation, secondary cell wall biosynthesis and maturation, leading to mature fibers. Ca^{2+} and ROS are two important factors involved in fiber cell growth [42]. Ca^{2+}/Calmodulin (CaM) is involved in plant growth and development through interaction with ROS signaling [43]. Based on gene expression profile analysis, Ca^{2+}/CaM is implicated in cotton fiber elongation. However, currently, there remains little direct evidence of the mechanism of Ca^{2+}/CaM on cotton fiber development. In our study, Ca^{2+} related genes were either down-regulated or up-regulated in Z128 compared with that in Z263.

Recent literature indicates that ethylene acts as a positive regulator of root hair, apical hook, and hypocotyl development [44–46]. Furthermore, Shi et al. [24] found that ethylene biosynthesis was one of the most significantly up-regulated biochemical pathways during fiber elongation. Exogenously applied ethylene promoted robust fiber cell expansion, whereas its biosynthetic inhibitor L-(2-aminoeth oxyvinyl)-glycine (AVG) specifically suppressed fiber growth. The ethylene biosynthesis pathways in our data were not shown in Table 3 as the "top 10 DEGs enriched pathways in KEGG analysis", however, a number of ethylene related genes were up-regulated in Z128 compared with that in Z263, such as Unigene61435_All, Unigene40770_All, and Unigene62869_All. This suggests that ethylene related genes may contribute to better fiber quality in Z128.

Author Contributions

Conceived and designed the experiments: WG SH XD. Performed the experiments: WG SH JT. Analyzed the data: WG SH GS XF. Contributed reagents/materials/analysis tools: JS ZP YJ. Contributed to the writing of the manuscript: WG.

References

1. Vreeland JM (1999) The revival of colored cotton. Sci Am 280: 90–96.
2. Dickerson DK, Lane EF, Rodriguez DF (1999) Naturally colored cotton: resistance to changes in color and durability when refurbished with selected laundry aids. California Agric Tech Inst, California State University, Fresno, California. 42p.
3. Murthy MSS (2001) Never say dye: the story of coloured cotton. Resonance 6, 29–35.
4. Feng HJ, Sun JL, Wang J, Jia YH, Zhang XY, et al. (2011) Genetic effects and heterosis of the fibre colour and quality of brown cotton (*Gossypium hirsutum*). Plant Breeding 130: 450–456.
5. Kohel R (1985) Genetic analysis of fiber color variants in cotton. Crop Sci 25: 793–797.
6. Xiao YH, Zhang ZS, Yin MH, Luo M, Li XB, et al. (2007) Cotton flavonoid structural genes related to the pigmentation in brown fibers. Biochem Biophys Res Commun 35: 873–78.
7. Zhan SH, Lin Y, Cai YP, Li ZP (2007) Preliminary deductions of the chemical structure of the pigment brown in cotton fiber. Chin Bull Bot 24: 99–104 (Chinese with english abstract).
8. Zhan SH, Lin Y, Cai YP, Li ZP (2007) Relationship between the pigment in natural brown cotton fiber and the condensed tannin. Cotton Sci 19: 183–188 (Chinese with english abstract).
9. Haas BJ, Zody MC (2010) Advancing RNA-seq analysis. Nat Biotechnol 28: 421–423.
10. Mortazavi A, Williams BA, McCue K, Schaeffer L, Wold B (2008) Mapping and quantifying mammalian transcriptomes by RNA-SNat Methods 5: 621–628.
11. Nagalakshmi U, Wang Z, Waern K, Shou C, Raha D, et al. (2008) The transcriptional landscape of the yeast genome defined by RNA sequencing. Science 320: 1344.
12. Hittinger CT, Johnston M, Tossberg JT, Rokas A (2010) Leveraging skewed transcript abundance by RNA-Seq to increase the genomic depth of the tree of life. Proc Natl Acad Sci USA, 107: 1476.
13. Guo S, Zheng Y, Joung JG, Liu S, Zhang Z, et al. (2010) Transcriptome sequencing and comparative analysis of cucumber flowers with different sex types. BMC Genomics 11: 384.
14. Tisserant E, Da Silva C, Kohler A, Morin E, Wincker P, et al. (2011) Deep RNA sequencing improved the structural annotation of the *Tuber melanosporum* transcriptome. New Phytol. 189: 883–891.
15. Ru Z, Wang G, He S, Du X (2013) The difference of fiber quality and fiber ultrastructure in different natural colored cotton. Cotton Sci 23: 184–188.
16. Wan CY, Wilkins TA (1994) A Modified Hot Borate Method Significantly Enhances the Yield of High-Quality RNA from Cotton (*Gossypium hirsutum* L.). Anal Biochem 223: 7–12.
17. Z Xia, H Xu, J Zhai, D Li, H Luo, C He, X Huang (2011) RNA-Seq analysis and de novo transcriptome assembly of *Hevea brasiliensis*. Plant Mol Biol 77: 1–10.
18. Li R, Zhu H, Ruan J, Qian W, Fang X, et al. (2010) De novo assembly of human genomes with massively parallel short read sequencing. Genome Res 20: 265–272.
19. Pertea G, Huang X, Liang F, Antonescu V, Sultana R, et al. (2003) TIGR Gene Indices clustering tools (TGICL): a software system for fast clustering of large EST datasets. Bioinformatics 19: 651–652.
20. Conesa A, Götz S, García-Gómez JM, Terol J, Talón M, et al. (2005) Blast2GO: a universal tool for annotation, visualization and analysis in functional genomics research. Bioinformatics 21: 3674–3676.
21. Ye J, Fang L, Zheng H, Zhang Y, Chen J (2006) WEGO: a web tool for plotting GO annotations. Nucleic Acids Res 34: 293–297.
22. Audic S, Claverie JM (1997) The significance of digital gene expression profiles. Genome Res 7: 986–995.
23. Mortazavi A, Williams BA, McCue K, Schaeffer L, Wold B (2008) Mapping and quantifying mammalian transcriptomes by RNA-SNat Methods 5: 621–628.
24. YH Shi, Zhu SW, Mao XZ, Feng JX, Qin YM, et al. (2006) Transcriptome profiling, molecular biological, and physiological studies reveal a major role for ethylene in cotton fiber cell elongation. Plant Cell 18(3): 651–664.
25. Chaudhary B, Hovav R, Flagel L, Mittler R, Wendel J (2009) Parallel expression evolution of oxidative stress-related genes in fiber from wild and domesticated diploid and polyploid cotton (*Gossypium*). BMC Genomics 10: 378.
26. Tang W, Tu L, Yang X, Tan J, Deng F, et al (2014) The calcium sensor GhCaM7 promotes cotton fiber elongation by modulating reactive oxygen species (ROS) production. New Phytol 202: 509–520.
27. Haigler CH, Singh B, Zhang D, Hwang S, Wu C, et al. (2007) Transgenic cotton over-producing spinach sucrose phosphate synthase showed enhanced leaf sucrose synthesis and improved fiber quality under controlled environmental conditions. Plant Mol Biol 63: 815–832.
28. Wang Z, Gerstein M, Snyder M. (2009) RNA-Seq: a revolutionary tool for transcriptomics. Nat Rev Genet 10(1): 57–63.
29. Zenoni S, Ferrarini A, Giacomelli E, Xumerle L, Fasoli M, et al. (2010) Characterization of transcriptional complexity during berry development in *Vitis vinifera* using RNA-SPlant Physiol 152: 1787–1795.

30. Wei W, Qi X, Wang L, Zhang Y, Hua W, et al. (2011) Characterization of the sesame (*Sesamum indicum* L.) global transcriptome using Illumina paired-end sequencing and development of EST-SSR markers. BMC Genomics 12: 451.

31. Dooner HK, Robbins TP, Jorgensen RA (1991) Genetic and developmental control of anthocyanin biosynthesis. Ann Rev Genet 25: 173–199.

32. Halloin JM (1982) Localization and changes in catechin and tannins during development and ripening of cottonseed. New Phytol 90: 651–657.

33. Grotewold E (2006) The genetics and biochemistry of floral pigments. Annu Rev Plant Biol 57: 761–780.

34. Yamazaki M, Gong Z, Fukuchi-Mizutani M, Fukui Y, Tanaka Y, et al. (1999) Molecular cloning and biochemical characterization of a novel anthocyanin 5-O-glucosyltransferase by mRNA differential display for plant forms regarding anthocyanin. J Biol Chem 274: 7405–7411.

35. Nishizaki Y, Matsuba Y, Okamoto E, Okamura M, Ozeki Y, et al. (2011) Structure of the acyl-glucose-dependent anthocyanin 5-O-glucosyltransferase gene in carnations and its disruption by transposable elements in some varieties. Mol Genet Genomics 286: 383–394.

36. Tu LL, Zhang XL, Liang SG, Liu DQ, LF Zhu, et al. (2007) Gene expression analyses of sea-island cotton (*Gossypium barbadense* L.) during fiber development. Plant Cell Rep 26: 1309–1320.

37. Tan J, Tu L, Deng F, Hu H, Nie Y, et al. (2013) A genetic and metabolic analysis revealed that cotton fiber cell development was retarded by flavonoid naringenin. Plant Physiol 162: 86–95.

38. Owens DK, Crosby KC, Runac J, Howard BA, Winkel B (2008) Biochemical and genetic characterization of Arabidopsis flavanone 3beta-hydroxylase. Plant Physiol Biochem 46: 833–843.

39. Gou JY, Wang LJ, Chen SP, Hu WL, Chen XY, et al. (2007) Gene expression and metabolite profiles of cotton fiber during cell elongation and secondary cell wall synthesis. Cell Res 17: 422–434.

40. Hovav R, Udall JA, Hovav E, Rapp R, Flagel L, et al. (2008) A majority of cotton genes are expressed in single-celled fiber. Planta 227: 319–329.

41. Rapp R, Haigler C, Flagel L, Hovav R, Udall J, et al. (2010) Gene expression in developing fibres of Upland cotton (*Gossypium hirsutum* L.) was massively altered by domestication. BMC Biol. 8: 139.

42. Qin YM, Zhu YX. (2011) How cotton fibers elongate: a tale of linear cell-growth mode. Curr. Opin. Plant Biol. 14: 106–111.

43. Lee SH, Seo HY, Kim JC, WD Heo, Chung WS, et al. (1997) Differential activation of NAD kinase by plant calmodulin isoforms. The critical role of domain I. J Biol Chem 272: 9252–9259.

44. Achard P, Vriezen WH, Van Der Straeten D, Harberd NP (2003) Ethylene regulates *Arabidopsis* development via the modulation of DELLA protein growth repressor function. Plant Cell 15: 2816–2825.

45. Seifert GJ, Barber C, Wells B, Roberts K (2004) Growth regulators and the control of nucleotide sugar flux. Plant Cell 16: 723–730.

46. Grauwe LD, Vandenbussche F, Tietz O, Palme K, Straeten DVD (2005) Auxin, ethylene and brassinosteroids: Tripartite control of growth in the *Arabidopsis* hypocotyl. Plant Cell Physiol 46: 827–836.

14

Pilon: An Integrated Tool for Comprehensive Microbial Variant Detection and Genome Assembly Improvement

Bruce J. Walker[1]*[◑]¤, **Thomas Abeel**[1,2◑], **Terrance Shea**[1], **Margaret Priest**[1], **Amr Abouelliel**[1], **Sharadha Sakthikumar**[1], **Christina A. Cuomo**[1], **Qiandong Zeng**[1], **Jennifer Wortman**[1], **Sarah K. Young**[1], **Ashlee M. Earl**[1]*

1 Broad Institute of MIT and Harvard, Cambridge, Massachusetts, United States of America, 2 VIB Department of Plant Systems Biology, Ghent University, Ghent, Belgium

Abstract

Advances in modern sequencing technologies allow us to generate sufficient data to analyze hundreds of bacterial genomes from a single machine in a single day. This potential for sequencing massive numbers of genomes calls for fully automated methods to produce high-quality assemblies and variant calls. We introduce Pilon, a fully automated, all-in-one tool for correcting draft assemblies and calling sequence variants of multiple sizes, including very large insertions and deletions. Pilon works with many types of sequence data, but is particularly strong when supplied with paired end data from two Illumina libraries with small *e.g.*, 180 bp and large *e.g.*, 3–5 Kb inserts. Pilon significantly improves draft genome assemblies by correcting bases, fixing mis-assemblies and filling gaps. For both haploid and diploid genomes, Pilon produces more contiguous genomes with fewer errors, enabling identification of more biologically relevant genes. Furthermore, Pilon identifies small variants with high accuracy as compared to state-of-the-art tools and is unique in its ability to accurately identify large sequence variants including duplications and resolve large insertions. Pilon is being used to improve the assemblies of thousands of new genomes and to identify variants from thousands of clinically relevant bacterial strains. Pilon is freely available as open source software.

Editor: Junwen Wang, The University of Hong Kong, Hong Kong

Funding: This project has been funded in part with Federal funds from the National Institute of Allergy and Infectious Diseases, National Institutes of Health, Department of Health and Human Services, under Contract No.:HHSN272200900018C. This project has been also been funded in part with Federal funds from the National Human Genome Research Institute, National Institutes of Health, Department of Health and Human Services, under grant U54HG003067. TA is a postdoctoral fellow of the Research Foundation-Flanders. The funders played no role collection, analysis, and interpretation of data; in the writing of the manuscript; and in the decision to submit the manuscript for publication.

Competing Interests: The authors have declared that no competing interests exist.

* Email: bruce@broadinstitute.org (BJW); aearl@broadinstitute.org (AME)

◑ These authors contributed equally to this work.

¤ Current address: Applied Minds, LLC, Boston, Massachusetts, United States of America

Introduction

Massively parallel sequencing technology has dramatically reduced the cost of genome sequencing, making the generation of large numbers of microbial genomes accessible to a wide range of biological researchers. For example, a single Illumina HiSeq2500 has the ability to generate the equivalent of 300 bacterial genomes of sequencing data in a single day using only one flow cell. Comparisons of whole genome sequence data from hundreds of microorganisms have provided unprecedented views on all aspects of microbial diversity, and there is growing recognition that 'hundreds' of genomes is the minimum scale needed to address pressing questions related to microbial evolution, diversity, pathogenicity and resistance to antimicrobial drugs [1–4]. As such, the methods needed to analyze these large volumes of data — including assembling and calling variants relative to a reference — must be robust, accurate, scalable, and able to operate without human intervention.

Several computational methods exist that make improvements to the quality of draft assemblies by recognizing and correcting errors involving (i) single bases and small insertion/deletion events (indels) [5], (ii) gaps [6], (iii) read alignment discontinuities [7] or by reconciling multiple *de novo* assemblies into an improved consensus assembly [8]. However, no single tool performs integrated assembly improvement of all error types. Computational tools for identifying sequence polymorphisms also exist [9,10], but focus primarily on identifying variants in the human genome [11], and particularly small events (SNPs and small indels) or structural rearrangements (chromosomal rearrangements) [11]. Furthermore, many of these tools require multiple steps to identify and subsequently filter variants to remove noise and false calls. In addition, for tools able to identify variants that exceed the length of the sequence reads (read-length) being evaluated, they generally indicate the approximate chromosomal location and estimated size of the predicted variant relative to the reference, but often do not provide exact coordinates [11]. For insertions that are longer than the read-length - particularly common in the microbial world - current tools do not assemble and report the inserted sequence.

We introduce Pilon, an integrated software tool for comprehensive microbial genome assembly improvement and variant detection, including detection of variants that exceed sequence read-length. Conceptually, Pilon treats assembly improvement and variant detection as the same process (Figure 1). Both start with an input genome — either an existing draft assembly or a reference assembly from another strain — and use evidence from read alignments to identify specific differences from the input genome supported by the sequencing data. Applying those changes to a draft genome assembly yields an improved assembly, while reporting the changes with respect to a reference genome yields variant calls.

In genomic regions where read alignments are poor, Pilon is capable of filling out and correcting sequence through an internal local reassembly process. This capability allows Pilon to further improve assemblies by filling gaps and correcting local mis-assemblies, and it also enables Pilon to capture many large insertion, deletion, and block substitution variants in their entirety. These larger events are often completely missed or inaccurately characterized by conventional variant calling tools that rely solely on read alignments. Pilon has built-in heuristics to determine which corrections and calls are of high confidence, so no separate filtering criteria need be specified. This allows for the automated processing of hundreds or thousands of data sets representing different microbial species with minimal human intervention.

We benchmarked Pilon both as an assembly refinement tool and variant caller. For assembly refinement, we used finished reference genome sequences from *Mycobacterium tuberculosis* F11, *Streptococcus pneumoniae* TIGR4 and *Candida albicans* SC5314 as benchmarks to evaluate the accuracy of Pilon in improving draft assemblies. Pilon-improved assemblies were more contiguous and complete than non-Pilon-improved assemblies and contained

improved sequences for genes implicated in pathogen-host interaction and virulence. We also evaluated Pilon's performance against tools specializing in assembly base quality improvement and gap filling, and, in each case, Pilon made more correct improvements while making far fewer mistakes than the other tools. For variant calling, we used read data from *M. tuberculosis* F11 to call polymorphisms against the finished *M. tuberculosis* H37Rv genome to evaluate Pilon's ability to accurately call polymorphisms. Pilon performed as well or better when compared with two state-of-the-art variant detection tools in calling small variants, and Pilon differentiated itself in its ability to identify large-scale variants.

Results

Assembly improvement evaluation

Assessing accuracy on bacterial assemblies. To test the accuracy of Pilon's improvements on bacterial assemblies, we sequenced and created draft assemblies for two bacterial strains with finished references: *S. pneumoniae* TIGR4 and *M. tuberculosis* F11 (see Methods). These strains were chosen because they represent different GC content (40% and 66% GC content, respectively) and both possess genomic features that are known to confound assemblers, leading to mis-assembled and/or incomplete genome sequences [12–14]. Sequence reads from both libraries were aligned back to their respective draft assemblies using *BWA* [15], and Pilon was run with those alignments.

To assess the benefits of running Pilon, we compared the original draft and Pilon-improved assemblies to each other and to their respective finished genome sequence. Pilon made significant improvements to the contiguity of both draft assemblies, increasing the contig N50 size by 443 Kbp for *S. pneumoniae* TIGR4 (see Table 1) and 196 Kbp for *M. tuberculosis* F11, even though the F11 draft assembly had been generated with assistance from a

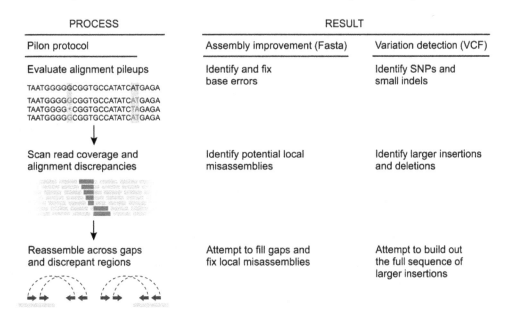

Figure 1. Simplified overview of the Pilon workflow for assembly improvement and variant detection. The left column depicts the conceptual steps of the Pilon process, and the center and right columns describe what Pilon does at each step while in assembly improvement and variant detection modes, respectively. During the first step (top row), Pilon scans the read alignments for evidence where the sequencing data disagree with the input genome and makes corrections to small errors and detects small variants. During the second step (second row), Pilon looks for coverage and alignment discrepancies to identify potential mis-assemblies and larger variants. Finally (bottom row), Pilon uses reads and mate pairs which are anchored to the flanks of discrepant regions and gaps in the input genome to reassemble the area, attempting to fill in the true sequence including large insertions. The resulting output is an improved assembly and/or a VCF file of variants.

Table 1. Summary assembly statistics before and after Pilon improvement.

Genome	M. tuberculosis F11	S. pneumoniae TIGR4	C. albicans SC5314
Contig N50 Increase	196 kb	443 kb	56 kb
Bases Added	11,516	9,608	33,804
Gaps Closed	9	9	54
Gaps Shrunk	7	7	102
Single-base Modifications	20	27	26,939
Mis-assembly fixes	3	1	44

In all cases the assemblies were more contiguous, contained more bases, and had fewer gaps and errors after Pilon improvement.

close reference. In addition, Pilon assemblies were more complete, with the *M. tuberculosis* F11 and *S. pneumoniae* TIGR4 Pilon-improved assemblies containing an additional 11,516 bp and 9,608 bp, respectively.

Observed gains in genome coverage and contig N50 were principally due to Pilon's ability to recognize and fill (or partially fill) by local assembly "captured gaps", *i.e.*, missing sequence between contigs within a scaffold. When run with default settings, Pilon does not introduce ambiguous bases or additional Ns during this process. Across the two draft assemblies, Pilon completely and accurately filled 17 of the 44 captured gaps (39% closure rate) including 8 gaps that represented more than 1 Kbp in sequence length (see Table S1). None of Pilon's gap closures were incorrect, though one was judged to be "no worse": the sequence used to bridge the gap was correct, but an error in the original assembly in one of the gap flanks was not detected by Pilon. An additional 14 gaps (32% of total captured gaps) were partially filled by Pilon, and 13 (93%) of those extensions were error-free. The one partial fill judged to be "Incorrect" involved a repetitive structure that Pilon extended into flanking sequence belonging to a different copy of the repeat.

We compared Pilon's ability to close gaps in these assemblies with two other tools commonly used for this purpose, IMAGE [16] and GapFiller [17] (see Table S1b). Pilon's overall gap closure rate was only somewhat higher than that of the other tools, but its accuracy was dramatically better. Across the two assemblies, IMAGE closed 13 captured gaps (30% closure rate) but only two of those closures were found to be correct by alignment with the reference (15% precision). Similarly, GapFiller closed 16 of captured gaps (36% closure rate) in the two assemblies, but only four of its closures were correct (25% precision). In addition to filling captured gaps, Pilon also corrected 43 single-bases and 4 small indels across both genomes, and all 47 changes were found to be correct by alignment against the reference (100% accuracy; see Table S2). By comparison, iCORN [18] made 47 single-base changes and 2 single-base deletions, but only 35 of the 49 (71%) of its changes were correct.

Optionally, Pilon can also make changes to genomic locations at which it finds significant evidence for more than one alternative, choosing the allele with the most support even where the evidence is insufficient to make a confident call. When run with this option on these assemblies, Pilon made 10 changes to ambiguous bases, but only 3 were verified to be correct. This option is turned off by default starting with Pilon version 1.8.

Pilon also detected and attempted to fix local mis-assemblies by reassembling contig regions that were suspected to be incorrectly assembled. Three of these regions were correctly fixed (see Table 1 and Table S1) and a fourth we classified as "No worse". For the latter, Pilon correctly identified a repetitive region within the

original *M. tuberculosis* F11 draft assembly that contained extra sequence with respect to the *M. tuberculosis* F11 reference. However, Pilon's change introduced a deletion with respect to the reference, underscoring the difficulty of accurately assembling repetitive regions with short read data [13].

For the *M. tuberculosis* F11 and *S. pneumoniae* TIGR4 Pilon improved assemblies, there were 13 and 4 regions, respectively, where Pilon detected a problem in the draft assembly, but was unable to provide solutions. In each of these cases, Pilon flagged the coordinates of the problematic region, and, in 10 of these cases, it also reported the length of the detected tandem repeat confounding resolution of the region. For example, Figure 2 shows scaffold00001 coordinates 3159800–3159898 of the *M. tuberculosis* F11 draft assembly, along with Pilon-generated genome browser tracks representing some of the internal metrics it used to identify this region as problematic. In this case, Pilon noted that it was unable to resolve a 57 bp tandem repeat, which enabled an experienced analyst to confirm the presence of a mis-assembly and accurately narrow the bounds of the unresolvable region. Manual comparison of the draft assembly with the reference revealed that there should have been three full and one partial copies of the 57 bp repeat in tandem, whereas the draft assembly only contained one full and one partial copy of the repeat.

Effect of assembly improvements on gene calls. To assess the impact of Pilon-improvement on gene calls (*i.e.*, functional interpretation of the genome), we examined Pilon improvements with respect to genes by investigating the regions that were affected by Pilon modifications and the effect of these modifications on coding sequences. Thirty-two genes and seven intergenic regions were impacted by Pilon changes to the *M. tuberculosis* F11 Pilon-improved assembly; of these, nearly all (95%; 37 of 39) were correct improvements. Nearly half (13) of the genes that were affected by a fix involved transposases that were completely or partially filled with sequence that perfectly matched the reference genome (see Table S3). One additional transposase had a single base pair corrected with perfect match to the reference. A particularly complex 13 Kbp region in *M. tuberculosis* F11 is highlighted in Figure 3. This region harbors three sets of transposases in close proximity that were not captured in the draft *M. tuberculosis* F11 assembly, but were accurately filled in by Pilon. Two of the gaps were completely closed, and the third transposase set was completely captured along with an additional gene. However, due to Pilon's conservative overlap requirement for closure (95 bp), that gap was not closed despite a 42 bp overlap in the extended flank sequences.

Of the remaining 19 genes, 6 were PE/PPE family protein encoding genes. Five corrections were perfect and, in one case (TBFG_11946), Pilon identified the problematic region, but could not completely resolve the problem. However, the correction that

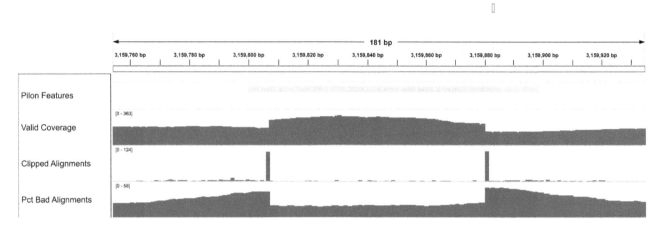

Figure 2. Example Pilon generated genome browser tracks. This region was flagged by Pilon as containing a possible local mis-assembly, but Pilon was unable to determine a fix due to a tandem repeat sequence. The tracks shown here include: *Pilon Features* track indicating the extent of the region flagged by Pilon as containing a potential mis-assembly, *Valid Coverage* track indicating the sequence coverage of valid read pair alignments excluding the clipped portions of the alignments, *Clipped Alignments* track indicating the number of reads soft-clipped at each location, *Pct Bad Alignments* track indicating the percentage of the total reads aligned to each location which are not part of *Valid Coverage*. These tracks are created with the '—tracks' command-line option. Together, these tracks reveal the true bounds of the mis-assembly, and indicate that there are likely missing copies of the tandem repeat in the draft assembly. In this case, manual analysis revealed the draft assembly was missing two of three full copies of a 57-base tandem repeat.

Pilon applied did not make the situation worse. Pilon also identified and accurately corrected a mis-assembly (highlighted in Figure S1) in which a gene had been truncated due to a collapsed repeat in the draft assembly.

In *S. pneumoniae* TIGR4, 20 genes and 12 intergenic regions were affected by fixes from Pilon. A majority (15 of 20) of the improved genes were transposases, of which Pilon was able to completely or partially fill 8 that matched completely and perfectly with the reference; the remaining 7 were individual base pair corrections. Pilon was also able to partially fill other genes encoding repetitive cell wall surface proteins - including choline binding protein A [19] and pneumococcal surface protein A [20] - both implicated in adhesion and virulence in *S. pneumoniae*.

Assessing accuracy on the assembly of the larger, polymorphic genome of *C. albicans*. To evaluate Pilon's ability to accurately improve assemblies of diploid genomes containing a high level of heterozygosity, we ran Pilon on an Illumina ALLPATHS-LG assembly of the SC5314 strain of *C. albicans* (Methods), for which there is a high quality reference curated by the Candida Genome Database (www.candidagenome.org). At 14.3 Mb, the *C. albicans* genome is 3- to 7-fold larger than the bacterial genomes evaluated here. It consists of 8 chromosomes that are present at diploid levels with an average of one SNP found at every 330–390 bases, although large regions of most chromosomes display loss of heterozygosity [21,22].

Pilon was capable of improving the assembly and added > 33 Kb of sequence (see Table 1). While the increase in contig N50 was relatively small (56 Kb), Pilon completely or partially filled 61% (156 of 256) of the total captured gaps including in both homo- and hetero-zygous regions of the genome. Homozygous

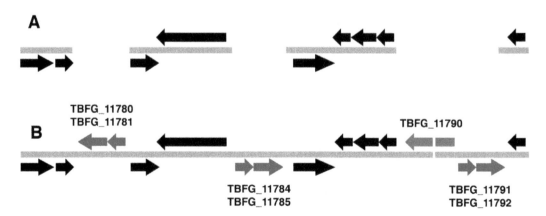

Figure 3. Comparative view of a transposase-rich region of the *M. tuberculosis* F11 genome (coordinates 1,991,000 to 2,006,300) obtained from the draft (A) and Pilon-improved (B) assemblies. In the draft assembly, three regions containing transposases (shown in blue) remained unassembled resulting in gaps. In the Pilon-improved assembly, all three sets of transposases were successfully assembled. The Pilon-improved assembly also contained a hypothetical gene, *TBFG_11790* (shown in red), missing from the draft assembly. Though *TBFG_11790* was not fully closed in the Pilon-improved version, closer inspection revealed that there was a 42 bp overlap in assembled sequence at this site. By default, Pilon will not close gaps unless there is at least 95 bp overlapping sequence to minimize spurious joins.

regions had a slightly higher fraction of completely closed gaps (33%; 8 of 24) as compared to heterozygous regions (20%; 46 of 232). Of completely filled gaps, 93% had full length alignment to the reference (including 300 bp of their flanking sequences) at 94% sequence identity or higher. Less than 100% identity is to be expected when comparing a heterozygous genome assembly against a flat reference. In several of the lower-identity cases, most of the base differences were in the flanks present in the draft assembly rather than the filled gap itself, suggesting that the gaps may have been caused by the original assembler's inability to assemble sequence through a highly polymorphic region.

Pilon also identified and corrected regions in the reference assembly where the read alignment evidence disagreed, including 44 regions that were likely mis-assembled. The nearly 27,000 corrected single-bases were mostly at heterozygous sites; Pilon identified these as potential bases to fix, as the majority of read-evidence favored an alternate allele from the reference base in the draft. These positions represented about half of the ~70,000 heterozygous SNP positions in this *Candida* genome [23]. While we did not investigate every change that Pilon made to this assembly, our results indicate that Pilon is suitable to be run on larger diploid genomes and can improve the quality of a draft assembly, resulting in fewer and longer contigs and an improved gene set.

Assembly improvements in a production environment. Given promising results from the benchmarking experiments, we implemented Pilon in the Broad Institute's *de novo* genome assembly production pipeline and assessed its performance by comparing assembly metrics from Pilon-improved assemblies of 50 representatives of the *Enterobacteriaceae* (including *Escherichia, Klebsiella, and Enterobacter*) to non-Pilon improved versions. Pilon reduced the mean number of contigs in the 50 assemblies from 33.7 to 20.9 (see Figure S2), a 38% reduction in total contigs representing closure of 47% of captured gaps. As a result, Pilon nearly doubled the contig N50 from 392 Kbp to 780 Kbp (99% increase; see Figure S3), capturing, on average, an additional 14,681 bp of sequence per assembly. This increase in genome size equates roughly to the addition of ~14 genes per genome (based on the average bacterial gene size of ~1 kb). Scaffold numbers were unchanged since, currently, Pilon does not attempt to join or break scaffold structures.

Variant detection evaluation

Assessing accuracy of polymorphism calls. To test the accuracy of Pilon's variant detection capability, we used BWA to align approximately 200-fold coverage of reads from the same *M. tuberculosis* F11 fragment and long insert libraries used in the assembly improvement assessment to the *M. tuberculosis* H37Rv finished reference genome. We generated two sets of variant calls with Pilon, one using both fragment and long insert reads as input, and one using only fragment reads. We also ran two popular variant detection tools, GATK UnifiedGenotyper (GATK-UG) and SAMtools/BCFtools (SAMtools), starting with the same aligned fragment BAM. All variant sites, including substitutions, deletions or insertions, were identified and two categories of variants were assessed: single nucleotide polymorphisms (SNPs) and multi nucleotide polymorphisms (MNPs) greater than 1 bp. Predicted polymorphisms were compared to a curated truth set of variants produced by comparing the *M. tuberculosis* F11 finished genome to the finished *M. tuberculosis* H37Rv genome (see Methods), resulting in a list of 1,325 events (summarized in Table 2) of which the majority were SNPs. We then compared Pilon's performance to that of the other two variant detection algorithms.

Overall, Pilon performed better in identifying SNPs, including single nucleotide insertions and deletions than did GATK-UG or SAMtools (Table 3). Pilon identified 8 to 11 percentage points (pp) more single nucleotide substitutions, 4 pp more single nucleotide deletions, and 4 to 8 pp more single nucleotide insertions from the curation set than did GATK-UG or SAMtools, respectively. Pilon's ability to precisely call single nucleotide substitutions was also high - only 3% of calls were not accounted for in the curation set - which was on par with the other two tools. Similarly, Pilon had perfect precision in calling single nucleotide insertions and only 5% of single nucleotide deletion calls were not accounted for in the curation set.

For MNPs, we allowed for a combination of two or more smaller events in the prediction set to contribute to a larger variant since there may be equivalent ways of representing the changes as a series of smaller edits (see Methods). Pilon greatly outperformed the other two variant callers in accurately identifying variants that involved more than one nucleotide (see Table 3; bottom three rows). Pilon identified three times as many multi nucleotide insertions as either GATK-UG or SAMtools (63% versus 17 or 21% of curated events), but made slightly more false predictions. For multi nucleotide deletions, Pilon identified two times as many events from the curation set than did GATK-UG or SAMtools and made fewer unsupported calls. In addition, Pilon identified all six curated multi-substitution events while the other two tools missed at least one, even when multiple SNPs were accounted for in these regions.

We next examined how overlapping the three tools were in either missing or overcalling variants. Panel A of Figure 4 summarizes the total number of variants appearing in the curation set that could not be detected by one or more of the variant callers. Pilon uniquely missed only one curated variant, while SAMtools and GATK-UG missed many more (32 and 13, respectively). The majority of variants that were missed by Pilon were also missed by SAMtools and GATK-UG (52 events). In addition, all three tools made predictions that were not supported by the curation set (summarized in Panel B of Figure 4), but, among unique unsupported events, Pilon and GATK-UG had ~3-fold fewer than SAMtools. Altogether, there were only 21 predictions where two or more of the tools agreed that there should be a variant called, most of which were SNPs, although four of the seven events shared by Pilon and GATK-UG were multi-nucleotide indels (5–15 nt in length).

Given the broad definition of 'multi' in Table 3 (>1 bp), we also evaluated how well Pilon performed for variants that were larger than 50 bp (see Table 4). We chose 50 bp since it is a length that is larger than the size of events for which short-read aligners are typically able to align, but shorter than the individual read length of the data used (101 bp). Overall, Pilon was able to accurately identify 74% of these large variants, including 100% of substitutions, 68% of insertions and 77% of deletions from the curated list. Of the eleven insertions that were missed by Pilon, eight involved a repetitive element (5 tandem insertions and 3 IS6110 insertions). Similarly, the six deletions not detected by Pilon involved deletion of one or more copies of a tandem repeat. Four of these tandem repeat regions were correctly reported by Pilon as possible tandem repeat variants in its standard output, but Pilon currently makes no attempt to provide a definitive copy number call in the presence of significant tandem repeat structures. Pilon also identified three events >50 bp that did not match variants in the curation set. These unsupported calls occurred within complex variable regions of the genome in which multiple nearby repeat structures prevented Pilon from correctly identifying the precise correct location or form of the events.

Table 2. Summary of variant types curated in the *M. tuberculosis* H37Rv and *M. tuberculosis* F11 finished genome comparison.

Type of variation between F11 and H37RV	Total found
Single substitution	1012
Single insertion	26
Single deletion	31
Multi substitution	13
Multi insertion	56
Multi deletion	47

The full list can be found in Table S8.

Since Pilon performed well in identifying and resolving MNPs and since GATK-UG and SAMtools were not explicitly designed to call these large variants, we sought to compare Pilon's MNP calls to that of methods specifically designed for MNP detection [11]. Though neither was described for use on microbial data, we evaluated how well BreakDancer [24] and CLEVER [25], two algorithms developed to detect large variants in eukaryotes, performed in calling MNPs on the *M. tuberculosis* test set. BreakDancer was unable to identify any MNP found in the curation set and CLEVER identified 21 multi nucleotide deletions, of which only 1 corresponded to a variant in the curated list. No large insertions or substitutions were predicted (data not shown).

Evaluating Pilon variant calls without long insert data. It is unsurprising that Pilon was better able to call larger variants since it is optimized to use both fragment (or small) and long (or large) insert libraries. Since many sequencing projects do not have access to long insert data and to also make a more direct comparison to existing variant callers that are not optimized to accept these data, we evaluated Pilon's performance using data from fragment insert libraries alone. To do this, we ran Pilon using the aligned fragment paired end reads from the *M. tuberculosis* F11 genome to the *M. tuberculosis* H37Rv finished reference genome. We then compared this output ("Pilon-frags") to the previously analyzed output from GATK and SAMtools and to Pilon output using data from both library types ("Pilon").

Pilon-frags performed well in identifying both single and multi nucleotide variants (see Table 3). Pilon-frags identified only 2 pp fewer single nucleotide substitutions, 4 pp fewer single insertions and 4 pp fewer single deletions as compared to the original Pilon output. Pilon-frag performance in calling SNPs was better or on par with both GATK-UG and SAMtools. Remarkably, Pilon-frags was also able to identify a large fraction of the MNPs, with nearly identical performance to Pilon with long insert read data. Pilon-frags also performed very well in calling variants larger than 50 bp (see Table 4), with one less insertion call and 4 fewer deletions calls as compared to Pilon.

To better understand the qualitative differences in what Pilon and Pilon-frags reported, we examined the concordance between results for each variant type. For SNP calls, we observed high concordance in the outputs from Pilon-frags and Pilon (95.2%; 871 of 915 events) (see Table S5). Discordance in SNPs often involved a position where a variant was found in both Pilon runs, but was considered high quality in one and low quality in the other. In fact, only 7 of 915 SNPs (0.8% of total) were confidently predicted to differ between the two Pilon run conditions, suggesting that the value of long insert library data when calling SNPs is small. However, for SNPs within repetitive regions of the genome, long insert data appeared to be very helpful in disambiguating these events (Table S6). Small indel variant calls

were also highly concordant for the two Pilon runs (93.3%; 56 of 60 events), and 78.5% concordance (73 of 93) for large indels.

For larger variants, the discordance between Pilon with and without long insert data was larger (see Table S5), particularly in regions of the genome encoding transposable IS6110 repeat elements. While Pilon-frags detected many of these events, the sequences that were assembled and reported at these sites were often incomplete, as illustrated in Table S7. Given the length of the IS6110 repeat (~1.3 Kbp), the fragment pairs — only ~180 bp apart — were unable to span the entire length of the IS6110 elements, leading to two large indels being called, one coming in from each side of the IS6110 *e.g.* Table S7, position 1,541,957. Pilon's improved ability to capture the full sequence of larger insertions is the primary value of including long insert read data for variant calling applications.

Assessing large-scale genome duplications. In addition to identifying substitutions and indels of various sizes, Pilon is able to identify areas in which the read evidence suggests additional copies of large genomic regions (>10 Kbp) compared with the input draft assembly or reference genome. These regions could indicate large collapsed repeats in an assembly improvement application or large genomic duplications in a variant detection application. To evaluate Pilon's ability to identify large duplications, we re-sequenced *M. tuberculosis* T67, a strain previously reported to harbor a large-scale duplication [26], using fragment and long insert libraries, and aligned the reads to the *M. tuberculosis* H37Rv finished reference. Pilon was then run to detect variants in T67 using H37Rv as a reference.

Pilon identified two duplication events that were >10 Kbp in size and separated by ~3 Kbp at *M. tuberculosis* H37Rv coordinates 3,494,063–3,551,070 (57 Kbp) and 3,554,192–3,712,284 (158 Kbp) resulting in a combined duplication of ~215 Kbp. The left gene boundary (Rv3128c) of the first predicted duplication and right gene boundary (Rv3427c) of the second predicted duplication corresponded to the upstream and downstream boundaries in the previously reported *M. tuberculosis* T67 duplication [26]. Upon closer inspection, the 3 Kbp intervening region contained two copies of the IS6110 element, which are routinely found in multiple copies within the *M. tuberculosis* genome (16 copies in H37Rv) [27]. Because these elements occur so frequently in the genome, the incremental coverage from the duplication was not sufficient for Pilon to identify them as part of the duplication event, breaking a true large duplication into two reported pieces.

Discussion

Pilon is an assembly improvement algorithm and variant caller that identifies differences between a draft assembly or closely

Table 3. Recall and precision metrics for *M. tuberculosis* F11 variants called against *M. tuberculosis* H37Rv by Pilon (with and without long insert library data), GATK UnifiedGenotyper and SAMtools.

	Pilon			GATK			SAMtools			Pilon-frags		
	R	P	F	R	P	F	R	P	F	R	P	F
Single substitution	0.96	0.98	0.97	0.85	0.98	0.91	0.88	0.93	0.90	0.94	0.98	0.96
Single insertion	0.83	1	0.91	0.75	1	0.86	0.79	1	0.88	0.79	1	0.88
Single deletion	0.91	0.95	0.93	0.87	0.9	0.86	0.87	1	0.93	0.87	0.95	0.91
Multi substitution	1	0.95	0.97	0.67	N/A	N/A	1	0.98	0.99	1	0.95	0.97
Multi insertion	0.63	0.73	0.68	0.17	0.79	0.28	0.21	0.5	0.30	0.63	0.76	0.69
Multi deletion	0.73	0.9	0.81	0.27	0.75	0.4	0.39	N/A	N/A	0.71	0.87	0.78

The three rows marked with 'Single' indicate single nucleotide variants. The three rows marked with 'Multi' indicate variants involving two or more nucleotides, which also include very large events that span several Kb. Recall (R) is the fraction of curated events that were called by the program. Precision (P) is the fraction of calls that the program made that were also described in the curation. The F-measure is the harmonic mean of recall and precision and provides measure of the trade-off between recall and precision. "N/A" indicates that all events of this type were captured in another variant category.

related reference assembly using evidence supported by the sequencing data, resulting in either an improved assembly or a list of variants in VCF format. We have demonstrated Pilon's performance on several microbial genomes with varying GC-content and different ploidy. Our results indicate that applying Pilon yields more contiguous and accurate assemblies. Furthermore, variant calls made by Pilon are of high quality when compared with other state-of-the-art tools, and Pilon's ability to find insertion and deletion events considerably larger than read-lengths sets it apart from traditional variant calling tools.

For many of the sub tasks that Pilon performs, there are a variety of existing tools that might be used in sequence to achieve similar output as Pilon. However, using existing tools to achieve the full complement of analyses performed by Pilon would require implementation of a complicated workflow that hands data around to various tools, and development of post-processing algorithms to ensure that results from the various tools are in agreement. Pilon is a single, benchmarked tool that performs comparably well, if not better, than other tools that do only a fraction of the work. In addition, Pilon is easy to use and is successfully being utilized for assembly and variant detection of thousands of data sets in Broad Institute's production pipelines.

We showed that the assembly improvements that were introduced by Pilon were both accurate and biologically relevant. In particular, several highly repetitive genes that were captured by Pilon are known to play a role in virulence and host-pathogen interactions [28,29]. To date, it has been difficult to study these genes comparatively, because they are often not captured or are only partially captured in draft assemblies. Furthermore, we were able to place more genes with repetitive features accurately in the genome. In the *M. tuberculosis* case, Pilon improved the sequence accuracy and placement of genes encoding transposases, which have an important role in genome reorganization in this species [27] and are used in strain typing schemes [30]. In addition, for *M. tuberculosis*, Pilon had a significant impact on the repetitive, GC-rich and not well understood PE and PPE genes, an expanded family of highly repetitive genes that account for about 10% of the gene repertoire in this species [31], and are implicated in pathogen-host interactions and virulence [32,33]. Pilon was able to resolve these repetitive structures because of its ability to use the long-distance mate pair information afforded by long insert libraries. In addition, data from long insert libraries often enable Pilon to completely fill in large sequence insertions and assemble across gaps.

For variant calling, the primary benefit of Pilon over other variant callers is its ability to capture large sequence polymorphisms and highly polymorphic local regions by performing a local assembly to generate complete sequences. By capturing intervening sequences in large variants, Pilon enables a more comprehensive view of the biological differences between strains *e.g.*, new genes that confer antibiotic resistance or virulence. In addition, by integrating the SNP and large sequence variant detection in a single tool, Pilon is less likely to erroneously call SNPs in regions that are affected by a large variant. Long insert libraries provide information that resolves larger, multi-nucleotide events, in particular by allowing Pilon to completely assemble inserted sequences. However, even without long insert libraries, most of these events are identified, albeit often with an incomplete alternative sequence.

Pilon was also able to perform comparably well in calling small variants, including SNPs and small indels. While all three variant callers benchmarked in this study had similar precision in their calls, Pilon demonstrated better recall (fewer false negatives) on single-base polymorphisms. There are two reasons for this.

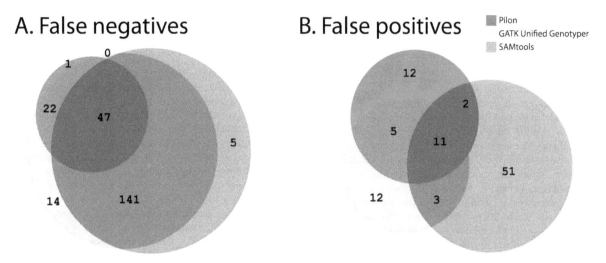

Figure 4. Venn diagram of the overlap in false negative (A) and false positive (B) calls by the three variant detection tools, Pilon, GATK UnifiedGenotyper and SAMtools. False negative calls are the number of unique events from the curation set that was missed by each tool. Overlaps in the Venn diagram show the number of variants that were missed by multiple tools. False positive calls are the number of predictions from *M. tuberculosis* F11 that were not supported by the curation set. Overlaps indicate predictions that were shared among tools.

First, there are a few local areas of the genome with very high polymorphism rates between F11 and H37Rv. When the local polymorphism rate is very high, the short-read aligners are unable to produce enough good alignments for any of the tools to call the base differences from the resulting pileups. However, Pilon was able to detect some of these problem areas and reassemble them into block substitutions, correctly capturing dozens of polymorphic locations the other tools were unable to resolve. Second, the use of long insert libraries allowed Pilon to make definitive calls inside some repeat regions by heavily weighting long insert reads which were unambiguously anchored to the flanks of the repeat area, capturing additional true SNPs.

We note that GATK contains a highly sophisticated collection of tools for variant calling. However, several of its tools and their associated best practices for human variant calling are not applicable to many microbial projects, because they rely upon a database of known variant locations (such as those found at dbSNP) to perform recalibration. For microbial variant calling, the extent of variation across the microbial species under investigation is typically unknown, so no such catalog of variation is available *a priori*. There is also a fundamental difference between GATK and Pilon's approach to variant calling. GATK's UnifiedGenotyper is designed to be aggressive in detecting possible variants, relying on a user-controlled VariantFiltration step to filter out calls of lower confidence or quality to minimize false positives. Pilon, on the other hand, relies exclusively on internal heuristics to make a

determination of which calls are confident. This makes Pilon easy to use "out of the box" in a highly automated environment, though it is less configurable than GATK for custom applications.

There are several areas where improvements could be made in future versions of Pilon with respect to assembly improvement and variant calling. First, it seems likely that there will be some benefit in iteratively applying Pilon in the assembly improvement and insertion variant calling process. Currently, Pilon builds out the gap/inserted sequence without re-aligning reads to build off the newly extended sequence. Using an iterative strategy has recently been used with some success by PAGIT [6]. Second, Pilon currently does not attempt to make fixes to larger structural issues within assemblies or make changes to scaffold architecture. With data from long insert libraries, it should be feasible to break and/ or join scaffolds accurately. Third, tandem repeats continue to be challenging and may require a more specific approach. These regions are inherently difficult with short read data because there is no unambiguous information in the data to determine how many copies are present. This challenge is true for any *de novo* assembly; in order to resolve a tandem repeat, reads are needed that anchor into unique sequence on either side and read through the entire tandem repeat sequence. Lacking this, tandem copy numbers can only be speculated from mate-pairing information, depth of coverage calculations and library insert sizes.

Currently, Pilon is rather conservative in its corrections: (i) it uses a large cut-off to merge overlaps, and (ii) it will not attempt to

Table 4. Pilon's performance in calling variants in *M. tuberculosis* F11 that were larger than 50 nt.

	Pilon		Pilon-frags	
	Missed	**Called**	**Missed**	**Called**
Insertion	11	23	12	22
Substitution	0	5	0	5
Deletion	6	20	10	16

Variants are divided by type across the rows. Missed variants are those that were annotated in the curation, but were not identified by Pilon. The called variants are those that were annotated in the curation that Pilon accurately identified.

resolve significant tandem repeats structures definitively. Notwithstanding the challenges encountered with tandem repeats, Pilon does an excellent job with other repetitive sequences and is able to fix many genes of known repetitive gene families and is able to fill in many transposable elements.

While we have evaluated Pilon's assembly improvements on both haploid and diploid genomes and obtained positive results for both, we acknowledge that there is still significant opportunity for future improvement in Pilon's handling of diploid genomes. Pilon could be enhanced to understand IUPAC ambiguity codes in its input genome and generate them in its output, and Pilon's heuristics for identifying insertions and deletions in diploid genomes could be improved, including its ability to recognize and report heterozygous indels. Finally, the local reassembly process could be improved to perform better in heterozygous regions. Even so, our results indicate that in its current form, Pilon is able to make valuable improvements to diploid genomes.

We have evaluated Pilon's performance using microbial genomes with finished references. However, there is no inherent limitation on the size of genomes to which Pilon can be applied. For example, we have used Pilon to improve assemblies of larger genomes, including 16 strains of the *Anopheles* genus (~200 Mbp diploid genome), but we were unable to verify the accuracy of Pilon's improvements since these genomes have not been finished. Pilon runs within minutes on small microbial genomes and will complete overnight on larger eukaryote genomes, such as *Anopheles*, which is similar to the tools included in our benchmarking.

Conclusion

Ultimately, Pilon has great utility and addresses an urgent need for better and more efficient methods to deal with the thousands of microbial genomes that are being produced. We have shown that Pilon performs well as compared to the state-of-the-art for both assembly improvement and variant detection, often outperforming these tools. Pilon is also unique in its user-friendly integrated approach to assembly improvement and is unique in its ability to identify large variants accurately in microbial genomes. As a recent addition to the production process for microbial genomes at Broad Institute, Pilon has been used to automatically improve the quality of over 8,000 prokaryote and eukaryote genomes prior to their submission to Genbank, and it has been used to call variants on over 6,000 genomes.

Material and Methods

Detailed algorithm description

Input requirements. Pilon requires an input genome in FASTA format and one or more BAM files containing sequencing reads aligned to the input genome. The BAM files must be sorted in coordinate order and indexed. For Illumina data, these BAM files are usually produced by an aligner such as BWA [15] or Bowtie 2 [34]. It is recommended that single best hit or random selection among equal best alignments is used as input into Pilon. Pilon can use three types of BAM files:

1. *Fragments*: paired read data of short insert size, typically < 1 Kbp. Reads should be in forward-reverse (FR) orientation;
2. *Long inserts*: paired read data of longer insert size, typically > 1 Kbp. Reads should be in forward-reverse (FR) orientation. Sequencing of long insert libraries that are generated using the standard Illumina mate-pair library preparation protocol

typically result in reverse-forward (RF) read orientation, so they will need to be reversed in the BAM file.
3. *Unpaired*: unpaired sequencing read data.

To use Pilon with default arguments, read length should be 75 bases or longer and total sequence coverage should be 50x or greater, though deeper total coverage of >100x is beneficial. Pilon can also make use of longer reads, such as those from Sanger capillary sequencing and circular-consensus or error-corrected reads from Pacific Biosciences (PacBio) sequencing. However, Pilon is not currently tuned to the error model of raw PacBio reads, and their use may introduce false corrections.

Pilon makes extensive use of pairing information when it is available, so paired libraries are highly recommended. Pilon is capable of using paired libraries of any insert size, as it scans the BAMs to compute statistics, including insert size distribution.

Improving local base accuracy and identifying SNPs. Pilon improves the local base accuracy of the contigs through analysis of the read alignment information. First, Pilon parses the alignment information from the input data and summarizes the evidence from all the reads covering each base position. Alignments can be less trustworthy near the ends of reads, especially in differentiating between indels and base changes, so Pilon ignores the alignments from a small number of bases at each end of the read, which is configurable at run time. For each base position in the genome, Pilon builds a pileup structure which records both a count and a measure of the weighted evidence for each possible base (A,C,G,T) from the read alignments. The contribution of base information from each read is weighted by the base quality reported by the sequencing instrument as well as the mapping quality computed by the aligner.

When Pilon is building pileups from paired alignment data, only reads from "valid" pairs (*i.e.*, those with the *PROPER_PAIR* flag set in the BAM by the aligner, indicating the reads of a pair align in proper orientation with a plausible separation) contribute evidence to the pileups. It is crucial that the PROPER_PAIR flag is set accurately by the tool that produced the BAM file. A count of non-valid alignments covering each position is also kept to help identify areas of possible mis-assembly. Pilon also keeps track of "soft clipping" in the alignments, which exclude sub-sections of a read which aligned poorly. A tally of soft-clip transitions is kept at each genomic location as another aid in identifying possible local misassemblies.

From the pileup evidence, Pilon classifies each base in the input genome into one of four categories:

1. *Confirmed*: the vast majority of evidence supports the base in the input genome;
2. *Changed*: the vast majority of evidence supports a change of the base in the input genome to another allele;
3. *Ambiguous*: the evidence supports more than one alternative at this position;
4. *Unconfirmed*: there is insufficient evidence to make a determination at this position due to insufficient depth of coverage by valid reads.

Ambiguous bases can occur for several reasons. If the genome is diploid, this is expected at heterozygous polymorphic locations. Difficult-to-sequence regions may result in a large enough fraction of sequencing errors to result in an ambiguous call. Finally, if the input genome has a smaller number of copies of a repeated genomic structure than occurs in the true genome (a "collapsed repeat"), the aligned reads may have originated from more than one instance of the repeat structure; where there are differences in the true instances of the repeat, the alignments can show mixed evidence.

Paired read information, especially information from long insert libraries which span a longer distance, is extremely valuable in helping resolve ambiguous locations due to collapsed repeats. Pairs for which one read lands inside a repeat element, but the other lands in unique anchoring sequence on the flanks of the repeat help to resolve the true base content of the repeat structure. Data from long inserts will typically have a higher alignment mapping quality than short-range fragment pairs that lie completely within the repeat because the fragment pairs may not be able to be placed uniquely among the repeats. Since Pilon uses mapping quality to weigh the evidence from each read, the long inserts can often pick the correct haplotype variations of the repeat structure.

Pilon includes corrections to single-base errors in its output genome, and optionally, it can also change ambiguous bases to the allele with the preponderance of evidence.

Finding and fixing small indels. While recording the base-by-base pileup evidence, Pilon also records the location and content of indels present in the alignments. Indel alignments which represent equivalent edits to the input genome may appear at different coordinates in the alignments. For instance, if the input genome has the sequence *ACCCCT*, but the read evidence suggests one of the Cs should be deleted (*ACCCT*), each individual read alignment might show a deletion at any of the four *C* coordinates. Pilon shifts alignment indels to their leftmost equivalent edit in the input genome, so that the evidence from all the equivalent edits is combined into evidence for a single event at a one location.

Pilon makes an insertion or deletion call if a majority of the valid reads support the change, though that threshold is lowered somewhat for longer events, as it is typically more difficult for aligners to identify longer indels in short read data. Called indels from the input genome are fixed in Pilon's output genome.

Fixing mis-assemblies, detecting large indels, and filling gaps. Pilon is capable of reassembling local regions of the genome when there is sufficient evidence from the alignments that the contiguity of the input genome does not match the sequencing data. For assembly improvement applications, this could be an indication of a local mis-assembly. For variant calling applications, this could be caused by insertions or deletions too large to be reflected in the short read alignments.

Pilon tries to identify areas of potential local read alignment discontinuity in the contigs of the input genome by employing four heuristics: (i) a large percentage of reads containing a soft-clipped alignment at a given base position, (ii) a large ratio of invalid pairs to valid pairs spanning a location, (iii) areas of extremely low coverage and (iv) rapid drops in alignment coverage over a distance on the order of a read length. Once Pilon has identified an area for local reassembly, it treats the suspicious region (which may be a single base or a larger region) as untrusted, using alignments to the trusted flanks on both sides to identify a collection of reads that might contribute evidence for the true sequence in the suspicious region.

Unpaired reads with partial alignments to the flanks are included in the collection. For paired data, Pilon identifies pairs in which one of the reads is anchored by proper alignment to one of the flanks (*e.g.*, with forward orientation on the left flank, or reverse orientation on the right flank), but whose mate is either unmapped or improperly mapped (*e.g.*, to a remote location in the genome). For fragment pairs, both reads of such pairs are included in the collection; for long inserts, only the unanchored read is included in the collection.

From the collected reads, Pilon builds a De Bruijn assembly graph (default K = 47). For each k-mer in the reads, it uses the same pileup structure to record the bases which follow that k-mer, including weighting by base quality. Then, the pileups are evaluated to determine the link(s) to the next k-mer(s); this results in either a single base call, resulting in one forward link to the next k-mer, or an ambiguous call, resulting in two links forward and a branch in the assembly graph. This process automatically prunes the assembly graph of most sequencing errors, as infrequent base differences are unlikely to present enough evidence to affect the forward links. A minimum coverage cutoff of five for each forward link also prunes the assembly graph of many false links that could appear because of sequencing errors.

Pilon then tries k-mers from the trusted flanks as starting points to walk into the untrusted region from each side, building all possible extensions with up to five branching points (2^5 possible extensions). Tandem repeats with combined length >K cause loops in the local assembly graph, and they are detected by noting when the assembly walk reaches an already-incorporated k-mer. Pilon currently does not attempt to determine the copy number of such tandem repeats; instead, it will report the length of the repeat structure encountered in its standard output, and it will not attempt to close the two sides.

When no tandem repeat is detected, the resulting extensions from each side are combinatorially matched for possible perfect overlaps of sufficient length (2K+1) to be considered for closure. If there is exactly one such closure and it differs from the input genome, the assembled flank-to-flank sequence will replace the corresponding sequence in the input genome. Since the default k-mer size is 47, an overlap of 95 bases is required for closure.

If there are no closures or more than one possible closure, Pilon will identify a consensus extension from each flank. If an optional argument is set to allow opening of new gaps, Pilon will replace the suspicious region with the consensus extensions from each flank and create a gap between them; otherwise, it simply reports that it was unable to find a solution. These reports identify areas that an assembly analyst might wish to investigate manually.

Pilon also attempts to fill gaps between contigs in a scaffold ("captured gaps") in the input genome. In order to fill captured gaps, Pilon employs the same local reassembly technique described above, treating the gap itself as the "untrusted" region. If there is a unique closure, the gap is filled; otherwise, consensus extensions from each flank are used to reduce the size of the gap. Pilon does not currently attempt to join or break scaffolds.

Large collapsed repeat (segmental duplication) detection. Pilon includes heuristics that attempt to flag areas indicative of large (>10 Kbp) collapsed repeats with respect to the input genome. These are characteristically large contiguous areas that appear to have double (or higher) read coverage compared to the rest of the genomic element being analyzed. Long insert data are excluded from this computation, as we have found long insert coverage to be far more variable across some genomes. Pilon does not attempt to fix these potentially collapsed regions, but it does report them in its standard output for further investigation.

In variant calling applications, large segmental duplications in the sequenced strain with respect to the reference have the same signature as large collapsed repeats in a draft assembly; a duplicated region of the genome will result in double the number of reads covering that sequence. Pilon's reporting of large collapsed repeat regions can be used to identify candidate segmental duplications.

Output files. Pilon generates a modified genome as a FASTA file, including all single-base, small indel, gap filling, mis-assembly and large-event corrections from the input genome. In the assembly improvement case, this is the improved assembly consensus. In variant detection mode, this is the reference sequence which has been edited to represent the consensus of the given sample more closely.

Pilon can optionally generate a Variant Call Format (VCF) [http://vcftools.sourceforge.net/specs.html] file, which lists copious detailed information about the base and indel evidence at every base position in the genome, including two scores regarding variant quality: the QUAL column, and a depth-normalized call quality (QD) field in the INFO column. For additional details on the VCF format, we refer to the VCF specification referred above. Changes generated by local reassembly, often triggered by larger polymorphisms in variant calling applications, are included as structural variant records (SVTYPE = INS and SVTYPE = DEL). Pilon can also, optionally, generate a "changes" file which lists the edits applied from input to output genome in tabular form, including source and destination coordinates and source and destination sequence. Finally, Pilon will optionally (with the — tracks option) output a series of visualization tracks ("bed" and "wig" files) suitable for viewing in genome browsers such as IGV [35] and GenomeView [36]. Tracks include basic metrics across the genome, such as sequence coverage and physical coverage, as well as some of the calculated metrics Pilon uses in its heuristics for finding potential areas of mis-assembly, such as percentage of valid read pairs covering every location.

Pilon's standard output also contains useful information, including coverage levels, percentage of the input genome confirmed, a summary of the changes made, as well as some specifically flagged issues which were not corrected, such as potentially large collapsed repeat regions, potential regions of mis-assembly which were not able to be corrected, and detected tandem repeats that were not resolved.

Data generation

All sequencing data used for these experiments were generated from an Illumina HiSeq 2000 machine. For sequencing *M. tuberculosis* F11 and T67, two libraries were generated: one PCR-free 180 bp insert paired fragment library [37] and large insert 3–5 Kbp long insert library [38]. *S. pneumoniae* TIGR4 data also consisted of two libraries: one robotically size selected 180 bp insert paired fragment library [37] and a large insert 3–5 Kbp long insert library [38]. The sequencing data for *C. albicans* SC5314 was generated from three libraries: a robotically size-selected 180 bp insert paired fragment library [37], a gel-cut 4 Kbp long insert library [39], and a 40 Kb Fosill library [40]. Sequencing data were submitted to the Sequence Read Archive with identifiers: SRX347313, SRX347312, SRX105400, SRX110130, SRX347317 and SRX347316.

Evaluation methods

Assembly improvement. All draft assemblies were generated using ALLPATHS-LG [41]. The draft assembly for *Mycobacterium tuberculosis* F11 utilized 100x of the 180 bp insert fragment library and 50x of the 3–5 Kb long insert library and was executed using ALLPATHS-LG v45395 utilizing the ASSISTED_PATCHING = 2.1 parameter and the *M. tuberculosis* H37RV reference genome for assisting (GenBank accession: CP003248). The draft assembly for *S. pneumoniae* TIGR4 was created using ALLPATHS-LG v45925 with default parameters and using 100x of the 180 bp insert fragment library and 50x of the 3–5 Kb long insert library. The *C. albicans* SC5314 utilized 100x of the 180 bp insert fragment library, 100x of the gel-cut 4 Kb long insert library and 50x of the Fosill library, and was assembled with ALLPATHS-LG v39846 utilizing the ASSISTED_PATCHING and HAPLOI-

DIFY options with the *C. albicans* SC5314 reference sequence as a reference for assisting.

We benchmarked Pilon's ability to close gaps in the draft bacterial assemblies against two tools built for this purpose, IMAGE v2.4.1 [16] and GapFiller v1.10 [17]. The same sets of sequencing reads used as input to Pilon were used for IMAGE (fragment library only) and GapFiller (fragment and long insert libraries). IMAGE was run in the manner implemented in the PAGIT [6] example scripts: 6 iterations, one with a kmer size of 61, three with a kmer size of 49, and two with a kmer size of 41. GapFiller was run for 10 iterations with a libraries.txt file specifying a ratio $r = 0.5$ and library insert sizes computed by Pilon from the aligned bams.

To evaluate the quality of Pilon's single base and small indel corrections to the draft assemblies, we also ran iCORN v0.97 [18], the consensus sequence improvement tool in PAGIT, on the same draft assemblies using the same sets of fragment reads. iCORN was run in the manner implemented in the PAGIT example scripts, only changing the library insert size mean and range parameters. For TIGR4, we used a mean of 180 and a range of 120–300. For F11, we used a mean of 226 and a range of 100–500, since the PCR-free library preparation resulted in a wider range of insert sizes.

Fixes to the assemblies (Table S1) made by Pilon and the other assembly improvement tools were assessed by extracting the changed region of sequence in the output genome along with 300 bp flanks on each side. These extracted sequences were aligned to their respective finished reference genomes with BLASTN [42], and the accuracy of the changes was assessed by manually inspecting the alignments for accuracy, judging each fix as "Correct" or "Incorrect". For larger block changes which resulted from local reassembly (gap filling and fixing of local mis-assemblies), a third category of "No worse" was established for situations in which: (i) the draft assembly contained a mis-assembly in the changed region, (ii) Pilon made a change attempting to fix the mis-assembly, and (iii) the fix was not entirely correct, but was no worse than the original problem.

For the assembly improvement statistics, *Bases added* was calculated by tallying bases added in locations where resulting fixes resulted in a net gain of bases during gap filling and mis-assembly correction processes, as reported in the Pilon standard output indicated by the "fix gap" or "fix break" lines.

Variant calling. Variant calls were made using *M. tuberculosis* H37Rv (GenBank accession: CP003248) as the reference and the *M. tuberculosis* F11 aligned read and long insert fragments as input data. From the sequenced fragment and long insert libraries, a random subset of read pairs was selected from each library to obtain an estimated 200x coverage of the *M. tuberculosis* H37Rv reference genome. Each library's reads were aligned to the *M. tuberculosis* H37Rv reference genome using BWA (version 0.5.9-r16) to generate BAM files suitable for input to the variant calling processes.

Pilon: Pilon was run with the —variant command line option, specifying the *M. tuberculosis* H37Rv reference genome and the above BAM file(s) as inputs. We evaluated two Pilon variant calling sets, one generated using both fragment and long insert library BAMs, and one using only the fragment library BAM.

GATK UnifiedGenotyper: Reads in the fragment library BAM were realigned by applying the Genome Analysis Toolkit (GATK version v3.2.2) RealignerTargetCreator and IndelRealigner tools on the fragment library aligned BAM file. Variants were then called from the realigned BAM file using UnifiedGenotyper run with the following settings: -nt 32 -A AlleleBalance -ploidy 1 -pnrm EXACT_GENERAL_PLOIDY -glm BOTH —output_mode

EMIT_ALL_SITES. Low confidence variants were then filtered using VariantFiltration (VF) run with the following settings: −filterExpression "((DP-MQ0)<10) || ((MQ0/(1.0*DP))>=0.8) || (ABHom <0.8) || (Dels >0.5)" −filterName LowConfidence.

These VariantFiltration settings filtered out variant calls at locations with less than 10 unambiguous read alignments, where 80% or more of the read depth had ambiguous mappings, where fewer than 80% of the reads supported the alternate allele, or more than half of the reads contained spanning deletions. This filter expression was based on one previously used to call variants from the European *Escherichia coli* O104:H4 outbreak [38], adjusting depth and allele balance thresholds to yield the best performance tradeoff between false negative and false positive results on these data.

SAMtools/BCFtools: The same aligned fragment library BAM file described above was used as input for variant calling using SAMtool/BCFtools v0.1.19 according to recommendations found on the SAMtools webpage (http://samtools.sourceforge.net/mpileup.shtml). samtools mpileup was used to generate pileups in bcf format, and variants were called using bcftools using the -bcg option. Finally, variants were filtered using vcfutils.pl varFilter -d 10 to filter out calls at locations where the aligned coverage was less than 10 reads. We chose the minimum depth of 10 to be consistent with the filtering used for GATK Unified-Genotyper.

CLEVER and BreakDancer: The aligned fragment and combined fragment and long insert library described above were used as input for CLEVER v2.0rc3 and BreakDancer 1.3.6. clever −sorted −use_xa was used to generate calls for CLEVER. bam2cfg.pl -g -h was used to generate the Break-Dancer config file, which was then used with breakdancer-max.

Curating differences between F11 and H37Rv. Differences between the finished *M. tuberculosis* F11 (GenBank accession: CP000717) and *M. tuberculosis* H37RV (GenBank accession: CP003248) references were curated by employing a banded Smith-Waterman algorithm to align syntenic regions of the two genomes. Alignments were run, separately, for each syntenic portion of the two sequences. When the alignment diverged significantly, the program was run again to pick up at the next syntenic block. The resulting alignments over syntenic regions identified coordinates of small blocks of mismatches, typically only a single-base long, but in some cases up to 289 bp. Areas where there was a significant break in synteny or where the banded Smith-Waterman alignment produced questionable results were analyzed using either Nucmer [43], ClustalW [44] or Blast2 [45] to verify the nature of the difference and to obtain more accurate coordinates. In some cases, the alignments proved too difficult to get accurate coordinates, but approximate definitions of these differences were obtained. The resulting table of differences between the two references (Table S8) has each difference annotated with most likely coordinates, with two exceptions where the variation between the strains was so high that it was impossible to know whether each difference was captured individually. The two highly variable regions corresponded to coordinates 1636857–1639600 and 3928967–3949709, which, together, account for less than 0.5% of the *M. tuberculosis* H37Rv genome. These regions were excluded from all variant analyses.

Variant Assessment. The resulting variant calls were compared to a manually curated set of differences between *M. tuberculosis* F11 and *M. tuberculosis* H37Rv as described above. Based on this comparison, recall and precision were calculated according to the strategy described in [46]. Briefly, recall is a measure of completeness of calls against the curated truth set; false

negatives lower the recall score. Precision is a measure of the accuracy of the calls made; false positives lower the precision score. Specifically, recall = $tp\_c/(tp\_c+fn)$ and precision = $tp\_p/(tp\_p+fp)$, where tp_c is the number of true positive calls based on the curation set, tp_p is the number of true positive calls from the set of predicted variants, fp is the number of false calls from the set of predicted variants, and fn is the number of missed calls from the predicted variants based on the curation set. The F-measure is the harmonic mean of the recall and precision rates, providing an "overall" metric that captures tradeoff between recall and precision.

True positives in the prediction set had at least one variant site called in the curation set. For variants that affected more than a single base in the curation set (*i.e.*, multi nucleotide polymorphisms), we allowed for a combination of two or more smaller events in the prediction set to be marked as correct, since tools may call a densely polymorphic region as a block substitution rather than a series of equivalent single-base changes. For example, for the multi nucleotide substitution in the curation set, ACCGT = > CCTGA, three SNP calls at the same location, A>C, C>T and T>A, would be counted as a true positive. In addition, predicted variants that affected more than 20 bases were allowed to match only partially with the curation set because there can be different ways to manually curate sites that vary among the *M. tuberculosis* F11 and H37Rv finished reference genomes. In particular, resolution of tandem repeats was challenging for both prediction and curation of variants since it was difficult to determine which copy of the repeat was inserted or deleted. In these cases, we counted the variant as correct if a similar event was predicted within 100 nucleotides.

Availability

Pilon is open source software available under the GNU General Public License Version 2 (GPLv2). Pilon is written in the Scala programming language, and it makes extensive use of the open source Picard Java libraries (http://picard.sourceforge.net) for parsing BAM and FASTA files. Pilon is compiled into a single Java Archive (JAR) file which runs inside a 64-bit Java Virtual Machine environment. Binary and source distributions can be obtained from GitHub (http://github.com/broadinstitute/pilon/releases/). Results in this paper were obtained with Pilon version 1.5. A summary of all command-line options is available in Table S9.

Online documentation, as well as two example data sets to test Pilon on the same data as was used in this manuscript, are available from the web site http://broadinstitute.org/software/pilon/. We provide the *Streptococcus pneumoniae* TIGR4 data set as an assembly improvement example and the *Mtb* F11 data set as a variant calling example.

Supporting Information

Figure S1 Muscle alignment of TB F11 gene TFBG_12611.

Figure S2 Contig count reduction in production.

Figure S3 Contig N50 increase in 50 production assemblies.

Table S1 Assessment of gap filling and local reassembly fixes. b: Comparison of assembly gap closures among Pilon, IMAGE, and GapFiller.

Table S2   Assessment of base corrections by Pilon and iCORN.

Table S3   Detailed information regarding the gene based assessment of F11 assemblies.

Table S4   Detailed information regarding the gene based assessment of TIGR4.

Table S5   Summary of SNP, small in-dels, and large in-dels in M. tuberculosis F11 relative to H37Rv.

Table S6   Example SNPs only found with regular Pilon.

Table S7   Example IS6110 insertion element variation.

Table S8   Curation of Mtb F11.

Table S9   Pilon Command Line Arguments.

Acknowledgments

We acknowledge the Broad Institute Genomics Platform for generating the Illumina libraries and sequence used here, and the Genome Assembly and Analysis Group and the A2E Group for their assistance in configuring and maintaining Pilon within the Broad Institute's production assembly infrastructure. We also thank Gustavo Cerqueira, Christopher Desjardins, Abigail McGuire, Jonathan Livny and, especially, Geraldine Van der Auwera for their helpful comments.

Author Contributions

Conceived and designed the experiments: BJW TA CAC QZ JW SKY AME. Performed the experiments: BJW TA TS SKY. Analyzed the data: BJW TA TS MP AA SS CAC QZ JW SKY AME. Wrote the paper: BJW TA TS MP AA SS CAC SKY AME.

References

1. Chewapreecha C, Harris SR, Croucher NJ, Turner C, Marttinen P, et al. (2014) Dense genomic sampling identifies highways of pneumococcal recombination. Nat Genet 46: 305–309. Available: http://www.ncbi.nlm.nih.gov/pubmed/24509479. Accessed 21 March 2014.

2. Comas I, Coscolla M, Luo T, Borrell S, Holt KE, et al. (2013) Out-of-Africa migration and Neolithic coexpansion of Mycobacterium tuberculosis with modern humans. Nat Genet 45: 1176–1182. Available: http://www.ncbi.nlm.nih.gov/pubmed/23995134. Accessed 19 March 2014.

3. Croucher NJ, Finkelstein J a, Pelton SI, Mitchell PK, Lee GM, et al. (2013) Population genomics of post-vaccine changes in pneumococcal epidemiology. Nat Genet 45: 656–663. Available: http://www.pubmedcentral.nih.gov/articlerender.fcgi?artid=3725542&tool=pmcentrez&rendertype=abstract. Accessed 21 March 2014.

4. Grad YH, Kirkcaldy RD, Trees D, Dordel J, Harris SR, et al. (2014) Genomic epidemiology of Neisseria gonorrhoeae with reduced susceptibility to cefixime in the USA: a retrospective observational study. Lancet Infect Dis 14: 220–226. Available: http://www.ncbi.nlm.nih.gov/pubmed/24462211. Accessed 21 March 2014.

5. Ronen R, Boucher C, Chitsaz H, Pevzner P (2012) SEQuel: improving the accuracy of genome assemblies. Bioinformatics 28: i188–96. Available: http://www.pubmedcentral.nih.gov/articlerender.fcgi?artid=3371851&tool=pmcentrez&rendertype=abstract. Accessed 20 January 2014.

6. Swain MT, Tsai IJ, Assefa S a, Newbold C, Berriman M, et al. (2012) A post-assembly genome-improvement toolkit (PAGIT) to obtain annotated genomes from contigs. Nat Protoc 7: 1260–1284. Available: http://www.pubmedcentral.nih.gov/articlerender.fcgi?artid=3648784&tool=pmcentrez&rendertype=abstract. Accessed 24 January 2014.

7. Hunt M, Kikuchi T, Sanders M, Newbold C, Berriman M, et al. (2013) REAPR: a universal tool for genome assembly evaluation. Genome Biol 14: R47. Available: http://www.pubmedcentral.nih.gov/articlerender.fcgi?artid=3798757&tool=pmcentrez&rendertype=abstract. Accessed 22 January 2014.

8. Vicedomini R, Vezzi F, Scalabrin S, Arvestad L, Policriti A (2013) GAM-NGS: genomic assemblies merger for next generation sequencing. BMC Bioinformatics 14 Suppl 7: S6. Available: http://www.pubmedcentral.nih.gov/articlerender.fcgi?artid=3633056&tool=pmcentrez&rendertype=abstract. Accessed 28 January 2014.

9. Li H, Handsaker B, Wysoker A, Fennell T, Ruan J, et al. (2009) The Sequence Alignment/Map format and SAMtools. Bioinformatics 25: 2078–2079. Available: http://www.pubmedcentral.nih.gov/articlerender.fcgi?artid=2723002&tool=pmcentrez&rendertype=abstract. Accessed 20 January 2014.

10. McKenna A, Hanna M, Banks E, Sivachenko A, Cibulskis K, et al. (2010) The Genome Analysis Toolkit: a MapReduce framework for analyzing next-generation DNA sequencing data. Genome Res 20: 1297–1303. Available: http://www.pubmedcentral.nih.gov/articlerender.fcgi?artid=2928508&tool=pmcentrez&rendertype=abstract. Accessed 21 January 2014.

11. Pabinger S, Dander A, Fischer M, Snajder R, Sperk M, et al. (2013) A survey of tools for variant analysis of next-generation genome sequencing data. Brief Bioinform 15: 256–278. Available: http://www.ncbi.nlm.nih.gov/pubmed/23341494. Accessed 19 March 2014.

12. Cubillos-Ruiz A, Morales J, Zambrano MM (2008) Analysis of the genetic variation in Mycobacterium tuberculosis strains by multiple genome alignments. BMC Res Notes 1: 110. Available: http://www.pubmedcentral.nih.gov/articlerender.fcgi?artid=2590607&tool=pmcentrez&rendertype=abstract. Accessed 28 January 2014.

13. El-Metwally S, Hamza T, Zakaria M, Helmy M (2013) Next-generation sequence assembly: four stages of data processing and computational challenges. PLoS Comput Biol 9: e1003345. Available: http://www.pubmedcentral.nih.gov/articlerender.fcgi?artid=3861042&tool=pmcentrez&rendertype=abstract. Accessed 21 January 2014.

14. Tettelin H, Nelson KE, Paulsen IT, Eisen J a, Read TD, et al. (2001) Complete genome sequence of a virulent isolate of Streptococcus pneumoniae. Science 293: 498–506. Available: http://www.ncbi.nlm.nih.gov/pubmed/11463916. Accessed 21 January 2014.

15. Li H, Durbin R (2009) Fast and accurate short read alignment with Burrows-Wheeler transform. Bioinformatics 25: 1754–1760. Available: http://www.pubmedcentral.nih.gov/articlerender.fcgi?artid=2705234&tool=pmcentrez&rendertype=abstract. Accessed 20 January 2014.

16. Tsai IJ, Otto TD, Berriman M (2010) Improving draft assemblies by iterative mapping and assembly of short reads to eliminate gaps. Genome Biol 11: R41. Available: http://www.pubmedcentral.nih.gov/articlerender.fcgi?artid=2884544&tool=pmcentrez&rendertype=abstract.

17. Nadalin F, Vezzi F, Policriti A (2012) GapFiller: a de novo assembly approach to fill the gap within paired reads. BMC Bioinformatics 13 Suppl 14: S8. Available: http://www.pubmedcentral.nih.gov/articlerender.fcgi?artid=3439727&tool=pmcentrez&rendertype=abstract. Accessed 10 July 2014.

18. Otto TD, Sanders M, Berriman M, Newbold C (2010) Iterative Correction of Reference Nucleotides (iCORN) using second generation sequencing technology. Bioinformatics 26: 1704–1707. Available: http://www.pubmedcentral.nih.gov/articlerender.fcgi?artid=2894513&tool=pmcentrez&rendertype=abstract. Accessed 10 July 2014.

19. Luo R, Mann B, Lewis WS, Rowe A, Heath R, et al. (2005) Solution structure of choline binding protein A, the major adhesin of Streptococcus pneumoniae. EMBO J 24: 34–43. Available: http://www.pubmedcentral.nih.gov/articlerender.fcgi?artid=544903&tool=pmcentrez&rendertype=abstract. Accessed 21 March 2014.

20. Tu AH, Fulgham RL, McCrory MA, Briles DE, Szalai AJ (1999) Pneumococcal surface protein A inhibits complement activation by Streptococcus pneumoniae. Infect Immun 67: 4720–4724. Available: http://www.pubmedcentral.nih.gov/articlerender.fcgi?artid=96800&tool=pmcentrez&rendertype=abstract. Accessed 21 March 2014.

21. Butler G, Rasmussen MD, Lin MF, Santos M a S, Sakthikumar S, et al. (2009) Evolution of pathogenicity and sexual reproduction in eight Candida genomes. Nature 459: 657–662. Available: http://www.pubmedcentral.nih.gov/articlerender.fcgi?artid=2834264&tool=pmcentrez&rendertype=abstract. Accessed 28 January 2014.

22. Jones T, Federspiel N a, Chibana H, Dungan J, Kalman S, et al. (2004) The diploid genome sequence of Candida albicans. Proc Natl Acad Sci U S A 101: 7329–7334. Available: http://www.pubmedcentral.nih.gov/articlerender.fcgi?artid=409918&tool=pmcentrez&rendertype=abstract.

23. Muzzey D, Schwartz K, Weissman JS, Sherlock G (2013) Assembly of a phased diploid Candida albicans genome facilitates allele-specific measurements and provides a simple model for repeat and indel structure. Genome Biol 14: R97. Available: http://www.ncbi.nlm.nih.gov/pubmed/24025428. Accessed 2014 January 28.

24. Chen K, Wallis JW, McLellan MD, Larson DE, Kalicki JM, et al. (2009) BreakDancer: an algorithm for high-resolution mapping of genomic structural variation. Nat Methods 6: 677–681. Available: http://www.pubmedcentral.nih.

gov/articlerender.fcgi?artid=3661775&tool=pmcentrez&rendertype=abstract. Accessed 2014 March 20.

25. Marschall T, Costa IG, Canzar S, Bauer M, Klau GW, et al. (2012) CLEVER: clique-enumerating variant finder. Bioinformatics 28: 2875–2882. Available: http://www.ncbi.nlm.nih.gov/pubmed/23060616. Accessed 10 July 2014.

26. Weiner B, Gomez J, Victor TC, Warren RM, Sloutsky A, et al. (2012) Independent large scale duplications in multiple M. tuberculosis lineages overlapping the same genomic region. PLoS One 7: e26038. Available: http://www.pubmedcentral.nih.gov/articlerender.fcgi?artid=3274525&tool=pmcentrez&rendertype=abstract. Accessed 2014 August 24.

27. Ioerger TR, Feng Y, Ganesula K, Chen X, Dobos KM, et al. (2010) Variation among genome sequences of H37Rv strains of Mycobacterium tuberculosis from multiple laboratories. J Bacteriol 192: 3645–3653. Available: http://www.pubmedcentral.nih.gov/articlerender.fcgi?artid=2897344&tool=pmcentrez&rendertype=abstract. Accessed 2014 January 22.

28. Kohli S, Singh Y, Sharma K, Mittal A, Ehtesham NZ, et al. (2012) Comparative genomic and proteomic analyses of PE/PPE multigene family of Mycobacterium tuberculosis H$_{37}$Rv and H$_{37}$Ra reveal novel and interesting differences with implications in virulence. Nucleic Acids Res 40: 7113–7122. Available: http://www.pubmedcentral.nih.gov/articlerender.fcgi?artid=3424577&tool=pmcentrez&rendertype=abstract. Accessed 2014 March 21.

29. Vordermeier HM, Hewinson RG, Wilkinson RJ, Wilkinson K a, Gideon HP, et al. (2012) Conserved immune recognition hierarchy of mycobacterial PE/PPE proteins during infection in natural hosts. PLoS One 7: e40890. Available: http://www.pubmedcentral.nih.gov/articlerender.fcgi?artid=3411574&tool=pmcentrez&rendertype=abstract. Accessed 2014 March 21.

30. Das S, Paramasivan CN, Lowrie DB, Prabhakar R, Narayanan PR (1995) IS6110 restriction fragment length polymorphism typing of clinical isolates of Mycobacterium tuberculosis from patients with pulmonary tuberculosis in Madras, south India. Tuber Lung Dis 76: 550–554. Available: http://www.ncbi.nlm.nih.gov/pubmed/8593378. Accessed 2014 January 28.

31. Karboul A, Mazza A, Gey van Pittius NC, Ho JL, Brousseau R, et al. (2008) Frequent homologous recombination events in Mycobacterium tuberculosis PE/PPE multigene families: potential role in antigenic variability. J Bacteriol 190: 7838–7846. Available: http://www.pubmedcentral.nih.gov/articlerender.fcgi?artid=2583619&tool=pmcentrez&rendertype=abstract. Accessed 2014 January 28.

32. Ford C, Yusim K, Ioerger T, Feng S, Chase M, et al. (2012) Mycobacterium tuberculosis—heterogeneity revealed through whole genome sequencing. Tuberculosis (Edinb) 92: 194–201. Available: http://www.pubmedcentral.nih.gov/articlerender.fcgi?artid=3323677&tool=pmcentrez&rendertype=abstract. Accessed 28 January 2014.

33. McEvoy CRE, Cloete R, Müller B, Schürch AC, van Helden PD, et al. (2012) Comparative analysis of Mycobacterium tuberculosis pe and ppe genes reveals high sequence variation and an apparent absence of selective constraints. PLoS One 7: e30593. Available: http://www.pubmedcentral.nih.gov/articlerender.fcgi?artid=3319526&tool=pmcentrez&rendertype=abstract. Accessed 2012 July 28.

34. Langmead B, Salzberg SL (2012) Fast gapped-read alignment with Bowtie 2. Nat Methods 9: 357–359. Available: http://www.pubmedcentral.nih.gov/articlerender.fcgi?artid=3322381&tool=pmcentrez&rendertype=abstract. Accessed 2014 January 20.

35. Thorvaldsdóttir H, Robinson JT, Mesirov JP (2013) Integrative Genomics Viewer (IGV): high-performance genomics data visualization and exploration. Brief Bioinform 14: 178–192. Available: http://www.pubmedcentral.nih.gov/articlerender.fcgi?artid=3603213&tool=pmcentrez&rendertype=abstract. Accessed 2014 January 20.

36. Abeel T, Van Parys T, Saeys Y, Galagan J, Van de Peer Y (2012) GenomeView: a next-generation genome browser. Nucleic Acids Res 40: e12. Available: http://www.pubmedcentral.nih.gov/articlerender.fcgi?artid=3258165&tool=pmcentrez&rendertype=abstract. Accessed 2012 March 15.

37. Ross MG, Russ C, Costello M, Hollinger A, Lennon NJ, et al. (2013) Characterizing and measuring bias in sequence data. Genome Biol 14: R51. Available: http://www.ncbi.nlm.nih.gov/pubmed/23718773. Accessed 2014 January 22.

38. Grad YH, Lipsitch M, Feldgarden M, Arachchi HM, Cerqueira GC, et al. (2012) Genomic epidemiology of the Escherichia coli O104:H4 outbreaks in Europe, 2011. Proc Natl Acad Sci U S A 109: 3065–3070. Available: http://www.pubmedcentral.nih.gov/articlerender.fcgi?artid=3286951&tool=pmcentrez&rendertype=abstract. Accessed 2014 January 20.

39. Ribeiro FJ, Przybylski D, Yin S, Sharpe T, Gnerre S, et al. (2012) Finished bacterial genomes from shotgun sequence data. Genome Res 22: 2270–2277. Available: http://www.pubmedcentral.nih.gov/articlerender.fcgi?artid=3483556&tool=pmcentrez&rendertype=abstract. Accessed 2014 January 23.

40. Williams LJS, Tabbaa DG, Li N, Berlin AM, Shea TP, et al. (2012) Paired-end sequencing of Fosmid libraries by Illumina. Genome Res 22: 2241–2249. Available: http://www.pubmedcentral.nih.gov/articlerender.fcgi?artid=3483553&tool=pmcentrez&rendertype=abstract. Accessed 28 January 2014.

41. Gnerre S, Maccallum I, Przybylski D, Ribeiro FJ, Burton JN, et al. (2011) High-quality draft assemblies of mammalian genomes from massively parallel sequence data. Proc Natl Acad Sci U S A 108: 1513–1518. Available: http://www.pubmedcentral.nih.gov/articlerender.fcgi?artid=3029755&tool=pmcentrez&rendertype=abstract. Accessed 21 January 2014.

42. Altschul SF, Gish W, Miller W, Myers EW, Lipman DJ (1990) Basic local alignment search tool. J Mol Biol 215: 403–410. Available: http://www.ncbi.nlm.nih.gov/pubmed/2231712. Accessed 23 January 2014.

43. Delcher AL, Phillippy A, Carlton J, Salzberg SL (2002) Fast algorithms for large-scale genome alignment and comparison. Nucleic Acids Res 30: 2478–2483. Available: http://www.pubmedcentral.nih.gov/articlerender.fcgi?artid=117189&tool=pmcentrez&rendertype=abstract.

44. Larkin M a, Blackshields G, Brown NP, Chenna R, McGettigan P a, et al. (2007) Clustal W and Clustal X version 2.0. Bioinformatics 23: 2947–2948. Available: http://www.ncbi.nlm.nih.gov/pubmed/17846036. Accessed 22 January 2014.

45. Tatusova TA, Madden TL (1999) BLAST 2 Sequences, a new tool for comparing protein and nucleotide sequences. FEMS Microbiol Lett 174: 247–250. Available: http://www.ncbi.nlm.nih.gov/pubmed/10339815. Accessed 28 January 2014.

46. Abeel T, Saeys Y, Rouzé P, Van de Peer Y (2008) ProSOM: core promoter prediction based on unsupervised clustering of DNA physical profiles. Bioinformatics 24: i24–31. Available: http://www.pubmedcentral.nih.gov/articlerender.fcgi?artid=2718650&tool=pmcentrez&rendertype=abstract. Accessed 8 July 2012.

15

Complete Chloroplast Genome Sequence of Omani Lime (*Citrus aurantiifolia*) and Comparative Analysis within the Rosids

Huei-Jiun Su[1], **Saskia A. Hogenhout**[2], **Abdullah M. Al-Sadi**[3], **Chih-Horng Kuo**[4,5,6]*

1 Institute of Ecology and Evolutionary Biology, National Taiwan University, Taipei, Taiwan, **2** Department of Cell and Developmental Biology, John Innes Centre, Norwich, United Kingdom, **3** Department of Crop Sciences, Sultan Qaboos University, Al Khoud, Oman, **4** Institute of Plant and Microbial Biology, Academia Sinica, Taipei, Taiwan, **5** Molecular and Biological Agricultural Sciences Program, Taiwan International Graduate Program, National Chung Hsing University and Academia Sinica, Taipei, Taiwan, **6** Biotechnology Center, National Chung Hsing University, Taichung, Taiwan

Abstract

The genus *Citrus* contains many economically important fruits that are grown worldwide for their high nutritional and medicinal value. Due to frequent hybridizations among species and cultivars, the exact number of natural species and the taxonomic relationships within this genus are unclear. To compare the differences between the *Citrus* chloroplast genomes and to develop useful genetic markers, we used a reference-assisted approach to assemble the complete chloroplast genome of Omani lime (*C. aurantiifolia*). The complete *C. aurantiifolia* chloroplast genome is 159,893 bp in length; the organization and gene content are similar to most of the rosids lineages characterized to date. Through comparison with the sweet orange (*C. sinensis*) chloroplast genome, we identified three intergenic regions and 94 simple sequence repeats (SSRs) that are potentially informative markers with resolution for interspecific relationships. These markers can be utilized to better understand the origin of cultivated *Citrus*. A comparison among 72 species belonging to 10 families of representative rosids lineages also provides new insights into their chloroplast genome evolution.

Editor: Tongming Yin, Nanjing Forestry University, China

Funding: Funding for this work was provided by research grants from Biotechnology and Biological Sciences Research Council (BB/J004553/1) and the Gatsby Charitable Foundation to SAH, Sultan Qaboos University (SR/AGR/CROP/13/01) to AMA, and the Institute of Plant and Microbial Biology at Academia Sinica to CHK. The funders had no role in study design, data collection and analysis, decision to publish, or preparation of the manuscript.

* Email: chk@gate.sinica.edu.tw

Introduction

Citrus is in the family of Rutaceae, which is one of the largest families in order Sapindales. Flowers and leaves of *Citrus* are usually strong scented, the extracts from which contain many useful flavonoids and other compounds that are effective insecticides, fungicides and medicinal agents [1–3]. *Citrus* is of great economic importance and contains many fruit crops such as oranges, grapefruit, lemons, limes, and tangerines. However, due to a long cultivation history, wide dispersion, somatic bud mutation, and sexual compatibility among *Citrus* species and related genera, the taxonomy of *Citrus* remains controversial [4,5] and the origination of many *Citrus* species and hybrids is still unresolved [6,7].

The chloroplast (cp) genome sequence contains useful information in plant systematics because of its maternal inheritance in most angiosperms [8,9] and its highly conserved structures for developing promising genetic markers. The only complete cp genome available in *Citrus* is sweet orange (*Citrus sinensis*) [10], which has provided valuable information to the position of Sapindales in rosids. Although a genome sequencing project is in progress for *C. clementine*, its complete chloroplast genome sequence is not available yet. To identify the cp genome regions that are polymorphic and may be used as molecular markers for resolving the evolutionary relationships among *Citrus* species, a second cp genome within the genus is necessary for comparative analysis. For this purpose, the major aim of this study is to determine the complete cp genome sequence of *C. aurantiifolia*.

C. aurantiifolia, which is commonly known as Key lime, Mexican lime, Omani lime, Indian lime, or acid lime, is native to Southeast Asia and widely cultivated in tropics and subtropics. Oman is known to be a transit country for lime, from which lime spread to Africa and the New World [11]. In Oman, Omani lime is considered the fourth most important fruit crop in terms of cultivated area and production. The products of Omani lime can be used for beverage, food additives and cosmetic industries [12]. Omani lime is sensitive to several biotic agents, the most serious of which is '*Candidatus* Phytoplasma aurantifolia', the cause of witches' broom disease of lime (WBDL). Recent studies on WBDL focused on effect of genetic diversity of Omani limes on the disease [13], transcriptome and proteomic analysis of lime response to infection by phytoplasma [14–16] and effect of phytoplasma on seed germination, growth and metabolite content in lime [17,18].

Here, we present the complete chloroplast genome sequence of Omani lime (*C. aurantiifolia*). To identify loci of potential utility for the molecular identification and phylogenetic analyses of *Citrus* cultivars and species, we compared the intergenic regions and SSRs in the cp genomes of *C. aurantiifolia* and *C. sinensis*. Furthermore, we performed phylogenetic analyses to infer the history of gene losses in the cp genome evolution among representative rosids lineages.

Materials and Methods

Sample Preparation and Sequencing

The Omani lime leaves were collected from a 5-year-old lime tree at a private farm located in the Omani territory of Madha (GPS coordinates: 25.276318, 56.318909). This farm is owned by one of the co-authors of this work, Dr. Abdullah M. Al-Sadi, whom should be contacted for future permissions. This study does not involve endangered or protected species and does not require specific permission from regulatory authority concerned with protection of wildlife. The sample was stored in a cool box and transported to the Plant Pathology Research Laboratory at Sultan Qaboos University (Al Khoud, Oman) for DNA extraction

Figure 1. Chloroplast genome map of *Citrus aurantiifolia*. Gene drawn inside the circle are transcribed clockwise, whereas those outside are counterclockwise. The within-genome GC content variation is indicated in the middle circles.

following a protocol of Maixner et al. [19]. The leaves were washed with clear water before the isolation procedure. 1 g of leaves were used and crushed in 3 ml CTAB extraction buffer (2% CTAB, 1.4 M NaCl, 500 mM EDTA pH8, 1 M Tris-HCl pH8 and 0.2% beta-mercaptol). 1.5 ml of the leave extract was transferred to a 2 ml tube and incubated in a water-bath at 65°C for 15 min. The tube was turned up and down twice during incubation, centrifuged at 960 g for 5 min, and the supernatant was subsequently transferred to a clean eppendorf tube. An equal volume of chloroform-isoamyl alcohol mix (24:1) was added and the tube was centrifuged at 21000 g for 20 min. The supernatant was transferred to a new tube and then 0.6 volume of isopropanol was added to the supernatant and incubated at −20°C for 30 min. The DNA pellet was collected by centrifugation at 21000 g for 20 min and then washed with 1 ml of 70% ethanol. The final DNA was resuspended in 100 µl TE (Tris 10 mM, EDTA 1 mM pH8) and was stored at −80°C until used.

The library construction and sequencing were done at the Genome Analysis Centre (Norwich, UK). The Illumina TruSeq DNA Sample Preparation v2 Kit was used to prepare an indexed library. The DNA sample was sheared to a fragment size of 500–600 bp using a sonicator, followed by end-repair and the addition of a single A base for binding of the indexed adapter. The appropriate sized library (500 bp) was selected by gel electrophoresis, followed by PCR enrichment. The 251 bp paired-end sequencing run was performed on an Illumina MiSeq instrument using the SBS chemistry and Illumina software MCS v2.3.0.3 and RTA v1.18.42. The raw reads were deposited at the NCBI Sequence Read Archive under the accession number SRR1611615.

Genome Assembly and Analyses

The procedures for genome assembly and annotation were based on our previous studies of cp genomes [20,21]. In addition to the standard *de novo* assembly approach by using Velvet v1.2.10 [22] with the k-mer size set to 243, a reference-based approach for assembly as described below was used in parallel. All

of the raw reads were initially mapped onto the published cp genome of *C. sinensis* [10] using BWA v0.6.2 [23]. The sequence variations were identified with SAMtools v0.1.19 [24] and visually inspected using IGV v2.3.25 [25]. The variants were corrected with the raw reads and the regions without sufficient coverage were converted into gaps. This corrected sequence was then used as the new draft reference for the next iteration of verification. Gaps were filled using the reads overhang at margins and the process was repeated until the reference was fully supported by all mapped raw reads. The final assembly, which was supported by our *de novo* and reference-based approaches, resulted in an average of 1,441-fold coverage of paired-end reads with a mapping quality of 60 and the region with the lowest coverage is 506-fold.

The preliminary annotations of the *C. aurantiifolia* cp genome were performed online using the automatic annotator DOGMA [26] and verified using BLASTN [27,28] searches (e-value cutoff = 1e-10) against other land plant cp genomes. Each annotated gene was manually compared with *C. sinensis* cp genome for start and stop codons or intron junctions to ensure accurate annotation. The codon usage was analyzed by using the seqinr R-cran package [29]. A circular map of genome was produced using OGDRAW [30].

To identify the differences between *C. aurantiifolia* and *C. sinensis*, the two sequences were aligned using Mauve v2.3.1 [31] and the result was analyzed using custom Perl scripts. Intergenic gene regions were parsed out from the two *Citrus* cp genomes and aligned using MUSCLE v3.8.31 [32] with the default settings. The pairwise distances were calculated using the DNADIST program in the PHYLIP package v3.695 [33].

The positions and types of simple sequence repeats (SSRs) in the two *Citrus* cp genomes were detected using MISA (http://pgrc. ipk-gatersleben.de/misa/). The minimum number of repeats were set to 10, 5, 4, 3, 3, and 3 for mono-, di-, tri-, tetra-, penta-, and hexanucleotides, respectively. For long repeats, the program REPuter [34] was used to identify the number and location of direct and inverted (i.e., palindromic) repeats. A minimum repeat

Table 1. Summary of the *Citrus* chloroplast genome characteristics.

Attribute	*C. aurantiifolia* (KJ865401)	*C. sinensis* (NC_008334)
Size (bp)	159,893	160,129
overall GC content (%)	38.4	38.5
LSC size in bp (% total)	87,148 (54.5%)	87,744 (54.8%)
SSC size in bp (% total)	18,763 (11.7%)	18,393 (11.5%)
IR size in bp (% total)[a]	26,991 (16.9%)	26,996 (16.9%)
Protein-coding regions size in bp (% total)	81,468 (51.0%)	79,773 (49.8%)
rRNA and tRNA size in bp (% total)	11,850 (7.5%)	11,850 (7.4%)
Introns size in bp (% total)	17,129 (10.7%)	18,252 (11.4%)
Intergenic spacer size in bp (% total)	49,446 (30.9%)	50,254 (31.4%)
Number of different genes	115	113[b]
Number of different protein-coding genes	81	79[b]
Number of different rRNA genes	4	4
Number of different tRNA genes	30	30
Number of different genes duplicated by IR	22	20
Number of different genes with introns	17	17

[a]Each cp genome contains two copies of inverted repeats (IRs).
[b]According to the original annotation, not including *orf56*.

Table 2. Differences between the *C. aurantiifolia* and *C. sinensis* cp genomes.

Indel				
	Length (bp)	Count		
	1	43		
	2–10	20		
	11–100	18		
	101–1,000	3		
Sum	1,780	116	Percentage[a]: 1.11%	
Substitution				
	Type	Count		
	A <-> T	34		
	C <-> G	15		
	A <-> C	81		
	T <-> C	64		
	A <-> G	51		
	T <-> G	85		
Sum		330	Percentage[a]: 0.21%	
10 most divergent intergenic regions				
	Region	Length[b] (bp)	Pairwise distance	
	rps3 - rpl22 (LSC)	234	0.027	
	ndhE - ndhG (SSC)	276	0.018	
	psaC - ndhE (SSC)	231	0.017	
	psbH - petB (LSC)	118	0.017	
	trnY-GUA-trnE-UCC (LSC)	59	0.017	
	trnH-GUG - psbA (LSC)	449	0.016	
	rpl32 - trnL-UAG (SSC)	1,141	0.015	
	psbT-psbN (LSC)	66	0.015	
	trnG-GCC-trnR-UCU (LSC)	204	0.015	
	trnD-GUC-trnY-GUA (LSC)	469	0.013	

[a]Relative to the length of *C. aurantiifolia*.
[b]Length in *C. aurantiifolia*.

size of 30 bp and sequence identity greater than 90% setting were used according to the study of *C. sinensis* cp genome [10]. The redundant or overlapping repeats were identified and filtered manually.

Phylogenetic Inference

Phylogenetic analysis of the representative rosids lineages with complete cp genomes available was performed using PhyML v20120412 [35] with the GTR+I+G model. A total of 72 rosids species were chosen as the ingroups and *Vitis venifera* was included as the outgroup, the accession numbers were provided in Table S1. The protein-coding and rRNA genes were parsed from the selected cp genomes and clustered into ortholog groups using OrthoMCL [36]. The presence/absence of orthologous genes in each genome was examined and further verified using TBLASTN [27,28] searches (e-value cutoff = 1e-10). The nucleotide sequences of the conserved genes were aligned individually by using MUSCLE with the default settings. The concatenated alignment was used to infer a maximum likelihood phylogeny as described above. The bootstrap supports were estimated from 1,000 resampled alignments generated by the SEQBOOT program in the PHYLIP package.

Investigations of *orf56* and *ycf68*

To investigate the presence/absence of *orf56* and *ycf68* in the selected cp genomes, the gene sequences from *C. aurantiifolia* was used as the queries to perform BLASTN [27,28] searches (e-value cutoff = 1e-10). The significant hits were examined to investigate the presence of intact open reading frames (ORFs). Phylogenetic analysis of the cp *orf56* genes and the homologous mitochondrial sequences was performed as described above. The final alignment contains 190 aligned nucleotide sites and a total of 70 sequences, including two sequences of *Amborella* as the outgroup.

Results and Discussion

General Features of the Omani Lime Chloroplast Genome

The complete cp genome of *C. aurantiifolia* (Christm.) Swingle (GenBank accession number KJ865401.1) is 159,893 bp in length, including a large single copy (LSC) region of 87,148 bp, a small single copy (SSC) region of 18,763 bp, and a pair of inverted repeats (IRa and IRb) of 26,991 bp each (Figure 1 and Table 1). A total of 137 different genes, including 93 protein-coding genes, 30 tRNA genes, and four rRNA genes, were annotated (Table S2). Among these, 12 protein-coding genes and 7 tRNA genes are duplicated in the IR regions. Most of the protein-coding genes are

Table 3. List of simple sequence repeats.

Repeat unit	Length (bp)	Number of SSRs	Start position[a]
A	10	6	4512; 47812; 53871; **72614**; 121748; **159288**
	11	6	**6866**; 10130; 69481; 71892; 117725; **134802**
	12	9	**8332**; 31399; 47307; 63928; **111804**; **113977** (*ycf1*); 118367; 140302 (*ycf68*); 144255
	13	2	10107; 84557
	14	1	**385**
	15	1	32360
	16	2	69965; 118302
	17	3	7620; 39139; 74176
	19	1	**12023**
	22	1	70289
T	10	10	**2424** (*matK*); 19786; **26964** (*rpoB*); **37622**; 46938; **63632**; **87731**; 117742; 117871; 118851
	11	11	9401; 10416; **17001**; 30912; 46021; 63530; 112216; 117988; 118224; **121703**; **131189** (*ycf1*)
	12	6	**14722**; 29024; 102773; 106715 (*ycf68*); **133040** (*ycf1*); 135213
	13	2	**73946**; 80423
	14	2	1776; 85274
	15	2	54209; 57817
	17	1	45965
	18	3	52748; 68339; 81409
	20	1	49202
	23	2	23694; **33282**
C	10	2	**28769; 104247**
G	10	1	**142772**
AT	10	4	**20631** (*rpoC2*); **33636**; 11817; **121517** (*ndhD*)
AAG	12	1	**97331**
AAT	12	2	**38604; 122629**
ATA	12	1	70220
ATT	12	3	**10283; 53810; 54088**
	18	1	**1760;**
CTT	12	2	**37353** (*psbC*); **149686**
TAA	12	2	**30250**; 61945
TAT	12	1	**83297**
TTC	12	1	**73084**
TAAA	12	2	4866; **45088**
AAAT	12	3	**30423; 32502; 71394**
ATAC	12	1	**51167**
ATTT	12	1	49193
	20	1	**117168**
TTAA	12	2	39175; 39188
TTAG	12	1	**61483**
TTTC	12	1	**14352**
TCTT	12	1	**46961**
AATAA	20	1	144226
TTTTA	20	1	102781
TTCAAA	18	1	63817

[a]The SSR-containing coding regions are indicated in parentheses. SSRs that are identical in the *C. sinensis* chloroplast genome are highlighted in bold; SSRs that are conserved but with different lengths are highlighted by underline.

Table 4. List of long repeat sequences.

Repeat size	Type[a]	Start position of 1st repeat	Start position the repeat found in other region	Location[b]	Region
30	D	1759	1762	IGS (*psbA-trnK-UUU*)	LSC
30	P	1771	12015	IGS (*psbA-trnK-UUU, atpA-atpF*)	LSC
30	P	8231	37726, 47606	**IGS (*trnS-GCU, trnS-UGA, trnS-GGA*),**	LSC
30	D	23226	85067	**intron (*rpoC1*), IGS (*rpl16-rps3*)**	LSC
30	D	23686	52733	intron (*rpoC1*), IGS (*ndhK-ndhC*)	LSC
30	P	23687	70291	intron (*rpoC1*), IGS (*trnP-UGG-psaJ*)	LSC
30	D	23692	33280	intron (*rpoC1*), IGS (*trnE-UUC-trnT-GGU*)	LSC
30	D	49192	117171	IGS (*psbA-trnK-UUU*), IGS (*atpA-atpF*)	LSC, IR
30	D, P	49197	102764, 144233	IGS (*trnT-UGU-trnL-UAA*), IGS (*rps12-trnV-GAC*)	LSC, IR
30	D, P	51215	102768, 144229	IGS (*trnF-GAA-ndhJ*), IGS (*rps12-trnV-GAC*)	LSC, IR
30	P	71344	71344	**IGS (*rpl33-rps18*)**	LSC
30	D, P	102768	102773, 144224	IGS (*rps12-trnV-GAC*)	IR
30	D	144225	144230	IGS (*trnV-GAC-rps12*)	IR
31	P	4492	117868	IGS (*trnK-UUU-rps16*), IGS (*rpl32-trnL-UAG*)	LSC, IR
31	P	10106	49188	IGS (*trnG-GCC-trnR-UCU, trnT-UGU-trnL-UAA*)	LSC
31	P	29811	29811	**IGS (*petN-psbM*)**	LSC
31	P	33281	70282	**IGS (*trnE-UUC-trnT-GGU, trnP-UGG-psaJ*)**	LSC
31	P	119977	119977	intron (*ccsA*)	IR
32	D	7615	74171	IGS (*psbK-psbI*), intron (*clpP*)	LSC
32	P	39166	39166	IGS (*trnG-GCC-trnfM-CAU*)	LSC
34	P	38774	38782	**IGS (*psbZ-trnG-GCC*)**	LSC
34	P	49186	70288	rps4, IGS (*trnP-UGG-psaJ*)	LSC
34	D, P	111432	111464, 135529, 135561	**IGS (*rrn4.5-rrn5*)**	IR
35	P	10097	49193	IGS *trnG-GCC-trnR-UCU, trnT-UGU-trnL-UAA*	LSC
36	P	27648	27648	IGS (*rpoB-trnC-GCA*)	LSC
40	P	77776	77776	IGS (*psbT-psbN*)	LSC
41	D	41294	43518	***psaB, psaA***	LSC
41	D, P	102353	124945, 144633	**IGS (*rps12-trnV-GAC*), intron (*ndhA*)**	IR
48	P	30626	30626	**IGS (*petN-psbM*)**	LSC
50	D	39020	39044	**IGS (*trnG-GCC-trnfM-CAU*)**	LSC
51	P	9984	9984	IGS (*trnG-GCC-trnR-UCU*)	LSC
53	P	8869	31095	**IGS (*trnS-GCU-trnG-GCC, psbM-trnD-GUC*)**	LSC
54	P	441	441	IGS (*trnH-GUG-psbA*)	LSC

[a]D: direct repeat; P: palindrome inverted repeat.
[b]IGS: intergenic spacer region. Sequences conserved in the *C. sinensis* chloroplast genome are highlighted in bold.

composed of a single exon, while 14 contain one intron and three contain two introns. The gene *rps12* was predicted to undergo trans-splicing, with the 5′ exon located in the LSC region and the other two exons located in the IR regions.

The protein-coding regions contain a total of 27,159 codons (Table S3). Isoleucine and cysteine are the most and least frequent amino acids and have 2,892 (10.7%) and 359 (1.2%) codons, respectively. The codon usage is biased towards a high ratio of A/T at the third position, which is also observed in many land plant cp genomes [37].

Sequence Comparisons with Sweet Orange

The general characteristics of the two *Citrus* cp genomes are summarized in Table 1, overall the compositions are quite similar. The GC content of these *Citrus* cp genomes is approximately 38.5%, which is slightly higher than the average of the 72 representative rosids lineages (36.7%). In these two *Citrus* cp genomes, the genic regions, introns, and intergenic regions account for ca. 58%, 11%, and 31%, respectively (Table 1).

The pairwise sequence alignment between the two *Citrus* cp genomes revealed approximately 1.3% sequence divergence (Table 2), including 1,780 indels (1.11%) and 330 substitutions (0.21%). The LSC region contains more sequence polymorphisms than expected by its size, including 1,360 (76.4%) indels and 235 (71.2%) substitutions. In contrast, the two IR regions account for ca. 34% of the cp genome yet contain only 16 (0.9%) indels and 12 (3.6%) substitutions. The size differences in the LSC and SSC regions between these two cp genomes are mostly explained by one large indel in each region. The LSC sizes differ by 596 bp and a 523-bp indel was found in the spacer between *rps16* and *trnQ-*

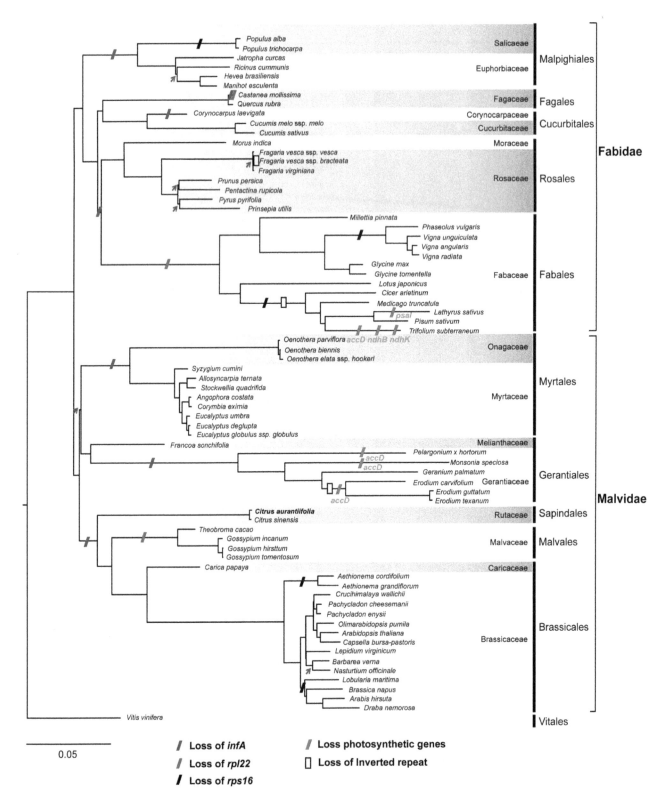

Figure 2. Maximum likelihood phylogeny of the representative rosids lineages. The common grape vine (*Vitis vinifera*) is included as the outgroup to root the tree. The concatenated alignment includes 62 conserved chloroplast genome genes and 54,689 aligned nucleotide sites. Nodes received <70% bootstrap support are indicated by gray arrows. The putative events of gene losses are inferred based on the most parsimonious scenario.

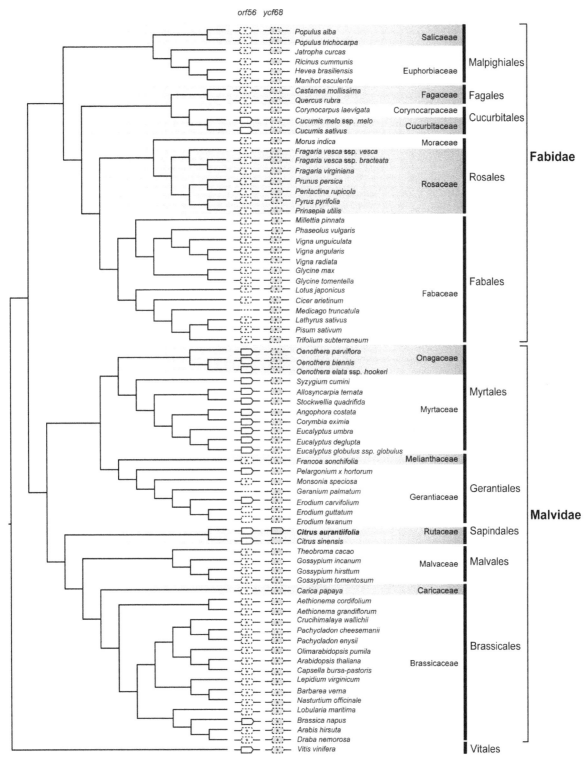

Figure 3. The phylogenetic distribution patterns of *orf56* and *ycf68*.

UUG. The SSC sizes differ by 370 bp and a 354-bp indel was found in the spacer between *rpl32* and *trnL-UAA*.

To identify the intergenic regions that may be useful for phylogenic analysis or molecular identification, we searched for the spacers that are >400 bp in length and exhibit above-average sequence divergence between the two *Citrus* species (i.e., >1.3%). A total of three regions satisfied these criteria, including the spacer between *trnH-GUG* and *psbA* (449 bp, 1.6% divergence), the spacer between *rpl32* and *trnL-UAG* (1141 bp, 1.5% divergence), and the spacer between *trnD-GUC* and *trnY-GUA* (469 bp, 1.3% divergence).

The junctions between the IR, LSC, and SSC regions in *C. aurantiifolia* are similar to that of *C. sinensis* except for the LSC-IRb boundary. A total of 23 indels and five substitutions were found at this region, resulting in one copy of *rpl22* spanning across the LSC-IRb junction in *C. aurantiifolia*. Comparing the IR junctions of *Citrus* with *Theobroma* and *Gossypium* in Malvaceae [38], it was found that the IRs in *Citrus* have expanded to include *rps19* and 252 nt of *rpl22*, whereas in Malvaceae, *rps19* is located in LSC and *rpl22* was missing [38–40].

Analyses of Repetitive Sequences

A total of 109 SSR loci were found in the cp genome of *C. aurantiifoliaa*, accounting for 1,352 bp of the total sequence (ca. 0.8%). Among these, 94 were also found in *C. sinensis* and 42 exhibit length polymorphism (Table 3). Most SSRs are located in intergenic regions, but some were found in coding genes such as *matK* and *ycf1*. Concerning the controversial status of *Citrus* taxonomy, the SSRs identified in this study may provide new perspective to refine the phylogeny and elucidate the origin of the cultivars. Furthermore, these SSRs may be used as molecular markers for population studies.

In addition, 62 large repeats that are longer than 30 bp were found in the *C. aurantiifolia* cp genome (Table 4). Most of these repeats are located in intergenic spacers, except for three that are located in the coding regions of *rps4*, *psaA* and *psaB*. Twelve of these long repeats were also found in *C. sinensis*, indicating that these repeats might be widespread in the genus.

Gene Content Analyses within the Rosids

A maximum likelihood phylogenetic analysis of 72 representative rosids lineages was conducted based on a concatenated alignment of four rRNA and 58 protein-coding genes with 54,689 sites (Figure 2). *Citrus* represents Sapindales and is sister to the clade containing Malvales and Brassicales. These relationships are congruent with the previous reports [10,41–43]. Based on this phylogeny and the gene content, we inferred the gene loss events during the cp genome evolution in rosids.

The translation initiation factor gene *infA* in cp has been lost independently at least 24 times in angiosperms and evidence provided from some cases suggested functional replacement by a nucleus copy [44]. Although the majority of *infA* in our selected cp genomes were found to be pseudogenized or completely lost, an intact *infA* was found in *Quercus*, *Francoa*, and two *Cuscumis* species.

The *rpl22* were found to be lost in Fabaceae [45] and *Castanea* of Fagaceae [46] following independent transfers to nucleus. Furthermore, another putative loss of *rpl22* was detected in *Passiflora* [46]. The *rpl22* in Malvaceae, including *Theobroma* and three *Gossypium* species, were found to be pseudogenized in our analysis. In *Citrus*, the ORF of *rpl22* was shortened to 252–264 nt compared to the typical length of 399–489 nt in other rosids [10,46]. However, compared with the pseudogenized *rpl22* found in Malvalvace, the *rpl22* homologs in *Citrus* still show high

sequence conservation. Additionally, the *rpl22* transcripts can be identified in the EST database for various *Citrus* species (data not shown). Taking account into the above consideration, we did not annotate *rpl22* as a pseudogene in *Citrus*.

The parallel losses of *rps16* were found in several rosids lineages (Figure 2), including one time in Salicaceae, two times in Fabaceae and another two times in Brassicaceae. The loss of *rps16* in *Medicago* and *Populus* was found to be substituted by a nuclear-encoded copy that transferred from the mitochondrion (mt) [47]. Because the nuclear-encoded RPS16 was found to target both mt and cp in *Arabidopsis*, *Lycopersicon*, and *Oryza* [47], it is possible that the cp genome-encoded *rps16* would not be maintained by selection and will eventually become lost in these lineages.

There are only a few gene loss events of photosynthetic genes found in rosids. In addition to the loss of *psaI* in *Lathyrus sativus* [48], the *accD* seems to be lost independently in *Trifolium subterraneum* and several Gerantiaceae species except for *Geranium palmatum*. In *Trifolium*, a nuclear-encoded *accD* copy has been reported [48], which presented another example of horizontal gene transfer from cp to nucleus. Successful gene transfers from cp to the nucleus in angiosperms are rare and have been only documented for four genes in rosids. Other than the three genes described above (i.e., *infA*, *rpl22*, and *accD*), the *rpl32* in *Populus* (Salicaceae) is the fourth example [49–51].

The IR has been reported to be independently lost at least five times among seed plants, two of which are within rosids [51]. In addition to the inverted repeat lacking clade (IRLC) of papilionoid Fabaceae [52] and *Erodium* of Gerantiaceae [53,54], the IR was found to be lost in two lineages of *Fragaria* (Rosaceae), which are *F. vesca* ssp. *bracteatea* and *F. mandschurica* (accession: NC_018767, not shown in Figure 2). Based on the *Fragaria* phylogeny shown in a previous study [55], it seems that IR loss was not a single event in *Fragaria*.

Molecular Evolution of *orf56* and *ycf68* within the Rosids

In the comparison of gene content between the two *Citrus* cp genomes, *C. aurantiifolia* was found to contain two additional protein-coding genes. The first gene, *orf56*, is located in the *trnA-UGC* intron that contains one sequence homologous to previously recognized mitochondrial *ACRS* (ACR-toxin sensitivity gene) in *Citrus* [56]. In addition to the 171-bp identical sequences between cp *orf56* and the ORF sequences of *ACRS* in mt, the full length of 355-bp region of *ACRS* that conferred sensitivity to ACR-toxin in *E. coil* are also identical. Furthermore, the whole *trnA-UGC* among two *Citrus* cp regions and *C. jambhiri* mitochondrial *ACRS* shared more than 96% identity (Figure S1), which highlight the conservation of this region between cp and mt.

The gene *orf56* has also been included in the annotation of complete cp genomes of *Calycanthus* [57] and *Pelargonium* [58]. Our BLAST search against the rosids genome database revealed that in addition to *Citrus* and *Pelargonium*, all of the species examined in Cucurbitaceae and Myrtales also contain an intact *orf56* (Figure 3). Moreover, an intact *ACRS* ORF is also present in the mt genomes of *Liriodendron* [59] and *Silene* [60] and the ORF sequences between cp and mt are identical. Goremykin et al. [57] suggested that the *ACRS* gene was relative recently transferred from cp to mt. Based on the phylogeny containing the cp *orf56* and the mt *ACRS* (Figure S2), it appears that *orf56* has been independently transferred from cp to mt in different lineages.

The second gene, *ycf68*, is located in the *trnI-GAU* intron. A nearly identical sequence was found in *C. sinensis* but an additional T insertion near the C-terminus abolished the stop codon at the corresponding position. The intact *ycf68* can be detected in several monocots and Nymphaeaceae [61,62]. However, in the majority of other rosids (Figure 3) and the rest of the eudicots [61], the *ycf68* homologs all contain premature stop codons. Although Raubeson et al. [61] argued that *ycf68* is not a protein-coding gene based on the lack of intron-folding pattern, the high levels of sequence conservation among the ORFs of identified homologs suggest that the true identity and functionality of this putative gene remains to be further investigated.

Conclusions

We reported the complete cp genome sequence of *Citrus aurantiifolia* (Rutaceae) in this study. The genome organization and gene content is typical of most angiosperms and highly similar to that of *C. sinensis* (i.e., 98.7% identical at the nucleotide level). The only difference in the gene content between the two *Citrus* cp genomes is the *C. aurantiifolia*-specific presence of a protein-coding gene (*ycf68*) in the *trnI-GAU* intron. Notably, three long intergenic spacers with high sequence divergence and 94 shared SSR regions were identified in the *C. aurantiifolia-C. sinensis* comparison. These regions may provide phylogenetic utility at low taxonomic levels and could be applied to the molecular identification of *Citrus* cultivars. Finally, our comparative analysis of gene content among 72 representative rosids lineages highlighted multiple events of gene losses within this group.

Supporting Information

Figure S1   Alignment of the *orf56*-containing sequences of two *Citrus* cp genomes and *C. jambhiri* mitochondrial *ACRS* sequences.

Figure S2   The maximum likelihood phylogeny of the cp *orf56* and mt *ACRS* ORF sequences.

Table S1   List of the complete chloroplast genome sequences included in the phylogenetic analysis.

Table S2   List of the genes found in the *C. aurantiifolia* cp genome.

Table S3   Codon usage of the *C. aurantiifolia* cp genome.

Acknowledgments

We thank Sam T. Mugford and Allyson M. MacLean for help with purification and quality controls of DNA samples.

Author Contributions

Conceived and designed the experiments: HJS CHK. Performed the experiments: SAH AMA. Analyzed the data: HJS CHK. Contributed reagents/materials/analysis tools: HJS AMA CHK. Contributed to the writing of the manuscript: HJS SAH AMA CHK.

References

1. Mabberley DJ (2004) *Citrus* (Rutaceae): a review of recent advances in etymology, systematics and medical applications. Blumea 49: 481–198.
2. Tripoli E, Guardia ML, Giammanco S, Majo DD, Giammanco M (2007) *Citrus* flavonoids: molecular structure, biological activity and nutritional properties: a review. Food Chem 104: 466–479.
3. Ezeabara CA, Okeke CU, Aziagba BO, Ilodibia CV, Emeka AN (2014) Determination of saponin content of various parts of six *Citrus* species. Int Res J Pure Appl Chem 4: 137–143.
4. Nicolosi E, Deng ZN, Gentile A, La Malfa S, Continella G, et al. (2000) *Citrus* phylogeny and genetic origin of important species as investigated by molecular markers. Theor Appl Genet 100: 1155–66.
5. Hynniewta M, Malik SK, Rao SR (2014) Genetic diversity and phylogenetic analysis of Citrus (L) from north-east India as revealed by meiosis, and molecular analysis of internal transcribed spacer region of rDNA. Meta Gene 2: 237–251.
6. Liu Y, Heying E, Tanumihardjo SA (2012) History, global distribution, and nutritional importance of citrus fruits. Compr Rev Food Sci Food Saf 11: 530–545.
7. Penjor T, Yamamoto M, Uehara M, Ide M, Matsumoto N, et al. (2013) Phylogenetic relationships of *Citrus* and its relatives based on *matK* gene sequences. PLoS ONE 8: e62574.
8. Corriveau JL, Coleman AW (1988) Rapid screening method to detect potential biparental inheritance of plastid DNA and results for over 200 angiosperm species. Am J Bot 75: 1443–1458.
9. Zhang Q, Liu Y, Sodmergen (2003) Examination of the cytoplasmic DNA in male reproductive cells to determine the potential for cytoplasmic inheritance in 295 angiosperm species. Plant Cell Physiol 44: 941–951.
10. Bausher MG, Singh ND, Lee SB, Jansen RK, Daniell H (2006) The complete chloroplast genome sequence of *Citrus sinensis* (L.) Osbeck var 'Ridge Pineapple': organization and phylogenetic relationships to other angiosperms. BMC Plant Biol 6: 21.
11. Davies FS, Albrigo LG (1994). Citrus. CABI International, Wiltshire, UK. 1–2.
12. Vand SH, Abdullah TL (2012) Identification and introduction of Thornless Lime (*Citrus aurantifolia*) in Hormozgan, Iran. Indian J Sci Technol 5: 3670–3673.
13. Al-Sadi AM, Al-Moqbali HS, Al-Yahyai RA, Al-Said FA (2012) AFLP data suggest a potential role for the low genetic diversity of acid lime (*Citrus aurantifolia* Swingle) in Oman in the outbreak of witches' broom disease of lime. Euphytica 188: 285–297.
14. Taheri F, Nematzadeh G, Zamharir MG, Nekouei MK, Naghavi M, et al. (2011) Proteomic analysis of the Mexican lime tree response to "*Candidatus* Phytoplasma aurantifolia" infection. Mol Biosyst 7: 3028–3035.
15. Zamharir MG, Mardi M, Alavi SM, Hasanzadeh N, Nekouei MK, et al. (2011) Identification of genes differentially expressed during interaction of Mexican lime tree infected with "*Candidatus* Phytoplasma aurantifolia". BMC Microbiol 11: 1.
16. Monavarfeshani A, Mirzaei M, Sarhadi E, Amirkhani A, Khayam Nekouei M, et al. (2013) Shotgun proteomic analysis of the Mexican lime tree infected with "*Candidatus* Phytoplasma aurantifolia." J Proteome Res 12: 785–795.
17. Faghihi MM, Bagheri AN, Bahrami HR, Hasanzadeh H, Rezazadeh R, et al. (2011) Witches'-broom disease of lime affects seed germination and seedling growth but is not seed transmissible. Plant Disease 95: 419–422.
18. Zafari S, Niknam V, Musetti R, Noorbakhsh SN (2012) Effect of phytoplasma infection on metabolite content and antioxidant enzyme activity in lime (*Citrus aurantifolia*). Acta Physiol Plant 34: 561–568.
19. Maixner M, Ahrens U, Seemüller E (1995) Detection of the German grapevine yellows (Vergilbungskrankheit) MLO in grapevine, alternative hosts and a vector by a specific PCR procedure. Eur J Plant Pathol 101: 241–250.
20. Ku C, Hu J-M, Kuo C-H (2013) Complete plastid genome sequence of the basal asterid *Ardisia polysticta* Miq. and comparative analyses of asterid plastid genomes. PLoS ONE 8: e62548.
21. Ku C, Chung W-C, Chen L-L, Kuo C-H (2013) The complete plastid genome sequence of Madagascar periwinkle *Catharanthus roseus* (L.) G. Don: plastid genome evolution, molecular marker identification, and phylogenetic implications in asterids. PLoS ONE 8: e68518.
22. Zerbino DR, Birney E (2008) Velvet: algorithms for *de novo* short read assembly using de Bruijn graphs. Genome Res 18: 821–829.
23. Li H, Durbin R (2009) Fast and accurate short read alignment with Burrows-Wheeler transform. Bioinformatics 25: 1754–1760.
24. Li H, Handsaker B, Wysoker A, Fennell T, Ruan J, et al. (2009) The Sequence Alignment/Map format and SAMtools. Bioinformatics 25: 2078–2079.
25. Robinson JT, Thorvaldsdottir H, Winckler W, Guttman M, Lander ES, et al. (2011) Integrative genomics viewer. Nat Biotechnol 29: 24–26.
26. Wyman SK, Jansen RK, Boore JL (2004) Automatic annotation of organellar genomes with DOGMA. Bioinformatics 20: 3252–3255.
27. Altschul SF, Gish W, Miller W, Myers EW, Lipman DJ (1990) Basic local alignment search tool. J Mol Biol 215: 403–410.
28. Camacho C, Coulouris G, Avagyan V, Ma N, Papadopoulos J, et al. (2009) BLAST+: architecture and applications. BMC Bioinformatics 10: 421.
29. Charif D, Lobry JR (2007) SeqinR 1.0–2: a contributed package to the R project for statistical computing devoted to biological sequences retrieval and analysis. In: Bastolla DU, Porto PDM, Roman DHE, Vendruscolo DM, editors.

Structural Approaches to Sequence Evolution. Biological and Medical Physics, Biomedical Engineering. Springer Berlin Heidelberg. 207–232.

30. Lohse M, Drechsel O, Bock R (2007) OrganellarGenomeDRAW (OGDRAW): a tool for the easy generation of high-quality custom graphical maps of plastid and mitochondrial genomes. Curr Genet 52: 267–274.

31. Darling ACE, Mau B, Blattner FR, Perna NT (2004) Mauve: multiple alignment of conserved genomic sequence with rearrangements. Genome Res 14: 1394–1403.

32. Edgar RC (2004) MUSCLE: multiple sequence alignment with high accuracy and high throughput. Nucl Acids Res 32: 1792–1797.

33. Felsenstein J (1989) PHYLIP - Phylogeny Inference Package (version 3.2). Cladistics 5: 164–166.

34. Kurtz S, Schleiermacher C (1999) REPuter: fast computation of maximal repeats in complete genomes. Bioinformatics 15: 426–427.

35. Guindon S, Gascuel O (2003) A simple, fast, and accurate algorithm to estimate large phylogenies by maximum likelihood. Syst Biol 52: 696–704.

36. Li L, Stoeckert CJ, Roos DS (2003) OrthoMCL: Identification of ortholog groups for eukaryotic genomes. Genome Res 13: 2178–2189.

37. Clegg MT, Gaut BS, Learn GH, Morton BR (1994) Rates and patterns of chloroplast DNA evolution. Proc Natl Acad Sci U S A 91: 6795–6801.

38. Kane N, Sveinsson S, Dempewolf H, Yang JY, Zhang D, et al. (2012) Ultra-barcoding in cacao (*Theobroma* spp.; Malvaceae) using whole chloroplast genomes and nuclear ribosomal DNA. Am J Bot 99: 320–329.

39. Lee SB, Kaittanis C, Jansen RK, Hostetler JB, Tallon LJ, et al. (2006) The complete chloroplast genome sequence of *Gossypium hirsutum*: organization and phylogenetic relationships to other angiosperms. BMC Genomics 7: 61.

40. Xu Q, Xiong G, Li P, He F, Huang Y, et al. (2012) Analysis of complete nucleotide sequences of 12 *Gossypium* chloroplast genomes: origin and evolution of allotetraploids. PLoS ONE 7: e37128.

41. Bremer B, Bremer K, Chase MW, Fay MF, Reveal JL, et al. (2009) An update of the Angiosperm Phylogeny Group classification for the orders and families of flowering plants: APG III. Bot J Linn Soc 161: 105–121.

42. Worberg A, Alford MH, Quandt D, Borsch T (2009) Huerteales sister to Brassicales plus Malvales, and newly circumscribed to include *Dipentodon*, *Gerrardina*, *Huertea*, *Perrottetia*, and *Tapiscia*. Taxon 58: 468–478.

43. Ruhfel BR, Gitzendanner MA, Soltis PS, Soltis DE, Burleigh JG (2014) From algae to angiosperms-inferring the phylogeny of green plants (*Viridiplantae*) from 360 plastid genomes. BMC Evol Biol 14: 23.

44. Millen RS, Olmstead RG, Adams KL, Palmer JD, Lao NT, et al. (2001) Many parallel losses of *infA* from chloroplast DNA during angiosperm evolution with multiple independent transfers to the nucleus. Plant Cell 13: 645–658.

45. Gantt JS, Baldauf SL, Calie PJ, Weeden NF, Palmer JD (1991) Transfer of *rpl22* to the nucleus greatly preceded its loss from the chloroplast and involved the gain of an intron. EMBO J 10: 3073–3078.

46. Jansen RK, Saski C, Lee S-B, Hansen AK, Daniell H (2011) Complete plastid genome sequences of three rosids (*Castanea*, *Prunus*, *Theobroma*): evidence for at least two independent transfers of *rpl22* to the nucleus. Mol Biol Evol 28: 835–847.

47. Ueda M, Nishikawa T, Fujimoto M, Takanashi H, Arimura S, et al. (2008) Substitution of the gene for chloroplast RPS16 was assisted by generation of a dual targeting signal. Mol Biol Evol 25: 1566–1575.

48. Magee AM, Aspinall S, Rice DW, Cusack BP, Sémon M, et al. (2010) Localized hypermutation and associated gene losses in legume chloroplast genomes. Genome Res 20: 1700–1710.

49. Cusack BP, Wolfe KH (2007) When gene marriages don't work: divorce by subfunctionalization. Trends Genet 23: 270–272.

50. Ueda M, Fujimoto M, Arimura S, Murata J, Tsutsumi N, et al. (2007) Loss of the *rpl32* gene from the chloroplast genome and subsequent acquisition of a preexisting transit peptide within the nuclear gene in *Populus*. Gene 402: 51–56.

51. Jansen RK, Ruhlman TA (2012) Plastid genomes of seed plants. In: Bock R, Knoop V, editors. Genomics of chloroplasts and mitochondria. Advances in photosynthesis and respiration. Springer Netherlands. 103–126.

52. Wojciechowski MF, Lavin M, Sanderson MJ (2004) A phylogeny of legumes (Leguminosae) based on analysis of the plastid *matK* gene resolves many well-supported subclades within the family. Am J Bot 91: 1846–1862.

53. Blazier JC, Guisinger MM, Jansen RK (2011) Recent loss of plastid-encoded *ndh* genes within *Erodium* (Geraniaceae). Plant Mol Biol 76: 263–272.

54. Guisinger MM, Kuehl JV, Boore JL, Jansen RK (2011) Extreme reconfiguration of plastid genomes in the angiosperm family Geraniaceae: rearrangements, repeats, and codon usage. Mol Biol Evol 28: 583–600.

55. Njuguna W, Liston A, Cronn R, Ashman TL, Bassil N (2013) Insights into phylogeny, sex function and age of *Fragaria* based on whole chloroplast genome sequencing. Mol Phylogenet Evol 66: 17–29.

56. Ohtani K, Yamamoto H, Akimitsu K (2002) Sensitivity to *Alternaria alternata* toxin in citrus because of altered mitochondrial RNA processing. Proc Natl Acad Sci U S A 99: 2439–2444.

57. Goremykin V, Hirsch-Ernst KI, Wölfl S, Hellwig FH (2003) The chloroplast genome of the "basal" angiosperm *Calycanthus fertilis* – structural and phylogenetic analyses. Plant Syst Evol 242: 119–135.

58. Chumley TW, Palmer JD, Mower JP, Fourcade HM, Calie PJ, et al. (2006) The complete chloroplast genome sequence of *Pelargonium* × *hortorum*: organization and evolution of the largest and most highly rearranged chloroplast genome of land plants. Mol Biol Evol 23: 2175–2190.

59. Richardson AO, Rice DW, Young GJ, Alverson AJ, Palmer JD (2013) The "fossilized" mitochondrial genome of *Liriodendron tulipifera*: ancestral gene content and order, ancestral editing sites, and extraordinarily low mutation rate. BMC Biol 11: 29.

60. Sloan DB, Müller K, McCauley DE, Taylor DR, Šorchová H (2012) Intraspecific variation in mitochondrial genome sequence, structure, and gene content in *Silene vulgaris*, an angiosperm with pervasive cytoplasmic male sterility. New Phytol 196: 1228–1239.

61. Raubeson LA, Peery R, Chumley TW, Dziubek C, Fourcade HM, et al. (2007) Comparative chloroplast genomics: analyses including new sequences from the angiosperms *Nuphar advena* and *Ranunculus macranthus*. BMC Genomics 8: 174.

62. Ahmed I, Biggs PJ, Matthews PJ, Collins LJ, Hendy MD, et al. (2012) Mutational dynamics of aroid chloroplast genomes. Genome Biol Evol 4: 1316–1323.

16

IL28B Polymorphism Cannot Predict Response to Interferon Alpha Treatment in Patients with Melanoma

Martin Probst[1], Christoph Hoeller[2], Peter Ferenci[3], Albert F. Staettermayer[3], Sandra Beinhardt[3], Hubert Pehamberger[2], Harald Kittler[2]*, Katharina Grabmeier-Pfistershammer[1]*

1 Division of Immunology, Allergy and Infectious Diseases, Department of Dermatology, Medical University Vienna, Vienna, Austria, 2 Division of General Dermatology, Department of Dermatology, Medical University Vienna, Vienna, Austria, 3 Division of Gastroenterology and Hepatology, Department of Internal Medicine III, Medical University Vienna, Vienna, Austria

Abstract

Background: Recent genome-wide association studies revealed the rs12979860 single nucleotide polymorphism (SNP) of the IL28B gene (CC genotype) to be the strongest pre-therapeutic predictor of therapy response to interferon alpha in patients with chronic hepatitis C infection. The favorable CC genotype is associated with significantly higher rates of sustained virologic response. No data exist on the role of IL28B polymorphism in interferon therapy of diseases other than viral hepatitis.

Methods: A retrospective study involving 106 patients with melanoma who received low- or high-dose interferon therapy was performed. The CC and non-CC genotype of IL28B rs12979860 SNP were correlated with progression-free and overall survival.

Results: 44 (41.5%) patients were CC and 62 (58.5%) non-CC. There was no statistically significant difference in age at diagnosis, melanoma type or localization, Breslow level or AJCC stage between CC and non-CC patients. During the observation period (6.43±4.66 years) disease progression occurred in 36 (34%) patients after 5.5±4.3 years. 43.2% (19) of patients with CC and 27.4% (17) of patients with non-CC genotype were affected (p = 0.091). Disease progression was more frequent in patients on high dose interferon therapy and with a worse AJCC stage.

Conclusion: In contrast to classical risk factors like tumor thickness and clinical stage, IL28B polymorphism was not associated with progression-free or overall survival in patients with melanoma treated with interferon alpha.

Editor: Andrzej T. Slominski, University of Tennessee, United States of America

Funding: The authors have no support or funding to report.

Competing Interests: The authors have declared that no competing interests exist.

* Email: harald.kittler@meduniwien.ac.at (HK); katharina.pfistershammer@meduniwien.ac.at (KGP)

Introduction

It is well established that melanoma represents a highly immunogenic tumor. Consequently various immune-modulatory approaches have been applied to cure melanoma or to delay disease progression. Immunotherapies aim at enhancing the immune response to malignant cells by increasing their immunogenicity or suppressing inhibitory pathways. Immunostimulatory mechanisms include interferon alpha treatment, interleukin-2 therapy, vaccination approaches and more recently blocking of inhibitory pathways with monoclonal antibodies directed against CTLA-4, PD-1 or PD-L1 [1].

Class I interferons enhance MHC and TAP expression, leading to improved antigen presentation in tumor cells [2,3]. Furthermore, they modulate STAT1/STAT3 relation and likely have an impact on signal transduction in T cells [3,4]. Interferon treatment has been associated with an increase in tumor infiltrating T cells and a reduction of regulatory T cells [5,6]. Thus interferon is broadly used either in a low-dose protocol in stage IB/II melanoma or in a high-dose protocol for stage III melanoma. In several independent studies both regimens have been shown to improve progression –free survival but not overall survival [7,8].

Interferon alpha treatment as adjuvant therapy of melanoma is a long-term therapy of 18 months and is often associated with severe side effects like liver, heart and bone marrow toxicity, fever and mood depression. This impairment of quality of life would make a predictive marker for the likeliness of therapy response very valuable and would also prevent unnecessary therapy costs.

Recently, a predictive marker has been identified for the probability of sustained virologic response after interferon alpha/ribavirin therapy in patients with chronic hepatitis C infection by genome wide association studies. A single nucleotide polymorphism (C/T dimorphism rs12979860) in the genomic region of IL28B, a type III interferon, is the strongest pre-therapeutic marker for success of therapy measured as sustained virologic response (SVR), i.e. negativity for HCV RNA 6 months post end of treatment. Whereas about 70% of patients with CC genotype reach this end point, SVR rates are about 32% and 23% in

Table 1. Baseline characteristics of patients.

		n (%)
Sex	male	62 (58.5)
	female	44 (41.5)
Age		50.3±14 a
Histology	NMM	44 (41.5)
	SSM	29 (27.9)
	other	33 (30.6)
Breslow		3.2±2.3 mm
AJCC Stage	I	21 (19.8)
	II	49 (46.2)
	III	32 (30.2)
	unknown	4 (3.8)
IFN Therapy	Low dose	92 (86.8)
	High dose	14 (13.2)

patients with CT or TT-genotype (i.e. non-CC genotype), respectively [9–11]. IL28B is involved in the induction of Interferon stimulated-genes (ISGs) and higher ISG activity in T allele carriers is thought to explain interferon resistance in hepatitis C [12,13]. The impact of IL28B genotype has not been investigated in patients on interferon alpha treatment for indication other than viral hepatitis.

In this study we have analyzed a possible correlation between IL28B polymorphism and overall and disease free survival of stage IB/II and III melanoma patients who have received interferon alpha adjuvant therapy.

Patients and Methods

Patients

106 caucasian patients were included in this monocentric retrospective study conducted at the dermato-oncologic outpatient clinics of the Department of Dermatology, Medical University of Vienna. The eligibility criteria for enrollment in this study were (i) histologically proven melanoma stage IB, II or III and (ii) adjuvant low-dose or high dose interferon alpha therapy.

Patients' history, clinical and histological data were collected from patients' charts. Parameters analyzed included age, sex, type and location of melanoma, clinical stage at diagnoses, and disease progression under interferon therapy, disease-free survival and overall survival.

Ethics

The study was approved by the Ethic committee of the Medical University Vienna (EK 1056/2011). Written informed consent was obtained from all patients who were alive at the time of enrollment in this study, but not of retrospectively analyzed deceased patients nor their relatives. Patients' data was anonymized and de-identified prior to analysis.

IL28B Genotyping

Genomic DNA was isolated either from EDTA whole-blood samples with the QIAamp DNA Blood Mini Kit or formalin-fixed, paraffin-embedded tissue samples using QIAamp DNA FFPE Tissue Kit (Qiagen, Hilden, Germany). The rs12979860 SNP was analysed using the StepOnePlus Real Time PCR System (Applied Biosystems, Foster City, USA) in conjunction with a Custom TaqMan SNP Genotyping Assay developed together with Applied Biosystems. Sequences were obtained from the NCBI Entrez SNPDatabase (http://www.ncbi.nlm.nih.gov/sites/entrez). For rs12979860, oligonucleotides with the sequences 5′-GCCTGTC-GTGTACTGAACCA-3′ and 5′-GCGCGGAGTGCAATTCA-A-3′ were used as forward and reverse primer, respectively.

Statistical analysis

Data were analyzed by IBM SPSS Statistics 22. We used the Mann-Whitney U test or Fisher's exact test for the comparison of groups as appropriate. Disease-free survival and overall survival were analyzed by the method of Kaplan-Meier, a Cox-regression

Table 2. Distribution of IL28B genotype in study cohort and correlation with expected distribution according to Hardy-Weinberg equilibrium.

polymorphism	genotype	observed	expected	p-value
	CC	44	46	
IL28B (SNP rs12979860)	CT	52	47	0.474
	TT	10	12	

Table 3. Risk factors like sex, age, type of melanoma, tumour thickness and AJCC stage did not differ between CC and non-CC genotype.

		CC	nonCC	p-value
Sex	Male	40.3%	59.7%	0.768
	Female	43.2%	56.8%)	
Age		48.8±14.4 a	51.3±14.4 a	0.379
Histology	NMM	38.6%	61.4%	0.099
	SMM	27.6%	72.4%	
Breslow		3.86±2.99 mm	2.75±1.45 mm	0.147
AJCC	I	20.9%	20.3%	0.496
	II	41.9%	52.5%	
	III	37.2%	27.1%	

model was used for multivariate analysis. For all statistical tests, a p-value <0.05 was considered statistically significant.

Results

Patient characteristics

62 (58.5%) male and 44 (41.5%) female patients were included in this study. The baseline characteristics are shown in table 1. Mean age at diagnosis was 50±14 years. The majority of patients were diagnosed with a nodular (41.5%) or superficial spreading melanoma (27.9%). In male patients melanomas were predominately located on the trunk (56.9%), while in female patients melanomas were equally distributed on the trunk and on the lower extremities (31.7% and 26.8% respectively). Mean Breslow index was 3.2±2.3 mm with a trend to higher levels in male and in older patients.

95 of 106 (89.6%) patients underwent sentinel lymph node biopsy which was positive in 26 (27.4%) patients. 8 patients showed additional lymph node metastasis in subsequent lymph node dissection. Thus, according to AJCC classification, 19.8%of the patients were staged IB, 29.2% IIA, 14.2% IIB, 2.8% IIC and 30.2% III (table 1). Patients with lymph node metastasis had a trend to thicker tumors (4.2±3.3 mm versus 2.9±1.8 mm, p=0.08) and a higher risk of disease progression (38.5% versus 27.5%, p=0.05).

92 (86.8%) patients received low-dose interferon alpha, 14 (13.2%) high-dose interferon alpha therapy (table 1). Low-dose interferon was applied for a mean of 25.4±15.4 months, high dose interferon for 12.6±3.9 months.

During the observation period (6.4±4.7 years) disease progression was seen in 36 of 106 (34%) patients. Disease progression occurred after 5.5±4.3 years and the mean overall survival was 7±4.7 years. Disease progression was seen more often in patients receiving high-dose interferon (8/14, 57.1% versus 28/92, 30.4%, p=0.05) and in patients with higher AJCC stages (2/21, 6.2% stage I versus 13/32, 40.6% stage III, p=0.05).

Distribution of IL28B polymorphism in the study population

44 (41.5%) patients had a CC and 62 (58.5%) a non-CC genotype. Distribution of C and T alleles in the study population was within the Hardy-Weinberg-equilibrium (table 2) and similar to the distribution previously described for the Austrian population (13). 25 (40.3%) male and 19 (43.2%) female patients had a CC genotype (p=0.768). Mean age in the CC group was 48.8±14.4

years and 51.3±14.4 years in the non-CC group (p=0.38). 17 (38.6%) patients with nodular melanoma and 8 (27.6%) patients with superficial spreading melanoma had a CC genotype (p=0.10). In CC and in non-CC patients tumors were mainly located to the trunk. There was no statistical significant association of IL28B polymorphism with sex, age at diagnosis, melanoma type, anatomic site or ulceration of the primary tumour. There was a trend towards a higher Breslow index in patients with CC genotype, however also this difference did not reach statistical significance (p=0.15). Distribution of AJCC stage did not differ between CC and non-CC patients (I-20.9%, II- 41.9%, III-37.2% versus I-20.3%, II-52.5%, III-27.1%, p=0.50, table 3).

Correlation of IL28B polymorphism and disease progression

43.2% of patients with CC genotype and 27.4% of patients with non-CC genotype experienced disease progression (p=0.09). Mean progression free survival was 5.3±5.1 years in the CC and 4.8±3.6 years in the non-CC group (p=0.92). Overall survival was 7.6±5.6 years for patients with CC and 6.5±3.9 in patients with non-CC genotype (p=0.45). During the observation period 4 patients in the CC and 4 patients in the non-CC group died of melanoma (p=0.61).

To assess an influence of IL28B polymorphism on survival of patients with melanoma treated with interferon alpha Kaplan-Meier analysis was done. There was no significant difference in progression free survival between CC and non-CC patients (Fig 1a). A similar result was seen when analyzing overall survival. Again, there was no impact of the IL28B polymorphism (Fig 1b). Applying univariate analysis, only tumor thickness and AJCC stage showed a statistically significant influence on progression and overall survival. There was no evidence for a correlation of IL28B polymorphism with progression free or overall survival (Fig 2). In multivariate Cox regression analysis only AJCC stage was significantly associated with disease-free survival (Fig 3a). Multivariate Cox analysis of overall survival did not reveal any statistically significant association when age, sex, IL28B polymorphism, Breslow index and AJCC stage were analysed together (Fig 3b). Since tumor thickness and AJCC stage are correlated multivariate Cox analysis was repeated separately with one of the two factors. In this analysis, a significant association of Breslow index and overall survival could be seen.

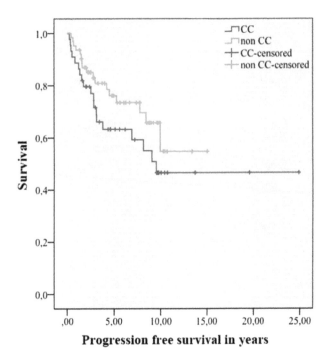

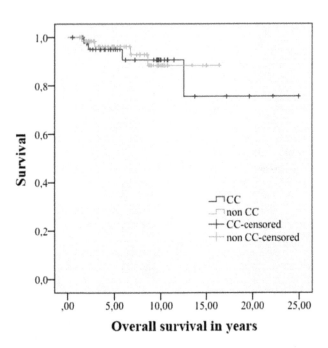

Figure 1. Correlation of IL28B polymorphism with progression-free and overall survival. A) Kaplan-Meier curve of progression-free survival according to IL28B polymorphism. No significant difference in progression-free survival could be observed for patients with CC or non-CC genotype (p = 0.176). Survival time on the x-axis is depicted in years. B) Kaplan-Meier curve of overall survival according to IL28B polymorphism. No significant difference in overall survival could be observed for patients with CC or non-CC genotype (p = 0.727). Survival time on the x-axis is depicted in years.

Discussion

Personalizing or tailoring health care to the individual patient's need and genetic makeup gains more and more importance in daily practice. It is especially rewarding in the setting of therapies associated with severe side effects but also with high costs. Reliable biomarkers, that either predict response to therapy or risk of side effects, are a prerequisite for this approach. In case of unfavorable predictors addition or the sole use of adiuvants with no or only marginal side effects but also cost may be a valuable approach [14,15].

Adjuvant interferon alpha treatment in melanoma fulfills the criteria for a therapeutic approach for which such a biomarker would be highly desirable: Several studies support the therapeutic efficacy in terms of disease-free survival and to a lower extent overall survival [16–21]. This potential benefit is counteracted by the fact that interferon alpha has to be taken for many months and is often associated with side effects that affect patients' quality of life.

Interferon alpha is used in several clinical settings. One main indication is the treatment of acute and chronic hepatitis C infection. Also in this application response to interferon is far below 100% and side effects are very common. However in this setting a biomarker – IL28 C/T dimorphism rs12979860 - predicting the probability to reach sustained virologic response (SVR) has recently been identified by genome-wide association studies [9–11]. Analysis of this SNP has become common practice and helps clinicians to either encourage their patients to undergo therapy or to defer treatment. Patients showing the favorable CC genotype have up to a twofold higher rate of SVR and higher rates of early virologic response than non-CC patients [22]. Another type III interferon, interferon λ4, was described recently as inducer of ISGs and has a similar prognostic value regarding treatment response [23].

The mechanism underlying the influence of IL28B polymorphism on the response to interferon alpha treatment is not known. IL28B is a type III interferon. Type I (interferon alpha) and type III interferon act via different receptors, but activate the same signal cascades and have similar effects [24]. IL28B polymorphism has been associated with IL28BmRNA expression levels and with activation levels of interferon responsive genes (ISG) [25,26]. More recently, additional effects of the IL28B polymorphism like higher levels of KIR expression on NK cells in non-CC genotype and enhanced caspase activity in CC genotype have been described [27].

To our knowledge this is the first study which analysis the influence of the IL28B polymorphism on the success rate of interferon therapy in a disease other than viral infection.

In contrast to the abundant literature showing a clear cut effect of IL28B polymorphism on the therapy-outcome and SVR rates in HCV infected patients, we did not find evidence for a significant association in melanoma patients on adjuvant interferon alpha therapy.

The collective presented here is representative of melanoma patients. Several prognostic factors have been identified including tumor thickness, clinical stage but also other factors like e.g. melanin content have been shown to influence treatment response and have been correlated with overall and disease free survival [28–29]. Thus – as expected - well established risk factors like tumor thickness and clinical stage correlate with disease free and overall survival also in our study population.

There may be several explanations for the differing results in HCV and melanoma patients.

A recent study showed the correlation of ISG levels and the IL28B polymorphism is inversely correlated in healthy and HCV infected livers [30]. While in healthy livers CC genotype is associated with high mRNA levels of ISG, in the case of HCV infection, patients with the TT genotype show the highest levels of

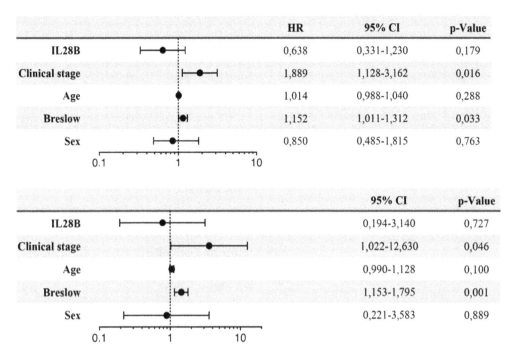

Figure 2. Univariate Cox regression analysis of risk factors associated with progression-free and overall survival. A) Risk factors associated with progression-free survival (univariate Cox regression analysis) B) Risk factors associated with overall survival (univariate Cox regression analysis).

ISG. Thus in the setting of acute or chronic HCV infection the virus seems to induce a shift in the regulation of interferon signaling pathways resulting in a situation where patients with TT genotype already show maximal stimulated ISG levels, that might not be further enhanced by exogenous interferon, i.e. Interferon alpha therapy [31]. This is supported by other studies showing HCV related changes in IFN signaling pathways [32–34]. Therefore, virally induced changes in the interferon signaling

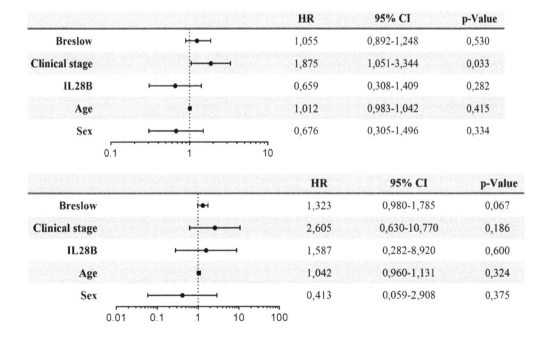

Figure 3. Multivariate Cox regression analysis of risk factors associated with progression-free and overall survival. A) Risk factors associated with progression-free survival (multivariate Cox regression analysis) B) Risk factors associated with overall survival (multivariate Cox regression analysis).

cascade could be a prerequisite to reveal an influence of IL28B polymorphism.

Although it is generally accepted that anti-viral and anti-tumor immune responses share common mechanisms, there are also significant differences. Interferons are one of the most important antiviral defense mechanism. Thus, most viruses have developed anti-interferon escape mechanism. Modulation of interferon associated signaling pathways may therefore have much stronger implication for anti-viral than anti-tumor responses masking the effect of IL28B genotype.

Furthermore, while anti-viral immune response is directed to non-self, anti-tumor responses must deal with altered self and thus recognize subtle differences between healthy cells and tumors. In addition, tumor epitopes vary much more inter-individually than viral antigens. Also these individual differences could lead to a higher variability of the efficacy interferon treatment, independently of the IL28B polymorphism, making appreciation of its influence more difficult.

Our study has certainly several limitations.

On one hand we observed only a relatively low number of patients with progressive disease in this cohort. Furthermore, in contrast to HCV infection in melanoma IFN alpha therapy is an adjuvant treatment. While in hepatitis C sustained virologic response defined as absence of HCV-RNA 6 months post treatment can be used as end point, surrogate markers have to be used for melanoma. Thus, due to the natural course of the disease the observation time of this study is short and much longer observation periods might be needed.

In conclusion, we did not find any evidence of an association of IL28B polymorphism and treatment success with interferon alpha in patients with melanoma. IL28B polymorphism cannot be recommended as predictor in decision guidance in the same way as in hepatitis C.

Acknowledgments

The authors would like to thank Claudia Willheim and Elisabeth Eder for performing the genetic assays.

Author Contributions

Conceived and designed the experiments: MP HK KGP. Performed the experiments: MP KGP CH. Analyzed the data: MP HK KGP. Contributed reagents/materials/analysis tools: PF AFS SB CH. Wrote the paper: KGP HK MP HP.

References

1. Kaufman HL, Kirkwood JM, Hodi FS, Agarwala S, Amatruda T et al. (2013) The Society for Immunotherapy of Cancer consensus statement on tumour immunotherapy for the treatment of cutaneous melanoma. Nat Rev Clin Oncol. 10: 588–98.

2. Kirkwood JM, Richards T, Zarour HM, Sosman J, Ernstoff M et al. (2002) Immunomodulatory effects of high-dose and low-dose interferon alpha2b in patients with high-risk resected melanoma: the E2690 laboratory corollary of intergroup adjucant trial E1690. Cancer 95: 1101–12.

3. Wang W, Edington HD, Rao UN, Jukic DM, Land SR et al. (2007) Modulation of signal transducers and activators of transcription 1 and 3 signaling in melanoma by high-dose IFNalpha2b. Clinical cancer research: an official journal of the American Association for Cancer Research 13: 1523–31.

4. Wang W, Edington HD, Jukic DM, Rao UN, Land SR et al. (2008) Impact of IFNalpha2b upon pSTAT3 and the MEK/ERK MAPK pathway in melanoma. Cancer immunology, immunotherapy: CII 57: 1315–21.

5. Ascierto PA, Napolitano M, Celentano E, Simeone E, Gentilcore G et al. (2010) Regulatory T cell frequency in patients with melanoma with different disease stage and course, and modulating effects of high-dose interferon-alpha 2b treatment. Journal of translational medicine 8: 76.

6. Ascierto PA, Kirkwood JM (2008) Adjuvant therapy of melanoma with interferon: lessons of the past decade. Journal of translational medicine 6: 62.

7. Ives NJ, Stowe RL, Lorigan P, Wheatley K (2007) Chemotherapy compared with biochemotherapy for the treatment of metastatic melanoma: a meta-analysis of 18 trials involving 2,621 patients. Journal of clinical oncology 25: 5426–34.

8. Mocellin S, Pasquali S, Rossi CR, Nitti D (2010) Interferon alpha adjuvant therapy in patients with high-risk melanoma: a systematic review and meta-analysis. Journal of the National Cancer Institute 102: 493–501.

9. Ge D, Fellay J, Thompson AJ, Simon JS, Shianna KV et al. (2009) Genetic variation in IL28B predicts hepatitis C treatment-induced viral clearance. Nature 461: 399–401.

10. Rauch A, Kutalik Z, Descombes P, Cai T, Di Iulio J et al. (2010) Genetic variation in IL28B is associated with chronic hepatitis C and treatment failure: a genome-wide association study. Gastroenterology 138: 1338–4511.

11. Tanaka Y, Nishida N, Sugiyama M, Kurosaki M, Matsuura K et al. (2009) Genome-wide association of IL28B with response to pegylated interferon-alpha and ribavirin therapy for chronic hepatitis C. Nature genetics 41: 1105–9.

12. Sarasin-Filipowicz M, Oakeley EJ, Duong FH, Christen V, Terracciano L, et al. (2008) Interferon signaling and treatment outcome in chronic hepatitis C. Proc Natl Acad Sci U S A 105: 7034–7039.

13. Beinhardt S, Aberle JH, Strasser M, Dulic-Lakovic E, Maieron A, et al. (2012) Serum level of IP-10 increases predictive value of IL28B polymorphisms for spontaneous clearance of acute HCV infection. Gastroenterology 142(1):78–85.

14. Slominski AT, Carlson JA (2014) Melanoma resistance: a bright future for academicians and a challenge for patient advocates. Mayo Clin Proc. 89(4):429–33.

15. Slominski A, Zbytek B, Slominski R (2009) Inhibitors of melanogenesis increase toxicity of cyclophosphamide and lymphocytes against melanoma cells. Int J Cancer. 124(6):1470–7.

16. Kirkwood JM, Strawderman MH, Ernstoff MS, Smith TJ, Borden EC, et al. (1996) Interferon alfa-2b adjuvant therapy of high-risk resected cutaneous melanoma: the Eastern Cooperative Oncology Group Trial EST 1684. Journal of clinical oncology 14: 7–17.

17. Kirkwood JM, Manola J, Ibrahim J, Sondak V, Ernstoff MS, et al. (2004) A pooled analysis of eastern cooperative oncology group and intergroup trials of adjuvant high-dose interferon for melanoma. Clinical cancer research: an official journal of the American Association for Cancer Research 10: 1670–7.

18. Kirkwood JM, Ibrahim JG, Sondak VK, Richards J, Flaherty LE et al. (2000) High- and low-dose interferon alfa-2b in high-risk melanoma: first analysis of intergroup trial E1690/S9111/C9190. Journal of clinical oncology 18: 2444–58.

19. Pehamberger H, Soyer HP, Steiner A, Kofler R, Binder M et al. (1998) Adjuvant interferon alfa-2a treatment in resected primary stage II cutaneous melanoma. Austrian Malignant Melanoma Cooperative Group. Journal of clinical oncology 16(4):1425–9.

20. Grob JJ, Dreno B, de la Salmoniere P, Delaunay M, Cupissol D et al. (1998) Randomised trial of interferon alpha-2a as adjuvant therapy in resected primary melanoma thicker than 1.5 mm without clinically detectable node metastases. French Cooperative Group on Melanoma. Lancet 351: 1905–10.

21. Mocellin S, Lens MB, Pasquali S, Pilati P, Chiarion Sileni V (2013) Interferon alpha for the adjuvant treatment of cutaneous melanoma, Cochrane Database Syst Rev. 6: CD008955. doi: 10.1002/14651858. CD008955.pub2.

22. Stättermayer AF, Stauber R, Hofer H, Rutter K, Beinhardt S et al. (2011) Impact of IL28B Genotype on the Early and Sustained Virologic Response in Treatment-Naive Patients with Chronic Hepatitis C, Clin Gastroenterol Hepatol 9(4):344–350.

23. Stättermayer AF, Strassl R, Maieron A, Rutter K, Stauber R et al. (2014) Polymorphisms of interferon-λ4 and IL28B- effects on the treatment response to interferon/ribavirin in patients with chronic hepatitis C, Aliment Pharmacol Ther 39(1):104–11.

24. Kotenko SV, Gallagher G, Baurin VV, Lewis-Antes A, Shen M et al. (2003) IFN-lambdas mediate antiviral protection through a distinct class II cytokine receptor complex. Nature immunology 4: 69–77.

25. Urban TJ, Thompson AJ, Bradrick SS, Fellay J, Schuppan D et al. (2010) IL28B genotype is associated with differential expression of intrahepatic interferon-stimulated genes in patients with chronic hepatitis C. Hepatology 52: 1888–96.

26. Honda M, Sakai A, Yamashita T, Nakamoto Y, Mizukoshi E et al. (2010) Hepatic ISG expression is associated with genetic variation in interleukin 28B and the outcome of IFN therapy for chronic hepatitis C. Gastroenterology 139: 499–509.

27. Naggie S, Osinusi A, Katsounas A, Lempicki R, Herrmann E et al. (2012) Dysregulation Of Innate Immunity In HCV Genotype 1 IL28B Unfavorable Genotype Patients: Impaired Viral Kinetics And Therapeutic Response. Hepatology 56(2): 444–454.

28. Slominski, Zmijewski MA, Pawelek J (2012) L-tyrosine and L-DOPA as hormone-like regulators of melanocyte functions, Pigment Cell Melanoma Res. 25(1):14–27.

29. Brożyna AA, Jóźwicki W, Carlson JA, Slominski AT (2013) Melanogenesis affects overall and disease-free survival in patients with stage III and IV melanoma. Hum Pathol. 44(10):2071–4.

30. Dill MT, Duong FH, Vogt JE, Bibert S, Bochud PY et al. (2011) Interferon-induced gene expression is a stronger predictor of treatment response than IL28B genotype in patients with hepatitis C. Gastroenterology 140: 1021–31.

31. Feld JJ, Nanda S, Huang Y, Chen W, Cam M, et al. (2007) Hepatic gene expression during treatment with peginterferon and ribavirin: Identifying molecular pathways for treatment response. Hepatology 46: 1548–1563.

32. Jilg N, Lin W, Hong J, Schaefer EA, Wolski D et al. (2013) Kinetic differences in the induction of interferon stimulated genes by interferon-alpha and IL28B are altered by Infection with Hepatitis C virus. Hepatology.

33. Thomas E, Gonzalez VD, Li Q, Modi AA, Chen W et al. (2012) HCV infection induces a unique hepatic innate immune response associated with robust production of type III interferons. Gastroenterology 142(4):978–88.

34. Weber F (2007) Interaction of hepatitis C virus with the type I interferon system. World J Gastroenterol 13(36):4818–23.

17

Identification of Allelic Heterogeneity at Type-2 Diabetes Loci and Impact on Prediction

Yann C. Klimentidis[1]*, Jin Zhou[1], Nathan E. Wineinger[2]

1 Mel and Enid Zuckerman College of Public Health, Division of Epidemiology and Biostatistics, University of Arizona, Tucson, Arizona, United States of America, **2** Scripps Translational Science Institute, La Jolla, California, United States of America

Abstract

Although over 60 single nucleotide polymorphisms (SNPs) have been identified by meta-analysis of genome-wide association studies for type-2 diabetes (T2D) among individuals of European descent, much of the genetic variation remains unexplained. There are likely many more SNPs that contribute to variation in T2D risk, some of which may lie in the regions surrounding established SNPs - a phenomenon often referred to as allelic heterogeneity. Here, we use the summary statistics from the DIAGRAM consortium meta-analysis of T2D genome-wide association studies along with linkage disequilibrium patterns inferred from a large reference sample to identify novel SNPs associated with T2D surrounding each of the previously established risk loci. We then examine the extent to which the use of these additional SNPs improves prediction of T2D risk in an independent validation dataset. Our results suggest that multiple SNPs at each of 3 loci contribute to T2D susceptibility (*TCF7L2*, *CDKN2A/B*, and *KCNQ1*; $p < 5 \times 10^{-8}$). Using a less stringent threshold ($p < 5 \times 10^{-4}$), we identify 34 additional loci with multiple associated SNPs. The addition of these SNPs slightly improves T2D prediction compared to the use of only the respective lead SNPs, when assessed using an independent validation cohort. Our findings suggest that some currently established T2D risk loci likely harbor multiple polymorphisms which contribute independently and collectively to T2D risk. This opens a promising avenue for improving prediction of T2D, and for a better understanding of the genetic architecture of T2D.

Editor: Shengxu Li, Tulane School of Public Health and Tropical Medicine, United States of America

Funding: YCK was supported by NIH grant K01DK095032. NEW was supported by NIH grant 1UL1TR001114. The funders had no role in study design, data collection and analysis, decision to publish, or preparation of the manuscript.

Competing Interests: The authors have declared that no competing interests exist.

* Email: yann@email.arizona.edu

Introduction

Approximately 65 loci have been shown to be associated with type-2 diabetes (T2D) through genome-wide association studies (GWAS). However, variation at these loci accounts for a small proportion of the expected heritability of T2D [1,2]. Among several potential strategies for identifying additional contributing genetic variation, one approach is to determine whether there are additional genetic markers near established loci that act independently or jointly with the reported marker (lead SNP).

Allelic heterogeneity is a feature of the genetic architecture of many traits, including common traits and diseases such as height, BMI, and T2D [3–6]. In the context of T2D genetics, both Morris et al. [2] and Yang et al. [7] have suggested that additional SNPs in established loci are associated with T2D risk. However, Morris et al. only considered SNPs in weak linkage disequilibrium ($r^2 < 0.05$) with the lead SNP, and that were not in the same recombination interval. Hence, without formal conditional analysis, they identified two loci as having multiple associations at genome-wide significance (*KCNQ1* and *CDKN2A/B*), and two more at suggestive levels (*DGKB* and *MC4R*). Yang et al. have recently developed a method for identifying additional associated SNPs based on conditional/joint (C/J) analysis using GWAS

summary statistics and linkage disequilibrium (LD) information from a reference sample [7]. They applied their method to only a single established T2D locus (*CDKN2A/B*), and identified two novel SNPs at that locus that were significantly associated with T2D when fitted jointly. Finally, on a smaller scale (1,924 cases and 5,380 controls), Ke [8] identified multiple associated loci at the *CDKN2A/B* and *TSPAN8* loci. Although higher power is afforded with the GWAS meta-analysis approach to identify associations with single SNPs, it does not allow for direct C/J analysis since the actual genotype data is not available. The advantage of the method developed by Yang et al. is that it takes advantage of the greater power of GWAS meta-analyses, while also testing for C/J associations, which would otherwise be impossible without individual level data.

Here, we comprehensively examine allelic heterogeneity based on the method developed by Yang et al. at 65 T2D loci discovered by the DIAGRAM consortium, using the summary statistics from their recent meta-analysis of T2D GWAS. We then examine the extent to which these newly identified SNPs increase the accuracy of T2D risk prediction in an independent validation dataset.

Methods

Datasets

We used 6,054 nominally unrelated European-American subjects (genomic relationship coefficient <0.025, based on approximately 2.5 million SNPs) from the Atherosclerosis Risk in Communities (ARIC) study [9] to obtain linkage disequilibrium (LD) estimates. According to Yang et al. [7], this sample size is sufficient for LD estimation with minimal error. In order to maximize the overlap of SNPs between the meta-analysis summary statistics (see below) and the ARIC study, we used IMPUTE2 software [10] along with 1000 Genomes reference data to impute millions of additional SNPs. Prior to imputation, we excluded individuals with a high genotype missing rate ($>10\%$). SNPs were excluded based on extreme minor allele frequency ($<0.5\%$), a high missing rate ($>10\%$), or failed Hardy-Weinberg equilibrium ($p<0.005$). After imputation, we excluded SNPs with 'info' <0.6 (measure of imputation quality), and SNPs with genotype dosage between 0.33 and 0.66, or between 1.33 and 1.66. Intermediate dosages outside of these specified ranges were rounded to the nearest integer. We did not use intermediate genotype dosages since this was not an option with the GCTA software, described below.

The validation dataset consisted of European-American subjects from the Multi-Ethnic Study of Atherosclerosis (MESA) [11], which included 225 T2D cases and 1,985 controls. T2D cases were defined as having a fasting glucose level ≥ 126 mg/dL, a self-report of taking diabetes medication, or a physician diagnosis of T2D. This dataset can be considered an independent validation dataset since it was not part of the DIAGRAM meta-analysis, whereas ARIC was a part of this meta-analysis, thus precluding it from any validation assessment. We implemented genotype QC and imputation as detailed above. However, we did not round or remove intermediate genotype dosages. The MESA and ARIC dataset were obtained from dbGaP (database of Genotypes and Phenotypes). IRB approval was obtained from the University of Arizona.

Conditional/Joint Analysis

Using the summary statistics from the discovery phase of the latest version (v3) of the DIAbetes Genetics Replication And Meta-analysis (DIAGRAM) consortium, available to the public through an online source [12] and LD estimates from the ARIC dataset as described above, we used GCTA software [13] to perform stepwise model selection. Briefly, SNPs were selected into the model based on p-values in the meta-analysis. An iterative scheme was adopted in which C/J analyses were alternatively performed with the stepwise selection procedure. SNPs with a re-estimated (i. e. through C/J estimation as opposed to marginal estimation) p-value under a certain threshold were selected. For a full description of the method, see Yang et al. [7]. We restricted our analysis to only the genomic regions within 1 Mb of the top SNP at the 65 established T2D loci as reported in Morris et al [2]. This filtering along with the QC filtering described above resulted in 112,329 SNPs being used in this analysis. For each SNP, we recorded the following information as input for the C/J analysis: effect allele, effect size (log of odds ratio), corresponding standard error, p-value, allele frequency of the effect allele (based on ARIC sample described above, as this was not available in the DIAGRAM summary statistic file), and sample size (sum of cases and controls). We used PolyPhen-2 [14] to determine whether any of the newly identified SNPs had any predicted functional effect, and RegulomeDB [15] to determine whether these SNPs may lie

in regulatory regions (e.g. transcription factor binding sites) or are associated with specific DNA features (e.g. DNAse sensitivity).

Validation/Prediction

We compared several prediction models. First we constructed a baseline model which only considered demographic information (sex and age). Then we added a weighted genetic risk score [16] based on only the set of lead SNPs with weights corresponding to the log odds ratios according to the DIAGRAM meta-analysis summary statistics. Lead SNPs were defined as those that had the lowest p-value in the respective 2 Mb region according to the DIAGRAM Stage 1 meta-analysis summary statistics. We then considered a weighted genetic risk score based on all SNPs identified by the C/J analysis with weights corresponding to the coefficients estimated from the C/J analysis. We conducted the above analyses at the following p-value thresholds based on the C/J results: 5×10^{-8}, 5×10^{-7}, 5×10^{-6}, 5×10^{-5}, and 5×10^{-4}. We examined the proportion of variance explained by these additional SNPs by calculating the variance explained on the liability scale, estimated through the odds ratios and allele frequencies of the SNPs, and assuming a disease prevalence of 10%, using the Mangrove package [17] in R [18]. We also calculated Nagelkerke's R^2 [19] of the models which include age and sex and each of the GRS, using the fmsb package [20] in R, and report the Akaike information criterion (AIC) [21] for each of these models. Prediction accuracy was estimated using the area under the receiver operating characteristic curve (AUC) as implemented in the pROC package [22] in R. Differences in AUC among models were compared by examining the change in AUC (ΔAUC) and assessed using the DeLong test [23] to determine statistical significance.

Results

Conditional/Joint analysis

We identified novel genome-wide significant ($p<5\times10^{-8}$) SNPs in the C/J analysis at the three following loci: *TCF7L2*, *CDKN2A/B*, and *KCNQ1* (see Table 1). In the *TCF7L2* region, we identified three SNPs (rs7917983, rs17747324, rs12266632) within a 32 kb region. The lead SNP (rs4506565) was not selected in this model, but is positioned in this region and is in moderate LD with each of the novel findings (r^2 between 0.18 and 0.70). For each of these novel findings, the marginal effect sizes and p-values in the meta-analysis are similar to those estimated in the C/J analysis. By relaxing the p-value threshold to $p<5\times10^{-4}$, we discovered an additional SNP in this region (rs10128255). In the *CDKN2A/B* region, the lead SNP (rs2383208) was not selected in the C/J analysis. Instead, two SNPs (rs10757282 and rs10811661) approximately 1.9 kb downstream of the lead SNP were discovered. These SNPs are only 110 bases apart. rs10757282 is in relatively low LD with the lead SNP ($r^2 = 0.29$). However, rs10811661 is in high LD with the lead SNP ($r^2 = 0.94$). It should be noted that the correlation between the respective risk alleles is negative ($r = -0.54$), suggesting that estimates obtained through the single marker association were underestimated for both SNPs, as evidenced by the larger effect sizes and lower p-values estimated in the C/J analysis compared to the meta-analysis marginal association results (see Table 1). In the *KCNQ1* region, we identified two SNPs in the C/J analysis ($p<5\times10^{-8}$). One SNP (rs462402) is in moderate LD with the lead SNP, rs231362 ($r^2 = 0.45$). The other SNP (rs163177) is approximately 121 kb upstream and is not in LD with the lead SNP ($r^2<0.01$). By relaxing the p-value threshold to $p<5\times10^{-6}$, we identified additional novel discoveries in the *DGKB* and *TP53INP1* genes.

Table 1. SNPs identified by conditional/joint analysis with p-value $<5\times10^{-5}$, and corresponding evidence of regulatory function from RegulomeDB.

Gene region	Chr	SNP	bp	refA	freq	b	se	p	n	bJ	bJ_se	pJ	LD (r)	RegulomeDB score
BCL11A	2	rs2192512	59854224	C	0.475	0.058	0.014	6.50E-04	110517	0.060	0.014	2.75E-05	-0.017	
	2	rs243019	60439310	C	0.451	0.086	0.014	2.70E-06	117602	0.087	0.014	3.24E-10	0.000	5
IGF2BP2	3	rs12233623	186712890	C	0.548	0.058	0.014	1.00E-03	111251	0.065	0.014	5.49E-06	-0.074	5
	3	rs6767484	187003272	G	0.305	0.122	0.018	3.90E-10	83684.8	0.127	0.018	7.89E-13	0.000	5
ANKRD55	5	rs9686661	55897543	T	0.189	0.095	0.023	9.20E-05	70923.6	0.100	0.023	1.05E-05	-0.063	4
	5	rs1895452	56096152	T	0.635	0.058	0.014	3.10E-03	118858	0.061	0.014	1.69E-05	0.000	
ZBED3	5	rs12522618	75491695	G	0.425	0.058	0.014	3.70E-03	112754	0.058	0.014	4.99E-05	0.009	
	5	rs7708285	76461623	G	0.300	0.122	0.022	1.10E-06	54523.4	0.122	0.022	3.70E-08	0.000	
DGKB	7	rs10282101	13867616	T	0.501	0.077	0.019	7.30E-05	64914.8	0.077	0.019	3.27E-05	-0.011	
	7	rs17168486	14864807	T	0.182	0.122	0.022	6.90E-07	76812	0.126	0.022	1.35E-08	-0.027	3a
TP53INP1	7	rs1974620	15031992	T	0.531	0.068	0.014	1.00E-04	112723	0.069	0.014	1.06E-06	0.000	
	8	rs4735337	96042641	T	0.490	0.077	0.014	6.90E-05	114412	0.078	0.014	2.01E-08	-0.022	5
CDKN2A/B	8	rs6991742	96533562	T	0.596	0.068	0.014	1.80E-04	116576	0.069	0.014	8.77E-07	0.000	5
	9	rs564398	22019547	T	0.579	0.077	0.014	1.70E-05	117268	0.059	0.014	2.48E-05	0.102	
	9	rs10757282	22123984	C	0.430	0.068	0.019	8.80E-04	65000.9	0.163	0.021	1.92E-14	-0.536	5
	9	rs10811661	22124094	T	0.825	0.166	0.021	1.50E-13	86410.9	0.247	0.024	1.08E-24	0.000	
ZMIZ1	10	rs3915932	80611942	G	0.592	0.095	0.018	4.70E-08	69629.2	0.098	0.018	7.10E-08	-0.040	5
	10	rs6480947	80906216	G	0.156	0.113	0.027	2.20E-04	59743.8	0.119	0.027	8.58E-06	0.000	
TCF7L2	10	rs7917983	114722872	C	0.479	0.148	0.013	1.50E-17	131838	0.082	0.014	2.17E-09	0.344	
	10	rs17747324	114742493	C	0.237	0.358	0.021	8.50E-55	70086.8	0.341	0.022	3.07E-54	-0.126	
	10	rs12266632	114754949	G	0.063	0.255	0.042	8.50E-11	53996.4	0.290	0.043	1.12E-11	0.000	5
KCNQ1	11	rs462402	2673869	C	0.507	0.077	0.014	1.10E-05	114391	0.076	0.014	6.49E-08	-0.002	
	11	rs163177	2794989	C	0.506	0.077	0.014	4.80E-05	114382	0.075	0.014	9.55E-08	0.040	5
	11	rs451041	3017301	G	0.496	0.068	0.014	8.20E-05	112303	0.063	0.014	8.89E-06	0.000	5
KCNJ11	11	rs7928810	17329019	C	0.396	0.068	0.014	1.30E-04	117423	0.068	0.014	1.50E-06	-0.004	
	11	rs757984	17576484	T	0.803	0.077	0.019	4.90E-04	102511	0.077	0.019	3.09E-05	0.000	
MTNR1B	11	rs10830962	92338075	G	0.408	0.104	0.014	1.50E-08	124966	0.101	0.014	1.61E-13	0.069	
	11	rs531573	92444531	C	0.182	0.086	0.018	1.70E-04	111150	0.077	0.018	2.75E-05	0.000	
TSPAN8	12	rs11178531	69694957	A	0.449	0.077	0.014	9.20E-06	115565	0.059	0.014	4.73E-05	0.271	
	12	rs1533104	69942814	T	0.338	0.086	0.014	2.80E-06	130115	0.072	0.014	4.26E-07	0.000	
SPRY2	13	rs1616547	78884037	C	0.512	0.058	0.014	3.00E-03	110308	0.058	0.014	4.84E-05	0.006	
	13	rs1327316	79607064	G	0.717	0.095	0.018	9.40E-07	82902.8	0.095	0.018	1.92E-07	0.000	
C2CD4A	15	rs6494307	60181982	C	0.572	0.077	0.014	1.80E-05	116797	0.077	0.014	3.65E-08	-0.001	
	15	rs2456936	60502334	C	0.785	0.077	0.019	1.10E-03	96195.2	0.077	0.019	3.32E-05	0.000	
ZFAND6	15	rs7176681	77714599	C	0.737	0.086	0.018	1.20E-04	85321.9	0.092	0.018	5.42E-07	-0.010	2a

Table 1. Cont.

Gene region	Chr	SNP	bp	refA	freq	b	se	p	n	bJ	bJ_se	pJ	LD (r)	RegulomeDB score
	15	rs1357335	78155918	A	0.684	0.086	0.018	1.60E-05	76497.6	0.086	0.018	2.92E-06	−0.006	5
	15	rs3848174	78531707	T	0.380	0.058	0.014	3.80E-03	117013	0.061	0.014	1.78E-05	0.000	5
HNF4A	20	rs387769	41745269	C	0.881	0.113	0.027	1.40E-05	75082.1	0.113	0.027	2.34E-05	0.008	1f
	20	rs6073708	43386291	A	0.562	0.058	0.014	2.40E-03	111949	0.058	0.014	4.81E-05	0.000	5

Abbreviations: Chr: chromosome, bp: base pair position, refA: reference allele, freq: frequency of the risk allele, b: regression coefficient from meta-analysis summary statistics, p: p-value from meta-analysis summary statistics, n: sample size in meta-analysis, LD (r): linkage disequilibrium between corresponding SNP and the following SNP at the same locus, bJ: regression coefficient estimated from conditional/joint analysis, bJ_se: standard error estimated from conditional/joint analysis, se: standard error from meta-analysis summary statistics, pJ: p-value from estimated from conditional/joint analysis.
(1f: eQTL+transcription factor (TF) binding/DNase peak; 2a: TF binding+matched TF motif+matched DNase Footprint+DNase peak; 3a: TF binding+any motif+DNase peak; 5: TF binding or DNase peak).

Continuing to relax this threshold, we identify 17 ($p<5\times10^{-5}$) (see Table 1) and 34 ($p<5\times10^{-4}$) regions with multiple associated SNPs. According to PolyPhen, none of the SNPs identified through C/J analysis had any predicted functional effect. According to our query in RegulomeDB, rs387769 near *HNF4A* shows evidence of being linked to expression of a gene target, affecting binding of a transcription factor, and shows evidence of a DNase footprint. SNP rs7176681 near *ZFAND6* also displays evidence of being a transcription factor binding site, and evidence of a DNase footprint. SNP rs17168486 in *DGKB* shows evidence of transcription factor binding and a DNase peak. Several other SNPs show evidence of transcription factor binding or a DNase peak (see Table 1).

Validation/Prediction

The AUC of the baseline prediction model which included only sex and age was 0.5702. For each of the three loci with additional SNPs that were significant at the $p<5\times10^{-8}$ threshold (*TCF7L2*, *CDKN2A/B*, and *KCNQ1*), the inclusion of the SNPs identified by the C/J analysis resulted in a higher AUC than a model including only the lead SNP (although not statistically significant) in all regions except for *KCNQ1* (see Figure 1).

Considering all three loci with additional SNPs at the $p<5\times10^{-8}$ threshold collectively, we found that the use of the seven SNPs identified by the C/J analysis resulted in a slightly higher AUC (0.5979) than when using only the three lead SNPs (0.5803). This represents a doubling in ΔAUC over the age+ sex model (see Figure 2), although this difference is not quite statistically significant (p = 0.055), according to the DeLong test. The inclusion of all SNPs (lead and from C/J analysis) results in a statistically significant (p = 0.049), yet small, increase in AUC (see Figure 2). At the $p<5\times10^{-6}$ threshold, the use of 11 SNPs at 5 loci (*TCF7L2*, *CDKN2A/B*, *KCNQ1*, *DGKB* and *TP53INP1*), slightly, but not significantly, increased prediction accuracy (AUC = 0.5885) over a model considering only the corresponding 5 lead SNPs (AUC = 0.5779; p = 0.126). At the $p<5\times10^{-5}$ threshold, we observe a small increase in prediction accuracy when using the 39 SNPs identified by the C/J analysis instead of the corresponding 17 lead SNPs (AUC = 0.5892 vs. 0.5724; p = 0.079). Finally, at the $p<5\times10^{-4}$ threshold, the use of 120 SNPs identified by the C/J analysis and the lead SNPs results in a slightly higher and nearly statistically significant increase in AUC over that of a model which includes only the 34 lead SNPs at the corresponding loci (AUC = 0.5965 vs. 0.5858; p = 0.067).

Table 2 shows the proportion of variance explained by the additional SNPs identified by the C/J analysis. For each of the three loci, the SNPs identified by the C/J analysis explained slightly more of the variance in T2D risk than the lead SNP. Similarly, for the collection of SNPs identified by the C/J analysis at various p-value thresholds, we observe an increase in the proportion of T2D variance explained by the SNPs and the GRSs, along with decreasing AIC values.

Discussion

Our analyses confirm previous findings regarding the allelic heterogeneity present at the *CDKN2A/B*, *KCNQ1*, *DGKB*, and *MC4R* loci. We provide novel evidence of allelic heterogeneity at genome-wide significance at the *TCF7L2* locus. We support our finding in *TCF7L2* by showing that the use of the three identified SNPs results in a small increase in AUC (albeit not statistically significant) compared to using the lead *TCF7L2* SNP (rs4506565) alone. We observe similar but much weaker trends at the *CDKN2A/B* and *KCNQ1* loci.

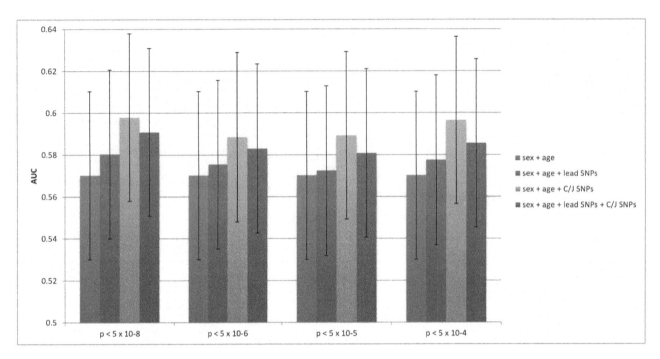

Figure 1. Prediction accuracy in MESA at 3 loci with additional detected SNPs at the 5×10^{-8} threshold.

At less stringent p-value thresholds, we observe additional putatively associated SNPs at up to 34 loci. Considering the collective set of loci in which additional associated SNPs were identified through C/J analysis, prediction accuracy appears to slightly improve with the addition of these additional SNPs in our validation dataset. At all p-value thresholds, the ΔAUC over the sex + age model is at least two-fold greater when using the C/J identified SNPs compared to using the lead SNPs alone.

The strength of the method developed by Yang et al. is well exemplified by the multiple associated SNPs identified at the *TCF7L2* locus, since the use of the three SNPs (which do not include the lead SNP) appears to be more informative than only using the lead SNP, rs4506565. Another example of the strength

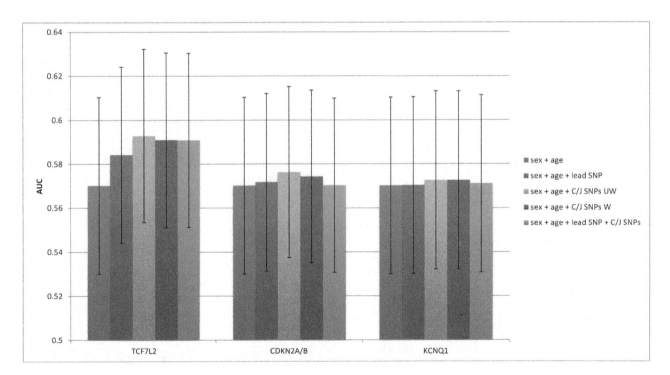

Figure 2. Prediction accuracy in MESA using lead SNPs vs. SNPs identified in C/J analysis at different p-value thresholds.

Table 2. Variance explained at various p-value thresholds in the MESA validation dataset by the collection of individual SNPs on the liability scale, variance explained by, and model fit of, the weighted GRS, using Nagelkerke's R^2, and AIC, respectively.

		Liability-scale variance	Nagelkerke R^2	AIC
age+sex			0.0115	1448.1
TCF7L2	1 Lead SNP	0.0097	0.0178	1443.3
	3 C/J SNPs	0.0189	0.0212	1439.7
	4 Lead+C/J SNPs	0.0282	0.0198	1441.1
CDKN2A/B	1 Lead SNP	0.0004	0.0118	1449.7
	2 C/J SNPs	0.0025	0.0127	1448.8
	2 Lead+C/J SNPs	0.0029	0.0117	1449.8
KCNQ1	1 Lead SNP	3.00E-5	0.0115	1450
	2 C/J SNPs	0.0029	0.0122	1449.3
	3 Lead+C/J SNPs	0.0014	0.0118	1449.8
<5.00E-08	3 Lead SNPs	0.0101	0.0153	1446
	7 GCTA SNPs	0.0225	0.0230	1437.8
	10 Lead+C/J SNPs	0.0322	0.0197	1441.2
<5.00E-06	5 Lead SNPs	0.0130	0.0134	1448.1
	11 GCTA SNPs	0.0258	0.0190	1442.1
	16 Lead+C/J SNPs	0.0381	0.0164	1444.8
<5.00E-05	17 Lead SNPs	0.0277	0.0134	1448
	39 GCTA SNPs	0.0648	0.0209	1440
	55 Lead+C/J SNPs	0.0865	0.0171	1444.1
<5.00E-04	34 Lead SNPs	0.0613	0.0158	1445.5
	91 GCTA SNPs	0.1443	0.0259	1434.6
	119 Lead+C/J SNPs	0.1801	0.0197	1441.2

of this method is the case in which two risk alleles are in negative LD. Without the C/J analysis, the additional SNPs in the *CDKN2A/B* region would not be identified when analyzed on their own.

The main limitation of this method is that associations are not tested directly, but rather through knowledge of marginal associations, and LD patterns in a different dataset (of the same ancestral background). A major limitation of the validation stage of our study is the relatively small sample size which limits the statistical power to detect differences in prediction accuracy between different GRSs. From this perspective, it will be important to continue validating these findings in larger datasets, and to combine actual genotype data across multiple datasets instead of using summary statistics. Furthermore, it will be important to dissect the allelic heterogeneity on a locus-by-locus basis to closely examine the patterns/existence of dependencies and additive or interactive effects. Finally, it will be important to functionally characterize these as well as all GWAS findings to more firmly establish causality and better understand molecular mechanisms leading to T2D.

Nevertheless, this approach is clearly promising for a greater understanding of the molecular basis of type-2 diabetes, and potentially for use in risk prediction scores. As additional loci are identified through GWAS, it will be important to systematically identify instances of allelic heterogeneity and to examine the extent to which additional SNPs can help to shed light on the functional basis of genetic variation.

Acknowledgments

The authors would like to thank the participants and organizers of all studies. Data from these studies was obtained from dbGaP through accession numbers: phs000280.v2.p1, and phs000209.v10.p2. We also thank Akshay Chougule for help with genotype imputation.

Atherosclerosis Risk in Communities

The Atherosclerosis Risk in Communities Study is carried out as a collaborative study supported by National Heart, Lung, and Blood Institute contracts (HHSN268201100005C, HHSN268201100006C, HHSN268201100007C, HHSN268201100008C, HHSN268201100009C, HHSN268201100010C, HHSN268201100011C, and HHSN268201100012C). Funding for GENEVA was provided by National Human Genome Research Institute grant U01HG004402 (E. Boerwinkle). The authors thank the staff and participants of the ARIC study for their important contributions.

Multi-Ethnic Study of Atherosclerosis

MESA and the MESA SHARe project are conducted and supported by the National Heart, Lung, and Blood Institute (NHLBI) in collaboration with MESA investigators. Support for MESA is provided by contracts N01-HC-95159, N01-HC-95160, N01-HC-95161, N01-HC-95162, N01-HC-95163, N01-HC-95164, N01-HC-95165, N01-HC-95166, N01-HC-95167, N01-HC-95168, N01-HC-95169 and CTSA UL1-RR-024156. MESA Family is conducted and supported by the National Heart, Lung, and Blood Institute (NHLBI) in collaboration with MESA investigators. Support is provided by grants and contracts R01HL071051, R01HL071205, R01HL071250, R01HL071251, R01HL071258, R01HL071259, UL1-RR-025005, by the National Center for Research Resources, Grant UL1RR033176, and the National Center for Advancing Translational Sciences, Grant UL1TR000124. This manuscript was not prepared in collaboration with MESA investigators and does not necessarily reflect the opinions or views of MESA, or the NHLBI. Funding for SHARe genotyping was provided by NHLBI Contract N02-HL-64278. Genotyping was performed at Affymetrix (Santa Clara, California, USA) and the Broad Institute of Harvard and MIT (Boston, Massachusetts, USA) using the Affymetric Genome-Wide Human SNP Array 6.0.

References

1. Voight BF, Scott LJ, Steinthorsdottir V, Morris AP, Dina C, et al. (2010) Twelve type 2 diabetes susceptibility loci identified through large-scale association analysis. Nat Genet 42: 579–589.
2. Morris AP, Voight BF, Teslovich TM, Ferreira T, Segrè A V, et al. (2012) Large-scale association analysis provides insights into the genetic architecture and pathophysiology of type 2 diabetes. Nat Genet 44: 981–990.
3. Lango AH, Estrada K, Lettre G, Berndt SI, Weedon MN, et al. (2010) Hundreds of variants clustered in genomic loci and biological pathways affect human height. Nature 467: 832–838.
4. Sim X, Ong RT, Suo C, Tay WT, Liu J, et al. (2011) Transferability of type 2 diabetes implicated Loci in multi-ethnic cohorts from southeast Asia. PLoS Genet 7: e1001363.
5. Williams AL, Jacobs SBR, Moreno-Macías H, Huerta-Chagoya A, Church-house C, et al. (2013) Sequence variants in SLC16A11 are a common risk factor for type 2 diabetes in Mexico. Nature.
6. Loos RJ, Lindgren CM, Li S, Wheeler E, Zhao JH, et al. (2008) Common variants near MC4R are associated with fat mass, weight and risk of obesity. Nat Genet 40: 768–775.
7. Yang J, Ferreira T, Morris AP, Medland SE, Madden PAF, et al. (2012) Conditional and joint multiple-SNP analysis of GWAS summary statistics identifies additional variants influencing complex traits. Nat Genet 44: 369–375.
8. Ke X (2012) Presence of multiple independent effects in risk loci of common complex human diseases. Am J Hum Genet 91: 185–192.
9. The Aric Investigators (1989) The Atherosclerosis Risk in Communities (ARIC) Study: design and objectives. The ARIC investigators. Am J Epidemiol 129: 687–702.
10. Howie B, Marchini J, Stephens M, Chakravarti A (2011) Genotype Imputation with Thousands of Genomes. G3 GenesGenomesGenetics 1: 457–470.
11. Bild DE, Bluemke DA, Burke GL, Detrano R, Diez Roux A V, et al. (2002) Multi-ethnic study of atherosclerosis: objectives and design. Am J Epidemiol 156: 871–881.
12. DIAGRAM website. Available: http://diagram-consortium.org/downloads.html. Accessed December 2013.
13. Yang J, Lee SH, Goddard ME, Visscher PM (2011) GCTA: A tool for genome-wide complex trait analysis. Am J Hum Genet 88: 76–82.
14. Adzhubei I, Jordan DM, Sunyaev SR (2013) Predicting functional effect of human missense mutations using PolyPhen-2. Curr Protoc Hum Genet.
15. Boyle AP, Hong EL, Hariharan M, Cheng Y, Schaub MA, et al. (2012) Annotation of functional variation in personal genomes using RegulomeDB. Genome Res 22: 1790–1797.
16. Purcell SM, Wray NR, Stone JL, Visscher PM, O'Donovan MC, et al. (2009) Common polygenic variation contributes to risk of schizophrenia and bipolar disorder. Nature.
17. Package Mangrove website. Available: http://cran.r-project.org/web/packages/Mangrove/index.html. Accessed May 2014.
18. R Development Core Team (2011) R: A language and environment for statistical computing. R Foundation for Statistical Computing.
19. Nagelkerke NJD (1991) A note on a general definition of the coefficient of determination. Biometrika 78: 691–692.
20. Nagelkerke R^2 function (R) website. Available: http://minato.sip21c.org/msb/man/Nagelkerke.html. Accessed June 2014.
21. Akaike H (1974) A new look at the statistical model identification. IEEE Trans Automat Contr 19.
22. Robin X, Turck N, Hainard A, Tiberti N, Lisacek F, et al. (2011) pROC: an open-source package for R and S+ to analyze and compare ROC curves. BMC Bioinformatics 12: 77.
23. DeLong ER, DeLong DM, Clarke-Pearson DL (1988) Comparing the areas under two or more correlated receiver operating characteristic curves: a nonparametric approach. Biometrics 44: 837–845.

Author Contributions

Conceived and designed the experiments: YCK NEW JZ. Performed the experiments: YCK. Analyzed the data: YCK. Wrote the paper: YCK NEW JZ.

18

Comparative Transcriptomic Analysis of the Response to Cold Acclimation in *Eucalyptus dunnii*

Yiqing Liu[1,2]*, Yusong Jiang[2], Jianbin Lan[2], Yong Zou[2], Junping Gao[1]

1 Department of Ornamental Horticulture, China Agricultural University, Beijing 100193, China, **2** College of Life Science & Forestry, Chongqing University of Art & Science, Yongchuan 402160, China

Abstract

Eucalyptus dunnii is an important macrophanerophyte with high economic value. However, low temperature stress limits its productivity and distribution. To study the cold response mechanisms of *E. dunnii*, 5 cDNA libraries were constructed from mRNA extracted from leaves exposed to cold stress for varying lengths of time and were evaluated by RNA-Seq analysis. The assembly of the Illumina datasets was optimized using various assembly programs and parameters. The final optimized assembly generated 205,325 transcripts with an average length of 1,701 bp and N50 of 2,627 bp, representing 349.38 Mb of the *E. dunnii* transcriptome. Among these transcripts, 134,358 transcripts (65.4%) were annotated in the Nr database. According to the differential analysis results, most transcripts were up-regulated as the cold stress prolonging, suggesting that these transcripts may be involved in the response to cold stress. In addition, the cold-relevant GO categories, such as 'response to stress' and 'translational initiation', were the markedly enriched GO terms. The assembly of the *E. dunnii* gene index and the GO classification performed in this study will serve as useful genomic resources for the genetic improvement of *E. dunnii* and also provide insights into the molecular mechanisms of cold acclimation in *E. dunnii*.

Editor: Turgay Unver, Cankiri Karatekin University, Turkey

Funding: 1. Natural Science Foundation of Chongqing Province, China (Grant NO. cstc2013jcyjA80035), 2. Key Program for forestry of Chongqing University of Art and Science (Grant NO. 201302), 3. National Science Foundation for Young Scientists of China (Grant NO. 31340016). The funders had no role in study design, data collection and analysis, decision to publish, or preparation of the manuscript.

Competing Interests: The authors have declared that no competing interests exist.

* Email: liung906@163.com

Introduction

Rapid population increase and the consequent increase in the requirement for different types of paper products, as well as the emphasis on paper as an environmentally friendly packaging material, have led to an increased demand for wood [1]. The imbalance between the supply and demand for forest products is growing. Eucalyptus is an economically important forest tree that grows in tropical and subtropical regions [2,3]. Eucalyptus trees can be highly productive over a short rotation period, tolerate a wide range of soils and commonly exhibit a straight stem form in those species utilized in production forestry. Furthermore, eucalypts, unlike many trees, do not have a true dormant period and retain their foliage, which enables growth during warm winter periods [4]. Nevertheless, in Eucalyptus plantations, low temperature stress limits their productivity and distribution. When the temperature drops to 8°C or below, Eucalyptus trees would exhibit various symptoms of cold injury due to their inability to adapt to the low temperature [5]. Cold stress also alters the physiological status, such as transient increases in hormone levels (e.g., ABA), changes in the membrane lipid composition, accumulates of compatible osmolytes (such as soluble sugars, betaine, and proline) and increases in antioxidant levels [6,7]. In contrast, temperate plants can withstand freezing temperatures following a period of low, but non-freezing temperatures, a process called cold acclimation. The mechanisms of cold acclimation have been extensively investigated in *Arabidopsis thaliana* [8] and other important crop species such as maize and barley [9,10]. Cold stress has been shown to induce changes in physiology and gene expression, and hundreds of cold-responsive genes have been identified so far [11]. However, in tropical and subtropical plants, especially *E. dunnii*, the molecular mechanisms of the cold response are not clear.

The physiological and biochemical changes that occur during plant cold acclimation result primarily from changes in the expression of cold-responsive genes. In general, Cold-responsive genes could be classified into two groups: 1) functional proteins, which directly protect plants against environmental stresses, and 2) regulatory proteins, which regulate the expression of downstream target genes in the stress response [4]. The first group mainly comprises enzymes involved in the biosynthesis of various osmoprotectants, such as late embryo genes is abundant (LEA) proteins, antifreeze proteins, chaperones, and detoxification enzymes[8,12]. The second group mainly includes transcription factors and protein kinases [12]. The best-characterized transcription factors (TFs) involved in the plant cold response are the class of AP2/ERF (APETALA2/ethylene-responsive element binding proteins), one kind of subfamily was known as CBF/DREB(C-repeat binding factor/dehydration resistance element binding protein), which regulate cold-responsive gene expression by binding to DRE/CRT cis-elements in the promoter region of cold-responsive genes [6,13]. Changes in the expression of cold-responsive contribute to

the differences in plant cold tolerance. For example, *Solanum.
commersonii* and *S. tuberosum*, which are closely related species
that differ in their cold acclimation abilities, exhibit considerable
differences in the expression levels of cold-responsive genes [6,14].
Chen *et al.* found that the activities of some detoxification
enzymes, such as catalase (CAT), superoxide dismutase (SOD),
peroxidase (POD) and esterase (EST) are increased in response to
cold stress, whereas the plant's metabolic activity is decreased [15–
17]. Some cold-induced genes have been cloned from Eucalyptus
plants. For example, four CBF paralogs were previously isolated
from *E. gunnii*, and qRT-PCR analysis demonstrated that they
exhibited complementary expression profiles in a range of natural
standard and cold conditions [18]. Navarro *et al.*found overexpression of *EguCBF1a* or *EguCBF1b* in the cold-sensitive *E.
urophylla·E. grandis* hybrid could enhance its freezing tolerance
[19].

Given the importance of cold-responsive genes in plant cold
tolerance, studying the cold response at the transcription level may
be a key step in identifying specific tolerance mechanisms. Next
generation sequencing (NGS) provides a high throughput
approach for analyzing genes involved a particular process at
transcription level. Compared to the traditional sequencing
techniques, NGS is more robust and demonstrates greater
resolution and inter-lab portability compared to several microarray platforms. NGS could detect millions of transcripts and is
beneficial to explore new genes and their expression profiling
independent of a reference genome [6,20,21]. For example, cDNA
libraries for *E. gunnii* have been constructed to identify genes
involved in cell protection (such as PCP, Lti6b and metallothio-
nein), LEA/dehydrin accumulation, and cryoprotection [22,23].
Despite its obvious potential, these next generation sequencing
methods have not been applied for *E. dunnii* yet.

The goal of this study was to construct a comprehensive
transcriptome to investigate the molecular mechanism of cold
tolerance in *E. dunnii*. The plants were exposed to low
temperature (4°C) for 0, 3, 6, 12, and 24 h, and the first two
expanded leaves below apical bud of *E. dunnii* were collected for
high throughput RNA-Seq analysis. Paired-end (PE) reads from
the RNA-Seq output were then assembled *de novo* to build an *E.
dunnii* transcriptome, which was subjected to a comparative
analysis. This analysis provides preliminary global insight into the
molecular mechanism of cold tolerance and a good base for future
basic research in *E. dunnii*.

Results

Physiological changes in *E. dunnii* in response to cold stress

Firstly, we detected the concentration of proline during the cold
treatment (at 4°C) from 0 to 48 h. The concentration of proline
decreased slightly from 0 to 3 h, but it increased rapidly as the cold
stress prolonging (Fig. 1A). The decrease in proline content at 0 to
3 h might be caused by a transient stress response of *E. dunnii* to
the low temperature shock. However, prolonged exposure to low
temperature (24 h) resulted in proline accumulation.

Plant cells could accumulate amounts of reactive oxygen species
under environmental stress, which result in severe damage of
proteins, membrane lipid, DNA and other cellular components
[15]. CAT could catalyze the decomposition of hydrogen peroxide
to water and oxygen, and it is important in protecting the cell from
oxidative damage by reactive oxygen species (ROS) [16–17]. The
activity level of CAT changed during cold acclimation in *E.
dunnii*. We observed an almost 25% increase in CAT activity after
3 h, and a nearly two-fold change after 24 h of cold stress

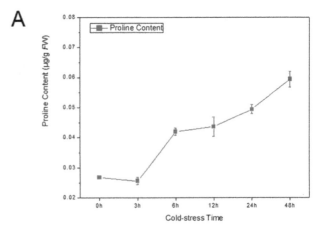

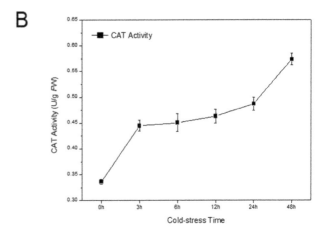

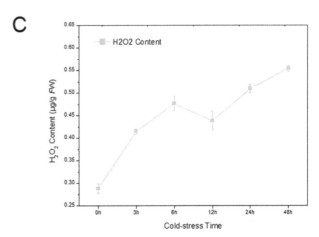

**Figure 1. Changes in proline content (A), CAT activity (B) and
H₂O₂ content (C) under low temperature (4°C) treatment over
time.**

(Fig. 1B). H_2O_2 is one kind of ROS molecule. The H_2O_2
concentration increased nearly 50% after 3 h, and then continued
to increase at a more moderate rate, remaining at high levels until
24 h (Fig. 1C). These results indicated that *E. dunnii* plants are
sensitive to the cold stress.

Table 1. Total number of reads for each treatment sample, as obtained by Illumina sequencing.

Duration of low temperature (4°C) treatment	Paired-end reads	Total length	Total number of contigs	Average length	N50 of contigs	Alignment rate (%)
0 h	25,407,247	120,616,917	118,761	1,015	1,817	90.3
3 h	24,817,373	179,471,852	149,467	1,200	2,136	91.7
6 h	25,703,824	215,673,491	161,627	1,334	2,343	90.6
12 h	34,870,702	302,614,952	195,733	1,546	2,630	92.4
24 h	33,846,411	217,801,712	160,461	1,357	2,367	91.2
TOTAL	–	349,381,021	205,325	1,701	2,827	94.5

RNA Sequencing, *de novo* assembly and functional annotation

To study the *E. dunnii* transcriptome in response to cold stress, we transferred plantlets with 10 leaves to a climate-chamber (4°C) and collected the first two expand leaves below apical bud at 0, 3, 6, 12, and 24 h time points, respectively. For the RNA-Seq analysis, we obtained 25,407,247, 24,817,373, 25,703,824, 34,870,702, and 33,846,411 clean paired-end reads, respectively (data not shown).

To obtain a more reliable and comprehensive transcriptome database, these five libraries were pooled together and then performed the *de novo* assembly. The pipeline for the bioinformatics analysis of the RNA-Seq data is shown in Fig. 2. The parameters of the contig databases assembled by each individual assembler, such as the alignment rate, sensitivity, accuracy and length distribution, were significantly different. Overall, the contig database produced by Trinity was significantly better than those from the other assemblers (Table S1). The optimal contig database contained 205,325 contigs ≥300 bp in length. The average length of these contigs was 1,701.6 bp, the N50 number was 2,827 bp, and the maximum length was 15,965 bp (Table 1). Additionally, there were 148,151 contigs with a length≥600 bp, 105,494 contigs with a length ≥1,200 bp, and 33,700 contigs with a length ≥ 3,000 bp (Fig. 3). The assembled contigs (≥300 bp) were deposited in the NCBI Transcriptome Shotgun Assembly (TSA) database under the accession number PRJNA208093.

Sequence similarity search against the NCBI non-redundant protein database (NR) was conducted using a locally installed BLAST program for functional annotation. Among all the assembled contigs (≥300 bp), 134,358 (65.4%) were annotated with BLASTx hits, matching 80,578 unique protein accessions (Table S2). For contigs longer than 600 bp, 80.9% had BLASTx hits, and for longer than 900 bp, the percentage increased to 88.8% (Fig. 3), indicating that most contigs, particularly the longer contigs, represent protein-encoding transcripts. As the completed genome information of *E. dunnii* was not available at this time, 70,967 contigs (34.6%) had no hits to any known proteins in the Nr database (Fig. 4A), suggesting that these contigs might be non-coding regions or potentially new genes [24].

In addition, among 134,358 contigs with BLASTx results, 52,265 (38.9%), 26,602 (19.8%), and 24,721 (18.4%) showed high sequence similarity to *Vitis vinifera*, *Populus* and *Glycine max*, respectively, but only 265 contigs (0.2%) shared homology with Eucalyptus (Fig. 4B). Alternatively, our results could indicate that *E. dunnii* is more closely to *V. vinifera* than *G. max* or *Arabidopsis* evolutionarily. Interestingly, some other plant transcriptomes, such as *Craterostigma plantagineum* [25] and *Fraxinus* spp. [26],

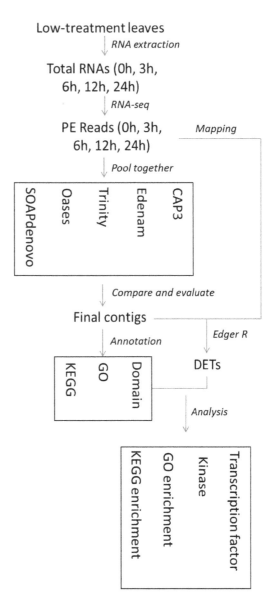

Figure 2. The pipeline for the bioinformatics analysis of the deep sequencing data.

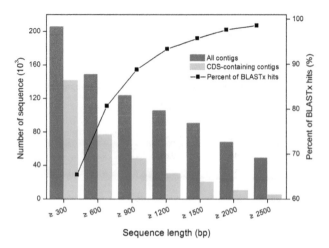

Figure 3. Annotation rate and proportion of long-CDS-containing sequences. A total of 205,325 contigs were used for the BLASTx search. The contig length is indicated on the X-axis. The size distributions of the final assembled contigs (red) and the number of long-CDS-containing contigs (green) are indicated on the left Y-axis. The percentage of BLASTx hits to size-grouped contigs is indicated by the diamond symbol.

display the same distribution pattern of top-hit species. These results could be simply explained by the number of genes deposited in Nr database. For example, by November 2013, the NCBI database contained 78,045 *V. vinifera* transcripts, 11,4590 *P. trichocarpa* transcripts, and 81,270 *G. max* transcripts, but only 7,146 Eucalyptus transcripts.

Differential expression between the groups and qRT-PCR validation

To characterize the digital gene expression profiles of the *E. dunnii* in response to low temperature, we performed a short-read alignment of each library using Perl script provided by the Trinity software package. For samples treated at 4°C for 0, 3, 6, 12, and 24 h, a total of 90.3%, 91.6%, 92.1%, 91.2%, and 91.5% of the reads could be aligned back to the contigs, and 64.2%, 63.8%, 65.1%, 62.7% and 61.5% aligned concordantly exactly once. To eliminate the effect of library size, edgeR (empirical analysis of digital gene expression in R) was used to create an effective library size for each sample. The number of aligned reads per transcript was normalized to FPKM based on an RESM-based algorithm. Differentially expressed transcripts (DETs) with FDR ≤ 0.05 and $\log_2$ fold-change ($\log_2$FC) ≥ 1 between pairs of samples were identified by edgeR [27]. The edgeR analysis generated 10 DET sets (0 h vs. 3 h, 0 h vs. 6 h, 0 h vs. 12 h, 0 h vs. 24, 3 h vs. 6 h, 3 h vs. 12 h, 3 h vs. 24 h, 6 h vs. 12 h, 6 h vs. 24 h, and 12 h vs. 24 h) with 11,395, 11,908, 11,901, 12,671, 8,935, 10,641, 11,843, 9,531, 11,489, and 10,230 DETs, respectively (Fig. 5). We also found that 7,059, 7,348, 7,479, and 7,636 DETs were up-regulated in the 0 h vs. 3 h, 0 h vs. 6 h, 0 h vs. 12 h, and 0 h vs. 24 h comparison sets, respectively. These results demonstrated that the number of up-regulated DETs increased as the duration of cold stress prolonged.

To validate the expression patterns of each DET obtained from the comparative RNA-Seq studies, we randomly selected 31 transcripts from the annotated DETs for qRT-PCR analysis. Noteworthy, qRT-PCR results are often affected by the choice of reference genes. Previously, a report explored the expression

stability of reference genes which are using in gene expression test in Eucalyptus in response to various abiotic stresses by qRT-PCR [28]. The authors found that expression of some genes, such as *PP2A-3/SAND*, *UPL7*, *UBC2* and *GAPDH*, are stable enough in all tested samples, while *ACT2* gene was not stable in response to environmental stimuli as expected. As mentioned in the paper of Cassan-Wang et al., GAPDH is a good choice as reference gene in qRT-PCR assay in Eucalyptus [28]. Therefore, we selected two most commonly used reference genes, beta-actin and GAPDH, because these two genes could be mutual support, mutual correction, and minimize the experimental errors. The results showed that the expression patterns of 25 DETs were compatible with the RNA-Seq analysis (Table S3), suggesting that the differential expression analysis based on high-throughput RNA sequencing produced reliable expression data.

Gene ontology (GO) and Kyoto Encyclopedia of Genes and Genomes (KEGG) enrichment analysis of DETs

GO (Gene Ontology) and KEGG (Kyoto Encyclopedia of Genes and Genomes) annotation were applied to the BLASTx results to provide comprehensive functional information for each transcript. In total, we obtained 198,528 GO annotations for 62,965 transcripts and 966 unique Enzyme Codes (ECs) for 28,295 transcripts (Table S2). Among the 62,965 transcripts with GO terms, 34,064 (54.1%) were assigned to the Biological process category, 19,959 (31.7%) to the Molecular function category, and 28,965 (46.0%) to the Cellular component category. In addition, 20,025 (31.8%) unique transcripts were assigned GO terms from all three categories (Fig. 6 and Table S4). To understand the mechanism of the cold stress response in *E. dunnii*, the DETs were subjected to GO and KEGG enrichment analysis. Under the GO category 'Biological process', the 'response to stress' and 'translational initiation' were the most highly enriched terms, with P.ad-values of 0 and 0.02, respectively. Under the category 'Molecular function', the 'quinolinate synthetase A activity' were the most highly enriched term, with a P.ad-values of 0.04. Under the category 'Cellular component', the 'cell part' was the most highly enriched term, with a P-value of 8.5E-11 (Table S4). KEGG analysis identified 27,688 contigs with pathway information were involved in 137 KEGG pathways. Among these 137 KEGG pathways, 'arginine and proline metabolism' and 'tropane, piperidine and pyridine alkaloid biosynthesis' were the two most significantly enriched KEGG pathways (Table S5).

Cold-responsive transcription factors and protein kinases in *E. dunnii*

Transcription factors and protein kinases are crucial upstream regulators that respond to various biotic and abiotic stresses in plants [29,30]. In this study, we identified a total of 586 contigs involving in transcription factor activity, which were classified into 65 types of transcription factors, including AP2, bZIP, JmjC, and SRF-TF. In order to verify the expression pattern of these transcription factors, an additional 5 transcripts were selected to carry out qRT-PCR analysis. The results displayed that the expression trend of these transcripts agreed with the results of RNA-seq analysis (Table S3). The START and bzip domains transcription factor families were the largest groups represented in the cold-responsive transcription factors, containing 64 (39 up-regulated and 25 down-regulated) and 62 (43 up-regulated and 19 down-regulated) unique transcripts, respectively. The next largest groups were the UDF (31 up-regulated and 24 down-regulated), AP2 (27 up-regulated and 24 down-regulated) and Sigma70 (24 up-regulated and 9 down-regulated) families (Table S6). In

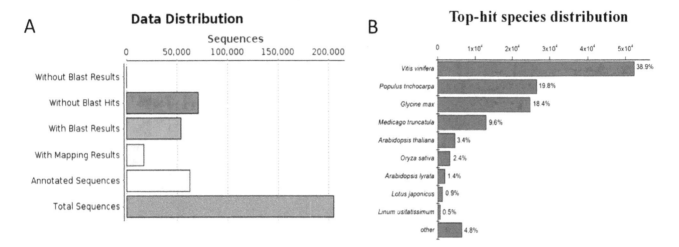

Figure 4. Distribution of the BLASTx results (A) and the top-hit species distribution of the *E. dunnii* transcriptome (B). A total of 205,325 contigs ≥300 bp in length were used for the sequence similarity searches, and 134,358 contigs produced BLASTx results. All of the contigs with BLASTx results were used for the species distribution analysis. Overall, 52,265 (38.9%), 26,602 (19.8%) and 24,721 (18.4%) contigs showed strong similarity to *Vitis vinifera*, *Populus* and *Glycine max*, but only 265 contigs (0.2%) shared the highest homology with Eucalyptus.

addition, we identified 169 contigs related to protein kinase activity, which were classified into 8 types of protein kinases based on their domains (Table S6).

Discussion

Improving *de novo* transcriptome assembly

The most critical step of an RNA-Seq study is the *de novo* assembly, especially for species without genome information [31–

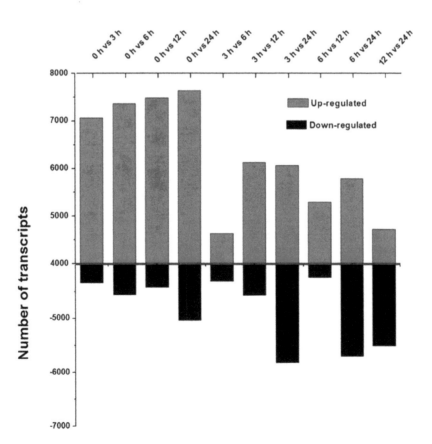

Figure 5. Transcripts that exhibited differential expression pattern. In total, 20,5325 contigs were used for the differential expression analysis, and the differential transcripts were identified by edgeR using the following parameters: FDR ≤0.05 and log$_2$fold-change ≥1.

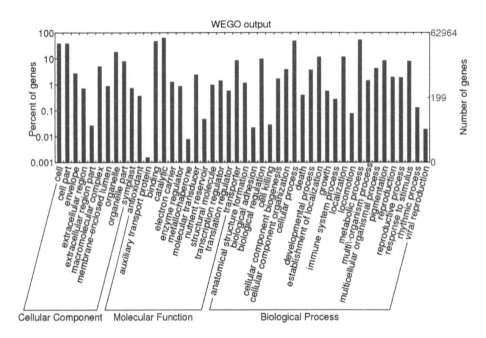

Figure 6. GO assignment of all contigs in the *E. dunnii* transcriptome. The contigs mapped to three main categories: Biological process, Cellular component and Molecular function. The right-hand y-axis indicates the number of annotated contigs.

34]. More and more genomes and/or transcriptomes sequences have been completed due to the development of high-throughput sequencing technologies. However, major published studies on transcriptome *de novo* assembly have typically used a single assembly program [35–37]. In this study, we compared the quality of 5 assemblers (Trinity, Osease, SOAP*denovo*, Edena, and Cap3) and then used the optimal combined strategy to construct the *E. dunnii* transcriptome database. When the reads were assembled using Trinity, Cap3, Edena, Oases, and SOAP *de novo*, the N50 (contig ≥300 bp) values were 2,827 bp, 2,551 bp, 1,368 bp, 1,838 bp, and 1,336 bp, respectively (Table S1). Although the accuracy and sensitivity of the contigs assembled by Trinity were the highest compared to the other assemblers, the assembly strategy still needs further optimization to obtain higher accuracy and sensitivity (Table S1). Different assembly software programs used different algorithms, such as the traditional OLC approach of the Edena assembler and the de Bruijn graph approach of the Oases and SOAP *de novo* assemblers [38]. For a particular species, these different algorithms have multiple advantages and disadvantages, which should be taken into account when selecting the most suitable assembler to complete the process of *de novo* assembly in different species. However, neither Trinity nor any other assembler is individually capable of assembling the results satisfactorily. When assembling the sweet potato transcriptome, Tao *et al.* [31] found that only 80% of the reads mapped back to contigs assembled by Trinity, implying that approximately 20% of the reads were not used effectively in the assembly process. In addition, sequencing quality, which is the foundation for obtaining an ideal assembly, should be improved. Xiao *et al.* [38] found that trimming all raw read sequences at the 3′-end and merging the assemblies from different assemblers significantly improved assembly outcome. Some researchers have also suggested that combining data produced by two or more sequencing methods, such as Illumina sequencing and 454 sequencing, could generate a more satisfactory assembly [39]. Combined assemblies use different assembling software and/or different assembling param-

eters, which means they benefit from the advantages of different software packages.

To date, there have been no standard criteria to evaluate the quality of transcriptome assemblies. Researchers appraise the quality of an assembly mainly by examining the data distribution of the assembly [40,41]. Besides the data distribution, we assessed the assembly quality using numerous metrics. Due to the lack of genomic resources for *E. dunnii*, we downloaded Eucalyptus genes with full-length from GenBank to use as reference sequences. The overlapping high-scoring segment pairs (HSPs) were only calculated once to determine the sensitivity. For each individual assembly, Trinity achieved higher sensitivity than Cap3, Oases and SOAP *de novo*. However, the final assembly generated by Edenam exhibited the highest sensitivity, which was slightly higher than that of Trinity (Table S1). To calculate the accuracy, we considered all unmatched components to be false positives, and Trinity exhibited a greatest accuracy. Taken together, the results from the above metrics indicate that our final assembly quality is optimal.

DETs involving in proline metabolism and quinoline alkaloid biosynthesis

Free proline in plant cells can significantly improve cold resistance [42], as it acts as a type of osmotic adjuster that can reduce the cell freezing point and stabilize intracellular water. Furthermore, free proline can also protect the cell from excessive dehydration and lipid peroxidation [42]. The accumulation of proline is frequently associated with whole plant tolerance to chilling and other stresses [6]. In this study, we observed that the free proline content was increased more than two-fold after 48 h cold treatment (Fig. 1), which were consistent with that accumulation pattern in Arabidopsis and cassava [6,43].

KEGG analysis showed that the 'arginine and proline metabolism' pathway was significantly enriched (Table S5) in *E. dunnii* during cold-stress. A total of 576 transcripts were involved in the 'arginine and proline metabolism' pathway, with 79 transcripts being up-regulated in response to cold stress at 24 h (Tables S6, S2).

In higher plants, proline can be synthesized via the glutamate (Glu) pathway or the ornithine (Orn) pathway, depending on the initial substrate [44,45]. P5CS (delta 1-pyrroline-5-carboxylate synthetase), a key enzyme in the Glu pathway, functions as a bifunctional enzyme to transform Glu to GSA [46]. The accumulation of free proline could improve the ability of stress resistance in many plants, which regulated by the expression of *p5cs* [47,48]. In *E. dunnii*, 3 transcripts were annotated as *p5cs*, and all three transcripts were up-regulated, particularly contig_6788, whose expression increased more than 10-fold when the plants were exposure to low temperature (Fig. 7 and Table S2). This transcriptome result correlated well with the change in free proline content, suggesting that at 4°C, the Glu pathway was activated to increase the free proline content to protect the plant against cold stress. δ-OAT (ornithine-oxo-acid transaminase) is a key enzyme in the Orn pathway that catalyzes the transformation of L-Orn into GSA. Because δ-OAT can catalyze arginine to glutamate, it could be involved in proline synthesis and accumulation [49]. However, we only identified two contigs (contig_6006 and contig_60065) annotated as δ-OATin the *E. dunnii* transcriptome, and neither was up- or down-regulated in response to cold stress (Fig. 7 and Table S2). Based on the expression profiles of these transcripts, we hypothesize that the Orn pathway may play a less important role than the Glu pathway during cold acclimation or that it may represent an alternative pathway for cold acclimation in *E. dunnii*.

Free proline accumulation is affected not only by the proline biosynthesis pathway but also by the proline degradation pathway. Under normal conditions, free proline functions as a feedback regulator to inhibit *p5cs* expression and concurrently induce ProDH (Proline dehydrogenase) gene expression. In contrast to the normal condition, *p5cs* expression is hyperactive during cold acclimation, whereas ProDH expression is inhibited, resulting in the accumulation of more and more free proline in plant cells. In *Arabidopsis* and other plants, proline levels are mainly determined

by balance of biosynthetic and catabolic pathways, controlled by P5CS and ProDH genes, respectively [6]. Nanjo *et al.* found that proline degradation was inhibited in *Arabidopsis* transformed with *At*ProDH [43], suggesting that free proline levels increased in leaves.

Secondary metabolism and its products are also involved in the response to various stresses in plants, representing a process that formed over a long evolutionary period [50–52]. There is some evidence that secondary metabolic products and environmental factors (biotic and abiotic) are closely linked, as in the case of alkaloids, which play an important role in resisting insects and herbivores via chemical defense mechanism [53]. In addition to 'arginine and proline metabolism', the DETs were significantly enriched in 'quinoline alkaloid biosynthesis' pathway during cold acclimation, based on KEGG pathway analysis (Table S5). Early in the cold stress period (0–6 h), 40% of transcripts related to quinoline alkaloid metabolism were up-regulated more than 2-fold compared to the 0 h time point (Table S2), including contig_65006 and contig_65485. This suggests that the up-regulation of transcripts in response to low temperatures may play a crucial role in plant stress tolerance. However, when the duration of cold stress exceeded 6 h, the expression levels of these up-regulated transcripts decreased gradually, dropping to their initial levels(i.e., comparable to their expression at 0 h) by 24 h (Table S2). This suggests that there may be a relationship between quinoline alkaloid biosynthesis and abiotic factors, although this relationship may not be as simple and direct as the relationship between the biological environment and chemical defense [54,55]. Further research is needed to explore this relationship in depth. Many researchers believe that plants produce secondary metabolites such as alkaloids at the cost of slower growth [56,57]. However, when biotic and abiotic stresses become severe enough to affect their survival, the plants have no choice but to produce some secondary metabolites for protection against such rigorous stress conditions.

'Response to stress' and 'translational initiation' response to low temperature

Under the GO category 'Biological process', the terms 'response to stress' and 'translational initiation' accounted for1.76% of the total 198,528 GOs, but the DETs accounted for 15.4% of the transcripts involved in these GO terms. Additionally, both of two GO terms were significantly enriched in four comparison sets (0 h vs.3 h, 0 h vs. 6 h, 0 h vs. 12 h, and 0 h vs. 24 h) according to the GO enrichment analysis (Table S4). The largest proportion of the 'Biological process' terms included the 'metabolic process' (30.07%), 'cellular process' (27.99%), and 'biological regulation' (5.75%) (Table S4), indicating comprehensive changes in *E. dunnii* gene expression before and after the cold stress. However, although only a few transcripts were identified as belonging to 'response to cold stress', as up-term of 'response to stress', these transcripts represented the most important components that are directly involved in protecting plants from cold stress. A total of 50 transcripts were annotated under this term based on GO categorization, and most were up-regulated in response to low temperature treatment. In particular, 26 transcripts involved in the 'response to cold stress' were not expressed under normal conditions but were induced by exposure to low temperature (Table S7 and Fig. S1). ROS scavenging enzymes, including catalase (CAT), superoxide dismutase (SOD), and glutathione transferase (GST), have been demonstrated to play key roles in the removal of ROS [17,58–61]. During exposure to low temperature, the CAT activity was increased (from 0.34 to 0.56 U/g Fw), which was in accordance with the expression level of the corresponding transcripts in the *E. dunnii* transcriptome (Figs. 1, and 7).

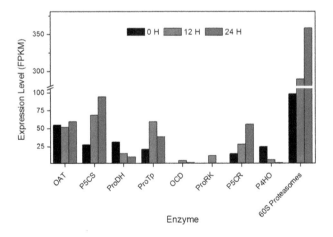

Figure 7. The expression level of some key enzymes involved in the 'arginine and proline metabolism' pathway during cold acclimation. Both the up-regulated expression of OAT (ornithine-oxo-acid transaminase), P5CS (pyrroline-5-carboxylatesynthase), ProTp (proline transporter), ProRK (proline-rich receptor protein kinase), P5CR (pyrroline-5-carboxylate reductase), and the down-regulated expression of ProDH (proline dehydrogenase), P4HO (prolyl 4-hydroxylase), OCD (ornithine cyclodeaminase) could result in proline accumulation.

Although the expression of peroxidases such as CAT and SOD increased significantly as the duration of cold exposure increased, these enzymes were still unable to completely clear the increased levels of H_2O_2, resulting in a significant increase in the amount of H_2O_2 during cold acclimation (Fig. 1). In this study, we found that some genes (e.g., MAP kinase and TCH2; Table S2) that are known to be involved in the response to other stresses (including salinity, heat and drought) in other plants are also involved in cold acclimation, which could support the hypothesis that the same gene have different functions in different plants.

The GO term 'translational initiation' was enriched in response to cold acclimation. A total of 254 transcripts were annotated under in this term, and 89 exhibited a greater than two-fold change in expression during the low temperature treatment (Table S8). Translation initiation in eukaryotes depends on many eukaryotic initiation factors (eIFs) that stimulate both the recruitment of the initiator tRNA, Met-tRNAiMet, and mRNA to the 40S ribosomal subunit and the subsequent scanning of the mRNA for the AUG start codon [62–64]. The largest of these initiation factors, the eIF-3 complex, organizes a web of interactions among several eIFs that assemble on the 40S subunit and participate in the different reactions involved in translation [62,65]. In plants, eIF-3plays the role of the central protein and interacts with many other translation initiation factors, such as eIF-4F, eIF-4G, eIF-4B, and eIF-1A [66]. Among the 89 contigs we identified that were annotated as 'translational initiation', 18 containedeIF-3 (eukaryotic translation initiation factor 3), and almost all were up-regulated during cold acclimation (Table S8). Daniel et al. [67] found that the expression level and phosphorylation state of these factors described above is subject to alteration during development, environmental stress (e.g., heat shock, and starvation), or viral infection. Tuteja [68] evaluated the roles of translation initiation, transcription factors, protein kinases, free proline, and reactive oxygen species in plant stress tolerance and found that these factors typically have synergistic effects in response to stress in plants. We also found that some transcripts encoding transcription factors, protein kinases (Table S6), translation initiation factors and antioxidant enzymes were up- or down-regulated in E. dunnii during cold acclimation, suggesting that the plant response to cold acclimation is a complex and global process.

Cold-responsive transcription factor genes in E. dunnii

In Arabidopsis, at least 5 transcription factor families have been reported to be involved in the cold stress response process, including AP2-EREBP, MYB, NAC, bHLH and WRKY family [29]. Wang et al. found there were many families of transcription factor, such as bHLH family, MYB family, WRKY family, NAC family and so on, responding to cold acclimation in C. sinensis [16]. Meanwhile, An et al. identified 6 AP2-EREBP and 5 Myb transcription factors participated in the process of cold stress in treated cassava [6]. In our study, many transcripts were annotated as AP2 transcription factor based on the domain analysis. Among theses, 27 transcripts were up-regulated and 24 down-regulated under cold stress (Table S6). In present work, we tested 5 AP2 TF genes by qRT-PCR and found four were up-regulated, one was down-regulated during cold-stress, and the changing trend of the two methods was accordant (Table S3). The AP2-EREBP family plays a major role in the early stages of the cold response and is the major regulator that functions in activating cold-regulated effectors in Arabidopsis and other plants [69,6]. In Eucalyptus plant, the CBF proteins, belonging to A-1 subfamily of ERF/AP2 TF family has been reported involved in response to cold stress in E. gunnii and E. globules [19,70–72].

Besides the AP2 family, the bZIP family has also been demonstrated to be involved in the cold response in Arabidopsis and C. sinensis [69,16]. In this study, we found that bZIP family was the most enriched TF family, containing 62 genes (43 up-regulated and 9 down-regulated). Differential expression of bZIP TFs implies that other environmental or hormonal pathways may be involved in cold response in E. dunnii.

In addition, four novel transcription factor families (JmjC, SRF-TF, and Sigma70-like) were also identified. Although their homologous genes in other plant species have not yet been reported in response to cold stress, the expression level of these genes were markedly changed before and after cold stress, suggesting they might be specific to E. dunnii or attractive targets for further functional characterization in plant.

Materials and Methods

Plant materials

Eucalyptus dunnii was used in this study. The plantlets of E. dunnii with 10 leaves were grew in a climate-chamber, with a temperature of $25°C$, $200\ \mu Em^{-2}s^{-1}$ illumination and a 14/10 h light/dark photoperiod. After eight weeks, the plants were moved into another climate-chamber with a temperature of $4°C$ and $200\ \mu Em^{-2}s^{-1}$ continuous illumination for cold-stress. For physiological measurement, we harvested the first two expanded leaves of these plantlets at 0, 3, 6, 12, 24 and 48 h after cold treatment, respectively. For RNA-seq, leaves from 6 plants treated by 0, 3, 6, 12, and 24 h were mixed for RNA isolation and sequencing. For test of physiological changes, leaves of plants treated by all time points were used. The harvested leaves were immediately frozen in liquid nitrogen for use.

Analysis of physiological parameters

The proline content of the leaves was analyzed using a free proline ELISA kit (Omega, Georgia, USA) according to the manufacturer's instructions. The CAT activity and H_2O_2 content were measured using a CAT ELISA kit (Omega, Georgia, USA) and a H_2O_2 ELISA kit (Omega, Georgia, USA), respectively. All measurements were performed on the platform of Epoch-ELIASA (Shmadzu, Tokyo, Japan), and all analysis were repeated three times in this study.

RNA extraction, library construction and RNA sequencing

Total RNA was isolated from the leaves by using Trizol reagent (Invitrogen, CA, USA) according to the manufacturer's instructions, and the RNAwas treated with RNase-free DNase I (TaKaRa, Dalian, China). The purity, concentration and RNA integrity number (RIN) were determined using a SMA3000 and/ or Agilent 2100 Bioanalyzer. The total RNA was then sent to Beijing Genomics Institute (BGI) -Shenzhen (Shenzhen, China) for RNA sequencing.

More than 20 μg of total RNA extracted from each group of plants exposed to low temperatures (n>3) was used to construct the cDNA libraries. First, the polyadenylated RNAs (mRNAs) were purified and retrieved using magnetic beads coated with a poly-T oligo. These mRNAs were then mixed with fragmentation media and fragmented. The fragmented mRNAs were subjected to reverse transcription using reverse transcriptase and random primers. The second-strand cDNA synthesis was performed using DNA polymerase I and RNase H. Finally, the resulting dscDNAs were repaired by adding a single 'A' base, and specific Illumina adapters were ligated to the repaired ends. Fragments of approximately 200 bp in size were purified and retrieved from the gels. To construct the fragmented cDNA library, these fragments, which served as the template, were enriched by PCR using two primers that annealed to the ends of the adapters. The cDNA libraries constructed above were sequenced using an Illumina Hiseq2000. The PE read information and quality values were generated using the Illumina sequencing-by-synthesis, image analysis and base-calling procedures.

Denovo assembly and functional annotation

Sequencing quality was assessed using fastQC software [http://www.bioinformatics.bbsrc.ac.uk/projects/fastqc/], and the PE reads were de novo assembled by five different assemblers: the Trinity software package (v2013-02-25) [73] with default parameters, the Oases software package (v0.1.21) [74] with a different K-value, the Edenam software package (v2013-07-15) [75] with a different M-value, the SOAP de novo software package (v2013-07-15) [76] with different K- and P-values, and the Cap3 software package (v12.07.21) [77] with default parameters. To evaluate the quality of the assemblies produced by the different assemblers, the PE reads were aligned back to the contigs assembled by a different assembler using Bowtie2 software (v2.0.0) [78], and the alignment rate was calculated. Subsequently, we analyzed the length distribution information of these contigs, such as the N50 number, average length, max length and total contig number, using common Perl scripts. Due to the lack of genomic information for Eucalyptus, 535 Eucalyptus sequences containing complete CDSs were downloaded from GenBank [http://www.ncbi.nlm.nih.gov/] and used as reference sequences to calculate the sensitivity and accuracy. Furthermore, we analyzed the best candidate coding sequence (CDS) for each contig from different assemblers and obtained the ratios of long CDS-containing transcripts to contigs with corresponding lengths.

All of the contigs (≥ 300 bp) produced by the Trinity software package were subjected to a similarity search against the NR database downloaded from GenBank utilizing local NCBI-BLAST software (v2.2.28+). The BLASTx searches were performed using a threshold E-value of $<10^{-3}$, max_target_seqs of 5, and an xml output file format. The BLASTx results were imported into Blast2GO software (v2.6.7) [79], and local functional annotation was performed. Enzyme codes, gene ontology (GO), and Kyoto Encyclopedia of Genes and Genomes (KEGG) pathways were retrieved from the KEGG web server (http://www.genome.jp/kegg/) [80]. GO classification [81] was performed using the WEGO program (http://wego.genomics.org.cn/cgibin/wego/index.pl) [82].

Differential expression profiling and enrichment

To investigate the expression level of each transcript at the five treatment time points, the PE reads for each sample were aligned back to the optimal assembly result (assembled by the Trinity assembler) using Perl scripts provided by the Trinity software package. Using these scripts, we obtained the digital expression levels of each transcript and normalized these data with a RESM-based algorithm to obtain the FPKM (Fragments per Kilobase per Million Mapped Fragments) values of each transcript. Based on the normalized expression profiles, the effect and bias introduced by library size and/or RNA composition were eliminated using edgeR [83], and significant differentially expressed transcripts (DETs) were identified with a P.ad-value ≤ 0.05 and $\log_2$ fold-change ($\log_2$ FC) ≥ 1.

The DET enrichment analysis was performed using the common Perl and R scripts. We first counted the number of transcripts involved in each KEGG pathway from the Trinity assembled contigs and/or DETs. Based on the transcript numbers in the contig database and DETs, we determined the enriched

KEGG pathway. Then, the P-value was adjusted by the Bernoulli equation, and a P.ad-value<0.05 was the threshold value for significant enrichment results. We applied a similar approach for the GO enrichment analysis.

Expression level verification

To verify the reliability and accuracy of the NGS-based expression level analysis, we randomly selected 31 transcripts from the contig database and evaluated the expression profiles among the five samples using quantitative real-time PCR. The primers for these transcripts are listed in Table S5. The first-strand cDNA was synthesized from 500 ng of total RNA using oligo (dT), random hexamers, and Moloney murine leukemia virus (M-MLV) reverse transcriptase (Invitrogen, CA, USA) according to the manufacturer's instructions. The real-time PCR was performed using the IQ5 Real-Time PCR System (Bio-Rad, CA, USA) in a total volume of 20 µL containing 100 ng of cDNA template, $1\times$ SYBR Premix Ex Taq^{TM}II (Perfect Real Time, TaKaRa), and 400 nM of each primer. Serial dilutions of each cDNA were used to generate a quantitative PCR standard curve to calculate the corresponding PCR efficiencies. The PCR conditions were as follows: initial denaturation at 95°C for 30 s, followed by 40 cycles of denaturation at 95°C for 5 s, primer annealing at 60°C for 30 s, and DNA extension at 72°C for 30 s. Two most commonly used reference genes, beta-actin and GAPDH, were selected for internal controls. Three biological replicates were used, and melting curve analysis was performed to check the amplification specificity. The relative expression levels were calculated using the BIO-RAD IQ5 standard edition Optical System software (version 2.1) and a normalized expression (ddCt) model.

Supporting Information

Figure S1 Differences of 'response to stimulus' (A), 'response to cold' (B), 'transcription factor activity' (C) and 'kinase regulator activity' (D) between each pair of samples. Overlap examinations were performed based on the resulting gene lists of four comparisons by VENNY. Overlap among four groups, D0 vs D3 (blue), D0 vs D6 (yellow), D0 vs D12 (yellow) and D0 vs D24 (red), were shown here.

Table S1 The characteristics of contig databases assembled by different assembler.

Table S2 Sequence annotations of E. dunnii transcripts and gene expression profiling of five samples.

Table S3 Comparison of expression patterns between RNA-Seq expression and qRT-PCR.

Table S4 GO classification of E. dunnii teanscriptome and differentially.

Table S5 KEGG classification of E. dunnii teanscriptome and differentially expressed transcripts indentified between each pairs comparisons.

Table S6 Transcription factor and kinase of E. dunnii response to low-temperature stress.

Table S7 Expression patterns of some transcripts involved in 'Response to cold'.

Table S8 Expression patterns of some transcripts involved in 'translational initiation'.

Acknowledgments

The authors are extremely grateful to Professor Yizheng Zhang (Sichuan University) for technical advice and assistance in data processing.

Author Contributions

Conceived and designed the experiments: YQL JPG. Performed the experiments: YQL YSJ JBL YZ. Analyzed the data: YSJ YQL JPG. Contributed reagents/materials/analysis tools: YQL YSJ. Wrote the paper: YSJ.

References

1. Leslie AD, Mencuccini M, Perks M (2012) The potential for Eucalyptus as a wood fuel in the UK. Appl Energ 89: 176–182.
2. Munoz F, Valenzuela P, Gacitua W (2012) Eucalyptus nitens: nanomechanical properties of bark and wood fibers. Appl Phys A-Mater 108: 1007–1014.
3. Brawner JT, Lee DJ, Meder R, Almeida AC, Dieters MJ (2013) Classifying genotype by environment interactions for targeted germplasm deployment with a focus on Eucalyptus. Euphytica 191: 403–414.
4. Gomat HY, Deleporte P, Moukini R, Mialounguila G, Ognouabi N, et al. (2011) What factors influence the stem taper of Eucalyptus: growth, environmental conditions, or genetics? Ann For Sci 68: 109–120.
5. Sands PJ, Landsberg JJ (2002) Parameterisation of 3-PG for plantation grownEucalyptus globulus. Forest Ecol Manage 163: 273–292.
6. An D, Yang J, Zhang P (2012) Transcriptome profiling of low temperature-treated cassava apical shoots showed dynamic responses of tropical plant to cold stress. BMC Genomics 13: 64.
7. Pennycooke JC, Cox S, Stushnoff C (2005) Relationship of cold acclimation, total phenolic content and antioxidant capacity with chilling tolerance in petunia (Petuniax hybrida). Environ Exp Bot 53: 225–232.
8. Zhao ZG, Tan LL, Dang CY, Zhang H, Wu QB, et al. (2012) Deep-sequencing transcriptome analysis of chilling tolerance mechanisms of a subnival alpine plant, Chorispora bungeana. BMC Plant Biol 12: 222.
9. Fernandes J, Morrow DJ, Casati P, Walbot V (2008) Distinctive transcriptome responses to adverse environmental conditions in Zea mays L. Plant Biotechnol J 6: 782–798.
10. Ziemann M, Kamboj A, Hove RM, Loveridge S, El-Osta A, et al. (2013) Analysis of the barley leaf transcriptome under salinity stress using mRNA-Seq. Acta Physiol Plant 35: 1915–1924.
11. He XD, Li FG, Li M, Weng QJ, Shi JS, et al. (2012) Quantitative genetics of cold hardiness and growth in Eucalyptus as estimated from E. urophylla x E. tereticornis hybrids. New Forest 43: 383–394.
12. Yang QS, Wu JH, Li CY, Wei YR, Sheng O, et al. (2012) quantitative proteomic analysis reveals that antioxidation mechanisms contribute to cold tolerance in plantain (Musa paradisiaca L.; ABB Group) seedlings. Mol Cell Proteomics 11: 1853–1869.
13. Tatusov RL, Koonin EV, Lipman DJ (1997) A genomic perspective on protein families. Science 278: 631–637.
14. O'Rourke JA, Yang SS, Miller SS, Bucciarelli B, Liu JQ, et al. (2013) An RNA-Seq transcriptome analysis of orthophosphate-deficient white lupin reveals novel insights into phosphorus acclimation in plants. Plant Physiol 161: 705–724.
15. Wang XC, Yang YJ (2003) Research progress on resistance breeding of tea plant. J Tea Sci 23: 94–98.
16. Wang XC, Zhao QY, Ma CL, Zhang ZH, Cao HL, et al. (2013) Global transcriptome profiles of Camellia sinensis during cold acclimation. BMC Genomics 14: 415.
17. Torres R, Teixidó N, Usall J, Abadias M, Mir N, et al. (2011) Anti-oxidant activity of oranges after infection with the pathogen Penicillium digitatum or treatment with the biocontrol agent Pantoea agglomerans CPA-2. Biol Control 57: 103–109.
18. Fernandez M, Villarroel C, Balbontin C, Valenzuela S (2010) Validation of reference genes for real-time qRT-PCR normalization during cold acclimation in Eucalyptus globulus. Trees-Struct Funct 24: 1109–1116.
19. Navarro M, Ayax C, Martinez Y, Laur J, El Kayal W, et al. (2011) Two EguCBF1 genes overexpressed in Eucalyptus display a different impact on stress tolerance and plant development. Plant Biotechnol J 9: 50–63.
20. Ponciano G, McMahan CM, Xie WS, Lazo GR, Coffelt TA, et al. (2012) Transcriptome and gene expression analysis in cold-acclimated guayule (Parthenium argentatum) rubber-producing tissue. Phytochemistry 79: 57–66.
21. Liu SH, Wang NF, Zhang PY, Cong BL, Lin XZ, et al. (2013) Next-generation sequencing-based transcriptome profiling analysis of Pohlia nutans reveals insight into the stress-relevant genes in Antarctic moss. Extremophiles 17: 391–403.
22. Fernandez M, Aguila SV, Arora R, Chen KT (2012) Isolation and characterization of three cold acclimation-responsive dehydrin genes from Eucalyptus globulus. Tree Genet Genomes 8: 149–162.
23. Fernandez M, Valenzuela S, Barraza H, Latorre J, Neira V (2012) Photoperiod, temperature and water deficit differentially regulate the expression of four dehydrin genes from Eucalyptus globulus. Trees-Struct Funct 26: 1483–1493.

24. Li XY, Sun HY, Pei JB, Dong YY, Wang FW, et al. (2012) De novo sequencing and comparative analysis of the blueberry transcriptome to discover putative genes related to antioxidants. Gene 511: 54–61.
25. Rodriguez MC, Edsgard D, Hussain SS, Alquezar D, Rasmussen M, et al. (2010) Transcriptomes of the desiccation-tolerant resurrection plant Craterostigma plantagineum. Plant J 63: 212–228.
26. Bai X, Rivera-Vega L, Mamidala P, Bonello P, Herms D A, et al. (2011) Transcriptomic signatures of ash (Fraxinus spp.) phloem. PLoS One 6: e16368.
27. Dussert S, Guerin C, Andersson M, Joet T, Tranbarger TJ, et al. (2013) Comparative transcriptome analysis of three oil palm fruit and seed tissues that differ in oil content and fatty acid composition. Plant Physiol 162: 1337–1358.
28. Hua CW, Marcal S, Hong Y, Eduardo L, Victor C, et al. (2012) Reference Genes for High-Throughput Quantitative ReverseTranscription-PCR Analysis of Gene Expression in Organs andTissues of Eucalyptus Grown in Various EnvironmentalConditions. Plant Cell Physiology 53(12): 2101–2116.
29. Feng BM, Lu DH, Ma X, Peng YB, Sun J, et al. (2012) Regulation of the Arabidopsis anther transcriptome by DYT1 for pollen development. Plant J 72: 612–624.
30. Ragusa M, Statello L, Maugeri M, Majorana A, Barbagallo D, et al. (2012) Specific alterations of the microRNA transcriptome and global network structure in colorectal cancer after treatment with MAPK/ERK inhibitors. J Mol Med-JMM 90: 1421–1438.
31. Tao X, Gu YH, Wang HY, Zheng W, Li X, et al. (2012) Digital gene expression analysis based on integrated de novo transcriptome assembly of sweet potato [Ipomoea batatas (L.) Lam.]. PLoS One 7: e36234.
32. Birol I, Jackman SD, Nielsen CB, Qian JQ, Varhol R, et al. (2009) De novo transcriptome assembly with ABySS. Bioinformatics 25: 2872–2877.
33. Wang L, Li PH, Brutnell TP (2010) Exploring plant transcriptomes using ultra high-throughput sequencing. Brief Funct Genomics 9: 118–128.
34. Wu HL, Chen D, Li JX, Yu B, Xiao XY, et al. (2013) De Novo characterization of leaf transcriptome using 454 sequencing and development of EST-SSR markers in tea (Camellia sinensis). Plant Mol Biol Rep 31: 524–538.
35. Schafleitner R, Tincopa LR, Palomino O, Rossel G, Robles RF, et al. (2010) A sweet potato gene index established by de novo assembly of pyrosequencing and Sanger sequences and mining for gene-based microsatellite markers. BMC Genomics 11: 604.
36. Shi CY, Yang H, Wei CL, Yu O, Zhang ZZ, et al. (2011) Deep sequencing of the Camellia sinensis transcriptome revealed candidate genes for major metabolic pathways of tea-specific compounds. BMC Genomics 12: 131.
37. Duan J, Xia C, Zhao G, Jia J, Kong X (2012) Optimizing de novo common wheat transcriptome assembly using short-read RNA-Seq data. BMC Genomics 13: 392.
38. Xiao M, Zhang Y, Chen X, Lee EJ, Barber CJS, et al. (2013) Transcriptome analysis based on next-generation sequencing of non-model plants producing specialized metabolites of biotechnological interest. J Biotechnol 166: 122–134.
39. Ong WD, Voo LYC, Kumar VS (2012) De novo assembly, characterization and functional annotation of pineapple fruit transcriptome through massively parallel sequencing. PLoS One 7: e46937.
40. Paszkiewicz K, Studholme DJ (2010) De novo assembly of short sequence reads. Brief Bioinform 11: 457–472.
41. Verma P, Shah N, Bhatia S (2013) Development of an expressed gene catalogue and molecular markers from the de novo assembly of short sequence reads of the lentil (Lens culinaris Medik.) transcriptome. Plant Biotechnol J 11: 894–905.
42. Bates LS, Waldren RP, Teare ID (1973) Rapid determination of free proline for water-stress studies. Plant Soil 39: 205–207.
43. Nanjo T, Kobayashi M, Yoshiba Y, Kakubari Y, Yamaguchi-Shinozaki K, et al. (1999) Antisense suppression of proline degradation improves tolerance to freezing and salinity in Arabidopsis thaliana. Febs Letters 461: 205–210.
44. Kishor P, Hong Z, Miao GH, Hu C, Verma D (1995) Overexpression of [delta]-pyrroline-5-carboxylate synthetase increases proline production and confers osmotolerance in transgenic plants. Plant Physiol 108: 1387–1394.
45. Delauney AJ, Verma DPS (2002) Proline biosynthesis and osmoregulation in plants. Plant J 4: 215–223.
46. Forlani G, Scainelli D, Nielsen E (1997) [delta]1-pyrroline-5-carboxylate dehydrogenase from cultured cells of potato (purification and properties). Plant Physiol 113: 1413–1418.
47. Verslues PE, Sharma S (2010) Proline metabolism and its implications for plant-environment interaction. Arabidopsis Book 8: e0140.

48. Ábrahám E, Rigó G, Székely G, Nagy R, Koncz C, et al. (2003) Light-dependent induction of proline biosynthesis by abscisic acid and salt stress is inhibited by brassinosteroid in Arabidopsis. Plant Mol Biol 51: 363–372.

49. Delauney AJ, Hu CA, Kishor PB, Verma DP (1993) Cloning of ornithine delta-aminotransferase cDNA from Vigna aconitifolia by trans-complementation in *Escherichia coli* and regulation of proline biosynthesis. J Biol Chem 268: 18673–18678.

50. Tena G, Boudsocq M, Sheen J (2011) Protein kinase signaling networks in plant innate immunity. Curr Opin Plant Biol 14: 519–529.

51. Goossens A, Hakkinen ST, Laakso I, Seppanen-Laakso T, Biondi S, et al. (2003) A functional genomics approach toward the understanding of secondary metabolism in plant cells. Proc Natl Acad Sci U S A 100: 8595–8600.

52. Vom Endt D, Kijne JW, Memelink J (2002) Transcription factors controlling plant secondary metabolism: what regulates the regulators? Phytochemistry 61: 107–114.

53. Constabel CP, Lindroth RL (2010) The impact of genomics on advances in herbivore defense and secondary metabolism in Populus. In: Jansson S, Bhalerao R, Groover A, editors. Genetics and genomics of Populus, vol 8. London: Springer. pp. 279–305.

54. Mithofer A, Boland W (2012) Plant defense against herbivores: chemical aspects. Annu Rev Plant Biol 63: 431–450.

55. Osbourn A (2010) Secondary metabolic gene clusters: evolutionary toolkits for chemical innovation. Trends Genet 26: 449–457.

56. Dixon DP, Skipsey M, Edwards R (2010) Roles for glutathione transferases in plant secondary metabolism. Phytochemistry 71: 338–350.

57. Wu B, Li Y, Yan HX, Ma YM, Luo HM, et al. (2012) Comprehensive transcriptome analysis reveals novel genes involved in cardiac glycoside biosynthesis and mlncRNAs associated with secondary metabolism and stress response in *Digitalis purpurea*. BMC Genomics 13: 15.

58. Heller J, Tudzynski P (2011) Reactive oxygen species in phytopathogenic fungi: signaling, development, and disease. Annu Rev Phytopathol 49: 369–390.

59. Reape TJ, McCabe PF (2010) Apoptotic-like regulation of programmed cell death in plants. Apoptosis 15: 249–256.

60. Baxter A, Mittler R, Suzuki N (2013) ROS as key players in plant stress signalling. J Exp Bot 65: 1229–1240.

61. Dey S, Ghose K, Basu D (2010) Fusarium elicitor-dependent calcium influx and associated ros generation in tomato is independent of cell death. Eur J Plant Pathol 126: 217–228.

62. Hinnebusch AG (2006) eIF3: a versatile scaffold for translation initiation complexes. Trends Biochem Sci 31: 553–562.

63. Sizova DV, Kolupaeva VG, Pestova TV, Shatsky IN, Hellen CU (1998) Specific interaction of eukaryotic translation initiation factor 3 with the 5' nontranslated regions of hepatitis C virus and classical swine fever virus RNAs. J Virol 72: 4775–4782.

64. Dever TE (2002) Gene-specific regulation by general translation factors. Cell 108: 545–556.

65. Maquat LE, Tarn WY, Isken O (2010) The pioneer round of translation: features and functions. Cell 142: 368–374.

66. Preiss T, Hentze MW (2003) Starting the protein synthesis machine: eukaryotic translation initiation. Bioessays 25: 1201–1211.

67. Gallie DR, Le H, Caldwell C, Robert LT, Nam XH, et al. (1997) The phosphorylation state of translation initiation factors is regulated developmentally and following heat shock in wheat. J Biol Chem 272: 1046–1053.

68. Tuteja N (2007) Mechanisms of high salinity tolerance in plants. Methods Enzymol 428: 419–438.

69. Lee BH, Henderson DA, Zhu J-K (2005) The Arabidopsis Cold-Responsive Transcriptome and Its Regulation by ICE1. Plant Cell17: 3155–3175.

70. Kayal WE, Navarro M, Marque G (2006) Expression profile of CBF-like transcriptional factor genes from Eucalyptus in response to cold. Journal Experiment Botany57: 2455–2469.

71. Navarro M, Marque G, Ayax C (2009) Complementary regulation of four Eucalyptus CBF genes under various cold conditions. Journal Experiment Botany 60: 2713–2724.

72. Fernandez M, Aguila SV, Arora R (2012) Isolation and characterization of three cold acclimation-responsive dehydrin genes from Eucalyptus globulus. TREE GENETICS & GENOMES, 8: 149–155.

73. Grabherr MG, Haas BJ, Yassour M, Levin JZ, Thompson DA, et al. (2011) Full-length transcriptome assembly from RNA-Seq data without a reference genome. Nature Biotechnol 29: 644–652.

74. Schulz MH, Zerbino DR, Vingron M, Birney E (2012) Oases: robust *de novo* RNA-seq assembly across the dynamic range of expression levels. Bioinformatics 28: 1086–1092.

75. Hernandez D, Tewhey R, Veyrieras JB, Farinelli L, Østerås M, et al. (2014) *De novo* finished 2.8 Mbp Staphylococcus aureus genome assembly from 100 bp short and long range paired-end reads. Bioinformatics 30: 40–49.

76. Luo R, Liu B, Xie Y, Li Z, Huang W, et al. (2012) SOAP*denovo*2: an empirically improved memory-efficient short-read *de novo* assembler. Gigascience 1: 18.

77. Huang X, Madan A (1999) CAP3: a DNA sequence assembly program. Genome Res 9: 868–877.

78. Langmead B, Trapnell C, Pop M, Salzberg SL (2009) Ultrafast and memory-efficient alignment of short DNA sequences to the human genome. Genome Biol 10: R25.

79. Conesa A, Gotz S (2008) Blast2GO: a comprehensive suite for functional analysis in plant genomics. Int J Plant Genomics 2008. doi: 10.1155/2008/619832

80. Kanehisa M, Goto S (2000) KEGG: kyoto encyclopedia of genes and genomes. Nucleic Acids Res 28: 27–30.

81. Arasan SK, Park JI, Ahmed NU, Jung HJ, Lee IH, et al. (2013) Gene ontology based characterization of expressed sequence tags (ESTs) of *Brassica rapa cv. Osome*. Indian J Exp Biol 51: 522–530.

82. Ye J, Fang L, Zheng H, Zhang Y, Chen J, et al. (2006) WEGO: a web tool for plotting GO annotations. Nucleic Acids Res 34: W293–297.

83. Robinson MD, McCarthy DJ, Smyth GK (2010) edgeR: a Bioconductor package for differential expression analysis of digital gene expression data. Bioinformatics 26: 139–140.

19

Vitamin D Insufficiency in Arabs and South Asians Positively Associates with Polymorphisms in GC and CYP2R1 Genes

Naser Elkum[1,5]*, Fadi Alkayal[2], Fiona Noronha[1], Maisa M. Ali[2], Motasem Melhem[2], Monira Al-Arouj[4], Abdullah Bennakhi[4], Kazem Behbehani[1], Osama Alsmadi[2], Jehad Abubaker[3]

1 Department of Biostatistics & Epidemiology, Dasman Diabetes Institute, Kuwait City, Kuwait, 2 Genetics & Genomics Unit, Dasman Diabetes Institute, Kuwait City, Kuwait, 3 Biochemistry & Molecular Biology Unit, Dasman Diabetes Institute, Kuwait City, Kuwait, 4 Clinical Services, Dasman Diabetes Institute, Kuwait City, Kuwait, 5 Clinical Epidemiology, Sidra Medical and Research Centre, Doha, Qatar

Abstract

Background: A number of genetic studies have reported an association between vitamin D related genes such as group-specific component gene (GC), Cytochrome P450, family 2, subfamily R, polypeptide 1 (CYP2R1) and 7-dehydrocholesterol reductase/nicotinamide-adenine dinucleotide synthetase 1 (DHCR7/NADSYN1) and serum levels of the active form of Vitamin D, 25 (OH) D among African Americans, Caucasians, and Chinese. Little is known about how genetic variations associate with, or contribute to, 25(OH)D levels in Arabs populations.

Methods: Allele frequencies of 18 SNPs derived from CYP2R1, GC, and DHCR7/NADSYN1 genes in 1549 individuals (Arabs, South Asians, and Southeast Asians living in Kuwait) were determined using real time genotyping assays. Serum levels of 25(OH)D were measured using chemiluminescence immunoassay.

Results: GC gene polymorphisms (rs17467825, rs3755967, rs2282679, rs7041 and rs2298850) were found to be associated with 25(OH)D serum levels in Arabs and South Asians. Two of the CYP2R1 SNPs (rs10500804 and rs12794714) and one of GC SNPs (rs1155563) were found to be significantly associated with 25(OH)D serum levels only in people of Arab origin. Across all three ethnicities none of the SNPs of DHCR7/NADSYN1 were associated with serum 25(OH)D levels and none of the 18 SNPs were significantly associated with serum 25(OH)D levels in people from South East Asia.

Conclusion: Our data show for the first time significant association between the GC (rs2282679 and rs7041), CYP2R1 (rs10741657) SNPs and 25(OH)D levels. This supports their roles in vitamin D Insufficiency in Arab and South Asian populations respectively. Interestingly, two of the CYP2R1 SNPs (rs10500804 and rs12794714) and one GC SNP (rs1155563) were found to correlate with vitamin D in Arab population exclusively signifying their importance in this population.

Editor: Masaru Katoh, National Cancer Center, Japan

Funding: This work was funded by Kuwait Foundation for the Advancement of Sciences. The funders had no role in study design, data collection and analysis, decision to publish, or preparation of the manuscript.

Competing Interests: The authors have declared that no competing interests exist.

* Email: nelkum@sidra.org

Introduction

Vitamin D deficiency is a common public health problem worldwide. It is associated with many medical outcomes, including osteoporosis [1], type 1 diabetes [2], cardiovascular diseases [3], asthma [4], and cancer [5]. Vitamin D plays a major role in calcium and phosphate homeostasis, both of which are essential in the mineralization of bone, muscle contraction, nerve conduction, and general cellular function in all cells of the body. Its active form, 1,25-dihydroxyvitamin D [1,25(OH)2D], control the expression of a vitamin D-dependent genes that code for calcium-transporting and bone matrix proteins [6]. The best

indicator of vitamin D status is the serum concentration of its main circulating metabolite, 25-hydroxyvitamin D [25(OH)D].

Factors that can potentially affect vitamin D status are dietary intake and exposure to ultraviolet-B (UVB) sunlight. Sunlight exposure catalyzes vitamin D photochemical synthesis from a cholesterol-like precursor in the skin, which is by far the most important source of vitamin D [7,8] and therefore limited exposure to sunlight is thought to be a key factor in vitamin D deficiency [9,10]. In Arabian Gulf countries where there is plentiful sunlight throughout the year, vitamin D could be expected to be adequate. Studies in Saudi Arabia and United

Arab Emirates have nevertheless highlighted a high prevalence of vitamin D deficiency in the local populations [10,11].

A study of a multi-ethnic population in the United Arab Emirates (UAE) [11] found serum 25(OH)D levels to be deficient in the overall population but sufficient among the Europeans contingent living in the same environment. Other studies have also reported wide ethnic differences. One study found a vitamin D insufficiency rate in Moroccans of 91% [12]. In the USA, Arab-American women living in southeast Detroit have been found to have dangerously low vitamin D levels [13]. These finding and others suggest potential genetic influences that predispose people of Arab backgrounds to vitamin D deficiency [14,15].

In this study, we look at which genetic variants underlie 25(OH)D status and how single nucleotide polymorphisms (SNPs) in key genes may be influencing vitamin D status. For example, SNPs in enzymes required for the production or secretion of 25(OH)D and 1,25-dihydroxyvitamin D [1,25(OH)D] could influence serum concentrations and polymorphism (e.g. if it results in more or less efficient enzyme, receptor, or binding protein), could either increase or decrease the concentration of 25(OH)D in sera.

Recent genetic studies have associated vitamin D deficiency with several candidate genes including Cytochrome P450, family 2, R, (CYP2R1), the group-specific component gene (GC) and 7-dehydrocholesterol reductase/NAD synthetase 1 (DHCR7/NAD-SYN1) [16–19]. These genes are involved in cholesterol synthesis (DHCR7/NADSYN1), hydroxylation (CYP2R1), and vitamin D transport (GC). The association between the polymorphisms of these genes and 25(OH)D has been previously studied in populations of European descent [18,20,21], African Americans [22,23], and Chinese [19,24]. Little is known about these associations in Arab populations. This study set out to (a) measure levels of 25(OH)D in Arabs, South Asians, and Southeast Asians; (b) to estimate the allelic frequencies of 18 SNPs from CYP2R1, GC, and DHCR7/NADSYN1 genes; and to investigate the relationship between these genetic polymorphisms and the level of 25(OH)D in Arab and Asian populations. In this study, we report for the first time the significant association between SNPs from GC (rs2282679 and rs7041), CYP2R1 (rs10741657) genes and 25(OH)D levels which clearly support their roles in vitamin D Insufficiency in Arab and South Asian populations. The lack of association between DHCR7/NADSYN1 SNPs and 25(OH)D levels minimize their role in controlling vitamin D levels in all three ethnic groups. Moreover, the fact that two of the CYP2R1 SNPs (rs10500804 and rs12794714) and one GC SNP (rs1155563) were found to associate with Vitamin D levels exclusively in Arabs signify their role in vitamin D insufficiency within this population.

Materials and Methods

Study population

A cross-sectional population-based survey was conducted with a random representative sample of adults ($\geq$18 years) from multi-ethnic origin across the six governorates (strata) of the State of Kuwait. A full description of the study population, design and data collection of this study have outlined previously [25]. Briefly, a stratified random sampling technique was used for the selection of participants from the computerized register of the Public Authority of Civil Information. This survey was carried out between June 2011 and August 2012. The study conformed to the principles outlined in the Declaration of Helsinki and was approved by the Scientific Advisory Council and Ethical Review Committee at the Dasman Diabetes Institute (DDI) IRB # 1– Biomedical. An informed written consent was obtained from all participants before their enrolment in the study.

Measurement of vitamin D levels

Serum 25(OH)D levels were measured by chemiluminescent competitive immunoassay (CLIA) using a DiaSorin LIAISON analyzer (DiaSorin Inc, MN, USA) and following company instruction. In brief, 25 OH Vitamin D was dissociated from its binding protein and bound to a specific antibody on the solid phase. Then the tracer (vitamin D linked to an isoluminol derivative) was added. Next, the unbound material was removed with a wash cycle. Finally, the starter reagents were added to initiate a flash chemiluminescent reaction. The light signal was measured by a photomultiplier as relative light units (RLU) and was inversely proportional to the concentration of 25(OH)D in the samples. The intra-assay coefficients of variations (CVs) were 5.5% and 4.0% at 10 and 25 ng/mL, respectively. The inter-assay CVs were 8% and 6% at 15 and 40 ng/mL, respectively. Body Mass Index (BMI) was calculated, using the standard BMI formula, as body weight (in kilograms) divided by height (in meters) squared. This study was conducted on adult male and female subjects comprising of Normal (BMI = 20–24.9 kg/m^2), overweight (BMI = 25–29.9 kg/m^2) and obese (BMI$\geq$30 kg/m^2) individuals. Other measurements such age, weight, height, gender and ethnicity were also obtained.

DNA collection, SNP selection, and genotyping

Blood samples were taken from consenting participants in accordance with DDI IRB approved consent form in 4 ml tubes containing EDTA anticoagulant. Gentra Puregene kit (Qiagen, Valencia, CA, USA) was used to extract DNA as per manufacturer's protocols. DNA was quantified, with a requirement that the A260/A280 ratio is in the range of 1.8–2.1, using Epoch Microplate Spectrophotometer. DNA stock aliquots were diluted to a concentration of 50 ng/μl and were frozen until needed for use in PCR assays.

We selected three candidate genes containing 18 SNPs (6 SNPs from each gene) that have been shown in previous GWAS reports [17,18] to have an association with vitamin D level, and are known to have a biological impact in vitamin D metabolism. The genes and SNPs included GC (rs17467825, rs2282679, rs3755967, rs2298850, rs7041, rs1155563), CYP2R1 (rs7116978, rs1993116, rs10500804, 4s12794714, rs10741657, rs206793), and DHCR7/NADSYN1 (rs7944926, rs12785878, rs4944957, rs12800438, rs3794060, and rs3829251).

We employed ready-to-use, manufacturer-validated, pre-designed allele discriminating TaqMan single nucleotide polymorphisms (SNP) assays. PCR amplification reactions were each carried out in clear optical 96-well plates on the Applied Biosystems (ABI) 7500 Real Time PCR system. Each reaction was performed in 20 μl volume, containing 20 ng of DNA template, 1X TaqMan pre-designed SNP assay master mix (Applied Biosystems, Carlsbad, CA, United States), and 1X HOT FIREPol EvaGreen qPCR master mix plus containing ROX reference dye (Solis BioDyne, Tartu, Estonia). Amplification cycling reactions were carried out under the following conditions: an initial incubation step at 95°C for 10 minutes followed by 35 cycles at 95°C for 1 minute, 58°C for 45 seconds, and 72°C for 45 seconds, then a final incubation step at 72°C for 7 minutes. An endpoint "plate read" was then performed using the ABI Sequence Detection System (SDS) Software. The Software detects fluorescence released from the bi-allelic TaqMan probes and plots fluorescence (Rn) signals (FAM & VIC) and generates the genotype calls. In every run, two previously validated DNA controls per genotype were included, in addition to two water (no template) negative controls.

Statistical analysis

We compared the baseline characteristics of the participants using analysis of variance tests (ANOVA) for continuous variables. Categorical variables were analyzed using the chi-square test. Mean serum 25(OH)D values were estimated within each group of homozygous referent (HR), heterozygous (HET), and homozygous variant (HV) genotypes for each SNP. The serum 25(OH)D levels were adjusted by age, gender and BMI using analysis of covariance (ANCOVA) approach. A P-value<0.05 was considered to be statistically significant. All analyses were performed using SAS (version 9.2; SAS Institute, Cary, NC). Genotype frequencies were determined according to the Hardy-Weinberg equilibrium (HWE).

Results

Clinical and biochemical characteristics of the participants

The study population (1549 in total) comprised subjects mainly of Arab and Asian background; 907 Arabs, 489 South Asians and 153 Southeast Asians. The descriptive characteristics of all participants are summarized in Table 1. There were substantial ethnic differences in BMI, age, and gender (p<0.0001); Arabs had a significantly higher mean BMI and tended to be older than Asians. Southeast Asians had higher mean levels of 25(OH)D compared to Arabs and South Asians.

Distribution of allele frequencies in different populations

Significant differences were found in the allele frequencies across all three study populations. The distribution of minor alleles frequencies of each SNP in the Arab, South Asia, and Southeast Asia populations are shown in Figure 1. Minor alleles (frequency below 0.20) were less common in the Arab population (5.3%) than in the South Asian (31.6%) and Southeast Asian (26.4%) populations. Hardy-Weinberg equilibrium (HWE) was not met for some SNPs across all three-populations (P<0.05). Three

CYP2R1 SNPs in the Arab population and four GC SNPs in the Southeast Asian population had HWE<0.05 (Table 2).

Association between genotypes and 25(OH) D serum concentrations

SNP genotype and mean serum 25(OH)D levels categorized by genotype are presented in Table 3. Only two CYP2R1 SNPs (rs10500804 and rs12794714) in Arab and one SNP (rs10741657) in South Asian populations were positively associated with vitamin D. No association was observed between the CYP2R1 gene and vitamin D in the South East Asian population. As for the GC gene, all the SNPs were significantly associated with serum 25(OH)D levels in the Arab population but not in the South East Asia population. In South Asian group, five SNPs in GC the gene (rs17467825, rs2282679, rs3755967, rs2298850, and rs7041) were significantly associated with vitamin D levels. SNPs in DHCR7/NADSYN1 gene showed no significant association with serum 25(OH)D in any of the three study populations. Table S1 shows unadjusted analysis of the association between genotypes and 25(OH)D serum levels.

Discussion

We conducted this research because different studies have recently shown ethnic differences in the allele frequency of "vitamin D associated SNPs" [22,26,27], but little information is available about the frequency of vitamin D SNPs in Arab and Asian populations. Our data show that the minor allele frequency (MAF<0.20) was higher in Asian populations (South Asians 31.6% and South East Asians 26.3%) compared to Arabs (5.3%); this can be attributed to an ascertainment bias as the SNPs examined were from studies representing European and African ethnicities. In the GC gene, MAFs for South Asians were significantly higher (P<0.05) than Arabs and South East Asians (Figure 1), whereas South East Asians have the lowest frequencies among all the SNPs. The Arab population has shown distinctive significant differences in the allele frequency of the NADSYN1/DHCR7 gene compared to other ethnic groups. Arabs showed the highest MAF among all the SNPs of the NADSYN1/DHCR7

Table 1. Characteristics of participants by ethnicity.

Factors	Ethnicities			P-value trend
	Arabs	South-East Asia	South Asia	
	n (%)	n (%)	n (%)	
Age (years)				
20–39	316 (20.3)	90 (5.8)	186 (11.9)	<0.0001
40–60	497 (31.9)	59 (3.8)	276 (17.8)	<0.0001
>60	98 (6.3)	5 (0.32)	28 (1.8)	<0.0001
Gender				
Female	270 (17.4)	104 (6.7)	104 (6.7)	<0.0001
Male	641 (41.2)	50 (3.2)	386 (24.8)	<0.0001
BMI				
Normal (18.5–24.9)	107 (6.9)	62 (3.9)	143 (9.2)	<0.0001
Overweight (25–29.9)	296 (19.0)	65 (4.2)	225 (14.5)	<0.0001
Obese (≥30)	508 (32.7)	27 (1.7)	122 (7.9)	<0.0001
*25 OHD (ng/l)	13.5±0.34	17.8±0.82	13.3±0.46	<0.0001

BMI, body mass index; *Data are presented as mean ± SD.

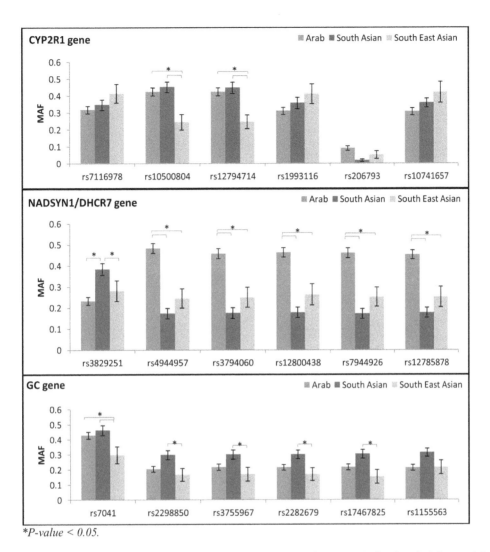

Figure 1. Minor Allele Frequency of SNPs from CYP2R1, DHCR7 & GC genes between Arabs, South Asian and South East Asian ethnicities. The standard deviation was estimated using the.

gene; only the rs3829251 SNP demonstrated higher frequencies in South Asians. Furthermore, analyses showed no significant differences in MAFs between Arab and South Asians populations across the CYP2R1 SNPs (rs206793 was an exception). These differences in allele frequencies are of significant importance in the design of association studies and selecting candidate vitamin D SNPs to investigate.

We were also interested in exploring the relationships between SNPs from the key Vitamin D genes (GC, CYP2R1 and DHCR7/ NADSYN1) and 25(OH)D levels. The GC and CYP2R1 genes were found to be significantly associated with 25(OH)D levels. The GC gene encodes the vitamin D binding protein (DBP) that is the key transporter of vitamin D and its metabolites (including 25(OH)D and 1,25(OH)₂D) in the circulation [28]. Recent studies have reported an association between SNPs in this gene and 25(OH)D concentrations [22,23,29,30]. Furthermore, other recent studies involving African Americans and Europeans have reported a significant association between SNP rs2282679 of the GC gene and vitamin D insufficiency [17,23]. It is notable that we found a strong association between SNP rs2282679 and 25(OH)D in Arabs and South Asians but not in South East Asians. Similar to the aforementioned studies on different ethnicities [24], rs7041

polymorphism was also associated with 25(OH)D in Arab and South Asian populations, but this was not the case with South East Asians. We also observed that polymorphisms rs17467825, rs3755967, and rs2298850 in the GC gene were associated with serum 25(OH)D levels in Arab and South Asian populations. Interestingly, rs1155563 SNP associated significantly with vitamin D level in Arabs, which suggests a possible involvement in Vitamin D secretion and transportation in this group. A previous study has found that the SNP rs1155563 was associated with vitamin D levels in men of non-Hispanic white background [30]. The significant associations between these GC SNPs and 25(OH)D levels further support the importance of the GC gene in Vitamin D insufficiency.

CYP2R1 is a microsomal vitamin D hydroxylase that hydroxylates vitamin D at the 25-C position for 25(OH)D synthesis (calcidiol) in the liver [31]. Subsequently, calcidiol is converted to calcitriol, the active form of vitamin D3 that binds to the vitamin D receptor (VDR) which arbitrates the majority of vitamin D physiological actions. Previous research has shown the gene CYP2R1 to be associated with several vitamin D related diseases such as type 1 diabetes [21], and in this study we found that rs10741657 SNP, which is a coding SNP that can change the

Table 2. Allele frequencies of 18 Vitamin D associated SNPs by ethnicity.

Gene & SNP	Region	Allele	MAF		Arab n=907			South Asian n=489			South East Asian n=153		
			All[a]	All[b]	MAF	HET	HWE	MAF	HET	HWE	MAF	HET	HWE
CYP2R1													
rs7116978	Ch 11	C/T	0.36	0.33	0.32	0.40	**0.0268**	0.35	0.43	0.2123	0.42	0.44	0.2248
rs10500804	Ch 11	T/G	0.43	0.42	0.42	0.45	**0.0164**	0.45	0.49	0.7610	0.24	0.39	0.6027
rs12794714	Ch 11	G/A	0.43	0.42	0.42	0.45	**0.0169**	0.45	0.48	0.5556	0.25	0.39	0.5849
rs1993116	Ch 11	G/A	0.40	0.33	0.31	0.42	0.6667	0.36	0.44	0.2498	0.41	0.44	0.3073
rs206793	Ch 11	T/C	0.40	0.07	0.09	0.16	0.0710	0.02	0.04	0.6613	0.05	0.10	0.4950
rs10741657	Ch 11	G/A	0.40	0.33	0.31	0.42	0.5537	0.36	0.43	0.1540	0.42	0.44	0.2067
DHCR7/NADSYN1													
rs3829251	Ch 11	G/A	0.24	0.28	0.23	0.36	0.9608	0.38	0.47	0.3126	0.28	0.41	0.9729
rs4944957	Ch 11	A/G	0.23	0.37	0.48	0.49	0.6672	0.17	0.40	0.4225	0.24	0.37	0.9334
rs3794060	Ch 11	C/T	0.23	0.35	0.46	0.49	0.6768	0.18	0.39	0.1186	0.25	0.38	0.8288
rs12800438	Ch 11	G/A	0.23	0.36	0.46	0.49	0.6117	0.18	0.39	0.1430	0.26	0.37	0.5376
rs7944926	Ch 11	A/G	0.23	0.35	0.46	0.49	0.6534	0.17	0.40	0.2068	0.25	0.39	0.7677
rs12785878	Ch 11	G/T	0.23	0.35	0.45	0.49	0.5725	0.18	0.42	0.5105	0.25	0.40	0.4687
GC													
rs7041	Ch 4	C/A	0.44	0.47	0.43	0.47	0.1302	0.46	0.47	0.3126	0.30	0.39	0.3394
rs2298850	Ch 4	G/C	0.28	0.23	0.20	0.32	0.6739	0.30	0.40	0.4225	0.17	0.20	**0.0008**
rs3755967	Ch 4	C/T	0.29	0.24	0.22	0.33	0.4866	0.30	0.39	0.1186	0.17	0.21	**0.0014**
rs2282679	Ch 4	T/G	0.29	0.24	0.21	0.33	0.6582	0.30	0.39	0.1430	0.17	0.21	**0.0014**
rs17467825	Ch 4	A/G	0.29	0.24	0.22	0.33	0.6172	0.30	0.40	0.2068	0.15	0.19	**0.0008**
RS1155563	Ch 4	T/C	0.30	0.25	0.21	0.32	0.2397	0.31	0.42	0.5105	0.22	0.31	0.3683

Allele major allele/minor allele; MAF minor allele frequency; HET heterozygosity; HWE *P*-values for Hardy-Weinberg Equilibrium test.
Bold numbers represent significant *P*-values.
[a]Based on GWAS by Wang, et al. 2010;
[b]based on our study.

Table 3. Single nucleotide polymorphisms in GC, CYP2R1, DHCR7/NADSYN1, and their association with serum 25(OH)D among Arabs, South Asian, Southeast Asian participants.

Gene & SNP	Region	HR, HET, HV	Arab n = 907 25(OH)D concentrations				South Asian n = 489 25(OH)D concentrations				South East Asian n = 153 25(OH)D concentrations			
			HR	HET	HV	P-values	HR	HET	HV	P-values	HR	HET	HV	P-values
CYP2R1														
rs7116978	Ch 11	CC, CT, TT	13.5	14.2	14.4	0.5747	13.7	14.4	16.2	0.1068	16.7	18.3	16.1	0.3397
rs1993116	Ch 11	GG, AG, AA	13.3	14.5	14.1	0.2736	13.4	14.6	16.2	0.0555	17.1	17.7	16.8	0.8459
rs1050804	Ch 11	TT, TG, GG	14.4	14.3	12.0	**0.0379**	14.5	14.7	13.3	0.3770	17.8	16.7	16.4	0.6780
rs12794714	Ch 11	GG, AG, AA	14.4	14.3	12.0	**0.0402**	14.5	14.8	13.1	0.2790	17.8	16.7	16.4	0.7009
rs10741657	Ch 11	GG, AG, AA	13.3	14.5	14.1	0.2873	13.4	14.6	16.2	**0.0437**	16.9	17.9	16.7	0.7224
rs206793	Ch 11	TT, CT, CC	13.9	13.7	9.6	0.3804	14.3	13.8	-	0.7915	17.4	16.4	-	0.6456
DHCR7/NADSYN1														
rs7944926	Ch 11	AA, AG, GG	13.2	14.1	14.2	0.5300	13.9	15.5	11.9	0.1096	16.4	18.1	21.9	0.1111
rs12785878	Ch 11	GG, GT, TT	13.2	14.2	14.0	0.5164	14.0	15.4	12.4	0.1660	16.2	18.2	22.2	0.0736
rs4944957	Ch 11	AA, AG, GG	13.0	14.1	14.4	0.3336	14.0	15.5	11.9	0.1234	16.6	17.8	21.9	0.1577
rs12800438	Ch 11	GG, AG, AA	13.3	14.1	14.2	0.6303	14.1	15.2	12.7	0.2761	16.3	18.9	16.0	0.1371
rs3794060	Ch 11	CC, CT, TT	13.3	14.1	14.3	0.5396	14.0	15.5	12.3	0.1364	16.4	18.1	21.9	0.1135
rs3829251	Ch 11	GG, AG, AA	14.4	12.9	14.1	0.1833	13.6	14.9	13.9	0.2599	18.1	16.8	15.0	0.3578
GC														
rs17467825	Ch 4	AA, AG, GG	14.4	13.4	10.3	**0.0165**	15.8	12.9	13.0	**0.0013**	17.7	17.4	12.8	0.2127
rs2282679	Ch 4	TT, GT, GG	14.4	13.3	10.4	**0.0377**	15.8	12.8	13.0	**0.0007**	17.8	17.0	12.8	0.1557
rs3755967	Ch 4	CC, CT, TT	14.4	13.3	10.5	**0.0368**	15.8	12.8	13.0	**0.0007**	17.8	17.0	12.8	0.1557
rs2298850	Ch 4	GG, GC, CC	14.4	13.3	10.3	**0.0374**	15.5	13.2	13.0	**0.0103**	17.9	16.7	13.4	0.2037
rs7041	Ch 4	CC, AC, AA	14.5	14.3	11.7	**0.0110**	15.8	14.4	12.3	**0.0072**	17.3	18.1	16.6	0.5594
rs1155563	Ch 4	TT, TC, CC	14.4	13.3	10.4	**0.0289**	14.8	14.2	12.8	0.3120	18.1	16.2	14.7	0.2541

HR: Homozygous referent, HET: Heterozygous, HV: Homozygous variant. Blue color indicates HV.
P-value for the association between the SNP and 25(OH)D levels from ethnic-stratified ANCOVA models adjusted for sex, age and BMI.

activity of the CYP2R1 enzyme and subsequently cause a relative lack of 25(OH)D, is significantly associate with vitamin D, but only in South Asians [31]. We also found two of the CYP2R1 SNPs (rs10500804 and rs12794714) to associate significantly with Vitamin D level in Arabs signifying their possible roles in vitamin D insufficiency in this population.

Gene DHCR7/NADSYN1 encodes the enzyme 7-dehydrocholesterol (7DHC) reductase, which catalyzes the production of cholesterol from 7 DHC, thereby removing the key substrate necessary for the vitamin D synthesis [18]. Recently, Zhang et al., 2012 have linked some of the DHCR7/NADSYN1 SNPs (rs3829251, rs12785878) to decreased serum 25(OH)D levels in northeastern Han Chinese children [24], while Cooper et al., 2011 has associated rs12785878 T allele carriers with vitamin D deficiency and type 1 diabetes [32]. In contrast, we found that none of the six variant genotypes of DHCR7/NADSYN1 was associated with serum 25(OH)D levels in any of the three population groups that were studied, suggesting minimal involvement of these SNPs or the DHCR7/NADSYN1 gene in mediating vitamin D insufficiency in these populations. However, it is worth noting that in our study the female participant from Arab and South Asian origins were underrepresented which might contribute to the lack of association between DHCR7/NADSYN1 gene and serum 25(OH)D levels. Interestingly, none of the 18 SNPs in this study associated significantly with serum 25(OH)D levels in South East Asians, minimizing the roles of these genes in mediating Vitamin D insufficiency and thus opening the door for the possible involvement of other genes, less common SNPs and/or predisposing factors for Vitamin D insufficiency.

Conclusion

Our study is one of the first to look at genetic determinant of vitamin D levels in Arabs and describes allele frequencies of 18 SNPs localized to three genes related to vitamin D deficiency in Arabs, South Asians and South East Asians. The significant associations between the GC (rs2282679 and rs7041), CYP2R1

(rs10741657) SNPs and 25(OH)D levels clearly support the idea of a role in vitamin D insufficiency in Arab and South Asian populations. The fact that none of the SNPs of the DHCR7/NADSYN1 gene associate with vitamin D levels in the three populations suggests minimized roles in controlling vitamin D release. The finding that GC SNP (rs1155563) and CYP2R1 SNPs (rs10500804 and rs12794714) associated exclusively with vitamin D level in Arabs, suggests the need for further study and possibly sheds light on their mechanism in the context of vitamin D insufficiency. Finally, the lack of association between the selected SNPs and vitamin D levels in South East Asians calls for larger population-based studies that include more genes linked to vitamin D and/or explore the less common SNPs within the existing genes.

Supporting Information

Table S1 Single nucleotide polymorphisms in GC, CYP2R1, DHCR7/NADSYN1, and their association with serum 25(OH)D among Arabs, South Asian, Southeast Asian participants.

Acknowledgments

We would like to thank our study team for their efforts and excellent work. We are grateful to the Clinical Laboratory and the Tissue Bank Core Facility at DDI for their contribution in performing the vitamin D profile analysis and handling samples, respectively. We would also thank Dr William Greer from the Sidra Medical and Research Centre for his proofreading and reviewing of the manuscript. Finally, we are grateful to the senior management, Scientific Advisory Board, and Ethical Review Committee of the Dasman Diabetes Institute for their unwavering support and valuable recommendations.

Author Contributions

Conceived and designed the experiments: NE. Performed the experiments: FA MMA MM OA. Analyzed the data: NE FN. Contributed reagents/materials/analysis tools: KB MA AB. Contributed to the writing of the manuscript: NE JA.

References

1. Simonelli C (2005) The role of vitamin D deficiency in osteoporosis and fractures. Minn Med 88: 34–36.
2. Borkar VV, Devidayal, Verma S, Bhalla AK (2010) Low levels of vitamin D in North Indian children with newly diagnosed type 1 diabetes. Pediatr Diabetes 11: 345–350.
3. Kilkkinen A, Knekt P, Aro A, Rissanen H, Marniemi J, et al. (2009) Vitamin D status and the risk of cardiovascular disease death. Am J Epidemiol 170: 1032–1039.
4. Bosse Y, Lemire M, Poon AH, Daley D, He JQ, et al. (2009) Asthma and genes encoding components of the vitamin D pathway. Respir Res 10: 98.
5. Heist RS, Zhou W, Wang Z, Liu G, Neuberg D, et al. (2008) Circulating 25-hydroxyvitamin D, VDR polymorphisms, and survival in advanced non-small-cell lung cancer. J Clin Oncol 26: 5596–5602.
6. Jones G, Strugnell SA, DeLuca HF (1998) Current understanding of the molecular actions of vitamin D. Physiol Rev 78: 1193–1231.
7. Gannage-Yared MH, Chemali R, Yaacoub N, Halaby G (2000) Hypovitaminosis D in a sunny country: relation to lifestyle and bone markers. J Bone Miner Res 15: 1856–1862.
8. Brustad M, Alsaker E, Engelsen O, Aksnes L, Lund E (2004) Vitamin D status of middle-aged women at 65–71 degrees N in relation to dietary intake and exposure to ultraviolet radiation. Public Health Nutr 7: 327–335.
9. Saliba W, Rennert HS, Kershenbaum A, Rennert G (2012) Serum 25(OH)D concentrations in sunny Israel. Osteoporos Int 23: 687–694.
10. Ardawi MS, Sibiany AM, Bakhsh TM, Qari MH, Maimani AA (2012) High prevalence of vitamin D deficiency among healthy Saudi Arabian men: relationship to bone mineral density, parathyroid hormone, bone turnover markers, and lifestyle factors. Osteoporos Int 23: 675–686.
11. Dawodu A, Absood G, Patel M, Agarwal M, Ezimokhai M, et al. (1998) Biosocial factors affecting vitamin D status of women of childbearing age in the United Arab Emirates. J Biosoc Sci 30: 431–437.
12. Allali F, El Aichaoui S, Khazani H, Benyahia B, Saoud B, et al. (2009) High prevalence of hypovitaminosis D in Morocco: relationship to lifestyle, physical

13. Hobbs RD, Habib Z, Alromaihi D, Idi L, Parikh N, et al. (2009) Severe vitamin D deficiency in Arab-American women living in Dearborn, Michigan. Endocr Pract 15: 35–40.
14. Hunter D, De Lange M, Snieder H, MacGregor AJ, Swaminathan R, et al. (2001) Genetic contribution to bone metabolism, calcium excretion, and vitamin D and parathyroid hormone regulation. J Bone Miner Res 16: 371–378.
15. Shea MK, Benjamin EJ, Dupuis J, Massaro JM, Jacques PF, et al. (2009) Genetic and non-genetic correlates of vitamins K and D. Eur J Clin Nutr 63: 458–464.
16. McGrath JJ, Saha S, Burne TH, Eyles DW (2010) A systematic review of the association between common single nucleotide polymorphisms and 25-hydroxyvitamin D concentrations. J Steroid Biochem Mol Biol 121: 471–477.
17. Ahn J, Yu K, Stolzenberg-Solomon R, Simon KC, McCullough ML, et al. (2010) Genome-wide association study of circulating vitamin D levels. Hum Mol Genet 19: 2739–2745.
18. Wang TJ, Zhang F, Richards JB, Kestenbaum B, van Meurs JB, et al. (2010) Common genetic determinants of vitamin D insufficiency: a genome-wide association study. Lancet 376: 180–188.
19. Zhang Z, He JW, Fu WZ, Zhang CQ, Zhang ZL (2013) An analysis of the association between the vitamin D pathway and serum 25-hydroxyvitamin D levels in a healthy Chinese population. J Bone Miner Res 28: 1784–1792.
20. Kurylowicz A, Ramos-Lopez E, Bednarczuk T, Badenhoop K (2006) Vitamin D-binding protein (DBP) gene polymorphism is associated with Graves' disease and the vitamin D status in a Polish population study. Exp Clin Endocrinol Diabetes 114: 329–335.
21. Ramos-Lopez E, Bruck P, Jansen T, Herwig J, Badenhoop K (2007) CYP2R1 (vitamin D 25-hydroxylase) gene is associated with susceptibility to type 1 diabetes and vitamin D levels in Germans. Diabetes Metab Res Rev 23: 631–636.
22. Engelman CD, Fingerlin TE, Langefeld CD, Hicks PJ, Rich SS, et al. (2008) Genetic and environmental determinants of 25-hydroxyvitamin D and 1,25-

performance, bone markers, and bone mineral density. Semin Arthritis Rheum 38: 444–451.

dihydroxyvitamin D levels in Hispanic and African Americans. J Clin Endocrinol Metab 93: 3381–3388.

23. Signorello LB, Shi J, Cai Q, Zheng W, Williams SM, et al. (2011) Common variation in vitamin D pathway genes predicts circulating 25-hydroxyvitamin D Levels among African Americans. PLoS One 6: e28623.

24. Zhang Y, Wang X, Liu Y, Qu H, Qu S, et al. (2012) The GC, CYP2R1 and DHCR7 genes are associated with vitamin D levels in northeastern Han Chinese children. Swiss Med Wkly 142: w13636.

25. Elkum N, Al-Arouj M, Sharifi M, Behbehani K, Bennakhi A (2014) Cardiovascular disease risk factors in the South Asian population living in Kuwait: a cross-sectional study. Diabet Med 31: 531–539.

26. Uitterlinden AG, Fang Y, Van Meurs JB, Pols HA, Van Leeuwen JP (2004) Genetics and biology of vitamin D receptor polymorphisms. Gene 338: 143–156.

27. Larcombe L, Mookherjee N, Slater J, Slivinski C, Singer M, et al. (2012) Vitamin D in a northern Canadian first nation population: dietary intake, serum concentrations and functional gene polymorphisms. PLoS One 7: e49872.

28. Speeckaert M, Huang G, Delanghe JR, Taes YE (2006) Biological and clinical aspects of the vitamin D binding protein (Gc-globulin) and its polymorphism. Clin Chim Acta 372: 33–42.

29. Sinotte M, Diorio C, Berube S, Pollak M, Brisson J (2009) Genetic polymorphisms of the vitamin D binding protein and plasma concentrations of 25-hydroxyvitamin D in premenopausal women. Am J Clin Nutr 89: 634–640.

30. Ahn J, Albanes D, Berndt SI, Peters U, Chatterjee N, et al. (2009) Vitamin D-related genes, serum vitamin D concentrations and prostate cancer risk. Carcinogenesis 30: 769–776.

31. Bu FX, Armas L, Lappe J, Zhou Y, Gao G, et al. (2010) Comprehensive association analysis of nine candidate genes with serum 25-hydroxy vitamin D levels among healthy Caucasian subjects. Hum Genet 128: 549–556.

32. Cooper JD, Smyth DJ, Walker NM, Stevens H, Burren OS, et al. (2011) Inherited variation in vitamin D genes is associated with predisposition to autoimmune disease type 1 diabetes. Diabetes 60: 1624–1631.

20

Genome-Wide Investigation and Expression Profiling of AP2/ERF Transcription Factor Superfamily in Foxtail Millet (*Setaria italica L.*)

Charu Lata[1,2], Awdhesh Kumar Mishra[3], Mehanathan Muthamilarasan[3], Venkata Suresh Bonthala[3], Yusuf Khan[3], Manoj Prasad[3]*

1 National Research Centre on Plant Biotechnology, New Delhi, India, 2 CSIR-National Botanical Research Institute, Lucknow, Uttar Pradesh, India, 3 National Institute of Plant Genome Research, New Delhi, India

Abstract

The APETALA2/ethylene-responsive element binding factor (AP2/ERF) family is one of the largest transcription factor (TF) families in plants that includes four major sub-families, namely AP2, DREB (dehydration responsive element binding), ERF (ethylene responsive factors) and RAV (Related to ABI3/VP). AP2/ERFs are known to play significant roles in various plant processes including growth and development and biotic and abiotic stress responses. Considering this, a comprehensive genome-wide study was conducted in foxtail millet (*Setaria italica* L.). A total of 171 *AP2/ERF* genes were identified by systematic sequence analysis and were physically mapped onto nine chromosomes. Phylogenetic analysis grouped *AP2/ERF* genes into six classes (I to VI). Duplication analysis revealed that 12 (~7%) *SiAP2/ERF* genes were tandem repeated and 22 (~13%) were segmentally duplicated. Comparative physical mapping between foxtail millet *AP2/ERF* genes and its orthologs of sorghum (18 genes), maize (14 genes), rice (9 genes) and *Brachypodium* (6 genes) showed the evolutionary insights of *AP2/ERF* gene family and also the decrease in orthology with increase in phylogenetic distance. The evolutionary significance in terms of gene-duplication and divergence was analyzed by estimating synonymous and non-synonymous substitution rates. Expression profiling of candidate *AP2/ERF* genes against drought, salt and phytohormones revealed insights into their precise and/or overlapping expression patterns which could be responsible for their functional divergence in foxtail millet. The study showed that the genes *SiAP2/ERF-069*, *SiAP2/ERF-103* and *SiAP2/ERF-120* may be considered as potential candidate genes for further functional validation as well for utilization in crop improvement programs for stress resistance since these genes were up-regulated under drought and salinity stresses in ABA dependent manner. Altogether the present study provides new insights into evolution, divergence and systematic functional analysis of *AP2/ERF* gene family at genome level in foxtail millet which may be utilized for improving stress adaptation and tolerance in millets, cereals and bioenergy grasses.

Editor: Swarup K. Parida, National Institute of Plant Genome Research (NIPGR), India

Funding: This work was financially supported by core grant of National Institute of Plant Genome Research, New Delhi, India, and Department of Science & Technology, Government of India through INSPIRE Faculty Award [IFA-11LSPA-01]. CL is the recipient of INSPIRE Faculty Award from Department of Science & Technology, while AKM and MM are the recipients of Research Fellowships from Council of Scientific and Industrial Research and University Grants Commission, New Delhi, respectively. The funders had no role in study design, data collection and analysis, decision to publish, or preparation of the manuscript.

* Email: manoj_prasad@nipgr.ac.in

Introduction

Plants frequently confront numerous environmental stresses which ultimately affect their growth and productivity. Therefore, in order to cope with these recurrent challenges, a plant species must acquire stress responsive and adaptive mechanisms that may assist in better survival and yield. Grass species of *Setaria* genus especially *S. italica* (foxtail millet) and *S. viridis* (green foxtail) prove to be excellent examples of stress adaptation and tolerance among graminaceous species [1]. Foxtail millet is a stress tolerant crop having a small genome (~515 Mb; 2n = 2x = 18) with relatively lower repetitive DNA, short life-cycle and inbreeding nature which makes it a perfect model for understanding various

biological aspects including architecture, phylogeny and physiology of related Panicoid crops, particularly potential bioenergy grasses which have closely related but relatively composite genomes [2]. Considering its significance as a model system for evolution and biological studies, its genome has recently been sequenced by Beijing Genomics Institute (BGI), China [3] and Joint Genome Institute (JGI), Department of Energy, USA [4] independently [5]. The availability of foxtail millet genome sequence has consequently encouraged plant biology researchers to work towards deciphering its structural and functional genomics that may give new insights for its stress response and adaptation mechanisms and eventually support crop improvement programmes to ensure sustainable food security [6]. However, stress response and adaptation is a complex process as stress may possibly occur at different stages of plant development with different intensities (moderate to severe) and often several stresses may act together, thus increasing the effects manifold. It is hypothesized that plants have evolved an intricate signaling network to survive stress conditions that begins with stress perception, initiation of signal transduction, modulation of stress responsive gene(s) expression and finally its manifestation at cellular and physiological levels. Stress response and adaptation entails differential gene expression which is controlled by specific transcription factors (TFs) that directly regulate majority of downstream multiple stress responsive gene expression in a synchronized manner. Hence TFs are attractive targets for application in plant molecular biology for gene manipulation and crop improvement. Among various TF families, the ethylene responsive TF (ERF) family plays an important role in plant growth and development and also enables them to adapt to changing environmental conditions, and therefore it is important to understand molecular functions of these genes in order to improve plant adaptability and productivity under varied ambiance/environmental changes.

The APETALA2/ethylene-responsive element binding factor (AP2/ERF) superfamily is a large group of TFs which is distinguished by the number of repetitions and the sequence of AP2/ERF DNA-binding domain based on which it is categorized into AP2, ERF and RAV families [7,8]. The AP2/ERF domain was first reported in the *Arabidopsis* homeotic *AP2* gene implicated in floral development [9]. This conserved DNA-binding domain usually consists of 60–70 amino acid residues and is known to interact directly with *cis*-acting elements namely GCC box and/or dehydration responsive element (DRE)/C-repeat element (CRT) present in the promoter regions of downstream target genes [10,11]. The homologous sequences of this domain have been found in homing endonucleases (HNH-endonucleases) of the cyanobacterium *Trichodesmium erythraeum*, the ciliate *Tetrahymena thermophila*, and the viruses Enterobacteria phage RB49 and Bacteriophage Felix 01 and hence it has been postulated that a horizontal transfer of an HNH-AP2 endonuclease from prokaryotes into plants resulted in evolution of the AP2/ERF superfamily [12,13]. Among the threeAP2/ERF families, the members of AP2 family contain two AP2/ERF domains connected by a 25 amino acid linker, whereas the members of ERF subfamily contain a single AP2/ERF domain. The RAV family members contain a single AP2/ERF domain and an additional B3 DNA-binding motif [14]. In addition, the AP2 family is again categorized into AP2 and AINTEGUMENTA (ANT) monophyletic groups in seed plants [15], while the ERF family is further subdivided into ERF and DREB subfamilies [6,7]. The ERF subfamily is characterized by the presence of conserved alanine and aspartic acid at 14th and 19th position respectively in the DNA-binding domain, while the DREB subfamily has valine and glutamine at respective positions [16].

A large number of AP2/ERF TFs have been identified and studied in various plants including *Arabidopsis*, rice, wheat, poplar, barley, castor bean, grape, cucumber, soybean, *Brassica* and *Malus* [8,17–25]. The genome-wide analyses of AP2/ERF TF superfamily have been performed in these crops, both to categorize each family member in an ordered nomenclature system as well as to investigate their expression profiles and chromosomal positions. As already mentioned, two full genome sequences of *Setaria italica* cv. Zhang gu and inbred Yugu 1 are available which have not only provided a useful genomic platform but have also paved pathway for researchers to carry out advanced genetic and genomic studies in this model crop. As AP2/ERF TFs

show wide diversity of functions including regulation of several developmental processes such as vegetative and reproductive development, cell proliferation, and responses to various abiotic and biotic stresses and plant hormones [8,10], their superfamily stands as one of the best candidates to examine important traits in foxtail millet. With this aim, a genome wide investigation of foxtail millet AP2/ERF TF superfamily and their expression profiling has been taken up in this study. Hence, this is the first comprehensive report on genome-wide survey, expression profiling and evolutionary analysis of AP2/ERF proteins in foxtail millet (named as *Setaria italica* AP2/ERF; 'SiAP2/ERF').

Materials and Methods

Sequence retrieval and identification of AP2/ERF proteins from *Setaria italica*

The Hidden Markov Model (HMM) profile of the AP2/ERF domain (PF00847) was obtained from Pfam v27.0 database (http://Pfam.sanger.ac.uk/) [26] and searched against the PHYTOZOME database of *Setaria italica* (www.phytozome.net/). All hits with expected (E) values less than 1.0 were retrieved and the non-redundant sequences were examined for the presence of conserved AP2/ERF domain by executing HMMSCAN (http://hmmer.janelia.org/search/hmmscan).

Chromosomal location, gene structure and genomic distribution of *AP2/ERF* genes

The identified AP2/ERF domain-containing proteins were BLASTP searched against *S. italica* genome of PHYTOZOME database with default settings, and physical map was constructed using MapChart [27]. Segmental duplications were calculated based on the method of Plant Genome Duplication Database [28] using MCSan v0.8 [29] and visualized using Circos v0.55 [30]. Tandem duplications were identified manually as described elsewhere [31,32] and marked on the physical map. The exonintron organizations of the genes were ascertained using Gene Structure Display Server (http://gsds.cbi.pku.edu.cn/) [33].

Phylogenetic analysis, Gene Ontology (GO) annotation, promoter analysis and identification of miRNAs targeting *SiAP2/ERFs*

The amino acid sequences of AP2/ERF proteins were imported into MEGA5 and an unrooted phylogenetic tree based on the Neighbor-joining method was generated after 1000 bootstrap replications [34]. The GO annotation of AP2/ERF protein sequences was performed using Blast2GO [35] and *cis*-regulatory elements were identified using PLACE (http://www.dna.affrc.go.jp/PLACE/) database. Further. the *S. italica* miRNAs reported by Khan et al. [36] were retrieved and searched for their targets in 171 *SiAP2/ERF* transcripts using psRNATarget tool (http://plantgrn.noble.org/psRNATarget/).

Comparative mapping and evolutionary analysis of paralogs and orthologs

The amino acid sequences of SiAP2/ERF proteins that were physically mapped onto foxtail millet genome were BLASTP searched against protein sequences of sorghum, maize, rice and *Brachypodium* (http://gramene.org/; www.phytozome.net). Reciprocal BLAST was also carried out to establish unique relationship between the orthologous genes. Hits with E-value≥ 1e-05 and at least 80% homology were considered significant. The comparative orthologous relationships of *AP2/ERF* genes among foxtail millet, sorghum, maize, rice and *Brachypodium* were finally

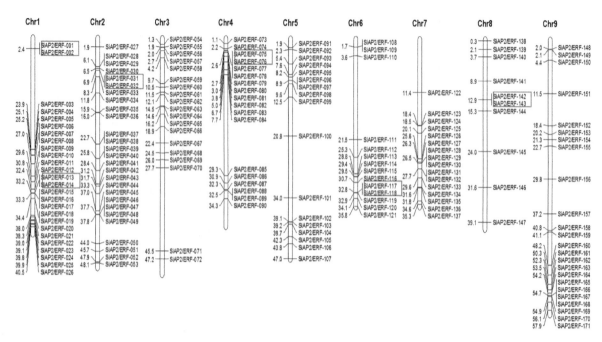

Figure 1. Physical map of 171 *SiAP2/ERF* genes. The bars represent the chromosomes with numbers at the left indicating the physical position (in Mb). The tandemly duplicated gene pairs are indicated within boxes.

illustrated using Circos [30]. For estimating the synonymous (Ks) and non-synonymous (Ka) substitution rates, the corresponding amino-acid as well as cDNA sequences of paralogous and orthologous SiAP2/ERF proteins were analyzed using PAL2NAL (http://www.bork.embl.de/pal2nal/) [37]. Time (million years ago, Mya) of duplication and divergence was calculated using a synonymous mutation rate of 1 substitutions per synonymous site per year as $T = Ks/2\lambda$ ($\lambda = 6.5 \times 10^{-9}$) [38,39].

Tissue-specific expression profiling using RNA-seq data

S. italica Illumina RNA-HiSeq reads from 4 tissues namely spica, stem, leaf and root, retrieved from European Nucleotide Archive [SRX128226 (spica); SRX128225 (stem); SRX128224 (leaf); SRX128223 (root)] [40] and were mapped onto the gene sequences of *Setaria italica* using CLC Genomics Workbench v.4.7.1 (http://www.clcbio.com/genomics). Normalization of the mapped reads was done using RPKM (reads per kilobase per million) method. The heat map for tissue-specific expression profile was generated based on the RPKM values for each gene in all the tissue samples using TIGR MultiExperiment Viewer (MeV4) software package [41,42].

Plant materials, growth conditions and stress treatments

Seeds of drought tolerant foxtail millet cultivar IC-403579 [8] were obtained from National Bureau of Plant Genetic Resources (NBPGR), Hyderabad, India. The seeds were sown in composite soil (peat compost: vermiculite: sand, 2:2:1) in glass house at National Phytotron Facility, Indian Agricultural Research Institute (IARI), New Delhi, India at $28 \pm 1°C$ day/$23 \pm 1°C$ night temperature with $70 \pm 5\%$ relative humidity and natural sunlight during June–July, 2013. For stress treatments, two week old seedlings were exposed to 20% polyethylene glycol (PEG 6000) (drought), 250 mM NaCl (salt), 100 µM abscisic acid (ABA), 100 µM salicylic acid (SA), 100 µM methyl jasmonate (MeJA) and 100 µM ethephon (Et) for 1 h (early) and 24 h (late) based on our previous studies (Lata et al. 2010; Lata et al. 2011a; Lata et al.

2011b). The plants were supplemented with water and Hoagland solution on alternate days. Unstressed plants were maintained as control. After stress treatments, whole seedlings were carefully harvested and immediately frozen in liquid nitrogen and stored at $-80°C$ until RNA isolation. Three independent experiments were conducted for precision and reproducibility, and for each experiment, ~100 mg seedling samples were collected by random sampling.

RNA extraction and expression analysis using qRT-PCR

Total RNA was isolated from the 14-day old unstressed and stressed (1 h and 24 h) foxtail millet cv. IC-403579 seedlings using TRIzol Reagent (Sigma, USA) following manufacturer's instructions. DNA contamination was removed from the RNA samples using RNase-free DNaseI (1 U µl^{-1}, Fermentas). The quality and purity of the RNA preparations were determined by measuring the OD_{260}/OD_{280} absorption ratio (1.9–2.0), and the integrity of the preparations was determined by electrophoresis in a 1.2% agarose gel containing formaldehyde as described in previous studies [43,44]. RNA concentrations were measured by a spectrophotometer (Nanodrop, USA). About 1 µg of total RNA was used to synthesize first strand cDNA primed with OligodT in a 20 µl reaction mix using 200 U/µl of PrimeScript M-MuLV reverse transcriptase (Takara Bio Inc., USA) following manufacturer's instructions. Quantitative real time (qRT) PCR was performed using SYBR Premix ExTaq II (Tli RNaseH Plus) (Takara Bio Inc., USA) on Mastercycler ep realplex system (Eppendorf) in triplicate. The constitutive gene RNA Polymerase II (RNA POL II; Accession No Si033113m) from foxtail millet was used as endogenous control which gave an amplification product of 146 bp [45]. The qRT-PCR primers of the *SiAP2/ERF* genes were designed from non-conserved regions of the corresponding genes using GenScript real-time PCR (TaqMan) Primer Design tool (www.genscript.com) using default parameters (Table S1). The PCR mixtures and reactions were used as detailed previously [45]. Melting curve analysis (60 to 95°C after 40 cycles) and

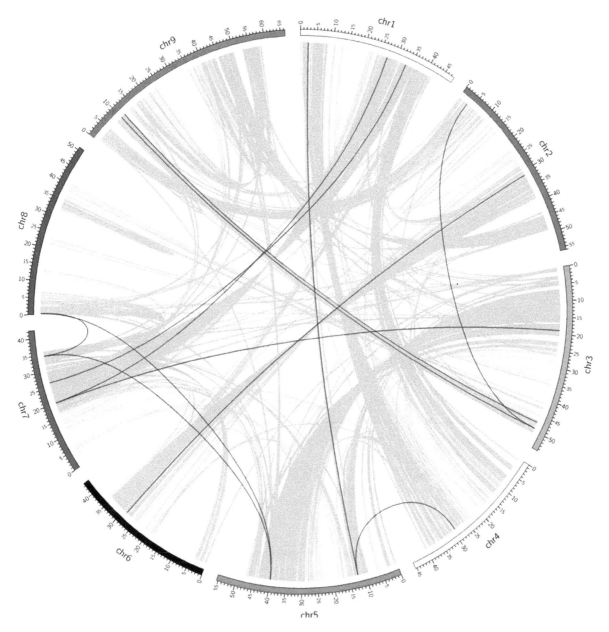

Figure 2. Distribution of segmentally duplicated *SiAP2/ERF* genes on foxtail millet chromosomes. Grey lines indicate collinear blocks in whole foxtail millet genome, and black lines indicate duplicated *SiAP2/ERF* gene pairs.

agarose gel electrophoresis were performed to check the amplification specificity of *AP2/ERF* genes normalized to the internal control RNA POL II and were analyzed using $2^{-\Delta\Delta Ct}$ method [46]. qRT-PCR data analysis was done according to previous studies [45,46]. The PCR cycling conditions were: initial denaturation at 95°C for 2 min, 95°C for 15 s, and 60°C for 1 min for 40 cycles followed by melting curve analysis using default parameters i.e. 95°C for 15 s, 60°C for 15 s, 95°C for 15 s with ramp time of 20 min.

Identification of molecular markers and homology modeling of SiAP2/ERF proteins

The presence of various types of DNA-based markers including simple sequence repeats (SSRs) [47], EST-derived SSRs (eSSRs) [48] and intron length polymorphic (ILPs) markers [49] retrived

from FmMDb (http://www.nipgr.res.in/foxtail.html) [50] were searched in the *SiAP2/ERF* genes using in-house perl script. For homology modeling, all the SiAP2/ERF proteins were queried against the Protein Data Bank (PDB) [51] to identify the best template with similar amino acid sequence and known 3D structure. The data was fed in Phyre2 server (Protein Homology/AnalogY Recognition Engine; http://www.sbg.bio.ic.ac.uk/phyre2) for predicting the three-dimensional structure of proteins by homology modeling under 'normal' mode [52]. Active site was predicted using COACH server (http://zhanglab.ccmb.med.umich.edu/COACH/) and highlighted using UCSF Chimera 1.8.

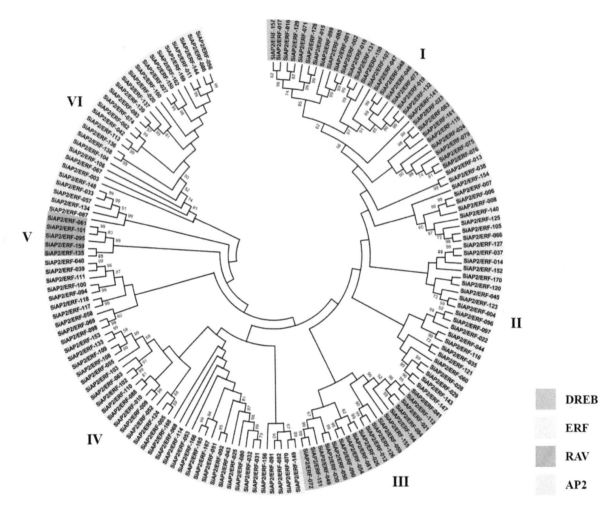

Figure 3. Phylogenetic tree of 171 *SiAP2/ERF* genes. Bootstrap confidence values from 1000 replicates are indicated at each branch. The different classes of SiAP2/ERF proteins are highlighted in different colors.

Results and Discussion

Identification of the AP2/ERF family transcription factors in foxtail millet genome

The HMM BLAST identified a total of 186 AP2/ERF protein sequences from foxtail millet. Fifteen proteins were found to be splice variants of primary transcripts, removal of which led to the identification of a total of 171 putative SiAP2/ERF proteins (Table S2) which represents approximately 0.4407% of all annotated genes (38801 genes total) in the *Setaria* genome [3] which is very similar to those present in poplar (0.4390%) and rice (0.4315%) however approximately 0.10% smaller than that of Arabidopsis (0.5481%) [7,8,18,53,54]. Among splice variants, Si022619m gene was found to encode a maximum of 8 alternate

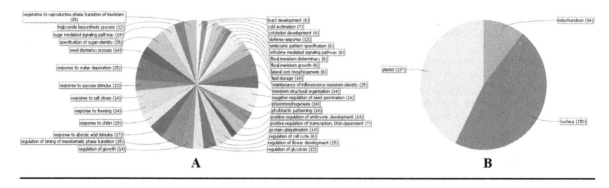

Figure 4. Gene Ontology annotation of SiAP2/ERF proteins. The Blast2GO output defining the (**A**) biological processes and (**B**) cellular localization of 171 SiAP2/ERF proteins.

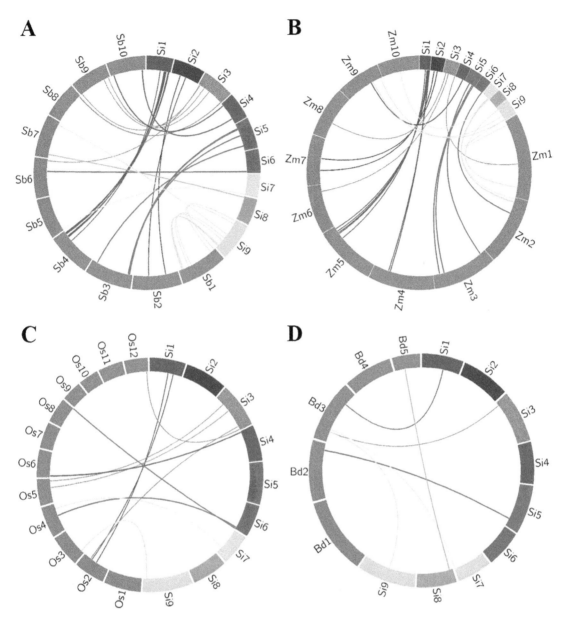

Figure 5. Comparative physical mapping showing the degree of orthologous relationships of *SiAP2/ERF* genes located on nine chromosomes of foxtail millet with (A) sorghum, (B) maize, (C) rice and (D) *Brachypodium*.

transcripts (Si022998m, Si022997m, Si022990m, Si022995m, Si022996m, Si022991m, Si022621m, Si022989m), followed by Si006802m and Si010289m which has 2 splice variants each (Si006941m, Si006905m in Si006802m and Si010292m, Si010301m in Si010289m) (Table S2). Genes Si030514m, Si036615m and Si036647m comprised of one alternate transcript Si030506m, Si036606m and Si036938m, respectively. The number conforms with the number of AP2/ERFs reported in Foxtail millet Transcription Factor Database [55]. Of note, the identification of pseudogenes among these 171 AP2/ERFs require further experimentations. In addition, the respective gene sequences encoding these proteins were retrieved and the presence of AP2/ERF domain was ascertained (Table S3). Due to lack of proper annotation, the existing identities of the genes were highly disordered and therefore for convenience, all 171 genes were assigned consecutive numbers from SiAP2/ERF-001 to SiAP2/

ERF-171 in the order of their chromosomal locations. All *SiAP2/ ERF* genes varied greatly in the size and sequence of their encoded proteins as well as in their physico-chemical properties (Table S2). Additionally all *SiAP2/ERF* genes were characterized by the presence of one or two highly conserved AP2/ERF DNA-binding domains and a B3 domain in case of RAV proteins (Table S3). The lengths of the identified proteins vary from 88 to 691 amino acids. ExPASy analysis revealed large variation in iso-electric point (pI) values ranging from 4.26 to 11.7 and molecular weight ranging from 9.92 to 72.31 kDa. Interestingly, this wide variation in pI and molecular weight revealed the presence of putative novel variants of SiAP2/ERFs and this is in accordance to previous genome-wide reports on DCL, AGO, RDR, C_2H_2 zinc finger and MYB genes in foxtail millet [56–58]. The characteristic features of SiAP2/ERF protein sequences are summarized in Table S2. Dual targeting is a term used to infer the ability of proteins to localize

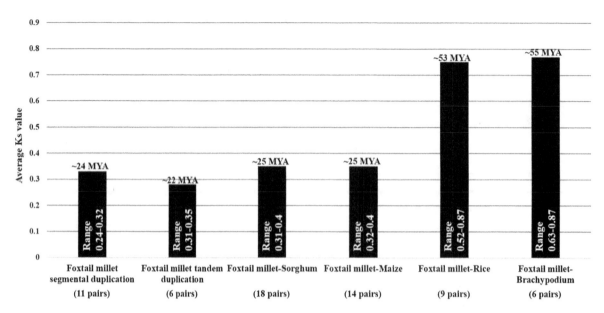

Figure 6. Time of duplication and divergence (MYA) based on synonymous substitution rate (Ks) estimated using paralogous and orthologous *SiAP2/ERF* gene pairs.

into more than one cellular compartment and it can also be viewed as a post translational regulatory mechanism [59]. Localization of 171 AP2/ERF proteins was determined using Blast2GO. Majority of the SiAP2/ERF proteins were predicted to be dual targeted or localized to nucleus, plastid and/or mitochondrion except SiAP2/ERF-025, SiAP2/ERF-032,

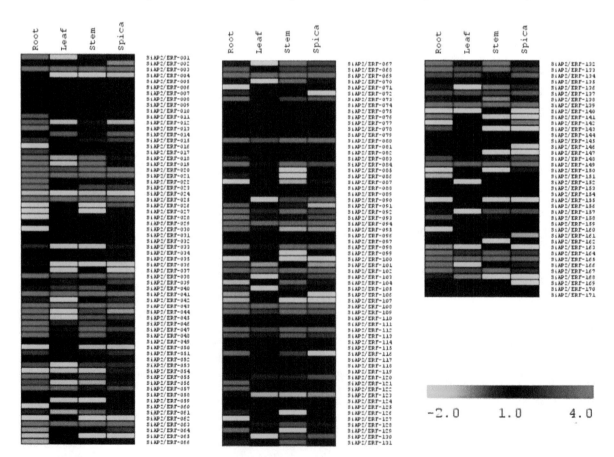

Figure 7. Heat-map showing the expression pattern of *SiAP2/ERF* genes in four tissues namely leaf, root, stem and spica. The color scales for fold-change values are shown at the bottom right.

SiAP2/ERF-035, SiAP2/ERF-040, SiAP2/ERF-051, SiAP2/ERF-055, SiAP2/ERF-063, SiAP2/ERF-065, SiAP2/ERF-091, SiAP2/ERF-100, SiAP2/ERF-108, SiAP2/ERF-121, SiAP2/ERF-122, SiAP2/ERF-153, SiAP2/ERF-165, SiAP2/ERF-166 (nucleus); SiAP2/ERF-024, SiAP2/ERF-059, SiAP2/ERF-066, SiAP2/ERF-078, SiAP2/ERF083 and SiAP2/ERF096 (plastid); and SiAP2/ERF-034, SiAP2/ERF-075, SiAP2/ERF-077, SiAP2/ERF-079, SiAP2/ERF-111 and SiAP2/ERF-159 (intracellular membrane-bound organelle) localized. Further, AP2/ERF superfamily was divided into four major families on the basis of nature and number of DNA-binding domains, namely AP2, ERF, DREB and RAV. The AP2/ERF proteins of *Setaria italica* were also classified into these families. Out of 171 genes, 28 belong to AP2, 90 to ERF, 48 to DREB and 5 to RAV (Table S3) indicating that foxtail millet genome supports large ERF and DREB subfamilies similar to Chinese cabbage genome [60].

Chromosomal distribution and structure of SiAP2/ERF proteins

Physical mapping of *SiAP2/ERFs* on all 9 chromosomes of foxtail millet revealed an uneven distribution of *SiAP2/ERF* genes in the genome (Figure 1). Among all chromosomes, chromosome 2 contained highest number of *AP2/ERF* genes (27; ~16%) followed by chromosome 1 (26; ~15%), while minimum number of genes were assigned on chromosome 8 (10; ~6%). The precise position (in bp) of each *SiAP2/ERF* on foxtail millet chromosomes is detailed in Table S2. Distribution pattern of these genes on individual chromosomes also pointed certain physical regions with a relatively higher accumulation of *AP2/ERF* gene clusters. As for example, *SiAP2/ERF* genes located on chromosomes 1, 6, 7 and 9 appear to congregate at the lower end of the arms as compared to chromosome 3 and 4 where these genes appear to cluster together at the upper end of the arm (Figure 1). It has recently been reported that foxtail millet genome underwent whole-genome duplication similar to other grass species about 70 million years

ago [3] and hence occurrence of such large number of *SiAP2/ERF* genes in foxtail millet genome suggests huge amplification of this gene family during the course of evolution. Hence, duplication of these genes was studied and found that 12 (~7%) *SiAP2/ERF* genes were tandem repeated (Figure 1) and 22 (11 pairs; ~13%) were segmentally duplicated (Figure 2). The tandem duplicated genes included six clusters (2 genes each) including two clusters on chromosome 1 and one each on chromosomes 2, 4, 6 and 8. The distance between these genes ranged from 6.2 kb to 32.2 kb. Among the segmentally duplicated gene pairs, three genes namely, SiAP2/ERF-001 (chromosome 1), SiAP2/ERF-013 (chromosome 1) and SiAP2/ERF-072 (chromosome 5) were duplicated twice in the genome forming six paralogs in chromosomes 2, 4, 5, 7 and 9 (Figure 2). Chromosomal localization study of *SiAP2/ERF* genes thus indicates that tandem- and segmental-duplication may be one of the contributing factors in evolution of new genes in foxtail millet genome. Moreover, analysis of *SiAP2/ERF* gene structures indicated highly diverse distribution of intronic regions (from 1 to 10 in numbers) among the exonic sequences suggesting significant evolutionary changes in the foxtail millet genome. Interestingly, 89 (~52%) *SiAP2/ERF* genes were found to be intronless (Figure S1). Similar results were also observed in case of *Arabidopsis* [7] and *Lotus corniculatus* [61]. Further, the shortest *SiAP2/ERF* gene was merely 266 bp (SiAP2/ERF-156), whereas the longest one was identified as SiAP2/ERF-150 with 3.7 kb genomic sequence.

Phylogenetic analysis of SiAP2/ERF proteins

Phylogenetic analysis is essential for understanding the evolutionary history of crop species. Therefore, to understand the evolutionary significance of domain architecture, a phylogenetic tree was constructed with 171 SiAP2/ERF proteins. The phylogenetic analysis clustered all the SiAP2/ERFs into distinct clades (AP2, ERF, DREB and RAV) comprising of 28, 90, 48, and 5 proteins, respectively, according to their domain composition (Figure 3). Interestingly, the DREB formed two clades intervened

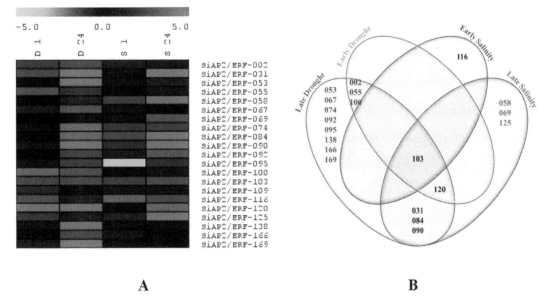

A   **B**

Figure 8. Expression profile of 21 *SiAP2/ERF* genes in response to dehydration and salinity stresses. (A) Heat map showing differential gene expression in response to dehydration (D) and salinity (S) stress across two time points (1 h and 24 h). The heat-map has been generated based on the fold-change values in the treated sample when compared with its unstressed control sample. The color scale for fold-change values is shown at the top. **(B)** Venn diagram showing stress-specific higher-expression of *SiAP2/ERF* genes during early and late stresses. The common subset of genes regulated by two stresses is marked by the overlapping circle. The numbers provided in the venn diagram corresponds to the *SiAP2/ERF* ID listed in Table S2.

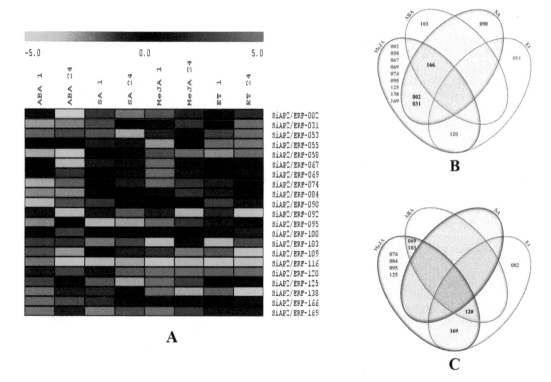

Figure 9. Expression profile of 21 *SiAP2/ERF* genes in response to Abscisic acid (ABA), salicylic acid (SA), methyl jasmonate (MeJA) and Ethephone (Et) treatments. (A) Heat map showing differential gene expression in response to ABA, SA, MeJA, Et treatments across two time points (1 h and 24 h). The heat-map has been generated based on the fold-change values in the treated sample when compared with its treated control sample. The color scale for fold-change values is shown at the top. **(B)** Venn diagram showing stress-specific higher-expression of *SiAP2/ERF* genes during early hormonal treatment. **(C)** Venn diagram showing stress-specific higher-expression of *SiAP2/ERF* genes during early hormonal treatment. The common subset of genes regulated by the hormonal treatments is marked by the overlapping circle. The numbers provided in the venn diagram corresponds to the *SiAP2/ERF* ID listed in Table S2.

by ERFs. Although similar observation was not reported in genome-wide studies of AP2/ERF conducted in other plants, in Chinese cabbage it has been found that the DREB clade was intervened by AP2 [60]. Further in-depth *in silico* analysis is requisite for finding the possible reason for such observation. The tree was divided into six groups based on the distribution of AP2, ERF, DREB and RAV. The derivation of statistically reliable pairs of possible homologous proteins sharing similar functions from a common ancestor was confirmed owing to high bootstrap values observed for a good number of the internal branches of the unrooted phylogenetic tree. Close association of SiAP2/ERF sub-families with their corresponding counterparts in other plant systems in terms of expression and/or biological and regulatory functions may be an implication of sequence conservation and also evidence to their similar *in planta* roles. Such phylogeny-based function prediction is obviously a rational systematic approach to facilitate identification of orthologous genes and has near-perfectly been applied for prediction of AP2/ERF proteins in other plant species such as rice, grapes and *Brassica* [20,21,60]. Thus, members of the sub-families of AP2/ERF superfamily in foxtail millet also have similar regulatory roles as those of their orthologs in other crop species.

Gene Ontology annotation

The GO analysis performed using rice protein sequences as reference showed the putative participation of SiAP2/ERF proteins in diverse biological, cellular and molecular processes (Figure 4; Table S4). The analysis of biological processes mediated

by SiAP2/ERF depicted that a predominant of SiAP2/ERF proteins were involved in stress responses, such as response to water deprivation, salt stress and freezing. In addition, SiAP2/ERF proteins were also evidenced to participate in regulation of timing of meristematic phase transition, specification of organ identity and maintenance of inflorescence meristem identity. The molecular processes of SiAP2/ERF proteins clearly showed that all the 171 proteins possess sequence-specific DNA binding transcription factor activity (Table S4). Further, cellular component analysis revealed the localization of SiAP2/ERF proteins in nucleus, plastids, mitochondria and other intracellular membrane-bound organelles. These are in concordance with the experimental findings reported earlier [8,10,61,62].

Promoter analysis and miRNA targets of *SiAP2/ERF* genes

Cis-regulatory elements are DNA sequences that are situated upstream of genes in the promoter region and act as TF-binding sites. These are known to play crucial roles in determining tissue-specific as well as stress-responsive gene expression [63]. There have also been reports that multi-stimuli genes are closely correlated with *cis*-regulatory elements in their promoter sequences [64]. Therefore, a comprehensive promoter analysis of all the 171 *SiAP2/ERF* genes was conducted in order to further understand transcriptional regulation and support functional prediction of the respective proteins (Table S5). A total of 300 cis-regulatory elements were found to be present in one or the other *SiAP2/ERF* gene. The cis-regulatory elements CACTFTPPCA1, CAATBOX1, EBOXBNNAPA, MYCCON-

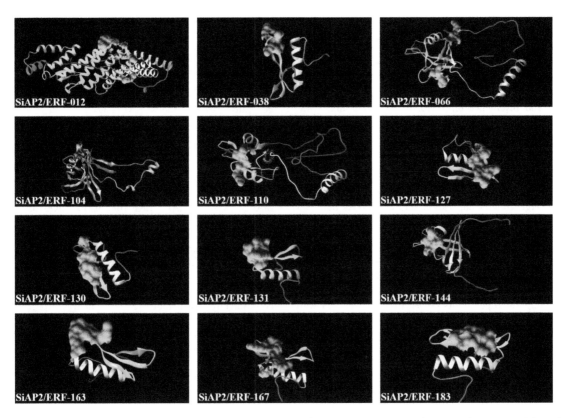

Figure 10. Predicated structures of SiAP2/ERF proteins. The structures of 12 SiAP2/ERF proteins with greater than 90% confidence level were shown along with its potential active site.

SENSUSAT, WRKY71OS, GT1CONSENSUS, ARR1AT, DOFCOREZM, GTGANTG10, RAV1AAT and GATABOX were present in all the 171 genes whereas HSE, VSF1PVGRP18, GMHDLGMVSPB, ABREAZMRAB28, PALBOXLPC, DR5GMGH3, SITEIIAOSPCNA, ABREBNNAPA, ABA-SEED1, ABAREG2, O2F3BE2S1, OPAQUE2ZM22Z, RGA-TAOS, CEREGLUBOX3PSLEGA, CPRFPCCHS, LREBOX-IIPCCHS1, AGL2ATCONSENSUS, TOPOISOM, NONAMERATH4, PALBOXPPC, WRECSAA01, SORLI-P4AT, NONAMERMOTIFTAH3H4, ABREMOTIFIOS-RAB16B, ACIIIPVPAL2, ABRE3OSRAB16, B2GMAUX28, AT1BOX, SPHZMC1, D3GMAUX28, INTRONUPPER, WINPSTPIIIK, JASE1ATOPR1, OCETYPEIINTHISTONE and MSACRCYM were unique to *SiAP2/ERF-020*, *SiAP2/ERF-023*, *SiAP2/ERF-024*, *SiAP2/ERF-026*, *SiAP2/ERF-029*, *SiAP2/ERF-036*, *SiAP2/ERF-038*, *SiAP2/ERF-059*, *SiAP2/ERF-063*, *SiAP2/ERF-066*, *SiAP2/ERF-067*, *SiAP2/ERF-069*, *SiAP2/ERF-071*, *SiAP2/ERF-072*, *SiAP2/ERF-073*, *SiAP2/ERF-073*, *SiAP2/ERF-082*, *SiAP2/ERF-083*, *SiAP2/ERF-091*, *SiAP2/ERF-091*, *SiAP2/ERF-091*, *SiAP2/ERF-093*, *SiAP2/ERF-094*, *SiAP2/ERF-099*, *SiAP2/ERF-110*, *SiAP2/ERF-112*, *SiAP2/ERF-128*, *SiAP2/ERF-132*, *SiAP2/ERF-140*, *SiAP2/ERF-152*, *SiAP2/ERF-157*, *SiAP2/ERF-159*, *SiAP2/ERF-166*, *SiAP2/ERF-169 and SiAP2/ERF-170*, respectively (Table S5). In addition, putative microRNAs (miR-NAs) targeting the *SiAP2/ERF* genes were also detected using psRNATarget server. The analysis showed five *SiAP2/ERF* genes to be targeted by *Setaria italica* miRNAs (Sit-miRs) (Table S6; Figure S2). The miRNAs identified in this study would be helpful in interpreting the post-transcriptional control of gene regulation during various physiological and stress-induced cellular responses in this otherwise naturally stress tolerant crop.

Orthologous relationships of *SiAP2/ERF* genes between foxtail millet and other grass species

To derive orthologous relationships of SiAP2/ERFs, comparative mapping approach was followed wherein the physically mapped *AP2/ERF* genes of foxtail millet were compared with those in the chromosomes of related grass genomes namely sorghum, maize, rice and *Brachypodium* (Figure 5). Maximum orthology of genes annotated on the foxtail millet chromosomes was exhibited with sorghum (18; ~11%) followed by maize (14; ~8%), rice (9; 5%) and least with *Brachypodium* (6; 4%). The extensive synteny among foxtail millet, sorghum and maize at gene level supports their close evolutionary relationships [3,4]. Intriguingly, most of *SiAP2/ERF* genes showed syntenic bias towards particular chromosomes of sorghum, maize, rice and *Brachypodium* and this suggests that the chromosomal rearrangement events like duplication and inversion predominantly shaped the distribution and organization of *AP2/ERF* genes in these grass genomes. The comparative mapping information thus offers a useful preface for understanding the evolution of *AP2/ERF* genes among grasses including foxtail millet. In addition, this study would be helpful in selecting candidate *SiAP2/ERF* genes and utilize them in genetic improvement of related grass family members. As for example, AP2-like ethylene-responsive transcription factors PLETHORA 1 and 2 are essential for QC specification and stem cell activity in roots of Arabidopsis [65]. It is thus likely that its orthologous foxtail millet gene (SiAP2/ERF-011; Phytozome ID: Si016558m) and rice gene (LOC_Os02g40070.1) may also be involved in similar function.

Duplication and divergence rate of the *SiAP2/ERF* genes

Whole genome duplications such as tandem and segmental duplications usually give rise to multiple copies of genes in a gene family. Such gene duplication events have been reported in various plant TF families including MYB, NAC and AP2/ERF [21,56–58,66–68]. Association of Darwinian positive selection in duplication and divergence of *AP2/ERF* in foxtail millet was explored by estimating the ratios of non-synonymous (Ka) versus synonymous (Ks) substitution rate (Ka/Ks) for 6 tandem and 11 segmentally duplicated gene-pairs as well as between orthologous gene-pairs of *SiAP2/ERF* with those of sorghum (18 pairs), maize (14 pairs), rice (9 pairs) and *Brachypodium* (6 pairs) (Figure 6) (Tables S7–S12). The Ka and Ks are a measure to examine the course of divergence after duplication, and the Ka/Ks ratio is a measure of the selection pressure to which a gene pair is subjected wherein Ka/Ks <1 means purifying selection, Ka/Ks = 1 stands for neutral selection, and Ka/Ks >1 signifies accelerated evolution with positive selection [38]. The Ka/Ks ratio for tandem duplicated gene-pairs in foxtail millet AP2/ERF genes ranged from 0.10 to 0.15 with an average of 0.13, whereas Ka/Ks for segmentally duplicated gene-pairs ranged from 0.03 to 0.14 with an average of 0.09 (Tables S7–S8). The data indicated that the duplicated *SiAP2/ERF* genes were under strong purifying selection pressure and had gone through substitution elimination and enormous selective constraint by natural selection during the course of evolution since their Ka/Ks ratios estimated as <1. Further, the duplication event of these *SiAP2/ERF* tandemly and segmentally duplicated genes may be estimated to have occurred around ~22 and ~24 Mya, respectively (Figure 6). Among the orthologous gene-pairs of *SiAP2/ERF* with those of other grass species, the average Ka/Ks value was maximum between *Brachypodium* and foxtail millet (0.4) and rice and foxtail millet (0.4), and least for sorghum-foxtail millet and maize-foxtail millet gene-pairs (0.2) (Tables S9–S12). The relatively higher rate of synonymous substitution between *Brachypodium*-foxtail millet and rice-foxtail millet *AP2/ERF* genes pointed their earlier divergence around 53–55 Mya from foxtail millet as compared to sorghum and maize *AP2/ERF* genes (Figure 6). Remarkably, the *AP2/ERF* gene-pairs between sorghum and foxtail millet, and maize and foxtail millet (average Ka/Ks = 0.2) appear to have undergone extensive intense purifying selection in comparison to foxtail millet-rice and foxtail millet-*Brachypodium* (average Ka/Ks = 0.4 for both) *AP2/ERF* genes. This is in agreement to their recent time of divergence, around 25 Mya. The estimation of tandem and segmental duplication time (average of 23 Mya) of foxtail millet *AP2/ERF* genes in between the divergence time of foxtail millet-rice (53 Mya), foxtail millet-*Brachypodium* (55 Mya) and foxtail millet-maize and foxtail millet-sorghum (both 25 Mya) orthologous *AP2/ERF* gene-pairs are comparable to evolutionary studies involving the protein-coding genes annotated from the recently released draft genome sequences of foxtail millet [3,4]. Interestingly, though the *SiAP2/ERF* gene-pairs showing segmental (Ka/Ks = 0.09) and tandem duplication (Ka/Ks = 0.13) events are not under similar evolutionary pressure, both set of gene pairs revealed that these events took place almost at similar time (22 Mya for tandem and 24 Mya for segmentally duplicated gene pairs). Therefore, overall, it can be concluded that the segmental and tandem duplication events including the divergence events of *SiAP2/ERF* genes with other grass species have played a major role in evolution of this gene family in foxtail millet. This is also in agreement with earlier genome-wide studies conducted for important gene families in foxtail millet [56–58,67,68].

In silico tissue-specific expression profiling of *SiAP2/ERF* genes

Tissue-specific expression data at a given developmental stage is helpful in identifying genes involved in defining precise nature of individual tissues. Therefore, in order to examine tissue-specific expression profiles of 171 *SiAP2/ERF* genes, a heat map was generated based on the RPKM values for each gene in all tissue samples using RNA-Seq data. A differential expression for all 171 transcripts was observed in 4 tissue samples namely root, leaf, stem and spica (Figure 7). A relative comparison of expression profiles of all 171 SiAP2/ERF showed a relatively higher expression of SiAP2/ERF-020, SiAP2/ERF-021, SiAP2/ERF-025, SiAP2/ERF-041, SiAP2/ERF-043, SiAP2/ERF-063, SiAP2/ERF-069, SiAP2/ERF-094, SiAP2/ERF-108, SiAP2/ERF-139 and SiAP2/ERF-165 in all the four tissues suggesting their importance as potential targets for further functional characterization. In general, majority of the SiAP2/ERFs exhibited root-specific expression (56; ~33%) followed by expression in stem (47; ~27%), then spica (43; ~25%) and least in leaves (26; ~15%). The results indicated that *AP2/ERF* genes in foxtail millet are mostly expressed in roots as confirmed by earlier studies [25,60]. The tissue-specific expression profiling of *SiAP2/ERFs* would further aid the combinatorial involvement of these genes in transcriptional regulation of various tissues, while ubiquitously expressed *SiAP2/ERFs* might control a broad set of genes at transcriptional level. The heat map data also facilitates the overexpression studies of *SiAP2/ERFs* across the tissues to impart stress tolerance to both foxtail millet as well as related grass species.

Expression profiling of *SiAP2/ERFs* during abiotic stresses and phytohormone treatments

Gene expression studies can provide essential indications regarding functions of a gene. In order to analyze the role of *AP2/ERF* genes in foxtail millet, we examined the expression profiles of 21 selected genes representing different sub-families using quantitative real-time (qRT) PCR analysis in response to drought (20% PEG 6000), salt (250 mM NaCl), 100 μM ABA, 100 μM SA, 100 μM MeJA and 100 μM Et during early (1 h) and late (24 h) durations of treatments. The heat map illustration of expression profiles of 21 selected *SiAP2/ERF* genes under drought and salinity is shown in Figure 8. The qRT-PCR analysis demonstrated an overall differential expression patterns to one or more stresses for the genes under study (Figs. 8, 9). The *SiAP2/ERF* genes, in general, were up-regulated by drought and salt treatments except SiAP2/ERF-116 which was down-regulated under drought stress and SiAP2/ERF-092 and SiAP2/ERF-095 which were down-regulated under salt stress at both time points. Only SiAP2/ERF-103 was co-regulated as it was induced by both stresses at all time points. However, 8 *SiAP2/ERF* genes were activated exclusively at late drought stress and 3 at late salinity stress suggesting their role in stress adaptation (Figure 8). The variability in gene expression patterns observed in this study indicated that *SiAP2/ERFs* might play an important role in regulating a complex web of stress responsive pathways for stress adaptation and tolerance towards multiple abiotic stresses.

Phytohormones or plant growth regulators not only play a crucial role in regulation of various plant processes including growth and development but also in signaling and gene expression during environmental stresses both abiotic and biotic. Therefore it was attempted to analyze the expression patterns of the selected 21 *SiAP2/ERF* genes under various hormone treatments. A hierarchical clustering demonstrated overlapping and specific gene expression patterns in response to various phytohormones

(Figure 9). No single gene was exclusively induced in all hormone treatments indicating their treatment-specific roles. However, several genes were exclusively repressed (SiAP2/ERF053, SiAP2/ERF-055, SiAP2/ERF-092, SiAP2/ERF-109 and SiAP2/ERF-116) in all hormone treatments indicating that these genes may be a part of general hormone response. Overall majority of the *SiAP2/ERF* genes were down-regulated in response to ABA except SiAP2/ERF-069, SiAP2/ERF-103 and SiAP2/ERF-120 confirming the previous reports that *AP2/ERF* genes (mostly ERFs and DREBs) are generally regulated in an ABA-independent manner with a few exceptions [11,43]. The regulation of certain *AP2/ERF* genes by SA, MeJA or Et suggests their potential roles in biotic stress responses. Several genes were found to be regulated exclusively by a specific or more than one hormone treatments (Figure 9). As for example, as many as 5 genes were specifically up regulated by MeJA at both early and late time points, while SiAP2/ERF-120 was induced by MeJA and Et at all time points. Phytohormones generally act synergistically or antagonistically to each-other thus influencing signaling response for maintaining cellular homeostasis [69]. Thus *SiAP2/ERF* TFs also act as important mediators of this signaling process. The differential expression patterns of *SiAP2/ERF* genes in this investigation again underlines the intimidating task of understanding the global milieu associated with any stress response. However, as an outcome of this study, we are able to compare their expression profiles during several environmental stress stimuli at early and late time points for precise identification of potential candidate genes for crop improvement programmes. In this regard, *SiAP2/ERF-069, SiAP2/ERF-103* and *SiAP2/ERF-120* may be considered as potential candidate genes for further functional validation as well for utilization in crop improvement programs for stress resistance since these genes were up-regulated under drought and salinity stresses in ABA dependent manner. It can thus be concluded that certain members of *AP2/ERF* gene family in foxtail millet exhibit stimulus-specific and temporal responses and hence expanding current knowledge on molecular basis of stress tolerance and adaptation conferred on plants by them.

Identification of markers in *SiAP2/ERF* genes

Marker-assisted selection (MAS) is a combination of conventional breeding and molecular biology and offers a methodology for accelerating the procedure of crop improvement. The tagging of useful genes, such as those involved in plant hormone synthesis, and those responsible for conferring stress resistance to plants, namely drought and salinity, has been a major target for improving crop growth and productivity [70,71]. With the use of molecular markers, it is now easy to trace important alleles either in segregating or natural populations. Some of the recent studies have shown the importance of *AP2/ERF* TFs, especially DREB TFs, in marker-aided breeding and crop-improvement strategies [16,71]. Considering this, the presence of previously reported DNA-based molecular markers such as SSRs [47], eSSRs [48] and ILPs [49] were searched for their presence in all the 171 *SiAP2/ERF* genes. The analysis identified 54 SSRs and 1 ILP marker in *SiAP2/ERF* genes (Table S13). These markers would be useful in genotyping and MAS for crop improvement.

Homology modeling of *SiAP2/ERF* proteins

Three dimensional protein models of twelve proteins were constructed by sequence similarity searching against the PDB database using BLASTP. These 12 proteins were selected owing to their higher homology to the known protein sequences in the PDB and Phyre2 was used for homology modeling of their predicted

structures. The protein structure of all the 12 SiAP2/ERFs were modelled at 90% confidence and the potential active sites were identified (Figure 10). The 3D structure revealed the presence of conserved AP2/ERF domain of about 60–70 amino acids in all the SiAP2/ERF proteins with a typical three-dimensional conformation ordered into a layer of three antiparallel β-sheets followed by a parallel α-helix. Further examination of the AP2/ERF domain showed the presence of two regions namely YRG and RAYD. The YRG region was 20-amino acid long N-terminal stretch rich in basic and hydrophilic residues and was reported to play a crucial role in establishing direct contact with the DNA [72]. Conversely, RAYD region comprises about 40 amino acids and this region was reported to mediate protein-protein interactions through α-helix. Moreover, reports also indicate that RAYD region is involved in DNA binding through interactions of hydrophobic face of the α-helix with the major groove of DNA [72]. The AP2 sub-family members possess two AP2/ERF domains separated by a linker sequence of 25 amino acids which is responsible for positioning of the DNA-binding domains [73]. The molecular modeling thus proved that all the predicted protein structures were highly consistent and this data would offer a preliminary foundation for comprehending the molecular functions of SiAP2/ERF proteins.

Conclusion

The AP2/ERF TFs are important regulators of various plant processes including growth, development and stress responses and thus have been subjected to intensive investigations in various crop plants (Figure S3). However, to the best of our knowledge, no such study has been taken up in otherwise naturally stress tolerant model panicoid C_4 crop *Setaria italica*. The present study identified 171 AP2/ERF TFs in the foxtail millet genome. Isolation and identification of these functional TF genes are expected to aid knowledge towards understanding the molecular genetic basis for foxtail millet stress adaptation and genetic improvement, and may also provide functional gene resources for genetic engineering approaches. To date, only one gene representing this TF superfamily has been characterized from foxtail millet [10]. Hence the present comprehensive study would assist in explicating AP2/ERF family gene function in regulations of stress signaling pathways, and defense responses as well as in providing new opportunities to discover foxtail millet stress tolerance and adaptation mechanisms. The *in silico* structure prediction might provide basic resources to study the molecular regulation of foxtail millet development and stress tolerance. However, extensive *in planta* characterization of putative candidate *SiAP2/ERF* genes is must to further explore its biological roles.

Supporting Information

Figure S1   Gene structures of 171 SiAP2/ERF proteins. Exons and introns are represented by green boxes and black lines, respectively.

Figure S2   Diagrammatic representation of alignment between the miRNA and the SiAP2/ERF targets.

Figure S3   Distribution of AP2/ERFs in sequenced plant genomes.

Table S1   List of primers used in quantitative real time-PCR expression analysis of *SiAP2/ERF* genes.

Table S2 Characteristic features of *SiAP2/ERF* Transcription factor gene family members identified in *Setaria italica*.

Table S3 Summary of functional domains present in the SiAP2/ERF proteins.

Table S4 Blast2GO annotation details of SiAP2/ERF protein sequences.

Table S5 Characteristics of the promoter region of *SiAP2/ERF* genes.

Table S6 List of putative *Setaria italica* miRNAs targeting *SiAP2/ERF* transcripts.

Table S7 The Ka/Ks ratios and estimated divergence time for tandemly duplicated *SiAP2/ERF* genes.

Table S8 The Ka/Ks ratios and estimated divergence time for segmentally duplicated *SiAP2/ERF* genes.

Table S9 The Ka/Ks ratios and estimated divergence time for orthologous SiAP2/ERF proteins between foxtail millet and sorghum.

Table S10 The Ka/Ks ratios and estimated divergence time for orthologous SiAP2/ERF proteins between foxtail millet and maize.

Table S11 The Ka/Ks ratios and estimated divergence time for orthologous SiAP2/ERF proteins between foxtail millet and rice.

Table S12 The Ka/Ks ratios and estimated divergence time for orthologous SiAP2/ERF proteins between foxtail millet and *Brachypodium*.

Table S13 Details of SiAP2/ERF transcription factor-based markers.

Acknowledgments

Timely assistance from Mr. Rohit Khandelwal is appreciated.

Author Contributions

Conceived and designed the experiments: MP CL. Performed the experiments: CL AKM MM VSB YK. Analyzed the data: MP CL. Contributed to the writing of the manuscript: CL MM MP.

References

1. Muthamilarasan M, Prasad M (2014) Advances in *Setaria* genomics for genetic improvement of cereals and bioenergy grasses. Theor Appl Genet DOI:10.1007/s00122-014-2399-3.
2. Lata C, Gupta S, Prasad M (2013) Foxtail millet: a model crop for genetic and genomic studies in bioenergy grasses, Crit Rev Biotechnol 33: 328–343.
3. Zhang G, Liu X, Quan Z, Cheng S, Xu X, et al. (2012) Genome sequence of foxtail millet (*Setaria italica*) provides insights into grass evolution and biofuel potential. Nature Biotechnol 30: 549–554.
4. Bennetzen JL, Schmutz J, Wang H, Percifield R, Hawkins J, et al. (2012) Reference genome sequence of the model plant *Setaria*. Nature Biotechnol 30: 555–561.
5. Lata C, Prasad M (2013) *Setaria* genome sequencing: An overview. J Plant Biochem Biotechnol 22: 257–260.
6. Muthamilarasan M, Theriappan P, Prasad M (2013) Recent advances in crop genomics for ensuring food security. Curr Sci 105: 155–158.
7. Sakuma Y, Liu Q, Dubouzet JG, Abe H, Shinozaki K, et al. (2002) DNA-binding specificity of the ERF/AP2 domain of Arabidopsis DREBs, transcription factors involved in dehydration- and cold-inducible gene expression, Biochem Biophys Res Commun. 290: 998–1009.
8. Nakano T, Suzuki K, Fujimura T, Shinshi H (2006) Genome-wide analysis of the ERF gene family in Arabidopsis and rice. Plant Physiol 140: 411–432.
9. Jofuku KD, Boer BGW, Montagu MV, Okamuro JK (1994) Control of Arabidopsis flower and seed development by the homeotic gene APETALA2. Plant Cell 6: 1211–25.
10. Lata C, Yadav A, Prasad M (2011) Role of plant transcription factors in abiotic stress tolerance. In: Shanker A. and Venkateshwarulu B. (eds) Abiotic Stress Response in Plants, INTECH Open Access Publishers, 269–296.
11. Mizoi J, Shinozaki K, Yamaguchi-Shinozaki K (2012) AP2/ERF family transcription factors in plant abiotic stress responses. Biochimica et Biophysica Acta 1819: 86–96.
12. Magnani E, Sjölander K, Hake S (2004) From endonucleases to transcription factors: evolution of the AP2 DNA binding domain in plants, Plant Cell 16: 2265–2277.
13. Shigyo M, Hasebe M, Ito M (2006) Molecular evolution of the AP2 subfamily, Gene 366: 256–265.
14. Saleh A, Pagés M (2003) Plant AP2/ERF transcription factors. Genetika 35: 37–50.
15. Shigyo M, Ito M (2004) Analysis of gymnosperm two-AP2-domain-containing genes. Development Genes Evoln 214: 105–14.
16. Lata C, Prasad M (2011) Role of DREBs in regulation of abiotic stress responses in plants. J Exp Bot 62: 4731–4748.
17. Riechmann JL, Meyerowitz EM (1998) The AP2/EREBP family of plant transcription factors. Biol Chem, 379: 633–646.
18. Zhuang J, Cai B, Peng RH, Zhu B, Jin XF, et al. (2008) Genome-wide analysis of the AP2/ERF gene family in *Populus trichocarpa*. Biochemical Biophysical Res Comm 371: 468–474.
19. Gil-Humanes J, Piston F, Martin A, Barro F (2009) Comparative genomic analysis and expression of the APETALA2-like genes from barley, wheat, and barley-wheat amphiploids. BMC Plant Biol 9: 66.
20. Dietz KJ, Vogel MO, Viehhauser A (2010) AP2/EREBP transcription factors are part of gene regulatory networks and integrate metabolic, hormonal and environmental signals in stress acclimation and retrograde signalling. Protoplasma 245: 3–14.
21. Licausi F, Giorgi FM, Zenoni S, Osti F, Pezzotti M, et al. (2010) Genomic and transcriptomic analysis of the AP2/ERF superfamily in *Vitis vinifera*. BMC Genomics 11: 719.
22. Zhuang J, Chen J-M, Yao Q-H, Xiong F, Sun C-C, et al. (2011) Discovery and expression profile analysis of AP2/ERF family genes from *Triticum aestivum*. Mol Biol Rep 38: 745–753.
23. Zhuang J, Yao Q-H, Xiong A-S, Zhang J (2011) Isolation, Phylogeny and Expression Patterns of AP2-Like Genes in Apple (Malus × domestica Borkh). Plant Mol Biol Rep 29: 209–216.
24. Zhang G, Chen M, Chen X, Xu Z, Guan S, et al. (2008) Phylogeny, gene structures, and expression patterns of the ERF gene family in soybean (*Glycine max* L.). J Exp Bot 59: 4095–4107.
25. Xu W, Li F, Ling L, Liu A (2013) Genome-wide survey and expression profiles of the AP2/ERF family in castor bean (*Ricinus communis* L.). BMC Genomics 14: 785.
26. Punta M, Coggill PC, Eberhardt RY, Mistry J, Tate J, et al. (2012) The Pfam protein families database. Nucleic Acids Res 40: D290–D301.
27. Voorrips RE (2002) MapChart: software for the graphical presentation of linkage maps and QTLs. J Hered 93: 77–78.
28. Lee TH, Tang H, Wang X, Paterson AH (2012) PGDD: a database of gene and genome duplication in plants. Nucleic Acids Res 41: D1152–D1158.
29. Tang H, Bowers JE, Wang X, Ming R, Alam M, et al. (2008) Synteny and Collinearity in Plant Genomes, Science 320: 486–488.
30. Krzywinski M, Schein J, Birol I, Connors J, Gascoyne R, et al. (2009) Circos: an information aesthetic for comparative genomics. Genome Res 19: 1639–1645.
31. Shiu S-H, Bleecker AB (2003) Expansion of the Receptor-Like Kinase/Pelle Gene Family and Receptor-Like Proteins in Arabidopsis. Plant Physiol 132: 530–543.

32. Du D, Zhang Q, Cheng T, Pan H, Yang W, et al. (2012) Genome-wide identification and analysis of late embryogenesis abundant (LEA) genes in *Prunus mume*. Mol Biol Rep 40: 1937–1946.

33. Guo AY, Zhu QH, Chen X, Luo JC (2007) GSDS: a gene structure display server. Yi Chuan 29: 1023–1026.

34. Tamura K, Peterson D, Peterson N, Stecher G, Nei M, et al. (2011) MEGA5: Molecular evolutionary genetics analysis using maximum likelihood, evolutionary distance, and maximum parsimony methods. Mol Biol Evol 28: 2731–2739.

35. Conesa A, Gotz S (2008) Blast2GO: a comprehensive suite for functional analysis in plant genomics. Int J Plant Genomics 2008: 619832.

36. Khan Y, Yadav A, Suresh BV, Muthamilarasan M, Yadav CB, et al. (2014) Comprehensive genome-wide identification and expression profiling of foxtail millet [*Setaria italica* (L.)] miRNAs in response to abiotic stress and development of miRNA database. Plant Cell Tiss Organ Cult 118: 279–292.

37. Suyama M, Torrents D, Bork P (2006) PAL2NAL: robust conversion of protein sequence alignments into the corresponding codon alignments. Nucleic Acids Res 34: W609–W612.

38. Lynch M, Conery JS (2000) The evolutionary fate and consequences of duplicate genes. Science 290: 1151–1155.

39. Yang Z, Gu S, Wang X, Li W, Tang Z, et al. (2008) Molecular evolution of the cpp-like gene family in plants: insights from comparative genomics of Arabidopsis and rice. J Mol Evol 67: 266–277.

40. Cochrane G, Alako B, Amid C, Bower L, Cerdeño-Tárraga A, et al. (2013) Facing growth in the European Nucleotide Archive. Nucleic Acids Res 41: D30–D35.

41. Saeed AI, Bhagabati NK, Braisted JC, Liang W, Sharov V, et al. (2006) TM4 microarray software suite. Methods Enzymol 411: 134–193.

42. Saeed AI, Sharov V, White J, Li J, Liang W, et al. (2003) TM4: a free, open-source system for microarray data management and analysis. Biotechniques 34: 374–378.

43. Lata C, Sahu PP, Prasad M (2010) Comparative transcriptome analysis of differentially expressed genes in foxtail millet (*Setaria italica* L.) during dehydration stress. Biochem Biophys Res Commun 393: 720–727.

44. Lata C, Bhutty S, Bahadur RP, Majee M, Prasad M (2011) Association of a SNP in a novel DREB2-like gene *SiDREB2* with stress tolerance in foxtail millet [*Setaria italica* (L.)]. J Exp Bot 62: 3387–3401.

45. Kumar K, Muthamilarasan M, Prasad M (2013) Reference genes for quantitative Real-time PCR analysis in the model plant foxtail millet (*Setaria italica* L.) subjected to abiotic stress conditions. Plant Cell Tiss Organ Cult 115: 13–22.

46. Livak KJ, Schmittgen TD (2001) Analysis of relative gene expression data using realtime quantitative PCR and the $2^{-\Delta\Delta Ct}$ method. Methods 25: 402–408.

47. Pandey G, Misra G, Kumari K, Gupta S, Parida SK, et al. (2013) Genome-wide development and use of microsatellite markers for large-scale genotyping applications in foxtail millet [*Setaria italica* (L.)]. DNA Res 20: 197–207.

48. Kumari K, Muthamilarasan M, Misra G, Gupta S, Subramanian A, et al. (2013) Development of eSSR-markers in *Setaria italica* and their applicability in studying genetic diversity, cross-transferability and comparative mapping in millet and non-millet species. PLoS ONE 8: e67742.

49. Muthamilarasan M, Venkata Suresh B, Pandey G, Kumari K, Parida SK, et al. (2014) Development of 5123 intron-length polymorphic markers for large-scale genotyping applications in foxtail millet. DNA Res 21: 41–52.

50. Suresh BV, Muthamilarasan M, Misra G, Prasad M (2013) FmMDb: a versatile database of foxtail millet markers for millets and bioenergy grasses research. PLoS ONE 8: e71418.

51. Berman HM, Westbrook J, Feng Z, Gilliland G, Bhat TN, et al. (2000) The protein data bank. Nucleic Acids Res 28: 235–242.

52. Kelley LA, Sternberg MJE (2009) Protein structure prediction on the Web: a case study using the Phyre server. Nature Protocols 4: 363–371.

53. Zhuang J, Peng R-H, Cheng Z-M, Zhang J, Cai B, et al. (2009) Genome-wide analysis of the putative AP2/ERF family genes in *Vitis vinifera*. Scientia Horticul. 123: 73–81.

54. Rashid M, Guangyuan H, Guangxiao Y, Hussain J, Xu Y (2012) AP2/ERF transcription factor in rice: genome-wide canvas and syntenic relationships between monocots and eudicots. Evol Bioinform Online 8: 321–355.

55. Bonthala VS, Muthamilarasan M, Roy R, Prasad M (2014) FmTFDb: a foxtail millet transcription factors database for expediting functional genomics in millets. Mol Biol Rep 41: 6343–6348.

56. Yadav CB, Muthamilarasan M, Pandey G, Prasad M (2014) Identification, characterization and expression profiling of Dicer-like, Argonaute and RNA-dependent RNA polymerase gene families in foxtail millet. Plant Mol Biol Rep DOI:10.1007/s11105-014-0736-y.

57. Muthamilarasan M, Bonthala VS, Mishra AK, Khandelwal R, Khan Y, et al. (2014) C_2H_2 type of zinc finger transcription factors in foxtail millet define response to abiotic stresses. Funct Integr Genomics 14: 531–543.

58. Muthamilarasan M, Khandelwal R, Yadav CB (2014) Identification and molecular characterization of MYB transcription factor superfamily in C_4 model plant foxtail millet (*Setaria italica* L.). PLOS ONE 9: e109920.

59. Karniely S, Pines O (2005) Single translation-dual destination: mechanisms of dual protein targeting in eukaryotes. Embo Reports 6: 420–425.

60. Song X, Li Y, Hou X (2013) Genome-wide analysis of the AP2/ERF transcription factor superfamily in Chinese cabbage (*Brassica rapa* ssp. pekinensis). BMC Genomics 14: 573.

61. Sun Z-M, Zhou M-L, Xiao X-G, Tang Y-X, Wu Y-M (2014) Genome-wide analysis of AP2/ERF family genes from *Lotus corniculatus* shows LcERF054 enhances salt tolerance. Funct Integr Genomics 14: 453–466.

62. Chen J, Xia X, Yin W (2009) Expression profiling and functional characterization of a DREB2-type gene from *Populus euphratica*. Biochem Biophys Res Commun 378: 483–487.

63. Le DT, Nishiyama R, Watanabe Y, Vankova R, Tanaka M, et al. (2012) Identification and expression analysis of cytokinin metabolic genes in soybean under normal and drought conditions in relation to cytokinin levels. PLoS ONE 7(8): e42411.

64. Fang YJ, You J, Xie KB, Xie WB, Xiong LZ (2008) Systematic sequence analysis and identification of tissue-specific or stress-responsive genes of NAC transcription factor family in rice. Mol Genet Genomics 280: 547–563.

65. Aida M, Beis D, Heidstra R, Willemsen V, Blilou I, et al. (2004) The PLETHORA genes mediate patterning of the Arabidopsis root stem cell niche. Cell 119: 109–120.

66. Cannon SB, Mitra A, Baumgarten A, Young ND, May G (2004) The roles of segmental and tandem gene duplication in the evolution of large gene families in *Arabidopsis thaliana*. BMC Plant Biol 4: 10.

67. Puranik S, Sahu PP, Mandal SN, B VS, Parida SK, et al. (2013) Comprehensive genome-wide survey, genomic constitution and expression profiling of the NAC transcription factor family in foxtail millet (*Setaria italica* L.). PLoS ONE 8: e64594.

68. Mishra AK, Muthamilarasan M, Khan Y, Parida SK, Prasad M (2014) Genome-wide investigation and expression analyses of WD40 protein family in the model plant foxtail millet (*Setaria italica* L.) PLoS ONE 9: e86852.

69. Zeller G, Henz SR, Widmer CK, Sachsenberg T, Ratsch G, et al. (2009) Stress induced changes in the *Arabidopsis thaliana* transcriptome analyzed with whole genome tiling arrays. Plant J 58: 1068–1082.

70. Lopez CG, Banowetz GM, Peterson CJ, Kronstad WE (2003) Dehydrin expression and drought tolerance in seven wheat cultivars. Crop Science 43: 577–582.

71. Lata C, Prasad M (2014) Association of an allele-specific marker with dehydration stress tolerance in foxtail millet suggests SiDREB2 to be an important QTL. J Plant Biochem Biotechnol 23: 119–122.

72. Okamuro JK, Caster B, Villarroel R, van Montagu M, Jofuku KD (1997) The AP2 domain of APETALA2 defines a large new family of DNA binding proteins in Arabidopsis. Proc Natl Acad Sci USA 94: 7076–7081.

73. Wolfe SA, Nekludova L, Pabo CO (2000) DNA recognition by Cys2His2 zinc finger proteins. Annu Rev Biophys Biomol Struct 29: 183–212.

21

Permanent Cardiac Sarcomere Changes in a Rabbit Model of Intrauterine Growth Restriction

Iratxe Torre[1], **Anna González-Tendero**[1], **Patricia García-Cañadilla**[1,2], **Fátima Crispi**[1], **Francisco García-García**[3,4], **Bart Bijnens**[6], **Igor Iruretagoyena**[1], **Joaquin Dopazo**[3,4,5], **Ivan Amat-Roldán**[1], **Eduard Gratacós**[1]*

1 BCNatal – Barcelona Center for Maternal-Fetal and Neonatal Medicine (Hospital Clínic and Hospital Sant Joan de Deu), IDIBAPS, University of Barcelona, and Centre for Biomedical Research on Rare Diseases (CIBER-ER), Barcelona, Spain, **2** Physense, Departament de Tecnologies de la Informació i les Comunicacions (DTIC), Universitat Pompeu Fabra, Barcelona, Spain, **3** Bioinformatics Department, Centro de Investigación Principe Felipe (CIPF), Valencia, Spain, **4** Functional Genomics Node, INB, CIPF, Valencia, Spain, **5** Centro de Investigación Biomédica en Red de Enfermedades Raras (CIBERER), CIPF, Valencia, Spain, **6** ICREA, Universitat Pompeu Fabra, Barcelona, Spain

Abstract

Background: Intrauterine growth restriction (IUGR) induces fetal cardiac remodelling and dysfunction, which persists postnatally and may explain the link between low birth weight and increased cardiovascular mortality in adulthood. However, the cellular and molecular bases for these changes are still not well understood. We tested the hypothesis that IUGR is associated with structural and functional gene expression changes in the fetal sarcomere cytoarchitecture, which remain present in adulthood.

Methods and Results: IUGR was induced in New Zealand pregnant rabbits by selective ligation of the utero-placental vessels. Fetal echocardiography demonstrated more globular hearts and signs of cardiac dysfunction in IUGR. Second harmonic generation microscopy (SHGM) showed shorter sarcomere length and shorter A-band and thick-thin filament interaction lengths, that were already present *in utero* and persisted at 70 postnatal days (adulthood). Sarcomeric M-band (GO: 0031430) functional term was over-represented in IUGR fetal hearts.

Conclusion: The results suggest that IUGR induces cardiac dysfunction and permanent changes on the sarcomere.

Editor: Christopher Torrens, University of Southampton, United Kingdom

Funding: This study was supported by grants from Ministerio de Economia y Competitividad PN de I+D+I 2008-2011 (ref. SAF2009_08815, SAF2012-37196, and BIO2011-27069); Instituto de Salud Carlos III (ref. PI11/00051, PI11/01709, PI12/00801) cofinanciado por el Fondo Europeo de Desarrollo Regional de la Unión Europea "Una manera de hacer Europa"; Centro para el Desarrollo Técnico Industrial (ref. cvREMOD 2009-2012) apoyado por el Ministerio de Economia y Competitividad y Fondo de inversión local para el empleo, Spain; The Cerebra Foundation for the Brain Injured Child (Carmarthen, Wales, UK); Obra Social "La Caixa" (Spain); Fundació Mutua Madrileña (Spain); Fundació Agrupació Mutua (Spain); AGAUR 2009 SGR grant n° 1099; and Red Temática de Investigación Cooperativa en cancer (ref. RD06/0020/1019). I.T. was supported by a post-doctoral fellowship from Carlos III Institute of Health, Spain (CD08/00176) during the time these studies were performed. P.G.C. acknowledges grant support to the Programa de Ayudas Predoctorales de Formación en Investigación en Salud del Instituto Carlos III, Spain (FI11/00362). A.G.T. was supported by an IDIBAPS (Institut d'Investigacions Biomèdiques August Pi i Sunyer) pre-doctoral fellowship. The funders had no role in study design, data collection and analysis, decision to publish, or preparation of the manuscript.

Competing Interests: The authors have declared that no competing interests exist.

* Email: gratacos@clinic.ub.es

Introduction

Intrauterine growth restriction (IUGR) is a major cause of perinatal mortality and long term morbidity [1] affecting up to 7–10% of pregnancies. IUGR results in low birth weight, which has been epidemiologically associated with an increased risk of cardiovascular disease in adulthood [2]. This association is thought to be mediated through fetal cardiovascular programming. Fetuses with IUGR suffer from chronic oxygen and nutrients restriction [3], which triggers an adaptive hemodynamic cardiovascular adaptation [4,5] associated with *in utero* volume and pressure overload [6]. Consequently, IUGR fetuses and newborns show signs of cardiovascular remodelling and dysfunction, including reduced annular peak velocities [7] and increased carotid intima-media thickness [8]. These fetal changes persist postnatally, as shown in human children [9] and in adult animal models [10,11].

Although the effects of IUGR on cardiac organ remodelling have been characterized, the features of cardiac fetal programming at subcellular scale are poorly documented. Identifying cellular and molecular pathways involved in the fetal cardiac programming may provide a better understanding into the pathogenesis of the disease and could be an opportunity to design therapeutic interventions reducing the burden of cardiovascular disease from early life.

The sarcomere is the basic functional unit of the cardiac contractile machinery. Changes in sarcomere structure and its key proteins have been observed in experimental models of cardiac dysfunction and failure [12–14]. In a previous study, we demonstrated that chronic pre-natal hypoxia induced permanent post-natal changes in the content and isoforms of sarcomeric proteins, including titin and myosin [10]. Interestingly, in another recent study in human hearts, we have demonstrated that severe

IUGR fetuses present signs of cardiac dysfunction associated with changes in sarcomere length [15]. Sarcomere length is strongly related to sarcomere function and contraction force, and has been described to be consistently altered in a substantial number of conditions associated with cardiac failure [16–18].

In the current study, we aimed to evaluate the long term impact of IUGR on sarcomere structure in an experimental rabbit model of IUGR previously described by our group [19] that reproduces the main biometric and cardiovascular features of human IUGR. Hearts from IUGR fetuses as well as hearts from young adult rabbits were evaluated in order to assess the postnatal persistence of the changes on sarcomere architecture. Additionally, gene expression analysis of the fetal hearts combined with the functional interpretation of the global gene expression profile were assessed to gain further insight on changes in pathways and proteins regulating the sarcomere.

Methods

Experimental model of IUGR

New Zealand White rabbits were provided by a certified breeder and housed for 1 week before surgery in separate cages on a reversed 12/12 h light cycle. Dams were fed a diet of standard rabbit chow and water *ad libitum*. Animal handling and all procedures were carried out in accordance to applicable regulations and guidelines and with the approval of the Animal

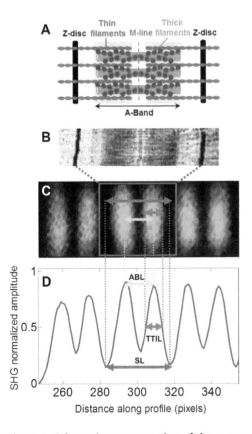

Figure 1. Schematic representation of the sarcomere cytoskeleton distances. A, an illustration of the sarcomere elementary parts; **B**, a sarcomere image by electron microscopy; **C**, a SHGM image of a myofibril; and **D**, a SHGM intensity profile along the myofibril showing the distances measured to characterize the sarcomere cytoskeleton: sarcomere length (SL), intra-sarcomeric A-band lengths (ABL) and thick-thin filament interaction length (TTIL).

Experimental Ethics Committee of the University of Barcelona (permit number: 310/11–5999) and all efforts were made to minimize suffering.

Ten New Zealand White pregnant rabbits were used to reproduce a previously described experimental model of IUGR [19,20]. Briefly, at 25 days of gestation dams were intramuscularly administered ketamine 35 mg/kg and xylazine 5mg/kg for anesthesia induction. Tocolysis (progesterone 0,9 mg/kg intramuscularly) and antibiotic prophylaxis (Penicillin G 300.000 UI endovenous) were administered prior to surgery. Rabbits have two uterine horns, which were exteriorized after a midline laparatomy. Randomly, one uterine horn was denominated as the IUGR horn, while the other as the control horn. In all gestational sacs from the IUGR horn, a selective ligature of the 40–50% of the uteroplacental vessels of each gestational sac was performed. No ligature was performed in the control horn in order to subject control animals to the same anaesthetic and surgical procedures than the IUGR animals. The abdomen was then closed and animals received subcutaneous meloxicam 0.4 mg/kg/24 h for 48 h, as postoperative analgesia. Five days after surgery, the same anaesthetic procedure was applied to perform a caesarean section. At this moment, fetal echocardiography was performed. Subsequently, rabbit kits were obtained and randomly assigned to two age study groups: fetal (30 days of gestation) and young adult (70 days post-natal) group. Hearts from rabbits included in the fetal group were obtained through a thoracotomy after anesthesia. Hearts for gene expression study were immediately snap frozen and stored at $-80°C$ until use. Hearts for multiphoton microscopy imaging were arrested in Ca^{2+}-free buffer and fixed in 4% paraformaldehyde in phosphate buffer for 24 hours at 4°C. Fetuses included in the young adult group were breast-fed by a wet-nurse rabbit until the age of 25 days. At the age of 70 postnatal days, animals were anesthetized and hearts were excised and fixed in 4% paraformaldehyde in phosphate buffer.

Fetal echocardiography

Echocardiography was performed in 10 paired control and IUGR rabbit fetuses at the time of the caesarean section by placing the probe directly on the uterine wall using a Vivid q (General Electric Healthcare, Horten, Norway) 4.5–11.5 MHz phased array probe. The angle of insonation was kept $<30°$ in all measurements and a 70 Hz high pass filter was used to avoid slow flow noise. Ultrasound evaluation included: (1) Ductus venosus pulsatility index obtained in a midsagittal section or transverse section of the fetal abdomen positioning the Doppler gate at its isthmic portion; (2) Aortic isthmus pulsatility index obtained in a sagittal view of the fetal thorax with a clear view of the aortic arch placing the sample volume between the origin of the last vessel of aortic arch and the aortic joint of the ductus arteriosus; (3) Left and right sphericity indices calculated as base-to-apex length/basal ventricular diameter measured from 2-dimensional images in an apical 4-chamber view at end-diastole; (4) Left ventricular free wall thickness measured by M-mode in a transverse 4-chamber view; (5) Left ejection fraction estimated by M-mode from a transverse 4-chamber view according to Teicholz formula; (6) Longitudinal systolic (S') peak velocities at the mitral annulus measured by spectral tissue Doppler from an apical 4-chamber view.

Second Harmonic Generation Microscopy (SHGM)

Fixed fetal and young adult hearts were dehydrated and embedded in paraffin. Transversal 30 μm heart sections were cut in a microtome (Leica RM 2135) and mounted onto silane coated thin slides. After deparaffination with xylene and hydration with

Table 1. Fetal biometric and echocardiographic results in IUGR and control fetuses.

	Control	IUGR	P-value
N	10	10	
Fetal Biometry			
Fetal weight (g)	48.97 (12.46)	29.94 (7.72)	0.000 *
Heart weight (g)	0.37 (0.10)	0.29 (0.09)	0.006 *
Heart weight/Fetal weight)*100	0.79 (0.11)	1.10 (0.31)	0.009 *
Fetal hemodynamics			
Ductus venosus pulsatility index	0.75 (0.25)	1.33 (0.75)	0.008 *
Aortic isthmus pulsatility index	3.05 (0.45)	3.85 (1.16)	0.009 *
Cardiac morphometry			
Left sphericity index	1.54 (0.34)	1.51 (0.26)	0.073
Right sphericity index	1.56 (0.24)	1.32 (0.23)	0.004 *
Left ventricle wall thickness (mm)	1.45 (0.37)	1.41 (0.31)	0.978
Systolic function			
Left ejection fraction (%)	89.1 (8.2)	82 (24.6)	0.39
Mitral annular systolic peak velocity (cm/s)	1.91 (0.27)	1.59 (0.33)	0.046 *

All values are median (interquartile range). P-value was calculated by t-test. *g:* grams; *mm:* millimeters; *cm/s:* centimetres/second. * P-value <0.05.

decreasing ethanol concentrations (100°/96°/70°), sections were covered with Mowiol 4–88 mounting medium (Sigma-Aldrich).

Detection of SHGM, from unlabelled cardiac tissue samples which were identified to contain mostly cardiomyocytes and no collagen, was performed with a Leica TCS-SP5 laser scanning

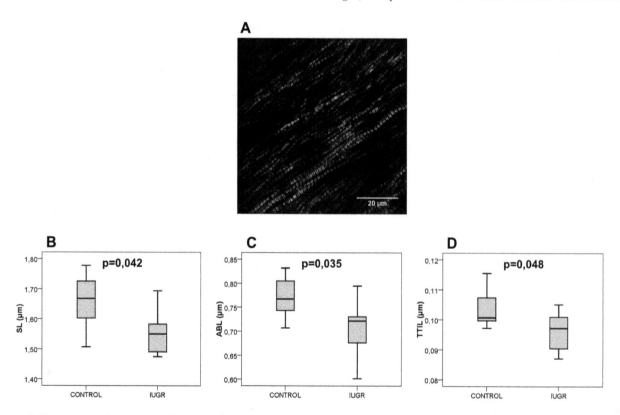

Figure 2. Ultrastructural sarcomere changes in fetal hearts from IUGR and controls. A, A representative SHGM image from unstained fetal rabbit left ventricle. The sarcomeres are clearly delimited by thick black lines (Z-discs) and the SHGM signal originates from the thick myosin filaments (in green). In the central region of the myosin filaments appears a thinner black line, identified as the M-band. Scale bar = 20 μm. **B** and **C,** show average distances between two consecutive Z-discs (SL) and intrasarcomeric A-bands (ABL), respectively. **D,** shows thick-thin filament interaction length (TTIL), as the mean width of the A-band-related peaks. Data are expressed as mean ± SD.

spectral confocal multiphoton microscope (Leica Microsystems Heidelberg GmbH, Manheim, Germany) equipped with a Near Infrared laser (Mai Tai Broad Band 710–990 nm, 120 fempto second pulse), at the Advanced Optical Microscopy Unit from Scientific and Technological Centres from University of Barcelona. 7 control and 7 IUGR hearts from both the fetal and the young adult group were included in the analysis. For each heart, between 8 and 10 SHGM images, randomly chosen from the left ventricular mid-wall, were acquired. Each image included an average of 15 cardiac muscle fibers containing 200 sarcomeres approximately. The image resolution was 40 nm/pixel. Left ventricular cardiac fibers, which are mostly oriented in a single axis in optical sections of approximately 1.5 microns, were aligned at 45 degrees to maximize Signal to Noise Ratio (SNR) of SHGM. All tissue sections consisted of almost parallel fibres and were contained in a single plane by the sample preparation to avoid any bias in measurements. This certified that the statistical differences are related to group differences and contributions from other source of experimental error are marginal.

Figure 1 shows the distinctive biperiodic pattern of sarcomeres, imaged by means of SHGM, and their characterization by two distances in unstained intact sarcomeres: resting sarcomere lengths (SL), measured as the distance between the two Z-discs delimiting each sarcomere; and intra-sarcomeric A-band lengths (ABL), defined as the distance between the two intra-sarcomeric segments of the A-band, divided by the M-band. Additionally, since SHGM arises from the thick-thin filament overlap in mature and developing sarcomeres [21], its length was calculated as the mean width of the A-band-related peak (namely in our study as thick-thin filament interaction length; TTIL). The three distinctive sarcomere distances (SL, ABL and TTIL) were measured automatically with a custom algorithm based on an optimal fitting of a parametric model of the autocorrelation function of the SHGM intensity profile in sarcomere fibers, as previously reported [22]. Briefly, the local orientation of the sarcomere fibers was estimated to perform automatically the following tracking of all the fibers within an image. After that, the SHGM intensity profile within each fiber was obtained and its autocorrelation function was computed. Finally, the autocorrelation function of the intensity profile was fitted with a parametric model to extract the average SL, ABL and TTIL of all the sarcomeres in the fiber. This calculation was performed for all the sarcomere fibers in the image. Additionally, a ratio between SL and ABL was calculated to further assess the quality of the acquired images and comparable physiological conditions between study groups. When this ratio was above 2.25, the sample was excluded from the analysis since the typical SHGM pattern of cardiac sarcomeres was lost.

Gene expression microarray

The gene expression profile was analyzed in 6 paired control and IUGR rabbit fetal hearts at 30 days of gestation as previously described [23]. This study focused on the expression profile of the cardiomyocyte contractile machinery – the sarcomere. Total RNA was isolated from left ventricle (RNeasy Mini kit, Qiagen), labelled with Quick Amp One-color Labelling kit (Agilent) and fluorochrome Cy3 and hybridized with a Rabbit Microarray (Agilent Microarray Design ID 020908). The hybridization was quantified at 5 μm resolution (Axon 4000B scanner) and data extraction was performed using Genepix Pro 6.0. Data obtained from the microarray were pre-processed and subjected to bioinformatics analysis from two viewpoints. Firstly, a differential gene expression analysis in order to identify up- or down-regulated individual genes associated with IUGR, carried out using the *limma* [24]

package from Bioconductor (http://www.bioconductor.org/). Differential gene expression of the samples in each group was assessed with the adjusted P value for every gene included in the microarray and with the fold change. Multiple testing adjustments of p-values were done according to Benjamini and Hochberg [25] methodology. All the pre-processing steps described can be carried out with the Babelomics software [26].

Secondly, a gene set analysis was carried out for the Gene Ontology (GO) annotations using FatiScan [27] algorithm, implemented in the Babelomics suite [28], as previously described [23]. Briefly, FatiScan is able to detect up or down-regulated blocks of functionally related genes within the lists of genes originated in the microarray experiment, which are ordered by differential expression. The algorithm uses various criteria to look for modules of genes that are functionally related, such as GO terms. Within the whole list of genes, FatiScan evaluates the distribution of functional terms and extracts GO terms that are significantly under- and over-represented. GO terms are grouped in three classes: i) cellular components; ii) biological processes and iii) molecular functions. In order to compare the two groups of genes and extract a list of GO terms whose distribution among the groups is significantly different, FatiScan uses a Fisher's exact test for 2×2 contingency tables and a multiple test correction to finally obtain an adjusted p-value.

GO annotations for the genes in the microarray where taken from Blast2GO Functional Annotation Repository web page (http://bioinfo.cipf.es/b2gfar/) [29].The raw microarray data is deposited in the Gene Expression Omnibus database under accession number GSE37860.

Statistical Analysis

Data were analyzed with the statistical package SPSS 15.0 (version 15.0; SPSS Inc, Chicago, IL). Data are expressed as mean ± standard deviation or median (Interquartile range (IQR)). Paired comparisons between the control and IUGR groups were done with t-test analysis. A classical parametric ANOVA test was carried out to compute significance between IUGR and control populations in experiments concerning SHGM analysis. Differences were considered significant with probability values of $p < 0.05$. Statistical methods related to the gene expression analysis have been detailed in its previous corresponding section.

Results

Fetal biometric and echocardiographic results

Table 1 details the fetal biometric and echocardiographic data. While absolute fetal body and heart weights were significantly lower in IUGR, heart to body weight ratio was significantly increased in IUGR fetuses as compared to controls. Fetal echocardiography showed a more globular cardiac shape with lower right sphericity index but similar wall thickness in IUGR as compared to controls. Despite similar results in ejection fraction, mitral annular peak velocities (S') were significantly lower in IUGR as compared to controls. Additionally, ductus venosus and aortic isthmus pulsatility indices were significantly increased in IUGR fetuses.

Fetal sarcomere morphometry

Seven paired control and IUGR rabbits were quantified demonstrating a readily detectable SHGM signal from unstained left ventricular sarcomeres. A representative image is displayed in **Figure 2A** and a magnification of it is shown in **Figure 1**.

As compared with controls, IUGR fetuses showed shorter sarcomere length (controls: 1.658 μm ±0.094 vs. IUGR 1.531 μm

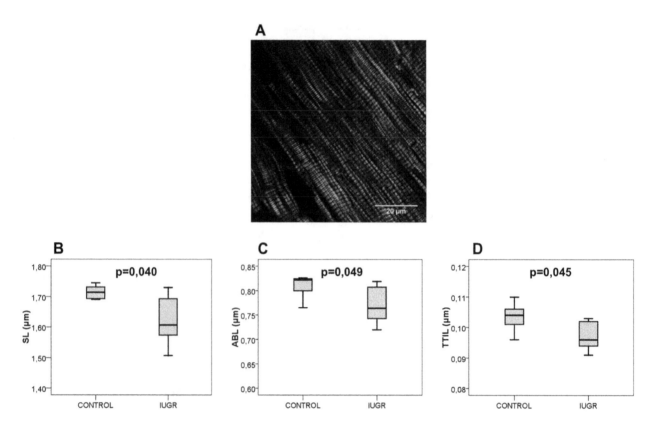

Figure 3. Ultrastructural sarcomere changes in adult hearts from IUGR and controls. A, Representative SHGM image from unstained adult rabbit left ventricle. Scale bar = 20 μm. **B** and **C,** show average distances between two consecutive Z discs (SL) and two intrasarcomeric A bands (ABL), respectively. **D,** shows the length of thick-thin filament interaction length (TTIL). Data are expressed as mean ± SD.

±0.114, p = 0.042) (**Figure 2B**). Intra-sarcomeric A-band length was also found to be decreased by IUGR (controls 0.772 μm ±0.044 vs. IUGR 0.705 μm ±0.060, p = 0.035) (**Figure 2C**). Additionally, thick-thin filament interaction length was shorter in IUGR (controls 0.104 μm ±0.006 vs. IUGR 0.096 μm ±0.007, p = 0.048) (**Figure 2D**). The ratio between sarcomere length and intrasarcomeric A-band length was similar in the study groups (controls 2.15±0.02 vs. IUGR 2.17±0.03).

Post-natal persistence of cardiac sarcomeric changes

Unstained young adult rabbit sarcomeres from left ventricular samples, in seven paired control and IUGR rabbits (70 postnatal days), produced a readily detectable SHGM signal (**Figure 3A**) with a pattern similar to the one observed in fetal sarcomeres, with similar ratios between sarcomere length and intrasarcomeric A-band length (controls 2.11±0.02 vs. IUGR 2.11±0.01). A significant decrease in sarcomere length (controls 1.720 μm ±0.068 vs. IUGR 1.626 μm ±0.084, p = 0.04) and intrasarco-meric A-band length (controls 0.817 μm ±0.036 vs. IUGR 0.772 μm ±0.041, p = 0.049) were observed in IUGR as compared to controls (**Figure 3B, C**). Additionally, sarcomeric thick-thin filament interaction length was shorter in IUGR (controls 0.103 μm ±0.005 vs. IUGR 0.097 μm ±0.005, p = 0.045) as compared to controls (**Figure 3D**).

Bioinformatic analysis of gene expression microarray data

Differential gene expression. All experiments showed a good level of labelling and hybridization onto the Agilent microarray. Differential gene expression (when analysing for a fold change higher than 0.5 and an adjusted p-value lower than 0.05– data not shown) was similar in both experimental conditions for all the cardiomyocyte sarcomere components included in the microarray.

Gene set analysis: functional interpretation of microarray data. A statistically significant enrichment in the group of genes composing the sarcomeric *M-band* (GO: 0031430) functional class (P raw <0.001; P adj = 0.069) was observed in fetal IUGR hearts. The GO cellular component M-band is represented in **Figure 4** as an acyclic graph. *M band* annotation was found in 1.6% of the most up-regulated genes in IUGR. On the other hand, only 0.69% of the most down-regulated genes in IUGR contained the annotation (p value <0.001; adjusted p-value <0.1). **Table 2** shows the most relevant genes that define the *M band* (GO: 0031430). The remaining sarcomeric functional terms identified by the gene set analysis were not found to be significantly modified due to IUGR.

Discussion

The present study shows an association between IUGR and fetal and postnatal permanent sarcomeric structural changes together with altered fetal cardiac function and sarcomere related gene expression changes. Since sarcomere length has been shown to have an important influence on cardiac contractility [30], the observed decrease in its length might be an important determinant of the currently described changes on cardiac shape and function. These findings provide important clinical and research suggestions that open future research to further characterize the molecular

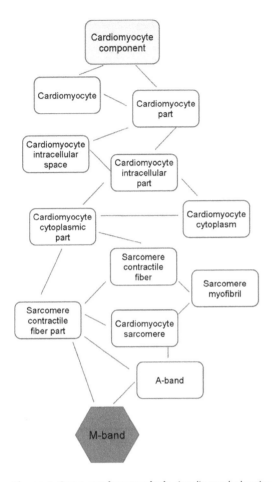

Figure 4. Gene ontology analysis. Acyclic graph showing the M-line cellular component significantly over-represented (in red) in IUGR compared to healthy control hearts. In the GO hierarchy, biological knowledge can be represented as a tree where functional terms near the root of the tree make reference to more general concepts while deeper functional terms near the leaves of the tree make reference to more specific concepts. If a gene is annotated to a given level, then is automatically considered to be annotated at all the upper levels up to the root.

basis of cardiac dysfunction and remodelling observed in IUGR individuals.

The experimental approach used in this study chronically reduces the blood supply to the fetuses by performing a selective ligature of the uteroplacental vessels in pregnant rabbits [19], and has been shown to reproduce the main biometric and fetal

hemodynamic features observed in human growth restriction [20]. Signs of fetal hypoxia are demonstrated by an increased pulsatility in the aortic isthmus flow as a reflection of the shift of blood towards the brain circulation in response to hypoxia, which is produced by a combination of brain vasodilation and systemic hypertension [31]. Additionally, the echocardiographic evaluation confirmed that IUGR results in more globular hearts, together with signs of systolic longitudinal dysfunction, which is similar to those changes observed in human fetuses and children [7,9,32].

Sarcomere length is important in myofilament force generation by different mechanisms since it has an effect on actin-myosin cross-bridge recruitment [33]. In this regard, sarcomere length is associated with the degree of thick-thin filament overlap, which affects the probability of actin-myosin cross-bridge formation and thus the capacity to generate force [34]. Shorter sarcomeres have been found in animal models of a variety of cardiac diseases, including ischemic contracture [16], diastolic dysfunction [17], dilated cardiomyopathy and heart failure [18]. Additionally, in a recent paper using human biopsies, passive force-length analysis suggested a shorter sarcomere length in pressure-overloaded myocardium compared to volume overload and control donors [35]. We recently described changes on the cardiomyocyte intracellular organization in the same experimental animal model of IUGR [23] that resemble to changes induced by pressure overload [36]. The permanent changes in sarcomere structure, as observed here in IUGR, could be as well a response to the known sustained increase in fetal blood pressure that occurs in IUGR. Supporting this notion, recent findings suggest that isolated neonatal cardiomyocytes undergo structural modifications within their myofibrils in response to changes in environmental stiffness, resulting in differences in resting sarcomere length [37]. Additionally, the observed changes in sarcomere length are consistent with previous research where we showed that chronic prenatal hypoxia led to a shift in the expression of titin isoform N2BA (larger and more compliant isoform) towards isoform N2B (smaller and stiffer isoform) [10]. This protein extends from the center of the sarcomere to the Z line, thus acting as a developmental template for sarcomere assembly [38]. Titin is thought to be a major determinant of the sarcomere length [39,17].

We recently published a work in which post-mortem cardiac samples from human fetuses that suffered severe IUGR were studied using the same SHGM methodology in order to assess changes in sarcomere morphometry [15]. Interestingly, results were strongly consistent with the ones described herein in the animal model: shorter sarcomere length in IUGR. This consistency strengthens the hypothesis that changes in sarcomere length could help to explain subclinical cardiac dysfunction previously described in human fetuses and children [9], and observed in IUGR rabbits in this study. The shorter sarcomere and thick-thin

Table 2. Results from most relevant sequences included in M-band functional class identified by FatiScan gene set analysis (M-band (GO: 0031430) block of genes).

Name	ID	Fold change
OBSCN (Obscurin)	ENSOCUT00000011554	0.266
OBSL (Obscurin-like protein 1)	ENSOCUT00000011142	0.171
Titin	ENSOCUT00000016899	0.265
Myopalladin	ENSOCUT00000009940	0.405
Myomesin-2	ENSOCUT00000013143	0.265

A fold change value above 0 indicates up-regulation in IUGR vs. controls.

filament interaction length might indicate a decrease in the number of binding events for cross-bridges between actin and myosin. Since cardiac muscle energy consumption depends on the number of recruited cross-bridges [40] shorter sarcomere length and thick-thin filament interaction length could be interpreted as an adaptive mechanism to cope with the oxygen and/or glucose restriction in IUGR. Importantly, in this study we show evidence of the postnatal persistence of the changes on sarcomere morphometry, which can not be assessed in humans. This observation might be relevant to explain the increased risk of cardiovascular disease and mortality in adult life related to IUGR as well as to understand the fetal cardiac programming.

Advanced bioinformatics analytic approaches were used in this study to complement the observations of SHGM and provide further evidence of the existence of functional gene expression differences encompassing structural changes. Fetal gene set analysis included a family of different tests designed to detect modules of functionally-related genes [41,42]. Among them, we used FatiScan [27], integrated in Babelomics [26], which has been employed to successfully detect coordinated variations in blocks of genes [43]. The results demonstrated functional differences in a basic structure for the sarcomeric cytoskeleton, the M-band, which plays an important organizational role during myofibrillogenesis by performing the regular packing of the nascent thick (myosin) filaments [44]. This finding is in line with previous observations suggesting that specific structural alterations at the M-band might be part of a general adaptation of the sarcomeric cytoskeleton to unfavorable working conditions in early stages of dilated cardiomyopathy, correlating with an impaired ventricular function [45]. Moreover, they suggest an underlying basis for the abnormal sarcomere cytoarchitecture observed in IUGR. Genes included in the M-band functional class involve a variety of associated molecules to this structure with key roles in sarcomere assembly and function, including titin, obscurin and myomesin. Interference with the physiological function of these proteins is of pathogenic relevance for human cardiomyopathies [46]. The findings warrant further investigation to clarify the individual role of sarcomere proteins in the generation of permanent structural changes of the contractile machinery under IUGR.

Several study limitations and technical considerations merit discussion. We used SHGM to measure sarcomere length. The technique has been validated to accurately visualize the SHGM signal produced by cardiac myosin thick filaments in unstained sarcomeres from different species with an accuracy of 20 nm [21,22,47]. SHGM offers comparable results to those provided by electron microscopy with the additional advantage of imaging

larger tissue areas [22]. Sarcomere length values in this study were consistent with measurements using electron and light microscopy in fixed rabbit cardiomyocytes, and showed increasing values with cardiac maturation [48,49]. Although the intrasarcomeric A-band length from the second harmonic generation signal changes little depending upon sarcomere contraction [21], we acknowledge the hearts not being arrested with potassium and fixed at a known transmural pressure as a potential limitation of this study. The bioinformatics analysis of gene expression data used herein has been shown to be useful for studying diseases like IUGR in which subtle differences are expected. This study illustrates the existence of sarcomere changes, but it provides a limited view of the mechanistic or molecular pathways underlying such changes. We acknowledge that with the present study is difficult to address the actual biological relevance of the gene expression and bioinformatics analysis and future studies are required to gain further insight on the relevance and to elucidate pathways that might be responsible for the observed changes. Additionally, results provided by bioinformatics gene set analysis strongly depend on the capabilities of bioinformatics tools and gene annotations, which are constantly evolving. Thus, gene pathways that are not well described yet might remain undetectable with current tools

In conclusion, this study provides new clues towards understanding the cellular and molecular mechanisms underlying cardiac remodelling through fetal programming and persisting in adulthood. Together, the findings presented here support that IUGR induces permanent changes in cardiac sarcomere morphometry, associated with functional changes in proteins involved in sarcomere function and assembly. These changes might help to explain the stiffer and less deforming hearts of fetuses and adults suffering from IUGR, and open new lines of research aiming at characterizing and interfering with the mechanisms of adaptation leading to cardiovascular remodelling in IUGR.

Acknowledgments

We thank Anna Bosch and Dr Maria Calvo from the Advanced Optical Microscopy Unit from Scientific and Technological Centres from University of Barcelona, for their support and advice with Second Harmonic Generation Microscopy techniques.

Author Contributions

Conceived and designed the experiments: IT FC BB JD IAR EG. Performed the experiments: IT AGT PGC FC FGG II. Analyzed the data: IT AGT PGC FC FGG IAR. Contributed reagents/materials/analysis tools: JD IAR. Wrote the paper: IT FC BB EG.

References

1. Alberry M, Soothill P (2007) Management of fetal growth restriction. Arch Dis Child Fetal Neonatal Ed 92: F62–F67.

2. Barker DJ (1999) Fetal origins of cardiovascular disease. Ann Med 1: 3–6.

3. Soothill PW, Nicolaides KH, Campbell S (1987) Prenatal asphyxia, hyperlacti-caemia, hypoglycaemia, and erythroblastosis in growth retarded fetuses. BMJ 294: 1051–1053.

4. Hecher K, Campbell S, Doyle P, Harrington K, Nicolaides K (1995) Assessment of fetal compromise by Doppler ultrasound investigation of the fetal circulation. Circulation 91: 129–138.

5. Tchirikov M, Schroder HJ, Hecher K (2006) Ductus venosus shunting in the fetal venous circulation: regulatory mechanisms, diagnostic methods and medical importance. Ultrasound Obstet Gynecol 27: 452–461.

6. Verburg BO, Jaddoe VW, Wladimiroff JW, Hofman A, Witteman JC, et al. (2008) Fetal hemodynamic adaptive changes related to intrauterine growth: the Generation R Study. Circulation 117: 649–659.

7. Comas M, Crispi F, Cruz-Martinez R, Martinez JM, Figueras F, et al. (2010) Usefulness of myocardial tissue Doppler vs. conventional echocardiography in the evaluation of cardiac dysfunction in early-onset intrauterine growth restriction. Am J Obstet Gynecol 203: 45.e1–45.e7.

8. Skilton MK, Evans N, Griffiths KA, Harmer JA, Celermajer D (2005) Aortic wall thickness in newborns with intrauterine restriction. Lancet 23: 1484–1486.

9. Crispi F, Bijnens B, Figueras F, Bartrons J, Eixarch E, et al. (2010) Fetal growth restriction results in remodeled and less efficient hearts in children. Circulation 121: 2427–2436.

10. Tintu A, Rouwet E, Verlohren S, Brinkmann J, Ahmad S, et al. (2009) Hypoxia induces dilated cardiomyopathy in the chick embryo: mechanism, intervention, and long-term consequences. PLoS One 4: e5155.

11. Ream M, Ray AM, Chandra R, Chikaraishi DM (2008) Early fetal hypoxia leads to growth restriction and myocardial thinning. Am J Physiol Regul Integr Comp Physiol 295: R583–595.

12. Haddad F, Bodell PW, Baldwin KM (1995) Pressure-induced regulation of myosin expression in rodent heart. J Appl Physiol 4: 1489–1495.

13. Warren CM, Jordan MC, Roos KP, Krzesinski PR, Greaser ML (2003) Titin isoform expression in normal and hypertensive myocardium. Cardiovasc Res 59: 86–94.

14. Falcão-Pires I, Palladini G, Gonçalves N, van der Velden J, Moreira-Gonçalves D, et al. (2011) Distinct mechanisms for diastolic dysfunction in diabetes mellitus and chronic pressure-overload. Basic Res Cardiol 106: 801–814.

15. Iruretagoyena JI, Gonzalez-Tendero A, Garcia-Canadilla P, Amat-Roldan I, Torre I, et al. (2014) Cardiac dysfunction is associated with altered sarcomere ultrastructure in intrauterine growth restriction. Am J Obstet Gynecol doi: 10.1016/j.ajog.2014.01.023.

16. Anderson PG, Bishop SP, Digerness SB (1987) Transmural progression of morphologic changes during ischemic contracture and reperfusion in the normal and hypertrophied rat heart. Am J Phatol 129: 152–167.

17. Radke MH, Peng J, Wu Y, McNabb M, Nelson OL, et al. (2007) Targeted deletion of titin N2B region leads to diastolic dysfunction and cardiac atrophy. Proc Natl Acad Sci U S A 104: 3444–3449.

18. Chen JF, Murchison EP, Tang R, Callis TE, Tatsuguchi M, et al. (2008) Targeted deletion of Dicer in the heart leads to dilated cardiomyopathy and heart failure. Proc Natl Acad Sci U S A 105: 2111–2116.

19. Eixarch E, Figueras F, Hernandez-Andrade E, Crispi F, Nadal A, et al. (2009) An experimental model of fetal growth restriction based on selective ligature of uteroplacental vessels in the pregnant rabbit. Fetal Diagn Ther 26: 203–211.

20. Eixarch E, Hernandez-Andrade E, Crispi F, Illa M, Torre I, et al. (2011) Impact on fetal mortality and cardiovascular Doppler of selective ligature of uteroplacental vessels compared with undernutrition in a rabbit model of intrauterine growth restriction. Placenta 32: 304–309.

21. Plotnikov SV, Millard AC, Campagnola PJ, Mohler WA (2006) Characterization of the myosin-based source for second-harmonic generation from muscle sarcomeres. Biophys J 90: 693–703.

22. Garcia-Canadilla P, Gonzalez-Tendero A, Iruretagoyena I, Crispi F, Torre I, et al. (2014) Automated cardiac sarcomere analysis from second harmonic generation images. J Biomed Opt doi: 10.1117/1.JBO.19.5.056010.

23. Gonzalez-Tendero A, Torre I, Garcia-Canadilla P, Crispi F, García-García F, et al. (2013) Intrauterine growth restriction is associated with cardiac ultrastructural and gene expression changes related to the energetic metabolism in a rabbit model. Am J Physiol Heart Circ Physiol 305: H1752–H1760.

24. Smyth GK (2004) Linear models and empirical bayes methods for assessing differential expression in microarray experiments. Stat Appl Genet Mol Biol 3: Article3.

25. Benjamini Y, Hochberg Y (1995) Controlling the false discovery rate a practical and powerful approach to multiple testing. J R Statist Soc B 57: 289–300.

26. Medina I, Carbonell J, Pulido L, Madeira SC, Goetz S, et al. (2010) Babelomics: an integrative platform for the analysis of transcriptomics, proteomics and genomic data with advanced functional profiling. Nucleic Acids Res 38: W210–W213.

27. Al-Shahrour F, Arbiza L, Dopazo H, Huerta-Cepas J, Mínguez P, et al. (2007) From genes to functional classes in the study of biological systems. BMC Bioinformatics 8: 114.

28. Al-Shahrour F, Carbonell J, Minguez P, Goetz S, Conesa A, et al. (2008) Babelomics: advanced functional profiling of transcriptomics, proteomics and genomics experiments. Nucleic Acids Res 36: W341–346.

29. Götz S, Arnold R, Sebastián-León P, Martín-Rodríguez S, Tischler P, et al. (2011) B2G-FAR, a species-centered GO annotation repository. Bioinformatics 27: 919–924.

30. Fukuda N, Sasaki D, Ishiwata S, Kurihara S (2001) Length dependence of tension generation in rat skinned cardiac muscle: role of titin in the Frank-Starling mechanism of the heart. Circulation 104: 1639–1645.

31. Fouron JC, Skoll A, Sonesson SE, Pfizenmaier M, Jaeggi E, et al. (1999) Relationship between flow through the fetal aortic isthmus and cerebral oxygenation during acute placental circulatory insufficiency in ovine fetuses. Am J Obstet Gynecol 181: 1102–1107.

32. Crispi F, Hernandez-Andrade E, Pelsers MM, Plasencia W, Benavides-Serralde JA, et al. (2008) Cardiac dysfunction and cell damage across clinical stages of severity in growth-restricted fetuses. Am J Obstet Gynecol 199: 254.e1–254.e8.

33. Wannenburg T, Heijne GH, Geerdink JH, Van Den Dool HW, Janssen PM, et al. (2000) Cross-bridge kinetics in rat myocardium: effect of sarcomere length and calcium activation. Am J Physiol Heart Circ Physiol 279: H779–H790.

34. Gordon AM, Homsher E, Regnier M (2000) Regulation of contraction in striated muscle. Physiol Rev 80: 853–924.

35. Chaturvedi RR, Herron T, Simmons R, Shore D, Kumar P, et al. (2010) Passive stiffness of myocardium from congenital heart disease and implications for diastole. Circulation 121: 979–988.

36. Schwarzer M, Schrepper A, Amorim PA, Osterholt M, Doenst T (2012) Pressure Overload Differentially Affects Respiratory Capacity in Interfibrillar and Subsarcolemmal Mitochondria. Am J Physiol Heart Circ Physiol 304: H529–H537.

37. Rodriguez AG, Han SJ, Regnier M, Sniadecki NJ (2011) Substrate stiffness increases twitch power of neonatal cardiomyocytes in correlation with changes in myofibril structure and intracellular calcium. Biophys J 101: 2455–2464.

38. Tskhovrebova L, Trinick J (2003) Titin: properties and family relationships. Nat Rev Mol Cell Biol 4: 679–689.

39. Labeit S, Kolmerer B, Linke WA (1997) The giant protein titin. Emerging roles in physiology and pathophysiology. Circ Res 80: 290–294.

40. Sela G, Yadid M, Landesberg A (2010) Theory of cardiac sarcomere contraction and the adaptive control of cardiac function to changes in demands. Ann N Y Acad Sci 1188: 222–230.

41. Mootha VK, Lindgren CM, Eriksson KF, Subramanian A, Sihag S, et al. (2003) PGC-1alpha-responsive genes involved in oxidative phosphorylation are coordinately downregulated in human diabetes. Nat Genet 34: 267–273.

42. Kim SY, Volsky DJ (2005) PAGE: parametric analysis of gene set enrichment. BMC Bioinformatics 6: 144.

43. Prado-Lopez S, Conesa A, Armiñán A, Martínez-Losa M, Escobedo-Lucea C, et al. (2010) Hypoxia promotes efficient differentiation of human embryonic stem cells to functional endothelium. Stem Cells 28: 407–418.

44. Agarkova I, Perriard JC (2005) The M-band: an elastic web that crosslinks thick filaments in the center of the sarcomere. Trends Cell Biol 15: 477–485.

45. Schoenauer R, Emmert MY, Felley A, Ehler E, Brokopp C, et al. (2011) EH-myomesin splice isoform is a novel marker for dilated cardiomyopathy. Basic Res Cardiol 106: 233–247.

46. Fukuzawa A, Lange S, Holt M, Vihola A, Carmignac V, et al (2008) Interactions with titin and myomesin target obscurin and obscurin-like 1 to the M-band: implications for hereditary myopathies. J Cell Sci 121: 1841–1851.

47. Boulesteix T, Beaurepaire E, Sauviat MP, Schanne-Klein MC (2004) Second-harmonic microscopy of unstained living cardiac myocytes: measurements of sarcomere length with 20-nm accuracy. Opt Lett 29: 2031–2033.

48. Nassar R, Reedy MC, Anderson PA (1987) Developmental changes in the ultrastructure and sarcomere shortening of the isolated rabbit ventricular myocyte. Circ Res 61: 465–483.

49. Wu Y, Wu EX (2009) MR study of postnatal development of myocardial structure and left ventricular function. J Magn Reson Imaging 30: 47–53.

PERMISSIONS

All chapters in this book were first published in PLOS ONE, by The Public Library of Science; hereby published with permission under the Creative Commons Attribution License or equivalent. Every chapter published in this book has been scrutinized by our experts. Their significance has been extensively debated. The topics covered herein carry significant findings which will fuel the growth of the discipline. They may even be implemented as practical applications or may be referred to as a beginning point for another development.

The contributors of this book come from diverse backgrounds, making this book a truly international effort. This book will bring forth new frontiers with its revolutionizing research information and detailed analysis of the nascent developments around the world.

We would like to thank all the contributing authors for lending their expertise to make the book truly unique. They have played a crucial role in the development of this book. Without their invaluable contributions this book wouldn't have been possible. They have made vital efforts to compile up to date information on the varied aspects of this subject to make this book a valuable addition to the collection of many professionals and students.

This book was conceptualized with the vision of imparting up-to-date information and advanced data in this field. To ensure the same, a matchless editorial board was set up. Every individual on the board went through rigorous rounds of assessment to prove their worth. After which they invested a large part of their time researching and compiling the most relevant data for our readers.

The editorial board has been involved in producing this book since its inception. They have spent rigorous hours researching and exploring the diverse topics which have resulted in the successful publishing of this book. They have passed on their knowledge of decades through this book. To expedite this challenging task, the publisher supported the team at every step. A small team of assistant editors was also appointed to further simplify the editing procedure and attain best results for the readers.

Apart from the editorial board, the designing team has also invested a significant amount of their time in understanding the subject and creating the most relevant covers. They scrutinized every image to scout for the most suitable representation of the subject and create an appropriate cover for the book.

The publishing team has been an ardent support to the editorial, designing and production team. Their endless efforts to recruit the best for this project, has resulted in the accomplishment of this book. They are a veteran in the field of academics and their pool of knowledge is as vast as their experience in printing. Their expertise and guidance has proved useful at every step. Their uncompromising quality standards have made this book an exceptional effort. Their encouragement from time to time has been an inspiration for everyone.

The publisher and the editorial board hope that this book will prove to be a valuable piece of knowledge for researchers, students, practitioners and scholars across the globe.

LIST OF CONTRIBUTORS

Wensheng Zhang, Andrea Edwards and Kun Zhang
Department of Computer Science, Xavier University of Louisiana, New Orleans, Louisiana, United States of America

Erik Flemington
Tulane Cancer Center, Tulane School of Medicine, New Orleans, Louisiana, United States of America

Newton Medeiros Vidal, Ana Laura Grazziotin and Thiago Motta Venancio
Laboratório de Química e Função de Proteínas e Peptídeos, Centro de Biociências e Biotecnologia, Universidade Estadual do Norte Fluminense Darcy Ribeiro, Camposdos Goytacazes, Rio de Janeiro, Brazil

Helaine Christine Cancela Ramos and Messias Gonzaga Pereira
Laboratório de Melhoramento Genético Vegetal, Centro de Ciências e Tecnologias Agropecuárias, Universidade Estadual do Norte Fluminense Darcy Ribeiro, Campos dos Goytacazes, Rio de Janeiro, Brazil

Erik Lysøe and Even S. Riiser
Department of Plant Health and Plant Protection, Bioforsk - Norwegian Institute of Agricultural and Environmental Research, Ås, Norway

Linda J. Harris
Eastern Cereal and Oilseed Research Centre, Agriculture and Agri-Food Canada, Ottawa, Canada

Sean Walkowiak and Rajagopal Subramaniam
Eastern Cereal and Oilseed Research Centre, Agriculture and Agri-Food Canada, Ottawa, Canada
Department of Biology, Carleton University, Ottawa, Canada

Hege H. Divon
Section of Mycology, Norwegian Veterinary Institute, Oslo, Norway

Carlos Llorens
Biotechvana, València, Spain

Toni Gabaldón
Bioinformatics and Genomics Programme, Centre for Genomic Regulation, Barcelona, Spain
Universitat Pompeu Fabra, Barcelona, Spain
Institució Catalana de Recerca i Estudis Avançats, Barcelona, Spain

H. Corby Kistler and Wilfried Jonkers
ARS-USDA, Cereal Disease Laboratory, St. Paul Minnesota, United States of Americ

Anna-Karin Kolseth
Department of Crop Production Ecology, Swedish University of Agricultural Sciences, Uppsala, Sweden

Kristian F. Nielsen, Ulf Thrane and Rasmus J. N. Frandsen
Department of Systems Biology, Technical University of Denmark, Lyngby, Denmark

Nicolas Tchitchek and Angela L. Rasmussen
Department of Microbiology, University of Washington, Seattle, Washington, United States of America

David Safronetz, Heinz Feldmann and Hideki Ebihara
Laboratory of Virology, Division of Intramural Research, National Institute of Allergy and Infectious Diseases, National Institutes of Health, Rocky Mountain Laboratories, Hamilton, Montana, United States of America

Craig Martens, Kimmo Virtaneva and Stephen F. Porcella
Genomics Unit, Research Technologies Section, National Institute of Allergy and Infectious Diseases, National Institutes of Health, Rocky Mountain Laboratories, Hamilton, Montana, United States of America

Michael G. Katze
Department of Microbiology, University of Washington, Seattle, Washington, United States of America

Washington National Primate Research Center, University of Washington, Seattle, Washington, United States of America

Carolina Mehaffy and Frances B. Jamieson
Public Health Ontario, Toronto, Canada
University of Toronto, Toronto, Canada

Jennifer L. Guthrie
Public Health Ontario, Toronto, Canada

David C. Alexander
Saskatchewan Disease Control Laboratory, Regina, Canada

Rebecca Stuart and Elizabeth Rea
Toronto Public Health, Toronto, Canada

Scott Schaeffer and Amit Dhingra
Department of Horticulture, Washington State University, Pullman, WA, United States of America
Molecular Plant Science Graduate Program, Washington State University, Pullman, WA, United States of America

Artemus Harper
Department of Horticulture, Washington State University, Pullman, WA, United States of America

Rajani Raja and Pankaj Jaiswal
2082 Cordley Hall, Department of Botany and Plant Pathology, Oregon State University, Corvallis, OR, United States of America

Xiaoyan Xu, Yubang Shen and Jianjun Fu
Key Laboratory of Exploration and Utilization of Aquatic Genetic Resources, Shanghai Ocean University, Ministry of Education, Shanghai 201306, PR China

Liqun Lu
National Pathogen Collection Center for Aquatic Animals, College of Fisheries and Life Science, Shanghai Ocean University, 999 Huchenghuan Road, 201306 Shanghai, PR China

Jiale Li
Key Laboratory of Exploration and Utilization of Aquatic Genetic Resources, Shanghai Ocean University, Ministry of Education, Shanghai 201306, PR China
E-Institute of Shanghai Universities, Shanghai Ocean University, 999 Huchenghuan Road, 201306 Shanghai, PR China

Marion Horsch, Helmut Fuchs and Valé rie Gailus-Durner
German Mouse Clinic, Institute of Experimental Genetics, Helmholtz Zentrum München GmbH, German Research Center for Environmental Health, Neuherberg, Germany

Johannes Beckers
German Mouse Clinic, Institute of Experimental Genetics, Helmholtz Zentrum München GmbH, German Research Center for Environmental Health, Neuherberg, Germany
German Center for Diabetes Research (DZD), Neuherberg, Germany
Experimental Genetics, Center of Life and Food Sciences Weihenstephan, Technische Universität München, Freising-Weihenstephan, Germany

Martin Hrabê de Angelis
German Mouse Clinic, Institute of Experimental Genetics, Helmholtz Zentrum München GmbH, German Research Center for Environmental Health, Neuherberg, Germany
German Center for Diabetes Research (DZD), Neuherberg, Germany
Experimental Genetics, Center of Life and Food Sciences Weihenstephan, Technische Universität München, Freising-Weihenstephan, Germany
German Center for Vertigo and Balance Disorders, University Hospital Munich, Campus Grosshadern, Munich, Germany

Birgit Rathkolb
German Mouse Clinic, Institute of Experimental Genetics, Helmholtz Zentrum München GmbH, German Research Center for Environmental Health, Neuherberg, Germany
Molecular Animal Breeding and Biotechnology, and Laboratory for Functional Genome Analysis (LAFUGA), Gene Center, LMU Mü nchen, Munich, German

Eckhard Wolf, Bernhard Aigner and Elisabeth Kemter
Molecular Animal Breeding and Biotechnology, and Laboratory for Functional Genome Analysis (LAFUGA), Gene Center, LMU München, Munich, German

Kim Vancampenhout and Claudia Spits
Research Group Reproduction and Genetics (REGE), Vrije Universiteit Brussel (VUB), Brussels, Belgium

Ben Caljon
Center for Medical Genetics, UZ Brussel, Vrije Universiteit Brussel (VUB), Brussels, Belgium

Katrien Stouffs, Willy Lissens and Sara Seneca
Research Group Reproduction and Genetics (REGE), Vrije Universiteit Brussel (VUB), Brussels, Belgium
Center for Medical Genetics, UZ Brussel, Vrije Universiteit Brussel (VUB), Brussels, Belgium

An Jonckheere
Department of Pediatric Neurology, UZ Brussel, Vrije Universiteit Brussel (VUB), Brussels, Belgium

Linda De Meirleir
Research Group Reproduction and Genetics (REGE), Vrije Universiteit Brussel (VUB), Brussels, Belgium
Department of Pediatric Neurology, UZ Brussel, Vrije Universiteit Brussel (VUB), Brussels, Belgium

Arnaud Vanlander, Joël Smet, Boel De Paepe and Rudy Van Coster
Department of Pediatrics, Division of Pediatric Neurology and Metabolism, University Hospital Ghent, Ghent University, Ghent, Belgium

Fan Fan, Richard J. Roman, Mallikarjuna R. Pabbidi and Stanley V. Smith
Department of Pharmacology and Toxicology, University of Mississippi Medical Center, Jackson, Mississippi, United States of America

Aron M. Geurts and Howard Jacob
Human and Molecular Genetics Center, Medical College of Wisconsin, Milwaukee, Wisconsin, United States of America

David R. Harder
Department of Physiology and Cardiovascular Research Center, Medical College of Wisconsin, Milwaukee, Wisconsin, United States of America
Daiva E. Nielsen and Ahmed El-Sohemy
Department of Nutritional Sciences, University of Toronto, 150 College St, Toronto, ON, M5S 3E2, Canada

Fang Deng
Department of Cell Biology, Third Military Medical University, Chongqing, 400038, China
Molecular Oncology Laboratory, Department of Orthopaedic Surgery and Rehabilitation Medicine, The University of Chicago Medical Center, Chicago, IL, 60637, United States of America

Xiang Chen, Sahitya Denduluri, Melissa Li, Nisha Geng, Guolin Zhou, Hue H. Luu and Rex C. Haydon
Molecular Oncology Laboratory, Department of Orthopaedic Surgery and Rehabilitation Medicine, The University of Chicago Medical Center, Chicago, IL, 60637, United States of America

Zhan Liao and Tong-Chuan He
Molecular Oncology Laboratory, Department of Orthopaedic Surgery and Rehabilitation Medicine, The University of Chicago Medical Center, Chicago, IL, 60637, United States of America
Department of Orthopaedic Surgery, the Affiliated Xiang-Ya Hospital of Central South University, Changsha, 410008, China

Zhengjian Yan, Zhongliang Wang, Youlin Deng, Qian Zhang, Jing Wang, Qiang Wei and Penghui Zhang
Molecular Oncology Laboratory, Department of Orthopaedic Surgery and Rehabilitation Medicine, The University of Chicago Medical Center, Chicago, IL, 60637, United States of America
Ministry of Education Key Laboratory of Diagnostic Medicine, and the Affiliated Hospitals of Chongqing Medical University, Chongqing, 400016, China

Zhonglin Zhang and Ruifang Li
Molecular Oncology Laboratory, Department of Orthopaedic Surgery and Rehabilitation Medicine, The University of Chicago Medical Center, Chicago, IL, 60637, United States of America
Department of Surgery, the Affiliated Zhongnan Hospital of Wuhan University, Wuhan, 430071, China

Jixing Ye
Molecular Oncology Laboratory, Department of Orthopaedic Surgery and Rehabilitation Medicine, The University of Chicago Medical Center, Chicago, IL, 60637, United States of America
School of Bioengineering, Chongqing University, Chongqing, 400044, China

Min Qiao
Molecular Oncology Laboratory, Department of Orthopaedic Surgery and Rehabilitation Medicine, The University of Chicago Medical Center, Chicago, IL, 60637, United States of America
Ministry of Education Key Laboratory of Diagnostic Medicine, and the Affiliated Hospitals of Chongqing Medical University, Chongqing, 400016, China

Department of Orthopaedic Surgery, the Affiliated Xiang-Ya Hospital of Central South University, Changsha, 410008, China

Lianggong Zhao
Molecular Oncology Laboratory, Department of Orthopaedic Surgery and Rehabilitation Medicine, The University of Chicago Medical Center, Chicago, IL, 60637, United States of America
Department of Orthopaedic Surgery, the Second Affiliated Hospital of Lanzhou University, Lanzhou, Gansu, 730000, China

Russell R. Reid
Molecular Oncology Laboratory, Department of Orthopaedic Surgery and Rehabilitation Medicine, The University of Chicago Medical Center, Chicago, IL, 60637, United States of America
The Laboratory of Craniofacial Biology, Department of Surgery, The University of Chicago Medical Center, Chicago, IL, 60637, United States of America

Tian Yang
Department of Cell Biology, Third Military Medical University, Chongqing, 400038, China

Wenfang Gong, Shoupu He, Jiahuan Tian, Junling Sun, Zhaoe Pan, Yinhua Jia and Xiongming Du
State Key Laboratory of Cotton Biology, Institute of Cotton Research, Chinese Academy of Agricultural Sciences, Anyang, China

Gaofei Sun
State Key Laboratory of Cotton Biology, Institute of Cotton Research, Chinese Academy of Agricultural Sciences, Anyang, China
Department of Computer Science and Information Engineering, Anyang Institute of Technology, Anyang, China

Bruce J. Walker, Terrance Shea, Margaret Priest, Amr Abouelliel, Sharadha Sakthikumar, Christina A. Cuomo, Qiandong Zeng, Jennifer Wortman, Sarah K. Young and Ashlee M. Earl
Broad Institute of MIT and Harvard, Cambridge, Massachusetts, United States of America

Thomas Abeel
Broad Institute of MIT and Harvard, Cambridge, Massachusetts, United States of America
VIB Department of Plant Systems Biology, Ghent University, Ghent, Belgium

Huei-Jiun Su
Institute of Ecology and Evolutionary Biology, National Taiwan University, Taipei, Taiwan

Saskia A. Hogenhout
Department of Cell and Developmental Biology, John Innes Centre, Norwich, United Kingdom

Abdullah M. Al-Sadi
Department of Crop Sciences, Sultan Qaboos University, Al Khoud, Oman

Chih-Horng Kuo
Institute of Plant and Microbial Biology, Academia Sinica, Taipei, Taiwan
Molecular and Biological Agricultural Sciences Program, Taiwan International Graduate Program, National Chung Hsing University and Academia Sinica, Taipei, Taiwan
Biotechnology Center, National Chung Hsing University, Taichung, Taiwan

Martin Probst and Katharina Grabmeier-Pfistershammer
Division of Immunology, Allergy and Infectious Diseases, Department of Dermatology, Medical University Vienna, Vienna, Austria

Christoph Hoeller, Hubert Pehamberger and Harald Kittler
Division of General Dermatology, Department of Dermatology, Medical University Vienna, Vienna, Austria

Peter Ferenci, Albert F. Staettermayer and Sandra Beinhardt
Division of Gastroenterology and Hepatology, Department of Internal Medicine III, Medical University Vienna, Vienna, Austria

Yann C. Klimentidis and Jin Zhou
Mel and Enid Zuckerman College of Public Health, Division of Epidemiology and Biostatistics, University of Arizona, Tucson, Arizona, United States of America

Nathan E. Wineinger
Scripps Translational Science Institute, La Jolla, California, United States of America

Yiqing Liu
Department of Ornamental Horticulture, China Agricultural University, Beijing 100193, China

College of Life Science & Forestry, Chongqing University of Art & Science, Yongchuan 402160, China

Yusong Jiang, Jianbin Lan and Yong Zou
College of Life Science & Forestry, Chongqing University of Art & Science, Yongchuan 402160, China

Junping Gao
Department of Ornamental Horticulture, China Agricultural University, Beijing 100193, China

Naser Elkum
Department of Biostatistics & Epidemiology, Dasman Diabetes Institute, Kuwait City, Kuwait
Clinical Epidemiology, Sidra Medical and Research Centre, Doha, Qatar

Fadi Alkayal, Maisa M. Ali, Motasem Melhem and Osama Alsmadi
Genetics & Genomics Unit, Dasman Diabetes Institute, Kuwait City, Kuwait
Monira Al-Arouj and Abdullah Bennakhi
Clinical Services, Dasman Diabetes Institute, Kuwait City, Kuwait

Kazem Behbehani and Fiona Noronha
Department of Biostatistics & Epidemiology, Dasman Diabetes Institute, Kuwait City, Kuwait

Jehad Abubaker
Biochemistry & Molecular Biology Unit, Dasman Diabetes Institute, Kuwait City, Kuwait

Charu Lata
National Research Centre on Plant Biotechnology, New Delhi, India
CSIR-National Botanical Research Institute, Lucknow, Uttar Pradesh, India

Awdhesh Kumar Mishra, Mehanathan Muthamilarasan, Venkata Suresh Bonthala, Yusuf Khan and Manoj Prasad
National Institute of Plant Genome Research, New Delhi, India

Iratxe Torre, Anna González-Tendero, Fátima Crispi, Igor Iruretagoyena, Ivan Amat-Roldán and Eduard Gratacós
BCNatal – Barcelona Center for Maternal-Fetal and Neonatal Medicine (Hospital Clínic and Hospital Sant Joan de Deu), IDIBAPS, University of Barcelona, and Centre for Biomedical Research on Rare Diseases (CIBER-ER), Barcelona, Spain

Patricia García-Cañadilla
BCNatal – Barcelona Center for Maternal-Fetal and Neonatal Medicine (Hospital Clínic and Hospital Sant Joan de Deu), IDIBAPS, University of Barcelona, and Centre for Biomedical Research on Rare Diseases (CIBER-ER), Barcelona, Spain
Physense, Departament de Tecnologies de la Informaciói les Comunicacions (DTIC), Universitat Pompeu Fabra, Barcelona, Spain

Francisco García-García
Bioinformatics Department, Centro de Investigación Principe Felipe (CIPF), Valencia, Spain
Functional Genomics Node, INB, CIPF, Valencia, Spain

Bart Bijnens
ICREA, Universitat Pompeu Fabra, Barcelona, Spain

Joaquin Dopazo
Bioinformatics Department, Centro de Investigación Principe Felipe (CIPF), Valencia, Spain
Functional Genomics Node, INB, CIPF, Valencia, Spain
Centro de Investigación Biomédica en Red de Enfermedades Raras (CIBERER), CIPF, Valencia, Spain

Index